Bioarchaeology of Ancient Northern Vietnam

Marc Fredrick Oxenham

BAR International Series 2781
2016

First Published in 2016 by
British Archaeological Reports Ltd
United Kingdom

BAR International Series 2781

Bioarchaeology of Ancient Northern Vietnam

ISBN: 978 1 4073 2421 0

Printed in England

All BAR titles are available from:

British Archaeological Reports Ltd
Oxford
United Kingdom
Phone +44 (0)1865 310431
Fax +44 (0)1865 316916
Email: info@barpublishing.com
www.barpublishing.com

ABSTRACT

This is a study of the health and behaviour of the past inhabitants of what is now the northern portion of the Socialist Republic of Vietnam. The thesis examines human skeletal evidence from two Holocene time periods in order to compare and contrast health profiles and behavioural implications. Skeletal and dental remains of ninety-four individuals excavated from a mid-Holocene Da But period site and ninety-six individuals from Metal period sites formed the samples.

Bio-archaeological macroscopic and radiological techniques were used to explore differences and similarities in the manifestations of health and behaviour within and between these two samples. All specimens were examined and assessed according to categories developed in a relational database using standard metric and non-metric skeletal and dental recording protocols. Analyses resulted in oral health profiles, insights into articular health, levels of physiological stress and case studies of traumatic and non-traumatic skeletal lesions.

The oral health of these samples was found to be relatively high by global and Southeast Asian standards. Nonetheless, there is evidence for slightly poorer oral health in the Metal period. Little evidence for serious articular disease was found for either sample. The temporally earlier Da But period sample displays slightly higher levels of articular change, consistent with the older age structure of this sample. There is evidence for a very high frequency of childhood anaemia in both samples, but without any obvious negative impact on adult health. Canine linear enamel hypoplasia is relatively high for both samples by Southeast Asian standards and has impacted on the health of adults in each dental series. A combination of factors associated with a sedentary lifestyle and marine economic focus are implicated. Low levels of traumatic injury in the Metal period and high levels in the Da But period implicate significant, albeit unknown, behavioural differences between each period. Infectious disease makes its first appearance in the Metal period sample and changes to the population-environment dynamic, including massive population movements into the region at the time, are implicated.

The evidence suggests both samples derive from sedentary communities with riverine, coastal and marine economic foci. For the Da But sample, there is no human biological data supporting the view that agriculture or horticulture was practised. As for the Metal period assemblage, the bioarchaeological information is consistent with historical evidence, albeit contrary to archaeological models, suggesting the local populace did not survive on the products of intensive agricultural pursuits. The Metal period data suggest a foraging or mixed agriculture/foraging subsistence base.

ACKNOWLEDGEMENTS

I am indebted to my supervisor and good mate Dr Ian Walters, for opening my eyes to the fun that can be had in playing with bones. Moreover, without Ian I would never had had the opportunity to visit Vietnam and make the contacts and friends in this exciting country, that I have since had the pleasure of doing. I also thank Dr Nguyen Lan Cuong, another close friend, co-supervisor and facilitator of all things from organising collection access to radiographs and much more. My work would have been impossible without his help.

Numerous people have helped me both personally and academically in my pursuit of this research. Thanks go to the Head of the Institute of Archaeology in Hanoi, Professor Ha Van Tan, who allowed me to visit Vietnam for an extended period of time on a research visa and gave me unhindered access to materials and resources at the Institute. I thank Dr Nguyen Kim Dung for making her time available to me and giving me an introduction to the complexities of Vietnamese archaeology. I thank Dr Nguyen Khach Su, Nguyen Kim Thuy and Nguyen Van Hao for their input and assistance. My thanks are also extended to the library staff at the Institute of Archaeology for their patience and help in locating various pieces of literature, including unpublished site reports.

I have been assisted with that most tedious of chores, being the reading of drafts, by Dr Kate Domett (University of Otago) and Dr Michele Toomay Douglas. For that I sincerely thank them. Furthermore, I am truly indebted to Michele for the enormous amount of red pen she spent on my draft chapters. Without her input my thesis would be less than it is.

I have been liberal with my pleas for help and assistance. I thank Dr Robert Hirsch (University of Adelaide) for his assistance with aspects of oral health; Dr George Dias (University of Otago) who also assisted with aspects of oral health; Dr Charlotte Roberts (University of Durham) for her input with regard to trauma; Professor Charles Merbs (Arizona State University) for his insights into vertebral trauma; Professor Rafe de Crespigny (Australian National University) for his generous gift of unpublished translations of ancient Chinese historical texts; Mr Steve Baddeley (orthopaedic specialist, Royal Darwin Hospital) for his input with regard to fractures; Dr Keith McGuinness (Northern Territory University) for his much needed statistical advise; and Dr Kevin Lee (formerly the Chief Forensic Pathologist of the Northern Territory, Australia) for his input with regard to taphonomic processes.

I also thank the Northern Territory University, for the infrastructure support as well as the all important scholarship, without which life would have been rather difficult. In particular I thank Elaine Glover and her staff in Interlibrary Loans, for speedy and helpful assistance in tracking down all manner of esoteric material that I always needed yesterday.

This thesis is dedicated to my spouse Shojun. Without her unflagging and unstinting support this research would never have happened. Thank you Shojun.

TABLE OF CONTENTS

LIST OF TABLES

LIST OF FIGURES

PREFACE

This monograph is a historical document in as much as it represents a reformatted, but otherwise unchanged, version of a PhD thesis submitted for examination at the then named Northern Territory University (now Charles Darwin University), Australia, in December 2000. A significant amount of the content, including data, presented here has not been published although the unpublished thesis has been frequently cited. Given the value, particularly with respect to the Bioarchaeology of Southeast Asia, of this monograph it was considered appropriate that it was (finally) published in its entirety.

While the majority of data was collected in Vietnam in the latter half of 1997, the recording protocols are still valid nearly 20 years later. Further, while obviously no cited literature post-dates the year 2000 much of the interpretations and conclusions are still as valid today as they were 15 years ago. Notwithstanding, my views, and the discipline as a whole, have developed extensively over the intervening period. To deal with changing interpretations and new data, an epilogue has been written that updates and visits new interpretations and findings where appropriate.

An issue that the reader should be aware of at the start concerns the culture-chronological framework that contextualises the two main skeletal series examined here. At the time of writing, the subsistence and economic focus or status of the mid-Holocene sample (referred to as the Da But period series) was unclear. Within the monograph the Da But period sample is referred to as Neolithic, while the other comparative sample belongs to the Metal (Bronze and Iron) period. It is now clear that the Da But represents a relatively widespread (including northern Vietnam and southern China) pottery using pre-Neolithic foraging culture complex. The Da But lacks any evidence for domesticated crops or animals, while making extensive use of pottery, edge ground lithics and a sophisticated bone tool repertoire. The Neolithic, proper, does not appear in Vietnam until approximately 4,000 BP (at least 2,000 years after the Con Co Ngua cemetery sample attributed to the Da But period). Efforts to bridge the bioarchaeology of the region, between the Da But and Metal periods, has focused on the c. 3,800 BP Neolithic Man Bac assemblage from northern Vietnam and is discussed in the epilogue.

PREFACE

This monograph is a historical document in as much as it represents a reformatted, but otherwise unchanged, version of a PhD thesis submitted for examination at the then named Northern Territory University (now Charles Darwin University), Australia, in December 2000. A significant amount of the content, including data, presented here has not been published although the unpublished thesis has been frequently cited. Given the value, particularly with respect to the Bioarchaeology of Southeast Asia, of this monograph it was considered appropriate that it was (finally) published in its entirety.

While the majority of data was collected in Vietnam in the latter half of 1997, the recording protocols are still valid nearly 20 years later. Further, while obviously no cited literature post-dates the year 2000 much of the interpretations and conclusions are still as valid today as they were 15 years ago. Notwithstanding, my views, and the discipline as a whole, have developed extensively over the intervening period. To deal with changing interpretations and new data, an epilogue has been written that updates and visits new interpretations and findings where appropriate.

An issue that the reader should be aware of at the start concerns the culture-chronological framework that contextualises the two main skeletal series examined here. At the time of writing, the subsistence and economic focus or status of the mid-Holocene sample (referred to as the Da But period series) was unclear. Within the monograph the Da But period sample is referred to as Neolithic, while the other comparative sample belongs to the Metal (Bronze and Iron) period. It is now clear that the Da But represents a relatively widespread (including northern Vietnam and southern China) pottery using pre-Neolithic foraging culture complex. The Da But lacks any evidence for domesticated crops or animals, while making extensive use of pottery, edge ground lithics and a sophisticated bone tool repertoire. The Neolithic, proper, does not appear in Vietnam until approximately 4,000 BP (at least 2,000 years after the Con Co Ngua cemetery sample attributed to the Da But period). Efforts to bridge the bioarchaeology of the region, between the Da But and Metal periods, has focused on the c. 3,800 BP Neolithic Man Bac assemblage from northern Vietnam and is discussed in the epilogue.

1
INTRODUCTION

This thesis investigates evidence for health and behaviour of the inhabitants of the northern portion of what is now the Socialist Republic of Vietnam, during the mid-Holocene and Metal periods. This study is based upon an analysis of the skeletal and dental remains of one hundred and ninety individuals excavated from archaeological sites in the region. Previous studies of human skeletal material recovered from Vietnam have focused upon late-Pleistocene and Holocene morphology, patterns of micro-evolutionary change and population movement within East Asia (Olivier 1966; Nguyen Lan Cuong 1985, 1992, 1996; Pietrusewsky 1987; Pietrusewsky *et al.* 1990; Nguyen Kim Thuy 1990, 1998).

These bones, stored for years in cardboard boxes in Hanoi, provide insights into the diseases, health and behaviour of people who lived in this region thousands of years in the past. Using the techniques of bioarchaeology, this study deciphers the skeletal record and for the first time reveals glimpses of their life and culture.

Why study the health of past communities? A study of health can contribute to a greater understanding of previous peoples and also furnish data relevant to modern health issues. With regard to the role studies of past health can play in modern epidemiological and clinical settings, Stirland (1991) has demonstrated how carefully designed bioarchaeological studies can contribute to the current understanding of the aetiology of osteoarthritis. Pfeiffer (1991) has shown how palaeopathological studies of tuberculosis have helped clarify the underlying environmental stresses associated with this disease. At the least, such studies contribute to a greater understanding of the history of disease, particularly in terms of the frequency of various conditions and potential aetiological factors involved.

A study of the health of past communities has an important role to play in understanding how people operated within a wide range of cultural and environmental contexts. For instance, a much greater understanding of the health effects of foraging versus agricultural subsistence bases, as well as the effects on the health of communities undergoing major shifts in subsistence orientation, has been effected in recent years (contributions in Cohen & Armelagos 1984; Swedlund & Armelagos 1990). The generally good correlation between certain markers of health and behaviours such as subsistence and residential mobility, appears to be global in nature. Another area where bioarchaeological studies of health have contributed, is in an increased appreciation of the health consequences of populations on the receiving end of invasions and unsolicited migrations (contributions in Larsen & Milner 1994). Moreover, studies of traumatic injury have helped to reconstruct past behaviour (Lovejoy & Heiple 1981), while studies of physiological stress have contributed to an understanding of past and present population stressors (Goodman & Armelagos 1988; Stuart-Macadam 1992;). This study of the ancient inhabitants of Vietnam was initially formulated in anticipation of what knowledge of their health could reveal of their lives.

The beginnings of this project extend back to the middle of 1995. While examining Pleistocene hominoid dental remains in Hanoi, as a research assistant, I was shown curated collections of Holocene human material. I sought permission from the Director of the Institute of Archaeology in Hanoi, Professor Ha Van Tan, to survey this material in 1996. After examining the collections, I began to define a research project based on a set of questions: What was the health of these past communities? Were there differences in health between different time periods in the region? Can specific diseases be discerned in this skeletal material? Is there any evidence for interpersonal or more generalised violence? What behaviours might be inferred from the health of these assemblages? With such questions in mind, and familiarity with the material gained by way of the survey, the project aims were developed. Following the successful application for a special research visa, to enable an extended stay in Vietnam, the vast bulk of the data collection for this study was carried out at the Institute of Archaeology in Hanoi. This was completed during a period of six months in the latter half of 1997. Data collected in Hanoi were transferred into a database that I developed using Microsoft Access. This is a form-based relational database that facilitates complex data interrogation. It was designed so that a virtually limitless number of variables could potentially be recorded for any single skeletal or dental element.

The Purpose of this Study

The paramount research aim of this study is to describe, infer and interpret the health of the people in each of the two temporal periods. The thesis explores health differences or similarities during the two time periods. It examines whether the skeletal and dental evidence for health is consistent with the archaeological evidence for behaviour, in particular subsistence and residential mobility. Further, with respect to the Metal period, inferences regarding health are assessed against historical evidence for subsistence and population dynamics. The theme of this study is the biological response to the sub-tropical, Southeast Asian context.

Nothing was previously known about the health of past inhabitants of this region. Skeletal series representing a mid-Holocene period site and several metal period sites, all from the same geographic locality, were available for analysis. There is archaeological and historical evidence suggesting possible differences between these time periods in terms of subsistence economies, culture and population dynamics. A bioarchaeological study can contribute to a better understanding of such issues. The samples provided an opportunity to examine: industrial uses of the dentition; skeletal signatures of diet, subsistence activities and residential mobility; high risk behaviours that might have led to elevated levels of traumatic injury; and changes in the population dynamics of these communities, that might lead to the introduction of new diseases.

A population approach to assessing health has been adopted in the main. Larsen (1997:3) has pointed out that the population approach is "critical for characterizing patterns of behavior, lifestyle, disease, and other aspects that form the fabric of the human condition". In the chapter dealing with evidence for traumatic and non-traumatic skeletal lesions a series of case studies, including detailed descriptions and differential diagnoses, complements the main, population based, study method.

Inferences regarding individual or population level health and behaviour is not an entirely straightforward process. Not only is bone a highly dynamic tissue that constantly remodels throughout the life of an individual, but health fluctuates over the life span of any individual. For some, much of their life may be spent in relative good health, while for others the converse is true. In inferring a population's health, bioarchaeologists necessarily work with samples that represent the individual end products of cumulative skeletal manifestations of health. In some instances, elevated levels of pathology in a sample can be interpreted in terms of defensive responses to disease and/or stress (Ortner 1991; Wood *et al.* 1992). A high frequency of skeletal lesions can be perceived as evidence for a healthy response to disease. As such, Wood *et al.* (1992: 356) have argued that "better health makes for worse skeletons". However, it must also be appreciated that these same *worse skeletons* represent periods of poor health, individuals or communities suffered in life. It is also arguable that a better immunological or physiological response to disease and/or stress is one that rids the body of the offending condition before recourse to a response requiring skeletal involvement (see Ortner 1991).

How health is defined, the pathological condition under consideration, environmental influences, cultural factors, including mode of subsistence and settlement patterns, all impact on how skeletal and dental evidence is interpreted. For this reason the cultural and environmental contexts for each sample need to be explored. However, environmental reconstruction has not been an archaeological research priority in Vietnam to date. This gap can be partially addressed by recourse to current geo-environmental and climatological data on northern Vietnam. The cultural context is developed by way of archaeological data. Further, for the Metal period contemporaneous and temporally near historical accounts exist which provide information about the people and geography of this area.

The Samples and Their Culture-Regional Context

The earliest accepted period of human occupation in northern Vietnam, Son Vi culture, dates back some 30,000 years ago (Nguyen Khac Su 1996). Moreover, the region may have been inhabited by hominids much earlier, if reports of teeth attributed to *Homo erectus* (Olsen & Ciochon 1990) are accepted. However, it is the inhabitants of the region during the mid to later Holocene that are the focus of study here. The temporally earlier sample of skeletal and dental remains derives from a single cemetery site, named Con Co Ngua, which is associated with the beginnings of the Neolithic Da But period. The Da But period began approximately 6000 years bp. The later sample is a composite of skeletal and dental remains from 11 separate sites. Over 80% of this sample is dated to between 2500 and 1700 years bp, while the remainder of the sample is dated to between 2600 and 3300 years bp. Combining it into a single sample allows discussion of what I will call the Metal Period. Figure 1.1 locates each sample spatially, while Table 1.1 provides dates and samples sizes for each individual site. At times, in the text and in illustrations, individuals are specifically referred to by their catalogue codes. These codes follow a similar format in all cases, the date excavated, the site abbreviation and the burial number. For instance, 78QCM31 refers to burial No. 31 excavated at Quy Chu in 1978.

The Da But Period

Insights from Archaeology

Origins and Prehistoric Context of the Da But Period

Da But period sites are located along the Ma River in Thanh Hoa province, northern Vietnam. The Da But is considered to be one of a number of emergent Neolithic cultures, with a coastal/marine economic focus, that developed out of the Hoabinhian culture approximately 6500 years bp (Bui Vinh 1991, 1994, 1996; Nguyen Kach Su 1997). Communities making up the Hoabinhian period, dated to between 18000 and 6000 years bp for the most part, were involved in hunting and gathering a wide variety of mammals, birds, fish and shellfish (Gorman 1971; Higham 1989). Apart from the possibility that a form of incipient horticulture was practised, there is also evidence for the remains of rice in at least one Hoabinhian site (Higham 1989). However,

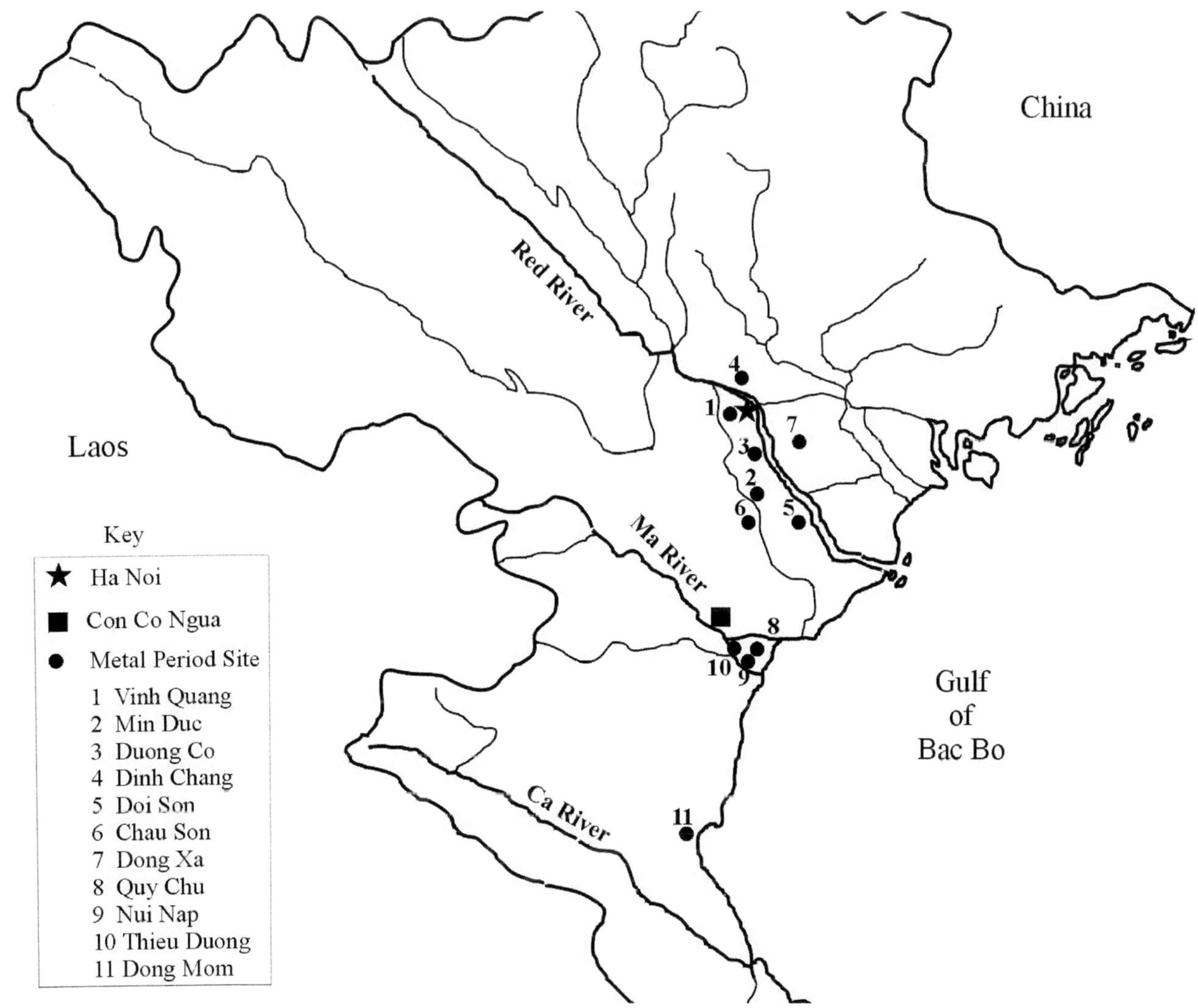

Figure 1.1. Northern Vietnam. Location of the Da But culture site Con Co Ngua and Metal Period sites

Table 1.1. Study Sample Summary

Period	Sites	Region	Individuals	Date[1]
Da But	Con Co Ngua	Ma River	94	~6kya
Metal	Vinh Quang	Red River	13	2200 to 1700 years bp
	Min Duc	Red River	6	2200 to 1700 years bp
	Duong Co	Red River	1	2200 to 1700 years bp
	Dinh Chang	Red River	8	2200 to 1700 years bp
	Doi Son	Red River	7	2200 to 1700 years bp
	Chau Son	Red River	1	2285+/-45 years bp[2]
	Dong Xa	Red River	1	2200 to 1700 years bp
	Quy Chu	Ma River	4	2450+/-45 and 2520+/-55 years bp[2]
	Quy Chu	Ma River	18	3300 to 2500 years bp
	Nui Nap	Ma River	2	3000 to 2500 years bp
	Nui Nap	Ma River	2	2400 to 2000 years bp
	Nui Nap	Ma River	30	1670+/-85 years bp[2]
	Thieu Duong	Ma River	1	3000 to 2800 years bp
	Dong Mom	Ca River	5	2200 to 1700 years bp

bp before present

[1] refers to date of skeletal sample

[2] radiocarbon date

the relationship between these emergent post-Hoabinhian cultures and the Hoabinhian itself is unclear. Hoabinhian cultures are generally restricted to inland caves and rock shelters.

Work by Geyh *et al.* (1979) in the Malacca Straits supports the view that between 36000 and 10000 years bp, the sea level was at least 40m to 60m below its present level. This would place most of the known Hoabinhian sites some considerable distance form the then coast. The sea level has been rising since then to reach a maximum of about 5m above the present level in the mid-Holocene (Geyh *et al.* 1979; Tjia 1996). However, research into old shorelines of the Malay-Thai Peninsula, indicate fluctuations in sea level following the maximum transgression (Tjia 1996). It would seem that these post-Hoabinhian Neolithic sites could represent the descendants of late-Pleistocene coastal dwellers who had been following the coast inland with the transgression of the sea (see Gorman 1971; Higham 1989). Alternatively, they may be the descendants of highland Hoabinhian peoples who came down to take advantage of the coast as it moved inland. Unfortunately, any evidence for Pleistocene coastal occupation lies beneath the waves now.

One intriguing Hoabinhian coastal site dated to between 5000 and 4500 years bp has been recently excavated (Ha Van Tan 1997). Both the date and location of the archaeological site Bau Du seems at odds with what is generally known of the Hoabinhian period in Vietnam. Bau Du even post-dates the so called post-Hoabinhian sites, such as Da But, thus complicating the timing of the terminal Hoabinhian. Moreover, the faunal remains from this midden site included shellfish such as *Anadara granosa* (an estuarine bivalve); mammals such as tortoise, deer, serow, monkey and rhinoceros; as well as crabs and fish (Ha Van Tan 1997). A mixed aquatic/terrestrial economy is thus indicated for this late Hoabinhian site.

Bau Du is not the only apparently anomalous Hoabinhian site in the region. Walters *et al.* (in prep.) have recently dated human skeletal material from what have hitherto been regarded as terminal Pleistocene Hoabinhian sites in western Thanh Hoa province, Mai Da Nuoc and Mai Da Dieu. The dates of this material indicated that the inhabitants of the shelters were present, in a Hoabinhian material culture context, between 5500 and 5000 years bp. As Walters *et al.* (in prep.) point out, "the Bau Du, Mai Da Nuoc and Mai Da Dieu dates render the Holocene expression of the Hoabinhian contemporaneous with these so-called post-Hoabinhian Neolithic culture periods". In the context of this study, one implication of the uncertainty surrounding the origin of mid-Holocene coastal oriented cultures, such as Da But, is that it is unclear if they represent a continuation of coastally oriented economies in the region from late-Pleistocene times. The alternative, that they represent an abrupt change and adaptation to coastal resources in the mid-Holocene by hitherto inland oriented Hoabinhians, would seem to be contraindicated from the mixed terrestrial and coastal orientation of Bau Du at least.

Emergent Post-Hoabinhian Cultures: The Da But Period

The Da But period, believed to have existed for some 1700 years, was far from static. The Da But begins between 6500 and 6000 years bp and is represented by shell midden sites that include polished stone artefacts, pottery and arguably domesticated pigs and buffalo (Bui Vinh 1991). The skeletal assemblage analysed in this study, from the basal layer of Con Co Ngua, represents the beginning of the Da But phase. The lower levels of this site also indicate the presence of salt-marsh plants and estuarine molluscs, thus indicating a close proximity to the coast. The second stage, represented by the upper level of Con Co Ngua and dated to around 5500 years bp, shows an absence of molluscs and salt marsh plants, suggesting the coast had started to prograde by this stage. But no skeletal material exists for this upper level at Con Co Ngua. The material culture also changes, with the first appearance of fishing net sinkers, spinning whorls, stone awls, stone chisels and technological changes in stone axes. Moreover, there is a large increase in the frequency of polished and curved stone artefacts that have been interpreted as being hoes (Bui Vinh 1991). There is also an increase in the number of domesticated buffalo and pig remains (Bui Vinh 1996). The final phase of the Da But period, beginning around 5000 years bp, termed the Go Trung, shows technological changes with respect to pottery and stone artefacts. Settlements are seen to spread along the coast and the development of sea travel and sea fishing has been suggested (Bui Vinh 1991,1996). This would indicate changes from a coastally oriented economy to a marine-based one, in the later phase of the Da But period.

The Da But Period Site: Con Co Ngua

The skeletal and dental remains of 94 individuals, representing the Da But period, derive from the single cemetery site Con Co Ngua. The following account of this site is based on the unpublished site report prepared by Bui Vinh (1980). Con Co Ngua lies in Ha Trung district, Thanh Hoa province some 30km from the present coast, approximately 3km north of the Ma River, latitude 20^0 North and longitude 1^0 5' 45" East (Figure 1.1). Con Co Ngua is situated on a low mound within a small valley at the foot of encircling hills. These hills range from 50 to 350m above sea level. It can thus be seen as situated within a transitional plains-upland region. However, it is still low enough to be constantly flooded during the rainy season. The occupants would have had access to riverine, estuarine, coastal and dense sub-tropical forest resources within close proximity to the site.

A rather large excavation, some 228.5m^2, comprising two adjacent pits on and by the mound was carried out in 1979/80. In the majority of cases, individuals were buried tightly flexed, both the upper and lower limbs, in a type of squatting or seated foetal position. Graves were generally only earthen pits with no sign of any other lining or covering material. However, at least five single graves were associated with a stone marker on top of or beside the grave. The vast majority of burials, for which an orientation could be discerned, were facing east or northeast. Signs of what appears to be ochre were spread over at least one clustered burial, while ochre solidified into the shape of a gourd was found in two separate graves. In three cases stag horns were found buried beside individuals. It was not uncommon to find in the graves bones of various vertebrates and invertebrates such as buffalo, pig, turtle, fish, oyster, mussel and snails.

Material cultural objects recovered from graves at Con Co Ngua include numerous grinding stones, ground axes and mat-impressed pottery fragments. I also recovered a worked bone artefact, possibly an ornament, from among the remains of one curated individual from the site.

While the stratigraphy of Con Co Ngua is not clear, the burials are essentially found within midden material containing a high concentration of fresh water molluscs and salt water oysters and mussels. A sterile capping layer including blue-grey clay, spread throughout the valley to a depth of 0.5-1.5m, has been interpreted as representing a major flood.

The two pits excavated are believed to be temporally discontinuous. However, it is not clear what period of time separates them. What does seem clear is that for the most part, the 28 burials from pit one are single or double burials. The 78 burials from pit 2, on the other hand, form mostly clustered inhumations. The skeletal remains of 94 individuals were available for examination. Of these, 25 were from pit 1, 64 from pit 2 and five individuals who could not be assigned confidently to a specific pit.

As noted, Vietnamese archaeologists, using material culture comparisons, associate Con Co Ngua with the Da But period that began approximately 6000 years bp and the Go Trung dating from around 4800 years bp. However, there is an absolute date for the site based on marine shell: Bln 2697: 3020+/-100 years bp. Vietnamese archaeologists refute this date and maintain it is too late given the nature of the pottery remains from Con Co Ngua (Nguyen Lan Cuong pers. comm.). There is also a recent radio carbon date based on a sample of human bone from the site: OZC324: 2740+/-1050 years bp. This date has an exceptionally large standard deviation and on its own does not help clarify the dating problems associated with this site.

The Metal Period

Insights from Archaeology

The earliest evidence for metal in the Red River delta region appears in late Phung Nguyen culture contexts. These sites are situated on relatively raised areas to the northern edge of the Red River delta. The Phung Nguyen culture, believed to have endured from 4000 to 3500 years bp, is notable for its rich and technologically sophisticated array of stone jewellery. Following this period the Dong Dau phase, 3500 to 3000 years bp, ushers in a major development in bronze working technology with the appearance of bronze weapons such as spears and arrow heads, as well as more utilitarian items such as fishing hooks, chisels and axes. The Dong Dau phase is followed by the Go Mun period, 3000 to 2500 years bp, which is characterised by the development of an enormous range of decorative, utilitarian, ritual and martial items (Higham 1996). Similar cultural sequences are reported for the Ma and Ca River regions immediately south of the Red River delta as well (Ha Van Tan 1985). Emerging from these early bronze phases is the Dong Son culture, flourishing in northern Vietnam from 2500 years bp until Han domination in the first centuries of the first millennium AD.

Higham (1989:30) has described the Vietnamese Dong Son as one of several Southeast Asian examples demonstrating social differentiation and a move "from village autonomy towards centralised chiefdoms". Apart from the diverse range and technological sophistication of material culture objects, richly decorated bronze drums exceeding 80kg for example, there is evidence for marked craft specialisation, a complex ritual life, the development of an aristocratic and centralised elite, maritime trade and sophisticated military skills and equipment. One notable example of the material achievements of this period can be seen in the size of the former capital Co Loa, built on the Red River plains some 2200 years bp. Higham (1996: 122) describes the site:

> There are two outer sets of ramparts and an inner, rectangular citadel. The outer rampart has a circumference of 8km [covering 600ha], and was punctuated by guard towers. The ramparts still stand up to 12m high, and are 25m wide at the base. It has been estimated that over 2 million m^3 of material were moved in constructing the defences, which also involved moats fed by the Hoang [Red] River.

Higham (1996:134) has gone on to suggest the Dong Son period:

> developed increasingly centralised social structures which in purely subsistence terms, were raised on the production and deployment of agricultural surpluses. These were used to sustain the central aristocracy and their retainers.

However, such a position must be examined in light of historical data spanning this period of time, and this is pursued presently. Moreover, there should also be discernible biological responses to such a dependence on agriculture, and investigating this is one of the expressed aims of this study. For now, to get a better idea of the people, their life-ways and also the geography of the region during the Metal period, it is necessary to investigate historical sources.

Insights from Historical Data

Historical accounts of the early inhabitants of northern Vietnam are of Chinese authorship. As with any historical evidence written by a conquering and colonising culture, they need to be treated with some circumspection. Be that as it may, much of this material represents the only extant descriptions of the customs and behaviours of people living in northern Vietnam from Han (206 BC-AD 220) through Tang (618-907 AD) dynasty times.

The Chinese commonly referred to the non-Chinese southern aboriginal populations collectively as the Man, a somewhat derogatory term implying barbarian status. Moreover, the iconography of the term reflected a reptilian flavour, reminiscent of the swamps, snakes and other reptiles inhabiting the south. What is now modern northern Vietnam, when not referred to specifically, came under the larger geographic construct Nam Viet. This area also included much of the southern and southwestern region of present day China such as Yun Nan, Guang Xi and Guang Dong provinces.

Politics and Geography

The Red River sites all fall within the later Han Chinese administrative unit, or commandary, Jiao Zhi. This unit encompassed the whole Red River delta area at the time, although the delta itself was much smaller than it is today (Holmgren 1980). Sites just south of the Red River region, the Ma and Ca Rivers, came under the administrative unit Jiu Zhen, which includes present day Thanh Hoa province. In Figure 1.1 the Han period administrative unit Jiao Zhi included the area occupied by the Metal period sites 1 to 7 above the Ma River. Jiu Zhen, on the other hand, encompassed the area occupied by sites 8 to 11 below the Ma River. Both Jiao Zhi and Jiu Zhen (known collectively then as Ou-Lo or Ou-Lak in Vietnamese) were originally conquered by Chao T'o (a former Qin dynasty general) in 186 BC. The Chinese names derive from later Han reconquering efforts in 111 BC (Schafer 1967).

The region, encompassing the vast majority of Metal period sites included in this study, moved in and out of direct Chinese administrative control from the late third century BC (De Crespigny 1990). Schafer (1967) suggests that initial military conquest and control of the region likely had a minimal cultural and social impact. It was not until concerted Chinese migration into the region occurred in the first century AD that significant socio-cultural effects were seen. Even with migration and the vagaries of central Chinese control of the region, De Crespigny (1996) argues that Chinese culture was maintained by the emigrants even during periods of independence from China. Moreover, De Crespigny (1996) talks of an insidious Confusianisation of the indigenes that had a significant influence on those social and cultural constructs crucial to the identity and viability of a society and way of life. This is aptly illustrated in a passage from the Hou Han Shu, or chronicles of the later Han:

> The people [of Jiao Zhi] are like animals, making no distinctions between the aged and young. They keep their hair in a knot on their necks, and they go barefoot wearing a cloth slipped over the head. In later times, many of the criminals of China proper were transported there and had to live among them, so they came to know something of the Chinese language and they gradually became acquainted with proper conduct and culture. After the restoration under Kuang-wu [AD 24], Hsi Kuang was Grand Administrator of Chiao-chih [Jiao Zhi], and Jen Yen governed Chiu-chen [Jiu Zhen], and they taught the people farming and had hats and shoes made for them. For the first time, the customs of the go-between and of betrothals were instituted and so they adopted the proper forms of marriage. Schools were established to teach good conduct and propriety.
> (HHS 86, 8a-9a translated by De Crespigny n.d.)

Nonetheless, subtlety was not always employed. A famous Vietnamese rebellion led by the now legendary Trung sisters (actually born of an immigrant Chinese family) was crushed by the Chinese general Ma Yuan in 43 AD. Ma Yuan immediately set about confiscating that local cultural icon, Dong Son drums, from the local chieftains and had them melted down and recast as a triumphal bronze horse (De Crespigny 1990).

Holmgren (1980) points out that early Chinese officials in northern Vietnam have traditionally been praised in the histories for their efforts to civilise these people. Such civilising being by way of the introduction of metal agricultural implements, agricultural techniques such as land reclamation and education as already noted. However, she goes on to suggest that:

> When the HHS speaks of the introduction of advanced methods of agriculture during Jen Yen's time, it means that desperate efforts were made to accommodate and provide for the influx of new settlers and refugees fleeing the north at the end of the former Han. Presumably large numbers of refugees arriving in Chiao-chih [Jiao Zhi] were also straining the resources of that commandery. The rapidly expanding pearl-fisheries off the coast of Ho-p'u in Kwangtung [Guang Dong province], and the increasing search for luxury items from Annam [northern Vietnam at this time], put additional pressure on Chiao-chih as a centre of supply for Ho-

> p'u, Chiu-chen [Jiu Zhen] and the merchant shipping in the far south. The agricultural improvements in Chiao-chih and Chiu-chen were thus aimed at feeding the new immigrant population and supplying coastal shipping, rather than supporting indigent native populations.
> Holmgren (1980:5)

The point regarding agricultural improvements is important in suggesting that prior to the intensification in Chinese colonising efforts, the region lacked an agricultural subsistence economy. While Chinese historical sources tell of local swidden agriculture, there is no mention of any Vietnamese communities having an agricultural economic focus. It would also appear that the introduction of intensive agricultural techniques and practices was essentially about Chinese exploitation of the region as a rice basket.

Holmgren (1980) says that migration into northern Vietnam from China proper did not have a radical impact on the region until after AD 140. This coincides with the more recent temporal extent of the Metal period human skeletal sample under investigation in this study. Of interest are census figures (from the Chinese histories Han Shu and Hou Han Shu cited in Holmgren 1980:70) for the Chinese defined administrative districts in northern Vietnam for both AD 2 and AD 140. The AD 2 census indicated that the number of households in Jiao Zhi was 92,440, or 8 members per household, and Jiu Zhen 35,743, or 4.6 members per household. The figures for AD 140 showed an eight-fold population increase for Jiao Zhi to 739,520. While in Jiu Zhen the population increased four or five fold to 164, 418.

It is highly probable that the strategic position of the Red River Delta and surrounds was a factor influencing Chinese efforts to gain a foothold and eventually to annex the region. In describing the region Taylor (1983:9) states:

> This plain lay astride the only lowland corridor between the Tibetan highlands and the sea; it was consequently in the path of peoples moving between East and Southeast Asia. Those who chose to make this place their home faced frequent challenges from all directions. They, of necessity, had to arm themselves with weapons equal to these challenges.

The Dong song period also had a close and extensive association with the sea. In describing the decorative designs and figures adorning Dong Son drums Taylor (1983:313) states: "Boats filled with oarsmen and warriors surrounded by seabirds and other forms of maritime life unmistakably testify to the ascendancy of sea-based power". The region was clearly politically and militarily volatile during much of the period from 200 BC onwards. Moreover, extensive and socio-culturally far reaching changes were being wrought by the Chinese, especially from the first century AD. So what were the local inhabitants of the region like? How did they live and what did they eat?

The People

At the time of the defeat of the Trung Sisters, the

> area south of the Red river delta [Jiu Zhen] is described as having been a wild place of swamps and forests, the abode of elephants, rhinoceros, and tigers. The inhabitants lived for the most part by hunting, using bows and arrows with bone arrowheads to kill beasts. They ate the flesh of pythons, fished, and also supplemented their diets with meager crops farmed in small swiddens. It was only around the administrative centres that some paddy fields were farmed, a cultivation technique learned from the Chinese colonists.
> (Hickey 1982:58)

With regard to the region encompassed by Jiao Zhi, more specific data are furnished by the early Chinese histories of the region.

> Lac society of the kingdom of Van Lang [capital near present day Hanoi] seems to have had a 'feudal' social organization with a structural hierarchy similar to that found among Muong, the Tai-speaking groups in the highlands of northern Vietnam. At the local level the Lac Marquis (Lac Hau) held hereditary positions of authority, while below them were the Lac chiefs, who were of noble village families. The Lac were sedentary people who lived in small communities, where they farmed the land with polished stone hoes (plows later were introduced by the Chinese). They knew how to use the tides to irrigate their paddy fields and produced two crops a year. They tattooed their bodies so as to 'avoid crocodiles', they chewed betel and areca, and they blackened their teeth…Religion was centred on agriculture with a large celebration at spring planting. Betel and areca were prescribed offerings for the bride at marriage, and the levirate was practiced. In warfare they used bows and arrows with bronze heads on which poison was smeared.
> (Maspero 1918, in Hickey 1982:60)

While difficult to associate with any particular group or area in Nam Viet, there are a plethora of Tang descriptions describing the Man in general. Several of these would appear to be specific to the southern part of Nam Viet (northern Vietnam).

> The southerners have always tattooed their bodies; they dwell on piles, as if they were birds in tree tops…they eat fish and rice; they drink through their noses; the lowlanders (especially Thai peoples, it seems) plant in water (that is in wet rice fields); uplanders cultivate the ashy soils of burned mountain slopes.
> (Schafer 1967:54)

The recurring theme of body tattooing is somewhat intriguing. Schafer (1967) says that coastal dwellers of Nam Viet have an extensive history of body tattooing. The principal reason for the tattooing being as a ward against the Kau, a Tang dynasty rendering of the word for dragon or sea monster. Schafer (1967) suggests that as such, it would have been a practical custom for divers and fisher people.

Diet

Health is to a considerable extent dependent on diet. Given the limited data available on the Metal period diet, the following account contributes to an understanding of what likely foods were consumed at this time. If Tang dynasty references to the foods consumed by the indigenes of Nam Viet are in any way reflective of the same peoples 300 to 500 years before, it would seem that the Metal period inhabitants enjoyed a varied diet. The following account is based on Schafer's (1967) work. These tropical coastal dwellers were observed to eat jellyfish prepared with various spices including ginger, cardamon, cassia and fagara. Various molluscs were consumed and oysters were a favourite eaten while drinking wine. In fact, at least one Tang observer found the wines of Nam Viet to be quite excellent. Squid was another popular food prepared in various ways such as fried with ginger and vinegar or as a dried and salted preserve. Various crustaceans were caught with the Horseshoe crab being used in a sauce or pickle. Another dish included living shrimp eaten with vegetables and sauces. Insects were not spared either with at least two favourites being cicadas and a sauce prepared with ant eggs. Other insects, such as various species of wasp were prized as ingredients in medicinal compounds. Various reptiles were also used in such a pharmacological fashion.

With respect to plants, both taro and yam were common in the region. Numerous varieties of palm were also grown in the area, with some supplying palm sugar. Sugar cane itself was known in Annam, and Schafer (1967) makes reference to the manufacture of cakes using sugar cane extracts in the third century AD.

A diverse range of mammals inhabited Nam Viet that were still very common in Tang times. Schafer (1967) makes reference to indigenous hunting of the elephant. Not only was this for the ivory (Annam was an important source of Tang ivory) but the trunk of the animal, at least, was regarded as a delicacy at feasts. Elephants were also trained and ridden by the Man. Another large animal, the rhinoceros, is also reported to have been hunted with poisoned arrows. Various species of deer, including sambar and muntjac, also populated Nam Viet, although Schafer (1967) makes no specific reference to these being hunted.

Health and Disease

Schaffer (1967) taps into a wealth of Tang dynasty material, much of it written by Chinese court officials banished to the south, giving some insight into issues of health and disease. A recurrent theme is the deadly disease carrying miasmas or fogs of Nam Viet. Many of these banished officials are recorded to have died after only short periods of time in the south (Schaffer 1967). Disease was such a problem in Nam Viet that military reprisals against local rebellions in the region, during the Tang dynasty at least, had been cancelled in the face of the predicted effects of malaria on the Chinese army (Schaffer 1967). One Tang insurgent, Huang Chao, was reported to have lost up to 40% of his army to disease during a campaign against Canton, a city a little north and east of Jiao Zhi near present day Hong Kong (Schaffer 1967).

Interestingly however, historical population statistics attest to a relative absence of the effects of malaria in both the Canton and Hanoi delta regions (Schaffer 1967). The higher ground and areas back from the coast do not seem to have been so lucky however. Lu & Needham (1993) note that the Chinese had some knowledge of malaria and were differentiating between the tertian (fever occurring every third day) and quartan forms by the sixth century BC.

The Chinese of the Tang also recorded indigenous perceptions and explanations for disease. The most striking conception of the causes of many diseases being black sorcery or Ku. Ku is still practiced in present day Guang Xi province and was known to the Chinese from Zhou dynasty (~1000-257 BC) times (Schaffer 1967). While Ku diseases were often seen as the result of sorcerous activities they could also be caused by the ubiquitous miasmas (Schaffer 1967). It has also been suggested that numerous oedematous conditions attributed to Ku, were in fact caused by intestinal parasites (Schaffer 1967). Another specific disease known to the Chinese in Nam Viet for a considerable period of time was beri-beri, or leg vapors (athiaminosis) (Schaffer 1967). Lu & Needham (1993) suggest the first Chinese reference to beri-beri occurs in 584 BC. A range of treatments for this condition were prescribed by Chinese pharmacologists including quince, gourd, almonds and betel nut (Schafer 1967). It is not clear what contribution indigenous knowledge had in developing these cures.

The severe gastrointestinal and fever symptoms of cholera were also known to the Chinese of the region (Schafer 1967). It has been suggested that cholera was being described in Chinese texts by around 200 BC (Lu & Needham 1993). A vast, diverse and at times bizarre range of treatments were prescribed for the condition (Schaffer

1967). Typhus is another disease that was well known and in fact a Han dynasty treatise was written on the condition (Schaffer 1967). No doubt the more generalised Chinese accounts of disease in Nam Viet, from very early on, drew on an extensive experience of all manner of colds, fevers, dysentery, skin lesions, parasitic infections and so forth.

The Metal Period Sites

As mentioned, the Metal period is represented by eleven sites, the vast majority of which are located either within the Red River delta region, or just south of the delta in the Ma River region (Figure 1.1). Much of the skeletal material from these sites is dated to between 2500 and 1700 years bp. This places the majority of the skeletal and dental sample within the periods of the Dong Son proper and early Han encroachments into the region. Very few radiometric dates exist for these sites, with the majority being dated by way of material culture associations. To better appreciate the material culture wealth of this period, descriptions of the burial goods associated with these skeletal samples are included in the following site descriptions. The available reports and published literature on these sites are somewhat variable in detail and also availability.

Red River Sites

Vinh Quang (VQ)

This site, situated in Vinh Quang Commune, Hoai District, Ha Tay Province was excavated three times between 1966 and 1969 (Institute of Archaeology HS 85). In 1966 an area of 100m^2 was excavated, in 1967 288m^2, with the final excavation extending through 1968 and 1969 being 200m^2. Two distinct cultural layers were discerned, a lower layer attributed to the Go Mun culture and a subsequent layer representing the Dong Son period. The skeletal material derives from the upper layer and is dated to between 2200 and 1700 years bp. Bronze burial goods from the three excavations included 47 spear-heads, 1 javelin, 5 arrow-heads, 51 axes, 13 fishhooks, 2 chisels, 1 needle, 1 bell, 8 bracelets and a chicken or cock figurine. Iron artefacts included 2 spear-heads and a further 2 unidentified objects. Stone goods included scale weights, 7 bracelets and 2 axes. Pottery objects included three spinning whorls, a number of moulds and also net sinkers. A single awl made from bone was also recovered (Nguyen Khac Su & Nguyen Lan Cuong 1997). While 62 burials were reported from the site, only the skeletal remains of 13 individuals either remain or were able to be recovered.

Minh Duc (MD)

Minh Duc has been dated by way of material cultural associations to between 2200 and 1700 years bp. The site is located within Sau Dinh pond, a 145m x 40m depression 1.5m-1.7m below the surrounding wet rice fields. It is situated within Minh Duc Commune, Ung Hoa District, Ha Tay Province (latitude 20^0 43' North and longitude 105^0 5' 20" East). It is some 30.5km south of the Ung Hoa provincial centre, 10km west of the Red River or 45km south of Hanoi. Seven well preserved boat coffins were excavated, of which five contained preserved human skeletal material (Institute of Archaeology HS 224; Bui Van Liem & Pham Lun Quan 1990). Such boat burials are quite common in this region (Bui Van Liem & Pham Quc Quan 1991). Artefacts recovered included wooden objects such as trays, bowls, disks and hoes. Bronze artefacts included basins, axes, spearheads and disks. In addition burial 89MDM7 had two iron hoes and a ceramic vessel. Some of these artefacts have been interpreted as suggesting occupations relating to textile manufacture and lacquer working (Bui Van Liem & Pham Quc Quan 1991).

Duong Co (DC)

Duong Co is dated by way of material cultural associations to between 2200 and 1700 years bp. This site is situated in Chi Minh Commune, Phu Xuyen District, Ha Tay Province and has been ascribed to the Dong Son Culture period. The coordinates of the site are latitude 20^0 44' 20" North and longitude 105^0 50' 40" East. It was first excavated in 1965 and again in 1970. The site itself is situated on a mound some 3.5m above sea level and is approximately 1600m^2. It is 30km from Hanoi and about 1km from the present course of the Red River. A total of 11 burials were discovered of which four contained preserved skeletal material. Currently, only one individual (70DCM1) is available for examination. Artefacts buried with these individuals included both indigenous and Han style ceramics. The Han style ceramics included glazed vessels, bowls and jars. Other artefacts included jade beads, stone ear-rings, bronze goods such as bracelets, axes, harpoons and also an unidentified iron artefact (Institute of Archaeology HS 63).

Dinh Chang (DT)

This site is dated by way of material cultural associations to between 2200 and 1700 years bp. It is situated in Dong Anh District, Hanoi Province. Dinh Chang was first surveyed in 1969 and a 108m^2 excavation was undertaken in 1970. The poorly preserved skeletal remains of one child and seven adults were recovered. Bronze artefacts recovered from this burial included spear-heads and javelins, an arrow head, two axes and a bracelet. An unidentified iron artefact, in addition to numerous pieces of pottery, was also recovered (Nguyen Khac Su & Nguyen Lan Cuong 1997).

Doi Son (DS)

This site is dated by way of material cultural associations to between 2200 and 1700 years bp. It is located in Nam Ninh District, Nam Dinh (formerly Nam Ha) Province. The eleven burials discovered at this site were interred using a variety of methods (Nguyen Lan Cuong 1996). Three of these were boat burials, two were stone lined graves and two were buried in pits without any clear sign of a coffin. The remaining four individuals were wrapped or rolled in bamboo mats. The skeletal remains of seven individuals were available for examination.

Chau Son (CHS)

This site has a radio carbon date of 2285+/-45 years bp. It

is located in Kin Bang District, Ha Nam (formerly Nam Ha) Province. It was excavated in 1976 and the skeletal remains of one individual were discovered. Bronze artefacts associated with this burial included 8 spear-points, 2 axes, 2 cylindrical jars and a pot. A wooden bowl, various wooden implements and bamboo artefacts were also recovered (Nguyen Khac Su & Nguyen Lan Cuong 1997).

Dong Xa (DX)
This site is dated by way of material cultural associations to between 2200 and 1700 years bp. Excavated in 1995, it is approximately 40km east of Hanoi in Luong Bang village, Kim Dong district, Hung Yen province. One individual buried within a bamboo mat was uncovered. Grave goods included a bronze axe and bronze necklet (Nguyen Lan Cuong pers. comm.).

Ma River Sites
Quy Chu (QC)
Quy Chu is situated in Hoang Hoa District, Thanh Hoa Province. Two temporally discrete groups of burials have been identified from Quy Chu. The later group has two radio carbon dates of 2520+/-55 and 2450+/-45 years bp, while the earlier group is thought to be situated between 2500 and 3300 years bp based on material cultural associations. The principal excavation, 108m^2, took place in 1978 with the discovery of 43 burials. The remains of 20 individuals were available for examination. Two of these included the remains of very well preserved neonates buried within pottery jars. A wide range of burial goods were recovered. Bronze material included 2 spear-points, 3 javelin-heads, 1 arrow, 2 tubular or cylindrical objects, 6 axes, 2 ploughshares, 1 chisel, 1 hammer, 2 fishhooks, 2 awls, 2 needles, 2 knives and 12 bracelets. A bone ornament and the remains of bronze casting crucibles were also found (Nguyen Khac Su & Nguyen Lan Cuong 1997).

Nui Nap (NN)
There are three separate groups of dates for this site. The earliest, for which only two individuals were available, is dated to between 3000 and 2500 years bp. The next group, for which again only two individuals are preserved, is dated to between 2400 and 2000 years bp. The last group, including the remaining 30 individuals, has a radio carbon date of 1670+/-85 years bp. Nui Nap is a high limestone mountain in Dong Son Commune, Dong Hieu District, Thanh Hoa Province. It is situated some 10km to the west of the eponymous Dong Son site. It was first discovered and surveyed in 1962, with a follow up survey in 1976 and excavation beginning in 1977 (Institute of Archaeology HS 251; Ha Van Phung & Pham Ninh Huyen 1978; Vo Hung & Nguyen Lan Cuong 1979; Ha Van Phung 1984). Two pits were excavated, pit A being 90m^2 and pit B 92m^2. The cultural layers identified for each pit are identical. The site is on a 13^0 incline with all the burials, long axis of the body, being fully extended an oriented north or northwest. Most of the burials were rich in artefacts. Bronze objects included spear-heads, arrow-heads, daggers, axes, harpoons, vessels, earrings, drums, and beads. A variety of pottery and ceramic vessels were also found. Other artefacts included a spindle whorl, stone earrings, glass beads and three burials that included minted Han coins.

Thieu Duong (TD)
There are two groups of dates for this site, approximately 2500 to 2000 years bp for the later and around 2800 to 3000 years bp for the earlier. Both sets of dates are based on material culture associations and the only individual available for skeletal assessment belongs to the earlier group. This site is within Thieu Duong Commune, Thieu Hoa District, Thanh Hoa Province and has been ascribed to the Dong Son culture period. The site was surveyed during 1959-1960 and initially excavated between 1960-1961. The site, 4m above sea level, is 4km north-northeast of Thanh Hoa Town and 2km from the Ma River (Institute of Archaeology HS 55). The excavation totalled some 10,800m^2. The two lower cultural layers have been ascribed to the Quy Chu period (temporally corresponding to the more northerly Go Mun period culture) while the upper two layers were Han period (Pham Minh Huyen & Diep Dinh Hoa 1981). The excavation uncovered 18 Han period tombs and 37 Dong Son style tombs. The vast majority of skeletal material was very poorly preserved and only one individual (61TDM17) from a Dong Son style tomb remains today. The Han Tombs were rich in artefacts such as pottery, ceramics, bronze and iron. Bronze artefacts included mirrors, seals and incense burners, while iron goods included knives and swords. In addition, silver and gold as well as stone items were recovered from these tombs. With one exception, all of the Dong Son tombs contained artefacts, for the most part pottery and bronze. Stone artefacts, such as earrings and also a single flaked stone item, were recovered (Nguyen Dong Chi 1965; Nguyen Van Buu & Diep Dinh Hoa 1987).

Ca River Site
Dong Mom (DM)
This site is dated by way of cultural associations to between 2200 and 1700 years bp. Dong Mom is the most southerly of the northern Vietnamese sites examined here, being in Dien Chau, Nghe An Province. It was first excavated in 1976, 115m^2, and then again in 1992 on a much smaller scale. The skeletal remains of two individuals are available from the earlier and three from the later excavation. Artefacts recovered from these burials included bronze spear-heads, knives, ploughshares, axes, vessels and bracelets. Iron artefacts included a spear-head, a sword, a hoe and a knife. There was also a glass or rock crystal bracelet and an earring (Nguyen Khac Su & Nguyen Lan Cuong 1997).

The Environmental Context

Some insights into the environment have already been touched on when reviewing the historical data for the Metal period. The following summary, drawn from Nguyen Trong Dieu (1995), of the geography of the study region, while not necessarily fully representative of conditions several thousand years ago, provides a generalised environmental context. Topographically, while about three-quarters of Vietnam can be described as mountainous, 85% of these are below 1000m in elevation. Vietnam has three plains systems that are still in the process of expansion. The archaeological sites of interest here are situated in the northern Bac Bo plain and fall between latitude 18^0 and 22^0 north. This northern plain is low lying, averaging between 10 to 0.3m above sea level, and at the mouth of the Red River is expanding at a rate of 80-100m per year. The Gulf of Bac Bo is very shallow at 40 m and the coast line is still dominated by mangrove forests today.

Vietnam presents two distinct climatic zones, a northern and a southern. Northern Vietnam has two seasons, cold and hot but with high levels of humidity occurring during both periods. Further, the north experiences marked climatic variability or instability that has restricted levels of ecological variation in comparison to southern Vietnam. The summer monsoon occurs between May and September and is dominated by southwesterly winds. In Hanoi the highest rainfall is between June and September with a peak during August. The highest average daily temperature in Hanoi is 28.9^{0}C in July. Average annual rainfall in the Hanoi region in modern times is 1,678mm while average daily humidity is 84%. In winter the winds are from the northeast with the lowest average daily temperature in Hanoi being 16.5^{0}C in January. Vietnam is prone to typhoons and has experienced over 400 during the last one hundred years. Storms are also very frequent with winds reaching up to 50 m/sec. This in combination with rainfall in excess of 600mm in a 24 hour period can lead to extensive agricultural and human disruption.

The Bac Bo plains are dominated by the Red River system, which empties into the Gulf of Bac Bo via four estuaries. The massive alluvial content of the Red River gives rise to both the rapidly extending delta and the river's name. Bac Bo experiences the greatest tidal variations in Vietnam with high tides moving up rivers as far as 50km. The alluvial plains of Bac Bo are rich and highly suited to rice cultivation

Because the Bac Bo region is essentially subtropical, tropical forests are found only below about 500m while more tropical flora is only found below some 300m. Around the coastal regions mangroves still predominate, while dense bamboo forests are common all over the northern plains. Moving up in elevation are the cold resistant dense tropical monsoon semi-deciduous forests. Dense tropical forests on limestone soil were formerly extensive but exist as remnants now. Moving higher in elevation, other forms of vegetation are differentially distributed such as evergreen humid subtropical broad-leaved forests and humid subtropical dense leafy gymnosperm forests.

Northern Vietnam is home to a diverse range of bird, riverine and marine life. Some 900 species of fish are recognised in the Gulf of Bac Bo alone. Terrestrial animals such as sambar deer, muntjac, chamois and numerous arboreal primates are still common in the region. In the past elephant, rhinoceros, tiger, and panther were also common.

Thesis Structure

Each chapter in this study investigates aspects of health and behaviour from a specific perspective. Chapter two is concerned with developing a description of the basic skeletal biology of the samples, including preservation, sex, age-at-death and stature. Chapter three is devoted to examining the oral health and oral behaviours inferred from these samples. Apart from investigating behaviours the oral cavity is sensitive to, one of the specific questions this chapter addresses is: what inferences with respect to the diet and general subsistence strategies of these samples is evident from their oral health profiles? Chapter four investigates the prevalence and distribution of articular changes in both samples. The specific aim of this chapter is not so much concerned with inferring past behaviours, an enterprise fraught with difficulty as will be demonstrated, but with determining if the level and distribution of articular changes had an obvious negative impact on the health of these samples. As such, the ancient Vietnamese evidence for articular change is assessed against such changes manifested globally in both past and contemporary populations. A further aim of this particular study of articular change is to contribute data to the global database on articular conditions, such as osteoarthritis. Chapter five explores the frequency and distribution of a skeletal and a dental marker of physiological stress that each manifest in childhood. An analysis of such stress markers is useful in inferring population health both in childhood and in later adulthood. Chapter six asks: what is the evidence for trauma and specific disease in these samples and what can this tell us with respect to issues of population health and behaviour? The method of this chapter departs from the purely population-based approach characterising much of this study and presents twenty-one case studies deriving from both temporal samples. Detailed descriptions and differential diagnoses are presented for each case study. Chapter seven, the final chapter, concludes and synthesises the evidence for health and behaviour in these samples within the biological, cultural and environmental contexts developed in chapters one and two.

2

THE SAMPLES: DESCRIPTIVE SKELETAL BIOLOGY

Health needs to be understood with respect to the age and sex profiles of the samples under examination. Age and sex are important aetiological components of the manifestation and frequency of many diseases and measures of health. Further, the level and pattern of skeletal and dental preservation will impact on how a study is designed and what questions can be asked. The purpose of this chapter is to describe the basic skeletal biology of the samples. Specifically, this chapter is concerned with describing the preservation, sex, age-at-death and stature of the samples. Following a discussion of the preservation of each assemblage, the literature on age and sex estimation is reviewed. Sex and age profiles are then developed for each sample. Methods for estimating stature are also reviewed and applied to individuals represented by these skeletal remains.

Sample Preservation

The preservation of these skeletal series has not been fully described to date and such a description forms one of the aims of this chapter. A further reason for a description of sample preservation is to assist in determining what materials and methods will be employed in developing the age and sex profiles of these assemblages. Furthermore, the potentially confounding factor of sample preservation can be factored into later inferences regarding the health of these samples. In the following discussion the cranium is treated as a unit and individual cranial bones are not referred to. Vertebral preservation is described with reference to the three main types of vertebrae rather than individual elements. For the most part, descriptions of hand and foot bone preservation uses element classes such as carpals, metacarpals and so forth as the unit of analysis.

Cranium, Maxilla and Mandible

Cranial preservation for both samples is quite good although Da But period post-cranial preservation is better than that for the Metal period (Figure 2.1a). While few crania or calvaria were preserved in the Da But period sample, more than 40% of the assemblage was represented by calvae. Over a quarter of the Metal period sample was represented by complete crania, maxillae and mandibles. In fact, for both samples there was good preservation of gnathic elements with over two thirds of the individuals in each sample being represented by some maxillary and/or mandibular component (Figure 2.1b & 2.1c).

Clavicle, Scapula and Ribs

There is quite a degree of variation in post-cranial bone preservation. For the most part this is related to variation in robusticity, including cortical thickness, of these bones. Moreover, bones also vary in the relative robusticity of their parts (Nawrocki 1995). Some 42% of Da But period individuals were represented by clavicular remains, while 16% of the Metal period sample of individuals were represented by clavicular remains. In the Da But period sample the acromial end of the clavicle fared better than either the diaphysis or sternal end (Figure 2.2a). In the Metal period sample the acromial and sternal portions of the clavicle were the best preserved sections.

Some 30% of the Da But period series exhibited scapula elements with the best preserved portion being the glenoid region (Figure 2.2b). No complete scapulae were present in the Da But sample. In the Metal period sample approximately 10% of individuals were represented by scapula elements and eight complete scapulae were available for examination. As with the Da But series, the robust glenoid region of the scapula was the most frequently preserved section of this bone (Figure 2.2b).

Almost 47% of individuals in the Da But period series were represented by ribs. However, for the most part this was in the form of small non-diagnostic fragments. Figure 2.2c illustrates the preservation of ribs, elements with an articular end only, as a percentage of the expected number of ribs. A little over 12% of the Metal period sample was represented by the remains of ribs. As was the case for the Da But sample, a low percentage of preserved rib articular ends were available for study in the Metal period sample. The fragile nature of the rib, including its thin cortex and large surface area to volume ratio, likely accounts for the levels of preservation recorded for these samples.

Vertebrae and Os Coxae

Over 43% of the Da But sample exhibited vertebral elements. Figure 2.3a shows the proportion of vertebrae preserved by each class of vertebra. It is presented as a ratio of the expected number of bones for each vertebral category. The relatively high level of sacral preservation in the Da But sample is due to good preservation of the more robust promontory and wings. Only one complete sacrum was present in the Da But series. Sacra aside, the best preserved vertebral elements were the lumbar vertebrae. This is probably due to their size and robusticity. In the Metal period series 25% of individuals have vertebral remains. Four complete sacra were available for assessment in the Metal period sample, while as with the Da But series, the lumbar vertebrae were the best preserved (Figure 2.3a).

Some 60% of the Da But period and almost 20% of the Metal period series are represented by components of the os coxae. Figure 2.3b illustrates preservation with respect to overall completeness of the os coxae. The os coxae are not as well preserved as other large post-cranial bones in the Da But series. A possible reason for this is the habit of the squatting burial position practised at the Con Co Ngua cemetery. Forensic pathologist Dr. Kevin Lee (pers. comm.) suggested to me the accumulation of body juices in the pelvic region, associated with this burial position, could have contributed to the dissolution of the pelvic bones. The best preserved components of the os coxae in each sample included the ischium, greater sciatic notch area, acetabular region and auricular area. The auricular area is useful in age assessment while the greater sciatic notch has a role in sex determination.

Small Bones: Hand, Knee and Foot

In the Da But series 51% of individuals were represented by hand bones while 27% of individuals in the Metal period had assessable hand elements. It is perhaps more meaningful to describe small bone preservation in terms of the number of bones preserved relative to the expected number if all individuals were perfectly preserved. Assessed in this manner, it can be seen that metacarpals are the best preserved small bones of the hand in both samples, while phalanges fair slightly better than carpals in terms of preservation. (Figure 2.4a).

With regard to the patella, almost 18% of the expected number of patellae were preserved in the Da But series, while 5% of the expected number were available in the Metal period sample (Table 2.4a).

Looking at bones of the feet, the talus stands out as being the best represented foot element in both samples, and almost half the expected number of tali are present in the Da But sample (Figure 2.4b). The calcaneus and then metatarsals are the next best represented foot elements. The fragile phalanges of the foot are not particularly well represented in either sample.

Long Bones

The femur is the best preserved long bone in both assemblages. In the Da But series over 75% of individuals are represented by a femoral element. In the Metal period sample some 22% of Individuals have preserved femoral elements. Examining the preservation of other long bones, over 60% of Da But period individuals have preserved humeral and tibial sections, 55% have radial and 64% have ulnar portions. In the Metal period series a little under 20% of individuals have tibial elements, 23% humeral, 18% radial and 19% ulnar elements.

There are two clear patterns of long bone preservation in the Da But period sample (Figure 2.5a). The femur, radius and ulna all show a trend to better preservation of the proximal end of the bone. The humerus, tibia and fibula all show a trend to better preservation of the distal end of the bone. This is likely related to relative robusticity of the respective ends of these bones. The overall pattern for the Metal period sample is for greater preservation of the shaft of the bone rather than any extremity (Figure 2.5b). With this description of the pattern and level of skeletal preservation of the samples, the most appropriate skeletal elements for sex and age estimation can now be examined.

Sex Estimation

> Normal sexual development may result from the functional and developmental integration of a number of different genes that play roles in sex determination, sexual differentiation, and sexual behavior.
> Vilain & McCabe (1998:74)

Moreover, this complexity can lead to ambiguous sexual differentiation (Haqq & Donahoe 1998) and this ambiguity is also encountered when estimating the sex of archaeological skeletal remains. However, for the most part it is possible to sex human skeletal remains with some level of certainty, although accuracy is dependent upon the completeness of the skeleton (Meindl *et al.* 1985a).

Before selecting particular bones and applying sex estimation techniques, a critical examination of a range of morphometric and non-metric morphological methods commonly employed in sex estimation is undertaken. Following this review, appropriate techniques are applied to the Vietnamese samples in order to develop a sex profile of the samples.

Non-Metric Morphological Traits and Sex Estimation

> Bones vary sexually in so many ways and often to such small degrees that it is infrequent, if not rare, to encounter a skeleton that is either typically male or female in absolutely every regard.
> (Wienker 1984:230)

St. Hoyme & Iscan (1989:59) suggest that when "we generalise that women are shorter than men, we forget that for most physical characteristics, both sexes share about 95% of the total range of variation". However, reasonably high levels of sex estimation accuracy have been reported for various morphological traits, and suites of traits, in the human skeleton. For the most part these traits are restricted to the skull and pelvis.

The Cranium and Mandible

Krogman (1962) notes a long held preference by researchers in using non-metric morphological traits for sexing the skull. He refers to early workers in this area such as Weisbach in the late 19th century and Pittard, Mobius and also Parsons in the early 20th century, and goes on to summarise the classic non-metric morphological criteria,

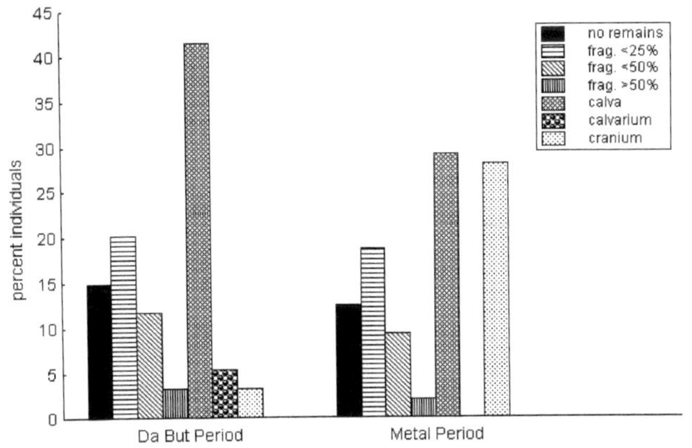

Figure 2.1a: Cranial Preservation

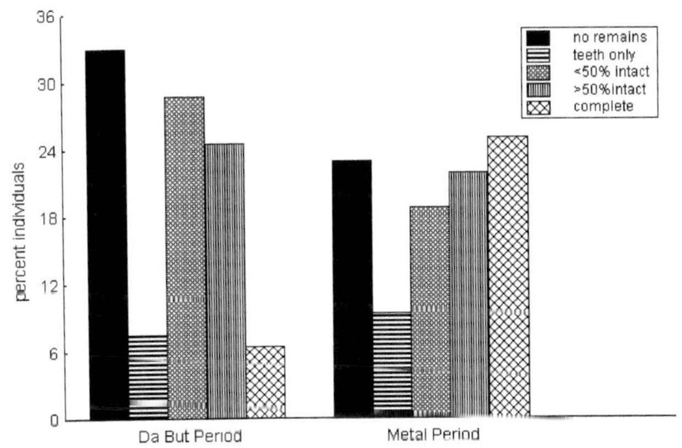

Figure 2.1b: Maxilla Preservation

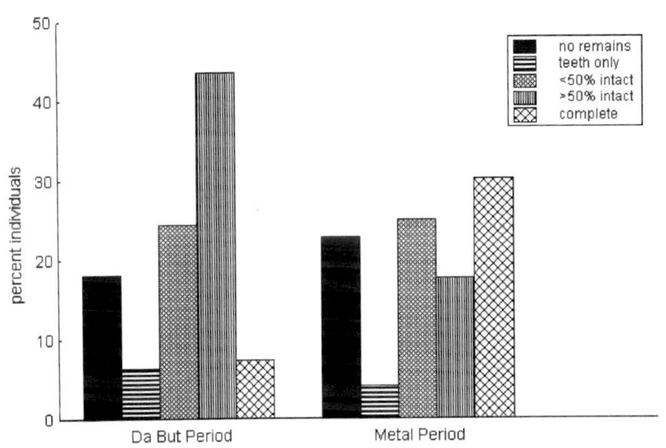

Figure 2.1c: Mandible Preservation

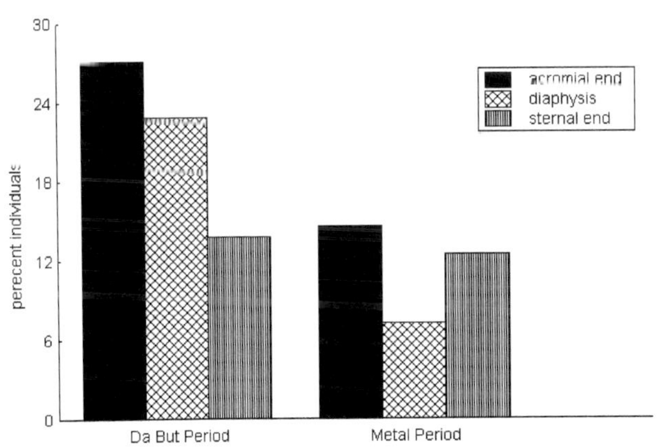

Figure 2.2a. Clavicle Preservation

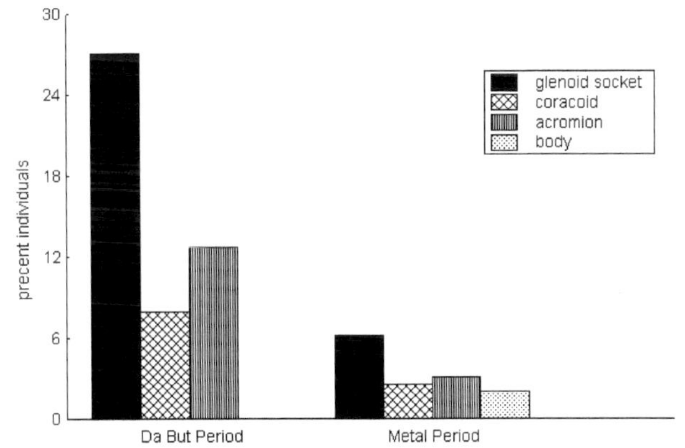

Figure 2.2b. Scapula Preservation

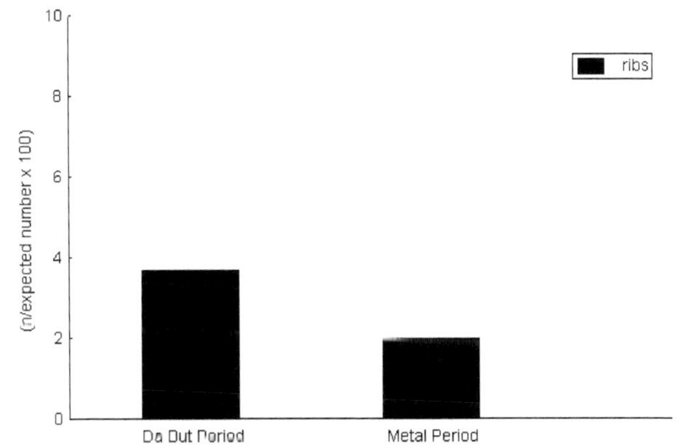

Figure 2.2c. Rib Preservation (articular ends only)

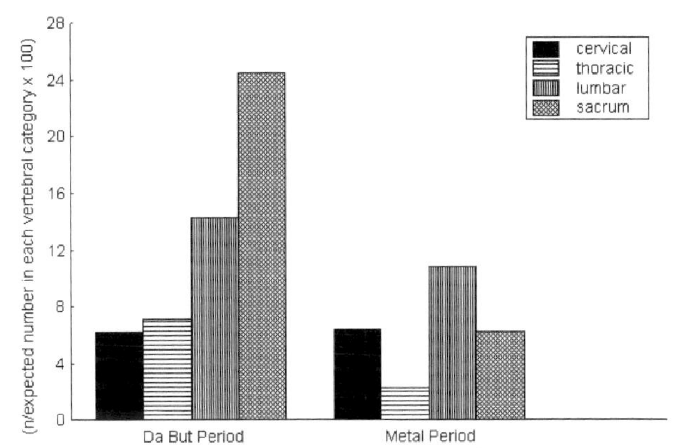

Figure 2.3a. Vertebral Preservation

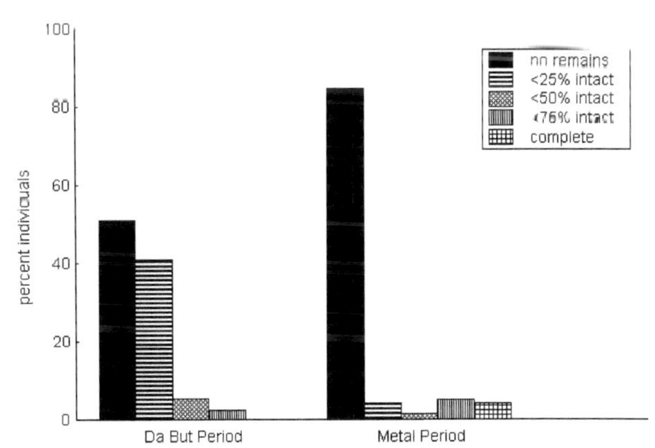

Figure 2.3b. Os Coxae Preservation

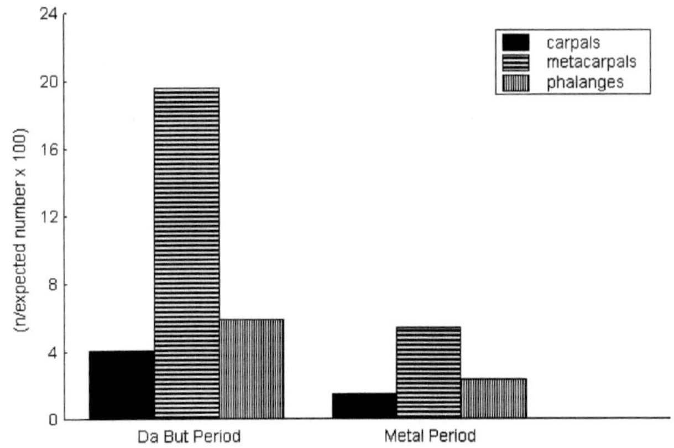

Figure 2.4a. Hand Bone Preservation

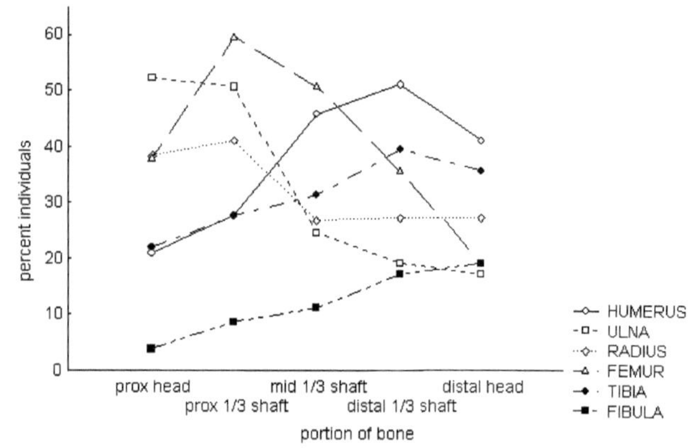

Figure 2.5a. Long Bone Preservation: Da But Period

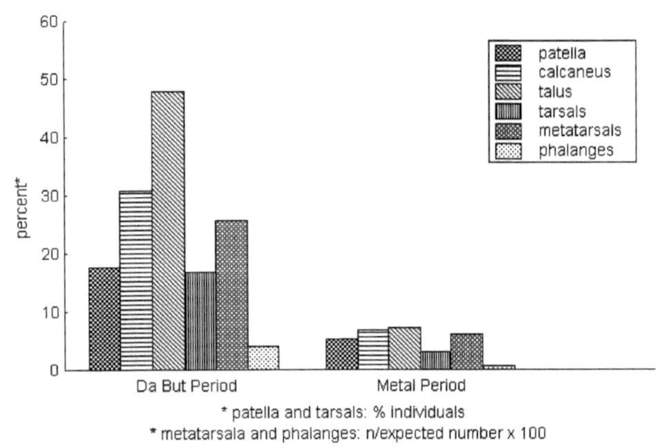

Figure 2.4b. Patella and Foot Bone Preservation

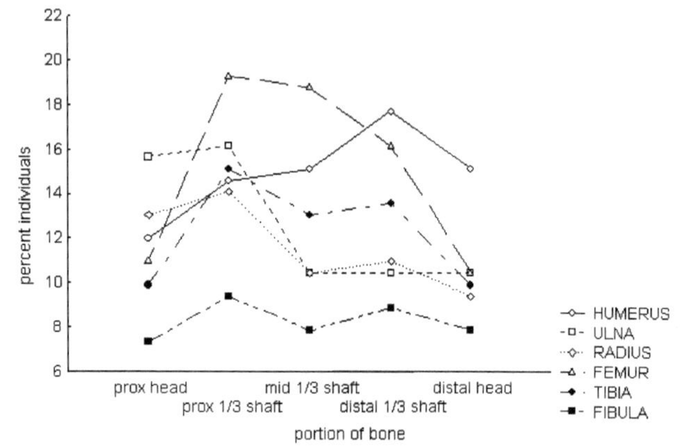

Figure 2.5b. Long Bone Preservation: Metal Period

still used today, in sexing the skull. Krogman (1962) cites the reported sexing accuracy of such techniques, as used by Stewart and Hrdlicka in the mid-twentieth century, as being consistently above 80% for the most part.

Standard osteological texts (e.g. Brothwell 1981; Bass 1987; Ubelaker 1989; White & Folkens 1991) and general summaries of sexing criteria (e.g. Buikstra & Mielke 1985) continue to list traditional non-metric morphological traits as very useful in sexing the skull. Nonetheless, there have been no recent detailed and controlled studies of the accuracy of cranial morphological traits used for sexing, especially with regard to inter-ancestral variation, in the presence or manifestation of these traits.

Meindl *et al.* (1985a) noted that discriminant function sexing of the crania using Giles & Elliot's (1963) functions were not as accurate as non-metric morphological assessment. While morphology seemed to have superior accuracy, they suggested part of this was attributable to the population specific nature of the discriminant functions. They also noted that it "is not that the linear discriminant functions in their present state of development fail to formalise what the eye discerns; rather, the two approaches [morphology and discriminant analysis] utilise overlapping but somewhat different sets of cranial information" (Meindl *et al.* 1985a:84). Nonetheless, researchers (Funayama *et al.* 1986; Inoue *et al.* 1992) have formalised and corroborated the use of several traditional cranial morphological traits. One additional factor to consider is Meindl *et al.'s* (1985a) discovery that age was a significant factor to consider in sexing the cranium. Increasing age is associated with an increasingly male morphology.

The addition of the mandible in sexing has been reported to increase the accuracy of sex estimation (Krogman 1962). The popular osteological texts referred to above also list traditional sets of morphological criteria for sexing the mandible. For example, St. Hoyme & Iscan (1989:74) note, amongst other traits, that "small round chins are typical of female sex; square, heavier chins ("Dick Tracy" type) are found in many males". However, a recent study comparing the morphological utility of the pelvis, cranium and mandible (Maat *et al.* 1997) suggested that the mandible in isolation was a very poor estimator of sex. Recently a new morphological trait has been added and has been proposed as the single most important trait for sexing the mandible. Loth & Henneberg (1996:475) describe a

> distinct angulation of the posterior border of the mandibular ramus at the level of the occlusal surface of the molars in adult males…in most females, the posterior border of the ramus was straight or, if flexure was observed, it was found to occur at a higher point near the neck of the condyle.

In a sample of 200 black African mandibles they developed a scoring system for the flexure based on presence, absence and ambiguous morphology and achieved 99% sexing accuracy. They tested the technique on a variety of different ancestral samples with accuracy ranging up from 91%. They caution that sex estimation efforts can be compromised because the mandible is susceptible to remodelling caused by a number of factors such as disease, masticatory stress and tooth loss. Nonetheless, as they point out, this places the mandible as at least as accurate, if not more, than the pelvis for the determination of sex. Moreover, the use of this trait surpasses the sex estimation accuracy of any metrical assessment of the mandible.

In a study exclusively examining juveniles and young adult females Koski (1996) argued that the presence of flexure was extremely common. Moreover, it was suggested that Loth & Henneberg's (1996) definition of flexure was somewhat vague (Koski 1996). Loth & Henneberg (1998) dismissed Koski's (1996) conclusions and suggested that Koski's (1996) sample was inadequate and the variable analysed was not mandibular flexure as defined in their original (Loth & Henneberg 1996) paper. In support of Loth & Henneberg's (1996) method is a study by Indrayana *et al.* (1998), which quantified mandibular flexure using Fourier analysis on a sample of Javanese (n=135) and Flores Islander (n=150) radiographs. Very high levels of accuracy were reported, ranging from 88% to 90% for the male samples and 94% to 95% for the female samples.

However, several studies have drawn attention to potential problems in using this technique. Donnelly *et al.* (1998) using a sample of 96 mandibles, the vast majority being an archaeological assemblage of native Americans, could only achieve an overall sex estimation accuracy of 62.5%. For males sexing accuracy was 80.4% compared with females at 50.0%. Most recently Haun's (2000) study, using an archaeological sample, reported 80.4% sexing accuracy when sexing adults. Moreover, whereas male adults were sexed with 96.3% accuracy females, on the other hand, were sexed with only 62.5% accuracy. Similar results were reported by Hill (2000) using a sample of adult mandibles of known age and sex from the Hamann-Todd collection. Hill (2000) also found ancestral differences in the accuracy of the mandibular flexure method. White males were sexed at 89.2% accuracy compared to 94.7% for black males, while white females were sexed with 60.7% accuracy compared to 51.9% for black females. The overall sexing accuracy for both ancestral groups and sexes was 79.1%. Hill (2000:575) states that "[t]hese discrepancies between the sexes were unexpected and indicate that flexure is more likely to be identified than its absence, leading to consistent misidentification of females".

It is unclear why Loth & Henneberg's (1996) and Indrayana *et al.'s* (1998) results are so encouraging while Donnelly *et al.* (1998), Hill (2000) and Haun's (2000) findings are so contraindicative. It is not improbable that recognition of flexure as originally defined by Loth & Henneberg (1996) is a major problem. If this is true, Indrayana *et al.'s* (1998) quantification of flexure may be a useful technique to deal with the issue of inter-observer error in assessment of this trait.

The Pelvis

St. Hoyme & Iscan (1989:76) note that "the adult pelvis has the best-known and most reliable sex identifiers". The use of the pelvis in sexing also has a long history (Hoyme1957). There are a large number of morphological traits and indices available for sexing this bone (Krogman 1962; Krogman & Iscan 1986; Schulter-Ellis *et al.* 1983; Schulter-Ellis & Hayek 1988). However, it is only recently that more vigorous testing of the usefulness of these characters has been undertaken.

Rogers & Saunders (1994) have examined both accuracy (% correctly sexed) and precision (intra-observer error in sexing) with regard to 17 pelvic traits traditionally used to sex this bone. Taking all pelvic traits together resulted in 95.9% accuracy but at the expense of 11.3% intra-observer error. The greatest levels of accuracy (98%) were obtained with the obturator foramen shape along with either true pelvis shape or ventral arc. Various other two and three trait combinations gave accurate allocation in excess of 95%. Moreover, 12 of their 17 traits performed at 80% or better accuracy but three of these, acetabulum, preauricular sulcus and ischiopubic ramus, exhibited very high levels of intra-observer error, thus mitigating against their use in sex estimation. The greater sciatic notch, as a single sex estimator, achieved 85.7% accuracy with an intra-observer error rate of 6.5%. The top six traits performing well in accuracy and precision include in order of utility: ventral arc, obturator foramen, true pelvis, sacrum shape, subpubic concavity and pubis shape. While this is an important and timely piece of work, the results need to be corroborated with a larger sample. Their sample comprised only 49 individuals. Further testing among a variety of ancestral groups also needs to be carried out.

There has also been a renewed interest in the utility of Phenice's (1969) technique for which 96% accuracy was originally reported for sexing the pelvis. Phenice's (1969) technique involves assessing the morphology of three aspects of the pubic bone: the ventral arc, subpubic concavity and medial aspect of the ishiopubic ramus. The results of these follow up studies have been mixed. At one end of the scale MacLaughlin & Bruce (1990), using a sample of 273 modern European dissecting room and 17-18th century cemetery burials, found quite low levels of accuracy overall; between 58.6% and 82.7% depending on the sub-sample. They also found differences in the usefulness of each of Phenice's sex markers when looked at individually. Subpubic concavity proved the most reliable with the other two adding little to the accuracy of sex estimation. They also found that the experience of the assessor played a significant role in correctly assigning sex. Phenice (1969), on the other hand, suggested that experience was not necessary to successfully use this technique.

MacLaughlin & Bruce (1990) suggest that, in part, lower accuracy probably related to their use of an indeterminate category and also not estimating sex unless at least two of the three traits gave the same value (male or female). They note that they had more luck with the English sub-sample because it was more sexually dimorphic than the other two samples. It would appear that levels of sexual dimorphism are important when interpreting the results of this technique.

Lovell (1989) tested Phenice's technique by having 12 researchers independently sex a sample of 50 pubic bones of known sex with an age distribution of 52-92 years. Lovell (1989) obtained an average accuracy of 83%. Replicating the test on two smaller series gave better results. Lovell (1989) suggests some of the difference in reported accuracy of the technique could be due to differing age distributions of samples used. In fact, she found that age could negatively affect sexing accuracy.

In Rogers & Saunders' (1994:1054) study they "reproduced the original rankings of the Phenice criteria for sex determination". However, their overall accuracy was lower than Phenice's at 88%. They also suggested the possible confounding affects of differential age distribution profiles in the various populations examined. However, they found no significant correlation of sexing accuracy with individual age in their own, as they admit, fairly age-restricted sample.

The most positive results testing any of Phenice's (1969) traits come from Sutherland & Suchey (1991). However, their work has limited value as they only examined the ventral arc, as the subpubic concavity could not be assessed due to their particular method of removing the pubis. Also, the ridge on the ischiopubic ramus was found to be too unreliable (only 70% of sample correctly sexed) to be reported on.

Sutherland & Suchey (1991) found that the percentage of mature females without any evidence of the ventral arc was rare (4%). They state that "the ventral arc does not usually appear in its easily recognisable form until the mid-20s" (Sutherland & Suchey 1991:504). They also note the presence of a precursor arch that "first appears at age 14 in the sample, and at age 20 it becomes the most frequent condition" (Sutherland & Suchey1991:504). They further point out that old age does not have a significant effect on the presence of the arc. The arc has a sexing utility from early adolescence through to advanced age. Notwithstanding this however, the absence of a precursor arc in an adolescent would be equivocal. Absence would mean sex could not be determined reliably. A sexing bias toward males would be introduced using the presence/absence of the arc in young adult cases. Sutherland & Suchey (1991) also suggest that there were not great differences in accuracy, using the ventral arc, between experienced and unexperienced assessors.

It is clear from these studies that more work needs to be done on sex estimation of the pelvis utilising as large a range of ancestral groups as possible. With regard to Phenice's three traits, MacLaughlin & Bruce (1990: 1391) say that no authors who have examined these traits have explained "fully the developmental or functional basis of the dimorphism of the three morphological traits of the os pubis". Moreover, "neither the ventral arc, subpubic concavity, or ischiopubic ridge is described in any detail in any standard anatomical reference text" (MacLaughlin & Bruce 1990: 1391). Further, such fundamental work should also be extended to include the range of characteristics examined by Rogers & Saunders (1994).

Morphometrics and Sex Estimation

While discriminant function analysis of human skeletal remains for the purposes of sexing has been used for some time (e.g. Pons 1955), it is only in the last couple of decades that the technique has become particularly popular. No doubt there is a non-fortuitous correlation with the increase in sophistication and ease of use of personal computers and statistical software. Discriminant function analysis studies have produced consistently high rates of sex estimation accuracy for a considerable variety of cranial and post-cranial bones. Here I concentrate on those skeletal elements and morphometric traits that show high accuracy, and are available for study in the Vietnamese samples.

Cranium

Recent work on Asian cranial material includes Song *et al.'s* (1992) analysis of a Chinese sample using traditional variables. They achieved in excess of 96% accuracy in sex assignment. Funayama *et al.* (1986) have looked at lateral cranial form and metrically described the development of the glabella eminence and nasal concavity in male Japanese skulls. Inoue *et al.* (1992) have looked at overall lateral cranial form in 100 Japanese skulls and quantitatively confirmed traditional qualitative morphological assessments that chief areas of sexual difference involve greater supraorbital development and stronger nasal bone development in males. Females, on the other hand, exhibit a "forehead contour" that is "higher, smoother, rounder and more vertical" (Inoue *et al.* 1992:112). Discriminant function sexing of their sample using variables describing lateral form ranged from 86% to 90% in their experimental sample and 85% to 88% in their control sample.

Mandible

Giles (1964, 1970) achieved good sexing accuracy with black and white ancestral group-specific discriminant functions derived from mandibular measurements. Hanihara (1959, cited in Giles 1970) developed a function derived from Japanese mandibles with 85.6% sexing accuracy. More recent work by Potsch-Schneider *et al.* (1985) looked at 17 mandibular variables from which functions were derived, giving sex allocation results of 79.5% (females) and 82.6% (males) from intact mandibles and 71.6% to 79.5% (females) and 72.5% to 81.7% (males) in fragmentary jaws.

Maxilla
A recent study of the maxilla (Burris *et al.* 1998) analysed casts from a living sample of American blacks and whites. The focus of the work was a number of palatal width and depth (length) variables, all taken from major buccal cusp landmarks. They had more success in estimating ancestry than sex. However, when ancestral group was known they could correctly sex two-thirds of their sample. In this study they found that "sex discrimination…depends largely on inter-molar widths, both in blacks and whites. This may simply reflect the greater overall breadth of the male skull in any given ethnic group" (Burris *et al.* 1998:962).

Scapula
Apart from earlier work looking at the scapula (Krogman 1962; Stewart 1979), Di Vella *et al.* (1994) have recently examined this bone for its sexing utility in a contemporary Italian population. They found the best single variable to be maximum scapula breadth. The size of the glenoid cavity alone also provided good sexing accuracy. This is important given the robust nature of this portion in fragmentary samples.

Humerus
The humerus has been favoured in quantitative sexing due to its general robusticity, and hence good preservation in archaeological and forensic contexts. Work by Dittrick & Suchey (1986) found the humeral head a better sex estimator than mid-shaft variables. Holman & Bennett (1991) looked at the three long bones of the arm, in particular long bone length and two variables looking at wrist robusticity, with promising results. They also noted differences in accuracy between blacks and whites in their samples.

Recent work by Iscan *et al.* (1998) on sexual dimorphism in the Thai, Japanese and Chinese humerus is important. They corroborate the notion that patterns of sexual dimorphism are quite population specific. Using broad racial group standards, such as Asian, will result in sex estimation errors.

Os Coxae
Apart from qualitative morphological sexing of the pelvis, metrical estimation of pelvic sex has also been quite successful. The ischium-pubis index (Washburn 1948) has long been used in sex estimation. More recently Schulter-Ellis *et al.* (1983) and Schulter-Ellis & Hayek (1988) have developed a number of pelvic indices that, particularly in combination with femoral head diameter sorting, achieve good rates of sexing accuracy. Di Bennardo & Taylor (1983) have developed functions with great accuracy from the Terry collection. Their study looked at discriminating both sex and ancestry. Gomez-Pellico & Fernandez-Camacho (1992) have found specific sex differences in the anterior border of the innominate that are useful for sexing both complete and fragmentary remains. Recently Luo (1995) detailed a study of Chinese material where 100% accuracy was achieved sexing the pubis from functions developed from four measurements.

Femur
The femur has attracted a lot of attention in the metrical sexing literature since Pons' (1955) early work which reported 95% sexing accuracy with a discriminant function analysis of this bone. A number of studies have found the femoral head to be exceptionally useful in sex estimation (Dittrick & Suchey 1986; Liu 1989; and see also Schulter-Ellis *et al.* 1983; and Schulter-Ellis & Hayek 1988). However, it is worth noting that Schulter-Ellis & Hayek (1988) found greater sexual dimorphism in femoral head diameter for American black and white samples (Terry collection) than in their native American and Eskimo samples. They refer to MacLaughlin & Bruce (1985) in suggesting this could be due to lower levels of sexual dimorphism in Asian populations. King *et al.* (1998) also found the femoral head to be of lesser importance in sexing Chinese material in a study that applied Thai derived functions to various ancestral groups. However, the femoral head was very important for sexing the Thai sample. Moreover, a recent study of Chinese femora (Liu 1989) found the best single variable for sex discrimination to be maximum femoral head diameter. Still another analysis of recent Chinese material (Iscan & Shi 1995) found the three variables, distal epiphyseal breadth, maximum length and antero-posterior diameter of the mid shaft, to be the most discriminating in sex estimation.

Black (1978) was the first to examine femoral diaphyseal width for its possible sexing utility. His reasoning was that the "dimorphism in bone diameters is due to sexual differences in bone remodeling in the tubular bones during adolescence" (Black 1978:227). The encouraging results of this study were followed up on by Di Bennardo & Taylor (1979). As well as corroborating Black's (1978) results they noted that age had a minimal affect on femoral circumference and can be essentially ignored as a factor. MacLaughlin & Bruce (1985) pointed out that Black's (1978) method could not be practically applied to fragmentary remains, as a complete femur was needed from which to measure mid-shaft width. They developed a technique for consistently establishing a landmark on damaged femora and achieved over 90% accuracy using their antero-posterior diameter variable. Recently, Seidmann *et al.* (1998) have achieved high levels of sexing accuracy, between 87% and 92%, using three functions describing femoral neck width.

Tibia
The tibia has proven as useful as the femur in sex estimation. Iscan & Miller-Shaivitz (1984a) looking at blacks and whites in the Terry collection found proximal epiphyseal breadth the best indicator of sex, regardless of ancestral group. In another publication using the same material, but only examining length and dimensions at the

nutrient foramen, they found that "sexual dimorphism was better assessed from the circumference than the length for whites. For blacks and American Indians, both lengths of the femur and tibia were better indicators of sex than the circumference" (Iscan & Miller-Shaivitz 1984b:56).

Steyn & Iscan (1997) found proximal and distal epiphyseal breadths, followed by breadth taken at the nutrient foramen (to the exclusion of tibial length), to be useful in sexing a South African white sample. Iscan *et al.* (1994) also found proximal and distal epiphyseal breadths to be the most discriminating morphometric traits in a recent Japanese sample. Using the Hamann-Todd collection, Holland (1991) found the dimensions of the articular facets of the proximal tibia very accurate in sex estimation. In a follow up study Kieser *et al.* (1992) corroborated Holland's (1991) technique on a sample of South African whites and blacks. They found bi-articular breadth to be the best single discriminating variable of sex.

Having reviewed biometrical work on specific bones, it must be further emphasised that functions derived from one population cannot generally be applied to ancestrally different populations (Giles 1970; Meindl *et al.* 1985a; Iscan *et al.* 1994; Iscan & Shi 1995). Further, it has been suggested that even short temporal hiatuses between ancestrally homogenous groups can result in difficulties in applying the same functions to each group (Iscan *et al.* 1994). For Vietnam there are no studies providing sexing functions for contemporary individuals. Even if there were, the applicability to archaeological populations up to 6000 years old, which may not be Vietnamese anyway, would be questionable.

Work has been done on contemporary Chinese femora (Liu 1989; Iscan & Shi 1995), pubic bones (Luo 1995), crania (Song *et al.* 1992) and humeri (Iscan & Shi 1995; King 1997). In addition to this is work on contemporary Thai femora (King 1997; King *et al.* 1998) and humeri (King 1997). Moreover, King *et al.* (1998) compared Thai sexing functions with Chinese, South African white, American white and American black femora. The results indicated that only the Chinese material was consistently correctly sexed with high levels of accuracy using the Thai formula. However, differences in the sexing utility of measurements were not identical for Thais and Chinese. They suggest that the reason for this is that although both are Asian populations, they are not homogenous and exhibit their own patterns of skeletal variability. Clearly the applicability of these functions to ancient Vietnamese remains would be problematic.

Estimating the Sex of the Vietnamese Skeletal Material

Methods and Materials

The level of preservation impacts on the number of available elements that can be employed in sex estimation. Effective sexing of these remains necessitated the use of both morphometric (quantitative) and non-metric morphological (qualitative) approaches.

Non-Metric Morphology and Sex Estimation

The sex of the cranium was estimated using five qualitative traits. These five traits were chosen both for their proven utility in sexing the cranium (Krogman & Iscan 1986; Bass 1987; Ubelaker 1989) and because they are preferentially preserved in the Vietnamese sample. These five variables included the morphological expression of the supra orbital ridges, orbital rim, occipital region, posterior zygomatic region and mastoid size. Consistent scores for at least two traits, with the exception of occipital morphology, were required to estimate the sex of any individual. Occipital morphology was difficult to assess with respect to sex unless there were very obvious nuchal markings. Moreover, while Funayama *et al.* (1986) and Inoue *et al.* (1992) have metrically confirmed the utility of supraorbital morphology in sex assignment, the evidence indicates that the occipital angle is not useful in sex estimation (Inoue *et al.* (1992). In the case of tied scores, for example two traits indicating male, and two female, the sex was recorded as indeterminate. In cases where there was a majority of only one trait indicating the sex (for example two male scores and one female; or only one trait could be assessed and sexed) then the sex was estimated to be either uncertain-male or uncertain-female.

Despite potential problems, mandibles were assessed for probable sex based on Loth & Henneberg's (1996) system of mandibular flexure. Indrayana *et al.* (1998) report a high degree of sexing accuracy with an Asian (Indonesian) sample. The morphology of the chin was also assessed for masculinity/femininity (Bass 1987). Disagreement between pelvic and cranial sexing was rare and when it occurred could be attributed to the fragmentary nature of the more traditional, pelvic and cranial, skeletal region used to assess sex.

The fragmentary nature of the sample limited the number of assessable pelvic sexing traits. Five traits were differentially available for assessment and included the pubic ventral arc, subpubic concavity, medial aspect of the ischiopubic ramus, greater sciatic notch and obturator foramen shape. Although there is some controversy over the accuracy of Phenice's (1969) traits, it would seem that they are still quite useful in sex determination and are used in this study. By far the most commonly preserved pelvic structure was the greater sciatic notch. While the human sciatic notch is the most sexually dimorphic among the hominoids (Hager 1996), it has been suggested that it is not a particularly reliable trait in isolation (MacLaughlin & Bruce 1986; St. Hoyme & Iscan 1989). Nonetheless, Rogers & Saunders' (1994) recent review of pelvic morphology and sex estimation reports 85.7% accuracy in using greater sciatic notch morphology in isolation. The level of preservation meant that sex estimation was recorded with one positive sex score for the pelvis. Nonetheless, a positive score for obturator foramen shape was not acceptable in isolation. The same criteria employed with the cranium for tied sex estimation scores was employed. However, a majority of one sex score was accepted for positive sex estimation, unless it included the obturator foramen. As the vast majority of os coxae were sexed using the greater sciatic notch in isolation, there were very few confounding cases. An indeterminate sex indicated that no pelvic traits clearly indicated male or female, or there was a tie.

A final non-metric morphological sex for each individual was recorded based on the results of the cranial, mandibular and pelvic scores. A single positive sex score (whether that be cranial, mandibular or pelvic) was necessary for assigning that case a sex of male or female. All conflicts in sex score were treated as indeterminate cases. For example, a score of male for cranial morphology and female for pelvis was treated as indeterminate sex. For cases where the original sex scores (cranial, mandibular, or pelvic) gave an uncertain result, such as uncertain-male or uncertain-female, this estimate was retained for the final non-metric morphological sex of the individual. Exceptions to this occurred where, for instance, the cranium was sexed as uncertain-male but the pelvis was sexed as male. In such cases the individual would be sexed as male.

Morphometrics and Sex Estimation

Use of discriminant function analyses in sexing human skeletal remains has been discussed. In developing useful sexing functions from single and multiple measurements of the skeleton, the sex of the individuals from which these functions are derived must be known. This is clearly a problem in an archaeological assemblage. Sex must be determined independently of the variables selected in a discriminant analysis. In such studies the sex is generally determined qualitatively by way of pelvic and or cranial morphology. In this study the individuals used for generating discriminant sexing functions were first sexed by way of the non-metric morphological criteria discussed above. Only adult remains, not displaying epiphyseal non-fusion, were used in developing the discriminant equations.

Due to the fragmentary nature of the samples, discriminant function equations for a number of the more commonly preserved skeletal elements were developed and assessed for sex estimation accuracy. The extremely fragmentary nature of Da But period cranial remains mitigated against the use of morphometric cranial variables for sex estimation. However, because of the good preservation of mandibles and maxillae, in both samples, they were investigated for sex estimation utility. Post-cranial elements such as the humerus, tibia and femur were selected for analysis due to their good representation in the Da But period sample and their demonstrated sexing utility even when fragmentary.

While articular facet dimensions have proven quite useful in sex estimation, they tended to be missing or too badly eroded post-mortem to provide a useful sample here. The glenoid cavity of the scapula was also potentially useful, but was not utilised as too few independently sexed scapulae were available. The talus and calcaneus, although well preserved, proved to be unreliable for sex estimation. This is not consistent with the findings of other studies such as Steele (1976) and Introna *et al.* (1997), who had better results. Much of the problem stems from the use of only one talar length and two calcaneal length variables in this study. However, Introna *et al.* (1997) found maximum calcaneal length to be the best univariate sex discriminator at 83.75% predictive accuracy. This variable was found to have no sex estimation value in this study with a 50% chance of correct sex assignment. This may reflect differences in the sexual dimorphism between Asian and European calcanea. Table 2.1 lists the variables utilised in this study for morphometric sex estimation.

Both the Da But period and Metal period samples were pooled for the purposes of sex allocation. The evidence is currently equivocal with respect to their biological distance. In a recent study of epigenetic characters in each population Parker (1998) suggests homogeneity for the pooled groups. On the other hand, Matsumura *et al.* (2001) suggest, on the basis of dental metrics and morphology, that the Da But period sample can be described as Australo-Melanesian in distinction to the Metal period sample which is seen as 'Mongoloid'. While this question cannot be resolved, at present it is not of great concern. The sample to benefit the most from discriminant function sexing is the Da But period assemblage. Moreover, the vast majority of post-cranial material used to generate these functions derives from this same sample. As will be seen, the Metal period sample did not greatly benefit from the use of discriminant function sexing. Additionally, the small sample size of Metal period post-cranial material mitigated against developing sample specific sexing functions.

Sex estimation equations were generated using the Discriminant Function Analysis facility in SPSS for Windows, version 8. The two principal assumptions of discriminant function analysis are multivariate normality and homogeneity of population covariances (Green *et al.* 1997). However, Green *et al.* (1997:331) point out the impracticality of ever achieving multivariate normality and suggest that the analysis is robust to departures from this given moderate to large sample sizes. The second assumption, the equality of group covariance matrices, can be tested using the Box's M statistic in the SPSS statistical package and this was done in the following analysis. However, Stojanowski (1999) notes the generally unrealistic nature of this second assumption as well. "It is possible for samples to exhibit nonnormality and unequal covariance matrices yet still be classified accurately with a linear discriminant function. In many cases these assumptions are unrealistic, and hence the tendency to ignore them in the literature" (Stojanowski 1999:249). Nonetheless, for this analysis Box's M was used to test the equality of group covariance matrices, even though negative findings do not necessarily refute the accuracy of

Table 2.1. Post-Cranial Measurements Used to Estimate Sex and Stature

Bone	Measurement	Abbreviation	Reference	Instrument
maxilla	arch length	(MX)AL	1	a
	arch length from infradentale superius[1]	(MX)INFDAL	2	a
	arch breadth	(MX)AB	1	a
	external (maxilloalveolar) length	(MX)EL	3	a
	external (maxilloalveolar) breadth	(MX)EB	3	a
mandible	arch length	(MN)AL	1	a
	arch length from infradentale[2]	(MN)INFDAL	2	a
	arch breadth	(MN)AB	1	a
	bigonial breadth	(MN)BB	3	a
	maximum height at M1	(MN)HTM1	4	a
	maximum breadth at M1	(MN)BM1	4	a
scapula	glenoid cavity length	GCL	3	a
humerus	maximum length	(H)MXL	3	b
	maximum diameter midshaft	(H)MSMXD	3	a
	minimum diameter midshaft	(H)MSMND	3	a
	maximum diameter head	(H)MDH	3	a
	vertical diameter head	(H)VDH	3	a
	transverse diameter head	(H)TDH	3	a
	biepicondylar width	(H)BEW	3	a
	articular width distal end	(H)AW	3	a
	shaft least circumference	(H)SLC	3	c
	maximum diameter at nutrient foramen	(H)NFMXD	2	a
	minimum diameter at nutrient foramen	(H)NFMND	2	a
Femur	maximum length	(F)MXL	3	b
	bicondylar length	(F)BL	3	b
	antero-posterior diameter mid shaft	(F)MSAPD	3	a
	medio-lateral diameter mid shaft	(F)MSMLD	3	a
	maximum diameter head	(F)MDH	3	a
	midshaft circumference	(F)MSC	3	c
	subtrochanteric antero-posterior diameter	(F)STAPD	3	a
	subtrochanteric medio-lateral diameter	(F)STMLD	3	a
	antero-posterior diameter at nutrient foramen	(F)NFAPD	3	a
	medio-lateral diameter at nutrient foramen	(F)NFMLD	2	a
	circumference at nutrient foramen	(F)NFC	2	c
Tibia	maximum length	(T)MXL	3	b
	antero-posterior diameter at nutrient foramen	(T)NFAPD	3	a
	medio-lateral diameter at nutrient foramen	(T)NFMLD	3	a
	circumference at nutrient foramen	(T)NFC	3	c
	biarticular breadth	(T)BB	5	a
	medial condyle articular breadth	(T)MCW	5	a
	medial condyle articular length	(T)MCL	5	a
	lateral condyle articular width	(T)LCAW	5	a
	lateral condyle articular length	(T)LCAL	5	a
Talus	maximum length	(TA)MTAL	6	a
Calcaneus	maximum length	(C)MCAL	6	a
	Posterior length	(C)PCAL	6	a

[1] measured in same manner as maxillary arch length, but taken from infradentale superius
[2] measured in same manner as mandibular arch length, but taken from infradentale

Reference:
1 Corruccini (1991)
2 This Study
3 Bass (1987)
4 Wood (1991)
5 Holland (1992)
6 Holland (1995)

Instrument:
a sliding calliper
b osteometric board
c cloth tape

the function in discriminating the groups being analysed.

For each bone examined the basic descriptive statistics of each trait is first tabulated and the male and female means examined for significant differences. However, because these traits behave differently in a multivariate universe, no variables were discarded based on the results of such univariate Student's t-tests. Unsexed elements were assessed using the most appropriate, or accurate, functions given their respective level of completeness (or number of traits available). The same method used for estimating a final non-metric morphological sex was applied to material sexed using discriminant functions. For example, in cases where sex estimates were in conflict, such as a femur being sexed male and a humerus female, the sex of the individual was considered to be indeterminate.

Sex Estimation Results

Non-Metric Morphological Sex Estimation

Table 2.2 summarises sex estimation results using qualitative morphological criteria for each individual. For the Da But period sample individuals sexed as male, the pelvis and cranium played an equal role in sex estimation while the mandible was utilised in half the cases. Moreover, the mandible did not play an exclusive role in estimating the sex of any male or uncertain-male. For those cases sexed as female there is an equal contribution to sex estimation from each of the three elements cranium, mandible and pelvis. The mandible played an exclusive role in estimating the sex of only two females and no uncertain-females. The majority of Da But period uncertain-males are represented by fragmentary cranial material without the benefit of mandibular or pelvic material suitable for sex estimation. While there were nine crania sexed as uncertain-female, a number of these were resolved as female with supporting mandibular or pelvic sex estimates or were assigned an indeterminate sex as individuals.

Of the individuals assessed in the Metal period sample the cranium and mandible played a major role in sex estimation, almost to the exclusion of the pelvis. The mandible played an exclusive role in estimating sex as male in four instances, uncertain-males once and females once. Few cases of uncertain-male or uncertain-female occurred in the Metal period sample. These cases were generally due to sexing conflicts within single elements (cranium, mandible or pelvis) or conflicts between the sex estimation of two elements.

Use of mandibular flexure in sex estimation has been discussed previously. Very few individuals were sexed based on the mandible exclusively. The two cases sexed as female based exclusively on the mandible in the Da But period sample have a high probability of being female. This is because the error in the system is biased toward sexing females as males based on the presence of flexure, and not the other way around. The four cases sexed as male based exclusively on the mandible in the Metal period sample do not benefit from the direction of error in sexing. Nonetheless, these cases make up a small part of the whole sample. Moreover, none of these cases had flexure scores of less than +1, and one had a score of +2.

For the Da But period only 42.5% of the adult sample could be sexed on non-metric morphological criteria alone. This figure goes up to 65.0% if uncertain cases are included. For the Metal period 57.5% of the adult sample could be assigned a sex, or 71.3% of individuals if uncertain cases are included.

Morphometric (Discriminant Function) Sex Estimation

Mandible and Maxilla

Descriptive statistics and t-tests of differences for male and female mandibles and maxillae are presented in Table 2.3a. Of the maxillary variables examined, only external breadth (MXEL) displayed a significant ($p<0.05$) difference between the sexes. The mandibular variables bigonial breadth (MNBB) and height at M1 (MNM1HT) were also significantly different ($p<0.005$) between males and females. As previously mentioned, variables used for multivariate analysis cannot be vetted using univariate significance tests of means. Variables displaying otherwise insignificant differences, when examined individually, may play an important role in separating groups in a multivariate universe. For this reason, all variables were utilised in the search for useful sex discrimination functions.

Discriminant function analysis indicated that the maxillary variables had little power to discriminate between the sexes. Even in combination with mandibular variables the resultant equations were poor separators of sex. Maxillary external length (MXEL) on its own had the greatest discriminatory power but resulted in only 64.0% of cases being correctly assigned. Correct assignation is defined as being assigned to the same sex as originally determined using non-metric morphological criteria.

Table 2.3b presents discriminant equations derived from the four most useful mandibular variables. Only those equations with a predictive accuracy in excess of 70% are listed in this study. Each equation in Table 2.3b includes

Table 2.2. Individual Non-Metric Morphological Sex Estimation Summary

Status	Da But Period Cranium	Mandible	Pelvis	Total No. Individuals	Metal Period Cranium	Mandible	Pelvis	Total No. Individuals
Male	17	9	17	19	17	21	2	23
Female	13	13	13	15	22	15	3	23
uncertain-Male	15	1	3	13	4	0	5	5
uncertain-Female	9	0	0	5	7	1	1	6
Subadult				13				14
Infant				1				2
Indeterminate Sex				28				23
TOTAL				94				96
% sexed adults				42.5%				57.50%
% sexed adults including uncertain cases				65.0%				71.30%

Table 2.3a. Descriptive Statistics for Independently Sexed Mandibles and Maxillae

	male			female					
Variable	mean	SD	n	mean	SD	n	t-test	df	p (sigma 2-tailed)
MXAL	44.92	2.27	11	42.86	1.95	5	1.751	14	0.102
MXINFDAL	42.02	1.19	13	41.23	1.95	12	1.233	23	0.230
MXAB	60.42	2.37	16	59.79	2.27	11	0.688	25	0.498
MXEL	55.48	4.25	14	51.92	3.83	11	2.169	23	**0.041**
MXEB	64.24	4.72	17	63.54	2.59	12	0.465	27	0.646
MNAL	40.14	1.08	11	39.74	1.80	5	0.553	14	0.589
MNINFDAL	39.11	1.64	22	37.94	2.02	15	1.944	35	0.060
MNAB	55.44	6.10	28	53.60	1.81	12	1.019	38	0.314
MNBB	103.01	5.78	17	91.07	5.98	12	5.405	27	**0.000**
MNM1HT	29.37	2.13	34	27.48	2.60	26	3.108	58	**0.003**
MNM1BR	14.58	1.92	34	14.39	1.45	26	0.430	58	0.669

MXAL	maxilla arch length
MXINFDAL	maxilla arch length from infradentale superius
MXAB	maxilla arch breadth
MXEL	maxilla external length
MXEB	maxilla external breadth
MNAL	mandible arch length
MNINFDAL	mandible arch length from infradentale
MNAB	mandible arch breadth
MNBB	mandible bigonial breadth
MNHTM1	mandible height at M1
MNBRM1	mandible breadth at M1

one or more coefficients and a constant. Sample size (n) refers to the number of independently sexed individuals used to generate the equation. The eigenvalue is a measure of group discrimination with a larger value meaning greater differentiation. Wilks' lambda, or the U statistic, is a significance test with a value between 1 and 0. Scores closer to 0 indicate greater group mean differences. The Chi-square test here assesses the significance of the U statistic. The Kappa statistic assesses accuracy of group membership prediction. Kappa ranges from –1 to 1 with a value of zero indicating random level group allocation and a value of 1 meaning perfect accuracy. The sectioning point for each group is calculated by determining the mid-point between each group's centroid. Predictive accuracy has been broken into groups in order to assess sex assignment for males as compared to females. For example, equation 1 in Table 2.3b reveals overall predictive accuracy as 82.8%, but female predictive accuracy was only 75.0% compared to 88.2% for males. In this case more females were being incorrectly assigned to the male group than males to the female group. A jackknife procedure is also included in the SPSS.8 package, called cross validation. The jack knife procedure recalculates the function, excluding any input from the case to be reclassified, and then reallocates this case based on the new function. It does this for all cases. Such jackknife procedures generally give lower percentages of correctly allocated cases (Manly 1994), as seen here.

Mandibular variables proved to be better than those of the maxilla in separating males from females. Five separate equations were derived (Table 2.3b) of which the most accurate in sex allocation is equation (5), which includes all four variables. The single most useful mandibular variable available was bigonial breadth (MNBB) with a sex allocation accuracy of 82.8%. Height at M1 (MNHTM1), the next most powerful single variable, had a sex allocation accuracy of only 66.7%. The assumption of equal covariance matrices was violated for equations (4) (Box's M 20.7, p=0.008) and (5) (Box's M 25.9, p=0.025). As noted, the importance of this is equivocal. In this case however, these two functions did not play a role in independently sexing any case.

The discriminant functions presented in Table 2.3b can be used to manually estimate the sex of any appropriate mandible. An appropriate bone is one where the requisite measurements are available. For example, the following steps are taken to determine the sex of a mandible with the following four variables: MNBB (88.4mm), MNHTM1 (29.9mm), MNBRM1 (12.0mm) and MNAB (55.2mm). The best equation to use in this case is equation (5). The measurements are entered into equation (5) in the following manner:

Equation (5):
sex = 88.4x0.207 + 29.9x-0.034 + 12.0x0.303 + 55.2x-0.061 + -20.493 = -2.942

The sectioning point for equation (5) is -0.292, which means that any value below this is female and any above is male. In this instance the discriminant function equation gives a value (-2.942) well below the sectioning point, suggesting the mandible is very probably female. For practical purposes the SPSS.8 program will assign all cases a sex for which the requisite measurements are available. Furthermore, the sex allocation output will also provide information on the level of probability for each case being assigned a given sex.

Humerus

Descriptive statistics and t-tests of differences for male and female humeri are presented in Table 2.4a. With the exception of transverse diameter of the head (TDH) and maximum diameter of the shaft at the nutrient foramen (NFMXD), all variables are significantly different

Table 2.3b. Discriminant Function Analysis Results for Mandibles

Variables Used	Discriminant Function Equation No:				
	1	2	3	4	5
MNBB	0.171	0.174	0.177	0.196	0.207
MNHTM1		-0.020	-0.005	-0.074	-0.034
MNBRM1			0.195		0.303
MNAB				-0.054	-0.061
constant	-16.738	-16.468	-20.075	-14.158	-20.493
n	29	29	29	25	25
eigenvalue	1.082	1.084	1.219	1.678	2.221
canonical correlation	0.721	0.721	0.741	0.792	0.830
Wilks' Lambda	0.480	0.480	0.451	0.373	0.310
Chi-Square	19.431	19.096	20.327	21.176	24.567
df	1	2	3	3	4
significance	0.000	0.000	0.000	0.000	0.000
Kappa	0.640	0.709	0.716	0.833	0.918
Assymp. Std. Error	0.146	0.134	0.132	0.113	0.080
Approx. T	0.346	0.386	3.854	4.167	4.606
Approx. Sigma	0.001	0.000	0.000	0.000	0.000
male group centroid	0.843	0.844	0.895	1.014	1.167
female group controid	-1.195	-1.196	-1.268	-1.522	-1.751
Predictive Accuracy					
male	88.2%	94.1%	88.2%	93.3%	93.3%
female	75.0%	75.0%	83.3%	90.0%	100.0%
total	82.8%	86.2%	86.2%	92.0%	96.0%
Cross Validated Accuracy					
male	88.2%	88.2%	88.2%	86.7%	86.7%
female	75.0%	75.0%	83.3%	90.0%	100.0%
total	82.8%	82.8%	86.2%	88.0%	92.0%

Note that for each set of variables there is only one discriminant function as there are only two groups (male and female) under consideration.

($p<0.05$).

Table 2.4b presents discriminant equations derived from the six most effective humerus variables. Twelve separate equations were derived of which the most accurate for sex allocation is equation (9) which includes vertical diameter of the head (VDH), midshaft maximum diameter (MSMXD) and shaft least circumference (SLC). The single most effective discriminating variable is vertical diameter of the head (VDH) with a Kappa of 0.766 and overall predictive accuracy of 88.9%. The assumption of the equality of covariance matrices was violated for functions (1) (Box's M 10.70, $p=0.028$) and (8) (Box's M 9.27, $p=0.046$). Again, this was not considered to be of great concern. Function (8) did not play an independent sexing role in the study. As for equation (1), the two dependent variables involved performed well when analysed independently.

Femur

Descriptive statistics and t-tests of differences between male and female means are presented in Table 2.5a. Of 11 variables examined maximum femur length (MXL), bicondylar length (BL), medio-lateral diameter of the midshaft (MSMLD), subtrochanteric antero-posterior diameter (STAPD) and medio-lateral diameter at the nutrient foramen (NFMLD) showed no significant difference between the male and female means. Moreover, with the exception of NFMLD these variables did not play a role in the discriminant function equations either.

Table 2.5b presents discriminant equations derived from the six most effective femoral variables. The single most effective discriminating variable is antero-posterior diameter of the midshaft (MSAPD). In combination with maximum diameter of the head (MDH), this function (equation 3) provided the most powerful discrimination with an overall predictive accuracy of 100.0%. Equations (4), (5), (6) and (8) also had good overall predictive accuracy (92.9% to 94.4%), but equation (4) fared poorly in the jackknife reclassification.

Table 2.4a. Descriptive Statistics for Independently Sexed Humeri

	male			female					
Variable	mean	SD	n	mean	SD	n	t-test	df	p (sigma 2-tailed)
TDH	40.66	2.33	8	38.00	2.45	4	1.84	10	0.096
VDH	35.79	1.54	11	32.04	2.38	7	4.08	16	**0.001**
MDH	44.02	2.08	11	39.92	3.46	6	3.08	15	**0.008**
AW	42.18	1.94	14	38.12	2.88	11	4.27	24	**0.000**
BEW	57.35	2.74	13	53.00	3.34	11	3.50	22	**0.002**
NFC	61.92	3.55	12	57.91	4.16	11	2.49	21	**0.021**
NFMND	16.98	0.82	12	15.11	1.56	11	3.63	21	**0.002**
NFMXD	20.25	1.45	12	19.83	2.20	11	0.55	21	0.589
SLC	60.71	2.61	14	55.38	5.08	13	3.47	25	**0.002**
MSMND	16.07	1.06	11	13.70	1.72	9	3.79	18	**0.001**
MSMXD	22.17	1.84	11	19.79	2.34	9	2.55	18	**0.020**
MXL	312.3	14.05	10	294.36	16.19	7	2.44	15	**0.028**

TDH	transverse diameter head
VDH	vertical diameter head
MDH	maximum diameter head
AW	articular width distal end
BEW	biepicondylar width
NFMND	minimum diameter at nutrient foramen
NFMXD	maximum diameter at nutrient foramen
SLC	shaft least circumference
MSMND	minimum diameter midshaft
MSMXD	maximum diameter midshaft
MXL	maximum length

Tibia
Descriptive statistics and t-tests of differences between male and female means are presented in Table 2.6a. With the exception of medial condyle articular length (MCAL), lateral condyle articular length (LCAL) and width (LCAW), all variables displayed significant ($p<0.05$) differences between the sexes. However, sample sizes for the articular variables were rather small and in fact preservation of the proximal tibia was poor with the facets being frequently eroded by post-mortem processes. Nonetheless, the single most useful sex discriminator was medial condyle articular width (MCAW) with a Kappa of 1.0 and predictive, and cross-validated, accuracy of 100.0%. (Table 2.6b) The three measurements taken at the nutrient foramen (equation 2) also displayed good discrimination based on a reasonable sample size. Due to the good preservation of this portion of the tibia more ungrouped cases could be assigned a sex than would be the case if articular variables alone were relied upon.

Final Sex Distribution

Table 2.7 summarises sex estimation, using discriminant functions, of individuals that could not be sexed using non-metric morphological criteria. For the Da But period sample an additional 19 individuals could be sexed, as compared to five additional individuals in the Metal period sample. Functions for the femur and humerus contributed to the sex estimation of 18/24 (75%) and 13/24 (54%) respectively, of previously unsexed individuals.

Table 2.8 summarises the distribution of sexed individuals based on non-metric morphology alone and the final sex distribution, based on the addition of sex estimation information gathered from the application of discriminant functions. Not only has the number of individuals for which a sex can be estimated increased, but the number of cases of uncertain male and female has decreased.

The somewhat smaller increase in Metal period sexed individuals is due to the small number of preserved post-cranial remains in this sample. Nonetheless, 67.5% of the Metal period sample can be sexed using a combination of non-metric morphological and metric criteria. If uncertain cases are included, 76.3% of the sample can be sexed. For the Da But period sample an increase from 42.5% to 72.5% can be sexed based on non-metric morphological and metric criteria. This figure climbs to 85.0% if uncertain cases of sex estimation are included. This is clearly a marked improvement over the results based on non-metric morphology alone, which is dependent on good preservation of specific areas of the skeleton.

Estimation of Age-at-Death

> [V]ariability rather than consistency is the rule in determining and individual's age at death.
> (Schwartz 1995: 185)

As with the literature on sex estimation, there has been an enormous amount of research directed to the problem of estimating the age-at-death of individuals based on skeletal and dental material alone. Again, as with sex estimation, only certain regions of the skeleton are consistently reliable in estimating age-at-death. Further, these particular areas of the skeleton are not generally the best preserved in the archaeological context. Only those techniques appropriate to the nature and level of preservation of the material in hand are examined here.

Estimating Age-at-Death from the Cranial Sutures

The popularity of using cranial sutures to estimate age-at-death has waxed and waned over the years. Kerley (1970), citing early work by Todd and Lyon in the 1920s, as well

Table 2.4b. Discriminant Function Analysis Results for the Humerus

Variables Used	Discriminant Function Equation No: 1	2	3	4	5	6	7	8	9	10	11	12
VDH	0.383		0.526		0.336		0.354		0.390	0.412	0.324	
AW	0.160	0.414			-0.028		0.097				0.038	
MSMND				0.717	0.361	0.885					0.355	0.885
MXL					0.012							
MSMXD						-0.153		-0.187	-0.419			-0.153
SLC							0.074	0.362	0.307	0.098		
constant	-19.608	-16.673	-18.066	-10.763	-19.635	-10.048	-20.358	-17.182	-22.348	-19.86	-18.034	-10.048
n	18	26	18	20	17	20	18	19	17	18	17	20
eigenvalue	1.134	0.76	1.04	0.797	1.3	0.828	1.191	0.622	1.584	1.162	1.279	0.828
canonical correlation	0.729	0.657	0.714	0.666	0.752	0.673	0.737	0.619	0.783	0.733	0.749	0.673
Wilks' Lambda	0.469	0.568	0.490	0.557	0.435	0.547	0.456	0.616	0.387	0.463	0.439	0.547
Chi-Square	11.37	13.287	11.05	10.254	10.827	10.251	11.375	7.742	12.814	11.566	11.119	10.251
df	2	1	1	1	4	2	3	2	3	2	3	2
significance	0.003	0.000	0.001	0.001	0.029	0.006	0.01	0.021	0.005	0.003	0.011	0.006
Kappa	0.766	0.455	0.766	0.694	0.757	0.596	0.766	0.776	0.876	0.766	0.757	0.596
Assymp. Std. Error	0.156	0.175	0.156	0.162	0.161	0.181	0.156	0.146	0.119	0.156	0.161	0.181
Approx. T	3.251	2.328	3.251	3.119	3.122	2.665	3.251	3.472	3.64	3.251	3.122	2.665
Approx. Sigma	0.001	0.020	0.001	0.002	0.002	0.008	0.001	0.001	0.000	0.001	0.002	0.008
male group centroid	0.801	0.776	0.767	0.766	0.896	0.781	0.821	0.636	0.989	0.811	0.889	0.781
female group centroid	-1.259	-0.905	-1.205	-0.936	-1.28	-0.954	-1.29	-0.875	-1.413	-1.274	-1.27	-0.954
Predictive Accuracy												
male	90.9%	78.6%	90.9%	90.9%	90.0%	81.8%	90.9%	100.0%	100.0%	90.9%	90.0%	81.8%
female	85.7%	66.7%	85.7%	77.8%	85.7%	77.8%	85.7%	75.0%	85.7%	85.7%	85.7%	77.8%
total	88.9%	73.1%	88.9%	85.0%	88.2%	80.0%	88.9%	89.5%	94.1%	88.9%	88.2%	80.0%
Cross Validated Accuracy												
male	90.9%	78.6%	90.9%	81.8%	90.0%	81.8%	90.9%	90.9%	90.0%	90.9%	90.0%	81.8%
female	71.4%	66.7%	85.7%	77.8%	71.4%	77.8%	85.7%	75.0%	85.7%	85.7%	85.7%	77.8%
total	83.3%	73.1%	89.9%	80.0%	82.4%	80.0%	88.9%	84.2%	88.2%	88.9%	88.2%	80.0%

Note that for each set of variables there is only one discriminant function as there are only two groups (male and female) under consideration.

Table 2.5a. Descriptive Statistics for Independently Sexed Femora

	male			female					
Variable	mean	SD	n	mean	SD	n	t-test	df	p (sigma 2-tailed)
MXL	455.50	13.82	4	420.00	32.36	3	2.013	5	0.100
BL	450.75	14.77	4	415.67	31.09	3	2.019	5	0.099
MSAPD	28.68	1.83	9	24.93	1.03	9	5.348	16	**0.000**
MSMLD	24.84	1.81	9	23.57	1.24	9	1.748	16	0.100
MDH	44.81	2.64	10	41.15	2.29	8	3.091	16	**0.007**
MSC	85.11	4.76	9	77.00	2.12	9	4.673	16	**0.000**
STAPD	22.22	1.78	14	21.04	1.71	11	1.680	23	0.107
STMLD	30.00	2.02	14	27.28	2.48	11	3.028	23	**0.006**
NFAPD	27.69	2.51	7	24.40	1.79	7	2.820	12	**0.015**
NFMLD	24.41	1.95	7	23.19	1.31	7	1.387	12	0.191
NFC	82.57	6.19	7	74.57	3.82	7	2.910	12	**0.013**

MXL	maximum length
BL	bicondylar length
MSAPD	antero-posterior diameter midshaft
MSMLD	medio-lateral diameter midshaft
MDH	maximum diameter head
MSC	midshaft circumference
STAPD	subtrochanteric antero-posterior diameter
STMLD	subtrochanteric medio-lateral diameter
NFAPD	antero-posterior diameter at nutrient foramen
NFMLD	medio-lateral diameter at nutrient foramen
NFC	circumference at nutrient foramen

as Singer in the early 1950s, notes the inadvisability of using suture closure in age estimation. Nonetheless, recent years have seen a rekindling of interest in using cranial suture closure for age-at-death estimation. Meindl & Lovejoy (1985) proposed a new technique in assessing suture closure. They assessed closure or synostosis at ten specific sites and scored these on a four-point scale from completely open to completely obliterated. The first seven sites fall on the coronal, sagittal and lambdoidal sutures and they refer to these as the vault system. The remaining three, referred to as the lateral-anterior system, include points at sphenofrontal, inferior sphenotemporal and superior sphenotemporal sutures. They found a correlation coefficient with actual age of 0.53, which they note was superior to the component pubic symphysis aging system they tested concurrently.

Their results indicated that the best combination of the ten points selected for age prediction were the five lateral-anterior sites (which includes two points on the coronal suture). They also stated that ectocranial sutures were superior to endocranial sutures in age estimation, but did not systematically test this statement. Another point to note is that although they state the sample size for the study was 236 crania from the Hamann-Todd collection, they do not mention how the crania were selected, nor do they describe the age distribution of the sample. Further, it is unclear if paired points, of which there were six of a total of ten points, were assessed for both sides of the skull and then averaged or if only one side of the skull was assessed. This is important given the results of Zivanovic's (1983) study of synostosis asymmetry. Zivanovic (1983) found coronal and parietal asymmetry in closure in more than 69% of a white male sample, 77% in a black male sample and 81% and 70% in white and black female samples respectively.

In a less positive study Hershkovitz et al. (1997) argue that they have sentenced age estimation from suture synostosis to a quiet death. They utilise a large sample of 3,636 skulls from the Hamann-Todd and Terry collections. Only the sagittal suture is assessed and this is done by assessing synostosis in four main categories: (1) totally closed - no sign of suture ectocranially; (2) partially closed - less than 10% of suture open; (3) totally open; (4) partially open - between 10% and 90% of suture open. They also include a 5th category for premature closure, which is carefully defined.

They found that for males the frequency distributions of all categories remain static and flat after age 35. This would indicate that there is no correlation between suture closure and age after age 35. Whereas for females, the pattern was more complex with the frequency of open sutures deceasing steadily with age and frequency of totally closed sutures increasing steadily with age. When looking at ancestry, the results indicated a "genetic predisposition for suture closure in general" (Hershkovitz et al. 1997:398).

Their conclusions included these two important points:

> 1. The sagittal suture cannot be used for aging the skeleton. 2. Although cross-sectional in nature, suture obliteration patterns (totally open, totally closed, partially open, partially closed) are not temporary progressive stages on an age scale, but rather independent permanent phenomena. (Hershkovitz *et al.* 1997:398)

However, there are a number of problems with this study that must impact on the verity of their conclusions. Most studies (Meindl & Lovejoy 1985 excluded) report that endocranial suture closure is more highly correlated with age than is ectocranial suture closure. From the start Hershkovitz *et al.* (1997) have chosen an aspect of suture closure that the literature has long reported as less reliable.

Table 2.5b. Discriminant Function Analysis Results for the Femur

Variables Used	Discriminant Function Equation No: 1	2	3	4	5	6	7	8	9	10	11
MSAPD	0.673		0.527	0.571	0.701	0.618		0.535			
MDH		0.401	0.215				0.295	0.107			
STMLD				0.160		0.309	0.153	0.132			
STAPD					-0.064	-0.311					
NFAPD									0.459		0.151
NFC										0.194	0.134
constant	-18.048	-17.299	-23.466	-19.905	-17.422	-18.744	-17.209	-22.825	-11.946	-15.277	-14.473
n	18	18	14	18	18	18	17	14	14	14	14
eigenvalue	1.788	0.597	3.156	2.060	1.812	2.502	0.864	3.306	0.663	0.706	0.718
canonical correlation	0.801	0.611	0.871	0.820	0.803	0.845	0.681	0.876	0.631	0.643	0.647
Wilks' Lambda	0.359	0.626	0.241	0.327	0.356	0.286	0.536	0.232	0.601	0.586	0.582
Chi-Square	15.890	7.257	15.669	16.774	15.506	18.175	8.719	15.331	5.846	6.141	5.956
df	1	1	2	2	2	3	2	3	1	1	1
significance	0.000	0.007	0.000	0.000	0.000	0.000	0.013	0.002	0.016	0.013	0.051
Kappa	0.778	0.667	1.000	0.889	0.889	0.889	0.764	0.851	0.714	0.714	0.714
Assymp. Std. Error	0.144	0.175	0.000	0.107	0.107	0.107	0.157	0.142	0.187	0.187	0.187
Approx. T	3.384	2.846	3.742	3.795	3.795	3.795	3.150	3.220	2.673	2.673	2.673
Approx. Sigma	0.001	0.004	0.000	0.000	0.000	0.000	0.002	0.001	0.008	0.008	0.008
male group centroid	1.261	0.652	1.899	1.353	1.269	1.491	0.823	1.944	0.754	0.778	0.785
female group centroid	-1.261	-0.815	-1.424	-1.353	-1.269	-1.491	-0.926	-1.458	-0.754	-0.778	-0.785
Predictive Accuracy	77.8%	80.0%	100.0%	88.9%	88.9%	100.0%	88.9%	83.3%	85.7%	85.7%	85.7%
male	100.0%	87.5%	100.0%	100.0%	100.0%	88.9%	87.5%	100.0%	85.7%	85.7%	85.7%
female	88.9%	83.3%	100.0%	94.4%	94.4%	94.4%	88.2%	92.9%	85.7%	85.7%	85.7%
total											
Cross Validated Accuracy											
male	77.8%	80.0%	83.3%	66.7%	77.8%	88.9%	88.9%	83.3%	85.7%	85.7%	71.4%
female	100.0%	87.5%	87.5%	77.8%	100.0%	88.9%	75.0%	87.5%	85.7%	85.7%	85.7%
total	88.9%	83.3%	85.7%	72.2%	88.9%	88.9%	82.4%	85.7%	85.7%	85.7%	78.6%

Note that for each set of variables there is only one discriminant function as there are only two groups (male and female) under consideration.

Table 2.6a. Descriptive Statistics for Independently Sexed Tibia

	male			female					
Variable	mean	SD	n	mean	SD	n	t-test	df	p (sigma 2-tailed)
MXL	370.17	19.03	6	326.00	9.90	2	3.033	6	**0.023**
NFAPD	33.38	2.17	11	29.36	1.81	12	4.851	21	**0.000**
NFMLD	23.39	1.88	11	20.81	1.96	12	3.224	21	**0.004**
NFC	90.55	5.91	11	81.38	7.01	12	3.338	21	**0.003**
BB	71.80	3.62	6	62.70	3.91	4	3.777	8	**0.005**
MCAW	30.80	2.28	7	23.90	2.48	5	4.994	10	**0.001**
MCAL	43.10	1.28	7	40.73	3.00	3	1.839	8	0.103
LCAW	30.59	4.44	7	26.50	2.22	5	1.877	10	0.090
LCAL	39.88	2.68	6	38.25	2.69	4	0.994	8	0.373

ML	maximum length	MCAW	medial condyle articular width
NFAPD	antero-posterior diameter at nutrient foramen	MCAL	medial condyle articular length
NFMLD	medio-lateral diameter at nutrient foramen	LCAW	lateral condyle articular width
NFC	circumference at nutrient foramen	LCAL	lateral condyle articular length
BB	biarticular breadth		

Table 2.6b. Discriminant Function Analysis Results for the Tibia

Variables Used	Discriminant Function Equation No:					
	1	2	3	4	5	6
NFAPD	0.503	0.962				
NFMLD		0.279				
NFC		-0.235				
BB			0.268			
MCAW				0.424	0.569	0.476
LCAW					-0.132	
LCAL						-0.081
constant	-15.742	-16.027	-18.259	-11.834	-12.096	-10.242
n	23	23	10	12	11	9
eigenvalue	1.120	1.612	1.783	2.494	4.070	2.867
canonical correlation	0.727	0.786	0.800	0.845	0.896	0.861
Wilks' Lambda	0.472	0.383	0.359	0.286	0.197	0.259
Chi-Square	15.408	18.725	7.676	11.885	12.987	8.115
df	1	3	1	1	2	2
significance	0.000	0.000	0.006	0.001	0.002	0.017
Kappa	0.738	0.826	0.783	0.375	1.000	1.000
Assymp. Std. Error	0.141	0.118	0.201	0.142	0.000	0.000
Approx. T	3.551	3.960	2.535	2.402	3.317	3.000
Approx. Sigma	0.000	0.000	0.011	0.016	0.001	0.003
male group centroid	1.056	1.267	0.975	1.218	1.666	1.336
female group centroid	-0.968	-1.162	-1.463	-1.706	-1.999	-1.670
Predictive Accuracy						
male	81.8%	90.9%	100.0%	100.0%	100.0%	100.0%
female	91.7%	91.7%	75.0%	100.0%	100.0%	100.0%
total	87.0%	91.3%	90.0%	100.0%	100.0%	100.0%
Cross Validated Accuracy						
male	81.8%	81.8%	100.0%	85.7%	100.0%	100.0%
female	91.7%	91.7%	75.0%	100.0%	100.0%	100.0%
total	87.0%	87.0%	90.0%	91.7%	100.0%	100.0%

Note that for each set of variables there is only one discriminant function as there are only two groups (male and female) under consideration.

Table 2.7. Individual Discriminant Function Sex Estimation Summary

	Da But Period		Metal Period	
	Male	Female	Male	Female
Mandible	2	1	0	3
Humerus	5	5	2	1
Femur	8	7	2	1
Tibia	5	4	0	0
Total No. Individuals Sexed	10	9	2	3

Table 2.8. Final Sex Distribution Summary

	Da But Period		Metal Period	
Status	Non-Metric Morphology	Non-Metric & Metric	Non-Metric Morphology	Non-Metric & Metric
Male	19	33	23	25
Female	15	25	23	29
uncertain-Male	13	7	5	3
uncertain-Female	5	3	6	4
Subadult	13	13	14	14
Infant	1	1	2	2
Indeterminate	28	12	23	19
TOTAL	94	94	96	96
% sexed adults	42.50%	72.50%	57.50%	67.50%
% sexed adults + uncertain cases	65.00%	85.00%	71.30%	76.30%

It would have been useful to assess both endocranial and ectocranial synostosis or have chosen endocranial synostosis given its better report in the literature.

They only assess the sagittal suture. The reasons for this were that "it is the only 'end-to-end' type suture in the calvaria (avoiding 'pseudo closure' due to overlapping of bone, as in frontal over parietal bone), and because its location at the midline neutralises it from biomechanical influences" (Hershkovitz *et al.* 1997:395). These points aside, it would have been useful to assess at least the other two most commonly reported on sutures, the coronal and lambdoidal.

Their selection of only four main categories of closure, and especially category #4 (10% to 90% of suture open) serves to swamp any possible variation in such an all encompassing category. It is perhaps not surprising that no correlation was found between age and level of synostosis when degrees of suture obliteration are so poorly resolved. Moreover, they do not pursue the evidence for a correlation between age and suture closure in the female sample. For some unexplained reason the negative male findings are given preference.

Meindl *et al.* (1983) note that only 512 of the total Hamann-Todd collection, of 3422 individuals, were sufficiently documented for age to be included in their own study. In other words, only 15% of the total Hamann-Todd sample can be considered reliably aged for the purposes of research requiring accurate information on known age-at-death. Hershkovitz *et al.'s* (1997) Hamann-Todd sample was 1958 individuals. In the event that these included all the 512 well documented cases, still only 25% of their Hamann-Todd sample had reliable information on actual age-at-death. Even if all the Terry individuals they used (n=1678) were reliably documented for age-at-death, still only 60% of their entire sample could be considered to be reliably documented for age-at-death. This is clearly of some concern, especially given their highly negative findings and conclusions. Reports on the quiet death of cranial suture analysis in age estimation are perhaps premature.

In a more productive study, Galera *et al.* (1998) compared the age assessment from cranial suture closure methods of Acsadi & Nemeskeri (1970), Masset (1982) and Meindl & Lovejoy (1985). They utilised a sample of 408 whites and 555 blacks from the Terry collection. Inaccuracy and bias were calculated (after Lovejoy *et al.* 1985a) and they found that in all sub-samples (by ancestry and sex) both the Acsadi & Nemeskeri (1970), and Masset (1982) methods were the most accurate. There were differences in accuracy when particular age categories were considered.

What these authors did not address is what Masset (1989) has described as systematic errors in using cranial suture closure for age estimation. All Galera *et al.* (1998) have pointed out is that different methods of scoring cranial suture closure will achieve differentially successful, but not inconsistent, results. Galera *et al.* (1998) also reproduced Masset's (1989) observation that the correlation between suture closure and age-at-death is quite small. The correlation coefficients, for all sub-samples and methods applied, range from 0.33 (white males using Acsadi & Nemeskeri and Meindl & Lovejoy) to 0.66 (black male endocranial of Acsadi & Nemeskeri). This means that between 89% and 56% of the variation in suture closure scores and indices is accountable by factors other than age. While this is a troubling finding, at least one researcher has devised a method to deal with the poor correlation between suture closure scores and age-at-death estimation that no one else, in this context, has dealt with before.

Masset (1989) took a fresh look at ageing the skeleton using cranial suture synostosis. He noted that if cranial sutures are to be used to age a population, a number of potential systematic errors must be addressed. These are summarised as sex differences in the timing of synostosis; differences in ectocranial and endocranial suture closure timing, the age composition of the reference population (used to develop equations to predict age from degree of synostosis); and the phenomenon of attraction to the middle. Masset (1989) points out that there is little indication in the literature that ancestral differences in suture closure timing are a confounding factor.

Differences in sex, as well as those between the outer and inner table of the cranium, are dealt with through the establishment of separate regression equations for the various permutations of these variables. Depending on the age composition of the reference sample, used to generate these regression formulae, a range of equations are available. Masset (1989:90, Table 4.2) provides equations for males and females separately for the internal and external table of the cranium for life expectancy at birth estimates of 25, 30, 40 and 60 years (a total of 16 regression equations). However, these equations cannot be used reliably to generate a true age distribution of the archaeological sample. The reason is that the poor correlation between age-at-death and suture closure, ranging between 0.42 and 0.63 in Masset's (1989) sample, leads to very high levels of error. Using these equations without regard to this problem leads to the phenomenon of attraction to the middle. That is, there is an accumulation of errors when assigning older or younger individuals an age. As the estimation error for young individuals can only be up and that for old individuals down, the estimation errors accumulate in the middle. This means the age distribution clumps in the middle, resulting in under representation in older and younger age classes. Masset (1989) suggests that correlation coefficients need to be greater than 0.9 before regression equations could reliably be utilised to build a composition by age distribution.

To deal with this problem of attraction to the middle, Masset (1989) has generated age probability matrices using a standardised population. This consisted of 914 crania with an even age distribution, that is, the same number of individuals in each of five age categories ranging from 18 to 85 years. Based on the level of suture closure, individual crania are assigned, or in reality distributed, over a range of age-at-death probabilities. For instance, a male skull with an ectocranial suture closure score representing Masset's (1989:93, Table 4.4) phase 3, would have a 35.2% probability of being aged between 18 and 29 years, 24.1% probability of being aged between 30 and 39 years, etc. This procedure is followed for the entire sample for which an age probability distribution is required. Departures from the reference population, represented by a flat age distribution curve, indicate the nature of the unknown sample.

If the objective of any study is to derive estimates of the age-at-death of individuals in a particular sample, then the use of cranial suture closure is not an attractive option. However, if the purpose is to assess the relative age structure of a sample, whether it be skewed to youth or old age, then Masset's (1989) method has merit.

Estimating Age-at-Death from Epiphyseal Closure

Epiphyseal information is generally restricted to that limited period of time between adolescence and adulthood. Differential fusion of the epiphyses can be used to narrow down probable age to within a couple of years within this age range. However, epiphyseal fusion is ancestral, sex and even geographic region specific (Banjeree & Agarwal 1998). Barret (1936) highlighted differences in timing of epiphyseal fusion between Asians and Europeans. In a sample of 101 female Burmese aged between 12 and 22 years, he found Burmese fusion consistently occurring earlier than for Europeans. In a recent study of epiphyseal fusion of the sternal end of the clavicle, Ji et al. (1994) noted earlier fusion times for modern Japanese as compared to American males. While not an epiphysis, but rather a synchondrosis, it is appropriate to note here that the basilar suture is generally regarded to remain open until 17 years (Schwartz 1995) and have fused by 20 to 25 years in most individuals (e.g. White & Folkens 1991). However, Sahni et al. (1998) have documented, in an Indian sample, partial male fusion beginning at 13 years with full fusion in all subjects by 19 years. Whereas partial female fusion begins at 12 years and full fusion is seen in all subjects by 17 years.

Several recent studies have looked specifically at the clavicle. Ji et al. (1994), using McKern & Stewart's (1957) five stage union system, assessed a recent autopsied Japanese sample of 54 individuals aged between 13 and 31 years. They found female union proceeded faster than males. Comparing McKern & Stewart's (1957) results with their own, they found Japanese male union was faster than for American males. For their sample no fusion at all (stage 0) equated with an age definitely less than 22 years, while stage 4 fusion indicated an age of more than 19 years.

Kreitner et al. (1997) and Kreitner et al. (1998), used a sample of 380 individuals below the age of 30 years, to document the appearance of a medial clavicular epiphyseal ossification centre between 11 and 22 years, partial union between 16 and 26 years, complete union beginning at 22 years. The entire sample displayed complete union by 27 years. They concluded:

> Compared to the experience recorded in the relevant literature, there are several landmarks that show no significant change between different ethnic groups and different periods of publication; these are the onset of

ossification, the time span of partial union, and the appearance of complete union. (Kreitner *et al.* 1998:1116)

Age and the Dentition

Non-invasive age determination from the teeth is generally either by way of an analysis of tooth wear or by way of radiological techniques. Costa (1986) provides a useful review of both invasive and non-invasive procedures for estimating age from the dentition. More recently, Smith (1991) has reviewed the literature on tooth formation, development and eruption studies. Smith (1991) highlights a wide range in the quality of such work, as well as the methods involved, and suggests the best studies to refer to given one's particular information needs.

When looking at sub-adult dentitions, Smith (1991) noted that the important study by Moorrees *et al.* (1963) is perhaps the best source for determining 'age of attainment of developmental stages' and in age prediction. Smith (1991) reproduced tables and interpolated further tables from Moorrees *et al.'s* (1963) work for estimating age-at-death in sub-adult samples. Saunders *et al.* (1993) have recently assessed the accuracy of the Moorrees *et al.* (1963) and Anderson *et al.* (1976) tooth formation standards on sub-adults from a cemetery population. They concluded that Moorrees' system should be used as it was more accurate and is based on a sample that began at birth (the Anderson sample begins at 3 years). They, as have others (e.g. Costa 1986; Ubelaker 1987), also stress the far greater accuracy in tooth formation chronologies as compared to tooth eruption standards. But they also point out that very little information or data is available on tooth formation standards for non-European populations. Other authors have also stressed the greater accuracy of development standards.

Ubelaker (1987) has argued that tooth eruption timing is variously affected by nutritional stress, disease, population variation and premature tooth loss. However, Smith (1991) maintains that both tooth development and eruption are relatively robust to environmental influences. Smith (1991) goes on to suggest that "in many cases, the dentition is the single best physiological indicator of chronological age in juveniles". Nonetheless, there is a certain degree of variability in dental eruption timing that needs to be appreciated when relying on this method for age estimation. A recent study by Lampl & Johnston (1996) corroborates the degree of variability in eruption timing demonstrated by Ubelaker (1987), for instance. Lampl & Johnston (1996) investigated differences in actual age, and age as assessed by standard osteological and dental standards. Their reference population consisted of several samples of Mexican children from a nutritionally deprived but otherwise healthy context. They highlight enormous amounts of variation in estimates of age based on standard reference chronologies to actual age. For example, in a sample of 183 children aged 6 years, it was found that using a standard osteological age estimator the 95% confidence interval was +/- 2 years. Or in other words, the range in estimated ages was between 4 and 8 years.

If tooth eruption timing is to be used to estimate age-at-death, the use of Ubelaker's (1987) modification of the Schour and Massler dental eruption chart is recommended. This chart has the twin advantages of factoring in variability in eruption timing and also including a large amount of data derived from non-European, particularly native American, populations.

Age estimation from measures of dental attrition has proven useful. The earliest systematic system for assessing age from levels of dental attrition was developed by Miles (1958; 1963a; 1963b), although Broca (1879 cited in Costa 1986) had developed a system for assessing wear in the late 19th century. The Miles system requires a sample of sub-adults from which age can be determined independent of wear. Functional ages of molars in the aged sample are then established based on years since eruption of the molar and level of wear. Functional age and time since eruption are summed for teeth from un-aged individuals in order to estimate age-at-death. The chief assumptions of the system include little variation in molar class eruption times and constant levels of wear for each molar class after eruption. Such assumptions present problems.

Tomenchuck & Mayhall (1979) devised a quantitative technique for establishing the predictive value of tooth wear in determining age. Their variable is the product of a measure of cusp height divided by mesio-distal diameter multiplied by bucco-lingual diameter, all multiplied by 100. They use least squares to model the relationship. They tested the applicability of their model on another population of Eskimos by comparing the predicted to actual mean ages of the population. The excellent results suggested that sample-specific age prediction models can be successfully applied across samples in certain circumstances.

Lovejoy (1985) detailed a complex system of estimating age, rates of wear and description of wear using the Libben population as an example. Rates of wear are assessed after Miles' (1963a, 1963b) system for the functional age of a tooth. Lovejoy's (1985) method involves seriating the sample by degree of wear, determining the rate of wear and assigning modal age phases. More incomplete specimens can then be aged in comparison to the seriated sample comprised of more complete specimens. The system is probably overly complex if the only information required is assessment of wear rates and age-at-death. However, if detailed descriptions of wear itself are required, the system is very useful.

Scott (1979a, 1979b) advocated the use of principal axis analysis for the description of rates of dental wear. Scott (1979a) argued that the use of the principal axis technique, rather than regression analysis, was appropriate because prediction of wear on one molar was not being sought from wear on the other molar. Further, principal axis analysis does not assume that the independent variable is measured without error, which is potentially a problem in this type of analysis. While the wear variable is ordinal, measured on a discrete scale between 4 and 40, the scale nonetheless approaches an interval level. This makes it more appropriate for some type of least squares analysis. While Scott (1979b) was not trying to estimate age-at-death, this wear scoring system is suitable for such an analysis.

The Pelvis and Age Estimation

The pelvis has been a traditional site for the estimation of age. The most well known work on aging the human pubic symphysis was that produced by Todd (1920, 1921) early last century. These standards are still reproduced in major human skeletal texts today (e.g. Brothwell 1981; Krogman & Iscan 1986; Bass 1987; White & Folkens 1991).

The next major development was McKern & Stewart's (1957) system, where three distinct pubic components were individually assessed and then summed to arrive at an estimate of age. However, independent testing of the male standards of the Todd and McKern & Stewart systems revealed problems in accounting for the range of variation in pubic morphology (Angel *et al.* 1986; Katz & Suchey 1989). Further, Suchey (1979) argued that the Stewart & McKern system was complicated and had low inter-observer replicability. A modification of the Todd system was suggested (Angel *et al.* 1986; Suchey *et al.* 1986; Katz & Suchey 1989) that lumped Todd's ten stages into six broad age phases, referred to as the Suchey-Brooks system.

Gilbert (1973) and Gilbert & McKern (1973) in noting sex differences in the timing of morphological change, developed a female standard three-component system. However, in an independent test of the system Suchey (1979) had 23 anthropologists independently age 11 pubes with appalling results. There was a wide range of variation in scores for each of the 3 components and importantly "only 51% of the assessments yielded age ranges which would include the known age of the specimen in question" (Suchey 1979:469). The chief problem with the system appeared to be that "it is often difficult to tell when the ventral rampart is building up or breaking down. Assessing Component III (the symphyseal rim) presents the same difficulty" (Suchey 1979:470). Suchey (Angel *et al.* 1986) noted that female morphological changes in the pubic symphysis were too variable for the application of a series of typical phases as was possible for male pubes. As such, three distinct morphological features for the assignment of female pubes to very broad age stages were highlighted (Angel *et al.* 1986).

A comparison of the McKern-Stewart/Gilbert-McKern and Suchey-Brooks systems (Klepinger *et al.* 1992) found the Suchey-Brooks system more accurate for forensic situations. However, Klepinger *et al.* (1992) suggest age estimates should be given +/-2 standard deviation scores.

At the same time Suchey was revising the Todd system, Meindl *et al.* (1985b) were also developing a revised Todd system. They evaluated the accuracy and bias of the McKern-Stewart/Gilbert-McKern, Hanihara-Suzuki and Todd ten phase pubic symphyseal systems. They found the Todd system to be superior but noted that it was similar to the other systems in that it was biased to higher age categories, a point also raised by Suchey (Katz & Suchey 1989). They then tested a revised Todd system which removed overall bias from all age categories. Their new system included five principal biological age phases in pubic symphyseal metamorphosis, which encompassed the various Todd stages. On the results of comparing the accuracy of their system they stated:

> Our results generally indicate that a subjective application based on the biological events recorded by the pubic symphyseal face is superior to a more typological–numerical approach. This is an interesting result because it implies that archetypal aging is less accurate than a more generalized and interpretive method.
> (Meindl *et al.* 1985b:33)

Meindl *et al.* (1990) further criticised the Suchey-Brooks system on the basis of it being typological, where a single dominant feature is used to assign a specimen to a particular age class. Meindl *et al.* (1990:365) maintained that:

> This practice restricts age estimation to a single, dominant feature of each specimen, despite the unquestionable existence of numerous other age-related changes which regularly occur in the symphysis. We continue to contend that knowledgeable, careful interpretation of a symphysis [Meindl's biological age phase system] will yield a more accurate age at death in both forensic and archaeological contexts than the mean age of stages used in comparison matching [Suchey's system].

Meindl and Lovejoy (Lovejoy *et al.* 1985b) have also reassessed the value of the auricular area of the ilium for estimating age-at-death. They identified five basic phases of change in the auricular area and a further series of five-year interval chronological stages from 20 years through to 60+ years. The penultimate stage is 10 years in duration. Blind tests of the method using individuals not included in developing the technique are reported in Lovejoy *et al.* (1985b), where a correlation of 0.71 between auricular age and actual age is recorded. A clear advantage of using the auricular area is its generally greater preservation, than the pubis, in archaeological contexts. An independent test of the method (Murray & Murray 1991) found no significant sex or ancestral factors on degenerative changes in the auricular area. However, they did note that "the rate of degenerative change is too variable to be used as a single criterion for the estimation of age" (Murray & Murray 1991:1162). Recognised problems with using single age indicators are addressed by Meindl and colleagues in their multi-factorial aging technique.

The Multi-Factorial Method

Rather than simply using one or two age estimators, it has been suggested (Meindl *et al.* 1983) that multiple age indicators can be examined in unison for a more accurate determination of actual age-at-death. Meindl *et al.* (1983) tested various single methods of age estimation and also assessed the multi-factorial approach to age-at-death estimation. Of the single age estimators they suggested that cranial sutures were problematic and the technique needs revision. They found traditional estimation of pubic age to slightly over-age in young adults and to significantly underage in older age groups, especially 50+. They recommend the use of dental attrition as it was the most highly correlated with actual age in their principal components analysis. They also recommend the use of the auricular area for age estimation due to its accuracy in age estimation and durability in the archaeological context.

They further developed their multi-factorial technique in a later paper where they employed revised pubic symphysis and auricular area aging techniques (Lovejoy *et al.* 1985a). Essentially, the technique requires the selection of a number of age estimators. These are independently serialised by increasing age and an estimate of age is given to each specimen. After which:

> All indicators are then used to generate an inter-correlation matrix. This matrix is then subjected to principal components analysis and the first factor is assumed to represent true chronological age. An indicator's correlation with the first principal component is then taken as its weight, and the final age of any individual is the weighted average of all available age indicators for that specimen.
> (Lovejoy *et al.* 1985a:4)

In fine tuning the technique they develop an algorithm to be applied to older individuals. This is to compensate for systematic under-aging of older individuals. A test of a sub-sample of the Hamann-Todd collection indicated very high levels of accuracy and low levels of bias when using the revised pubis and auricular area estimators. Moreover, the multifactorial system was far superior to the use of individual age estimators.

Additionally, Mensforth & Lovejoy (1985) using a range of traits independent of the multifactorial approach, compared the results of these age estimators on an archaeological sample with a modern assemblage of known age structure. High levels of corroboration were found between these independent estimates and the multifactorial estimates of the age profile of the archaeological sample. Bedford *et al.* (1993) have further corroborated this method in showing inter-observer error levels to be quite low. Bedford *et al.* (1993) found mean accuracy of age estimation to be greatest for summary age, or in other words, greater than any single age estimator. The pubic symphysis proved to be the single most accurate estimator for younger individuals while auricular area was more accurate for older adults.

This corroborative analysis has been criticised by Fairgrieve & Oost (1995). Their chief concerns were a lack of appropriate statistical testing and small sample size (only 55 individuals in test), thus casting doubt on the validity of the results and the utility of the aging method. While Meindl *et al.* (1995) reply to this criticism, the issue of further testing with a much greater sample size has not been addressed to date.

Summary of Age Estimators

It is clear that there is no single trait that can be utilised to estimate the age-at-death of skeletal material without some degree of error. Further confounding the problem are the vagaries of preservation in archaeological samples. This latter problem affects the potential to employ a multifactorial technique in age estimation. Seldom are multiple traits available. The use of cranial sutures in estimating age is inadvisable unless a probabilistic approach is adopted, which means that the actual age composition of the sample is unknown. Epiphyseal closure can only be used in adolescents and young adults for isolating age with any level of accuracy. This is also the case for tooth eruption timing, which is restricted to sub-adult cases. The pubic symphysis and auricular areas of the pelvis are probably the most useful areas for estimating the age of adult remains. However, morphological changes to the symphysis are inaccurate after about 50 years of age. Moreover, this portion of the skeleton is generally very poorly preserved, whereas the auricular area is very robust and frequently preserved. Tooth attrition scores have a great deal of potential to age otherwise fragmentary material. Nonetheless, a sample of individuals aged by way of other methods, such as pubic symphyseal morphology, is needed in order to generate equations for estimating age-at-death from tooth wear data.

Methods and Materials

Multifactorial aging was not possible in this study due to the level and pattern of preservation. What is important, for inter-sample comparative reasons, is an idea of the general age structure of each sample. This is clearly necessary in the context of discussions concerning the manifestations and frequencies of various pathological conditions. Two independent approaches to age estimation were employed. In the first instance age probability profiles were developed for both the Da But period and Metal period samples. The relatively large number of individuals with assessable cranial sutures in both samples facilitated this approach. However, as discussed, the actual age composition of the samples cannot be revealed through this type of analysis. The importance of age probability profiles is that the relative age structure of each sample can be assessed.

The second approach to age estimation involves the development of equations to predict age-at-death based on tooth wear scores. A sample of individuals for which age-at-death could be estimated by way of tooth eruption, epiphyseal status, pubic and auricular area morphology provided the basis for developing these equations. However, due to the poor representation of Metal period individuals, for which such independent age estimates could be performed, the Da But period sample was solely used to generate age estimation equations based on tooth wear. Tests of the suitability of applying these equations to the Metal period sample are carried out and discussed later in the chapter.

Cranial Sutures

Masset's (1989) method for assessing cranial suture synostosis was employed. Age probability profiles were generated using Masset's probability matrices, based on his standardised population (Masset 1989:93, Table 4.4). Probability profiles were generated for each sample, males and females separately and combined for both ectocranial and endocranial sutures. The method of obtaining the suture 'obliteration coefficient' from the sagittal, coronal and lambdoidal sutures also follows that of Masset (1989). However, due to preservation problems an average synostosis score was used where five or more of the potential ten synostosis scores were available.

Tooth Wear

Unfortunately, when dealing with tooth wear the method used for cranial suture synostosis is not applicable. There is no universal biological relationship between tooth wear and age-at-death. Other factors, such as diet, play a significant role in determining tooth wear for any given age. Moreover, there cannot be an appropriate standard reference population from which to generate equations relating wear to age-at-death. The relationship between tooth wear and age is population specific. However, because of the generally good correlation between age-at-death and tooth wear (Lovejoy 1985; Richards & Miller 1991; Molnar *et al.* 1983; Xu *et al.* 1992; but see Santini *et al.* 1990), it may be possible to generate an age distribution that approaches actual age composition.

As previously mentioned, Scott (1979a, 1979b) originally used these wear scores for a principal axis analysis to develop age independent rates of dental wear. In this case it is age-at-death that is to be predicted, for which regression analysis is more appropriate. Age category, the dependent variable, was derived from independent age assessment of individuals with appropriately preserved pelvic areas, or by way of epiphyseal fusion and tooth eruption for sub-adult individuals.

Preservation of the os coxae was poor for both populations and this, coupled with recording circumstances, prevented initial seriation of ageable components. Where possible, age-at-death was estimated using Lovejoy *et al.'s* (1985b) age phase system for the auricular area. Where sufficiently preserved pubic symphyses were available, Meindl *et al.'s* (1985b) biological age phase system was employed.

In lieu of a comprehensive published corpus of data on Asian epiphyseal fusion times, the standards tabulated by Bass (1987) and Brothwell (1981) were followed. Ubelaker's (1987) modification of the Schour and Massler dental eruption chart was used to estimate age based on dental eruption status. While problems have been highlighted with respect to using such charts (Lampl & Johnston 1996), Ubelaker's (1987) modifications to the chart included a large sample of Asian (native American) material. The error ranges associated with using such charts are not unacceptable for the purposes at hand.

The age predictor variables, in the regression analysis, were ordinal wear scores, based on Scott's (1979b) scoring method, of mandibular and maxillary first and second molars. To maintain an effective sample size it was not a condition that first and second molars be paired. The method of selection was simply that the left molar was used whenever possible with the substitution of the right molar when necessary. Ideally it would be preferable to use left and right averaged wear scores for each of the four molar types. However, to do this would have rendered the resultant sample sizes impractically small for least squares modelling.

Intra-observer error with regard to scoring the teeth was also examined. Scott's (1979b) system involves dividing the occlusal crown into four equal quadrants and assessing wear on a scale from 1 to 10. The scores from each quadrant are summed to give a wear score of between 4 and 40. A total of 79 teeth from the Da But period and Metal period samples, both first and second molars, were reassessed for wear. At least one month and a maximum of four months separated these re-sampled cases from the original assessments of wear.

Results

(1) Cranial Sutures

Figures 2.6a to 2.6f and 2.7a to 2.7f show the age probability profiles for each sample based on cranial suture closure. Figure 2.8 represents the shape of the standardised reference population, and it is deviations from this that reveal the nature of the unknown populations of interest. The clear overall pattern that emerges is that for both the Da But period and Metal period samples there is a strong bias toward youth. Moreover, a slightly greater proportion of the Da But period sample falls into the youngest age class.

Masset (1989) notes that endocranial scores are more

highly correlated with true age than ectocranial. Figure 2.6a indicates a male population highly skewed toward youth. This may be due to sample size, as the skewness, although still apparent, is spread across more age classes in the ectocranial distribution (Figure 2.6d). A similar picture is evident with females looked at endocranially and ectocranially (Figures 2.6b, 2.6e). Combining males and females still clearly reveals the youth of the population (Figures 2.6c, 2.6f).

The Metal period distributions also indicate the youthful composition of this population. However, there tends to be a better representation of the middle age categories than for the Da But period. The greatest skewing to youth appears in the endocranial distributions (Figures 2.7a, b & c). The ectocranial distributions display greater middle age class representation (Figures 2.7d, e & f), which may be due to slightly better sample sizes.

There is not a great deal of point in trying to gain further information from these distributions. They are not 'real' representations of the Da But and Metal period populations. They are probability distributions for which departure from the reference population (Figure 2.8) is

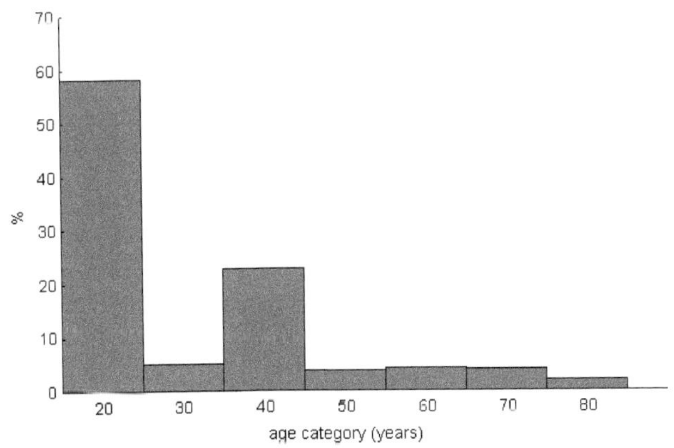

Figure 2.6a. Da But Period: Age Probability Profile Males, Endocranial Suture Closure (n=22)

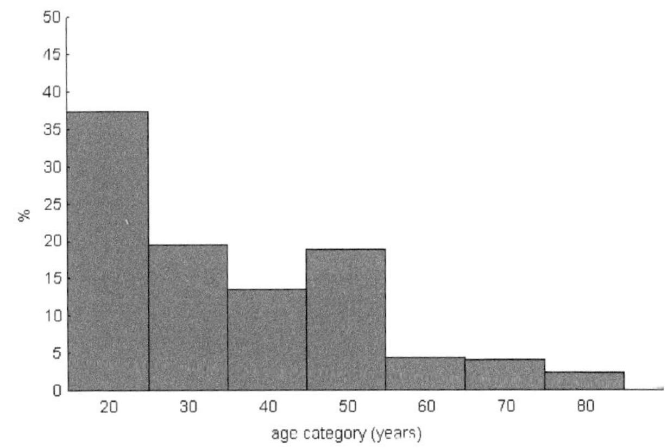

Figure 2.6d. Da But Period: Age Probability Profile Males, Ectocranial Suture Closure (n=26)

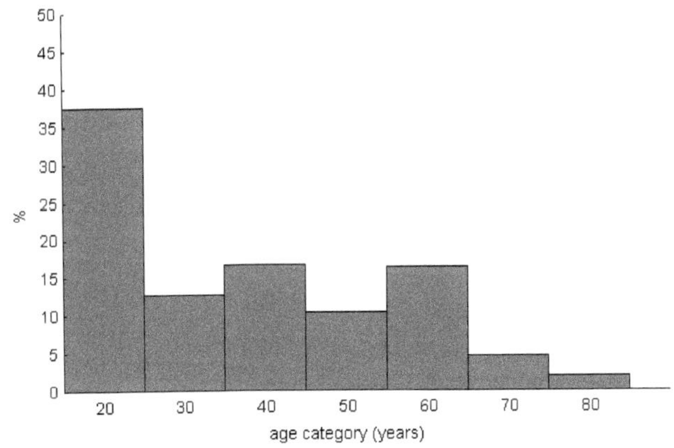

Figure 2.6b. Da But Period: Age Probability Profile Females, Endocranial Suture Closure (n=18)

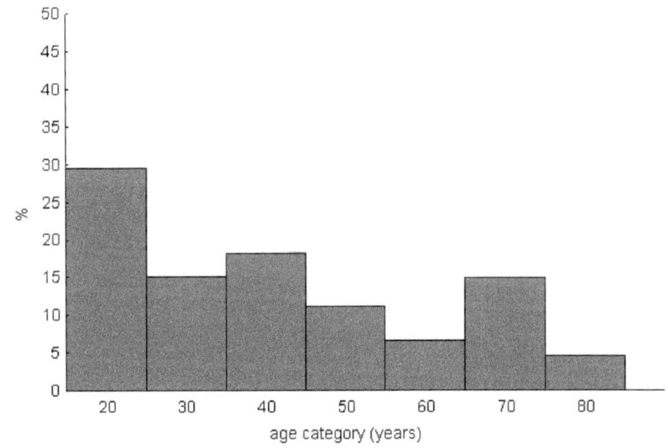

Figure 2.6e. Da But Period: Age Probability Profile Females, Ectocranial Suture Closure (n=18)

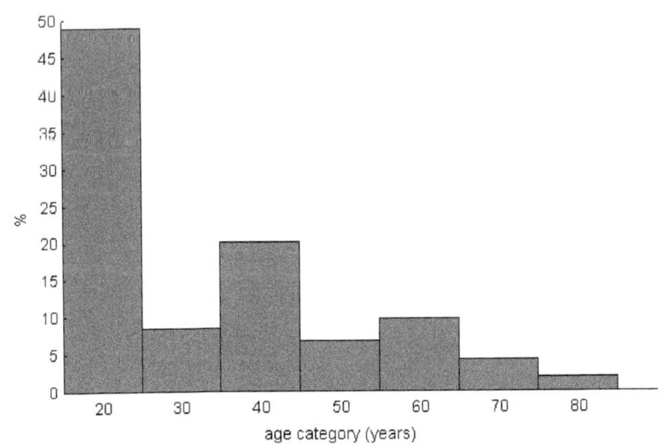

Figure 2.6c. Da But Period: Age Probability Profile Males and Females, Endocranial Suture Closure (n=40)

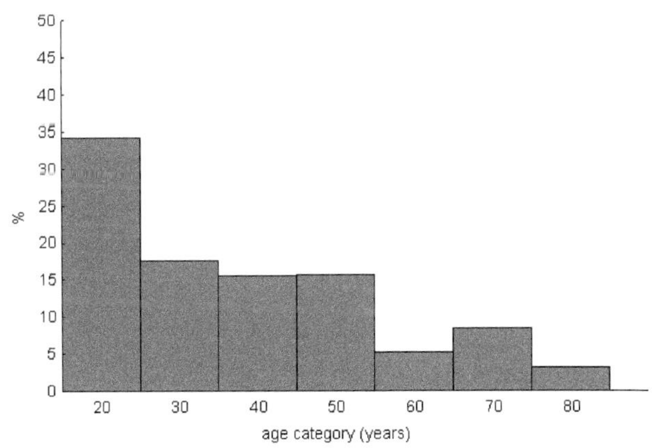

Figure 2.6f. Da But Period: Age Probability Profile Males and Females, Ectocranial Suture Closure (n=44)

of interest. Further, only about half of each adult sample was well enough preserved to permit inclusion in these distributions. The principal point is that both the Da But period and Metal period cemetery samples display similar age probability distributions.

(2) Tooth Wear

(a) Wear Score Intra-Observer Error

A paired sample t-test was performed (Table 2.9) comparing wear scores at time 1 (original wear assessment) and time 2 (subsequent wear assessment). The means are not significantly different and the standard deviation is similar in both instances. There is a high Pearson's Correlation Coefficient of 0.988. Scott (1979b) argued that this was an indicator of precision in analyses of intra-observer error although it could not be so used in looking at inter-observer error. The mean difference between time 1 and time 2 was only 0.23 'Scott wear score units' (Scott 1979b), further indicating a low level of intra-observer error.

(b) Regression Analysis

Regression analysis assumes several conditions of the data are not violated. Departures from these assumptions can be visually inspected by way of residual analysis

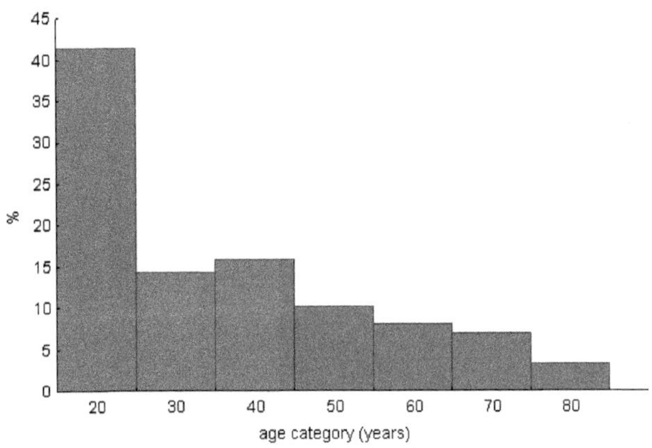

Figure 2.7a. Metal Period: Age Probability Profile Males, Endocranial Suture Closure (n=17)

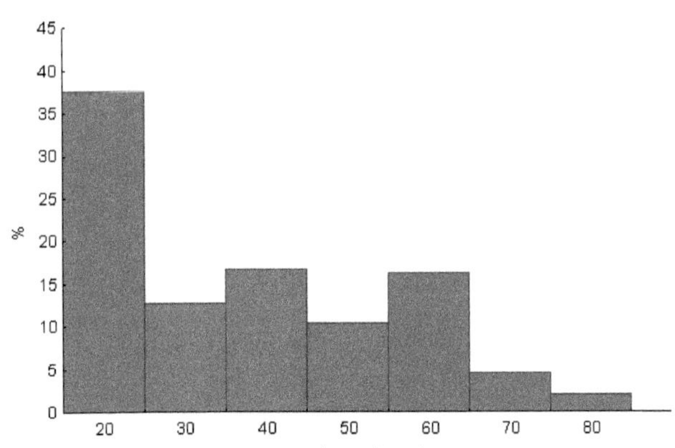

Figure 2.7b. Metal Period: Age Probability Profile Females, Endocranial Suture Closure (n=18)

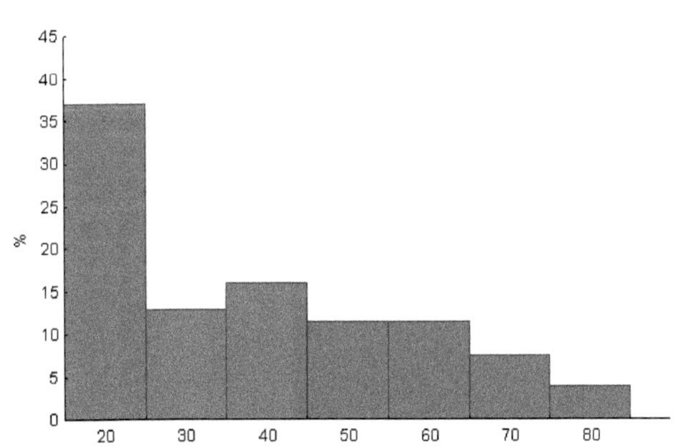

Figure 2.7c. Metal Period: Age Probability Profile Males and Females, Endocranial Suture Closure (n=35)

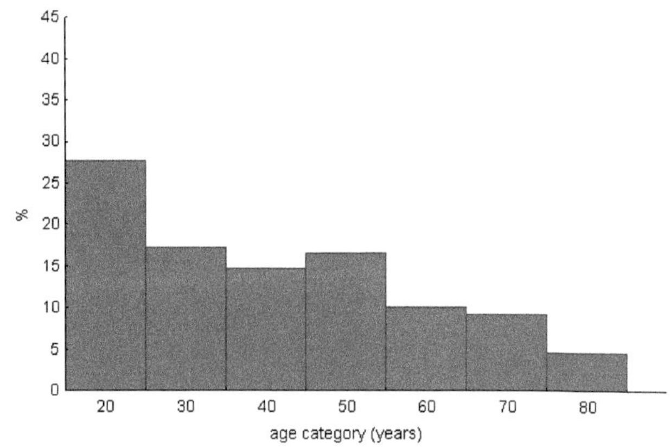

Figure 2.7d. Metal Period: Age Probability Profile Males, Ectocranial Suture Closure (n=20)

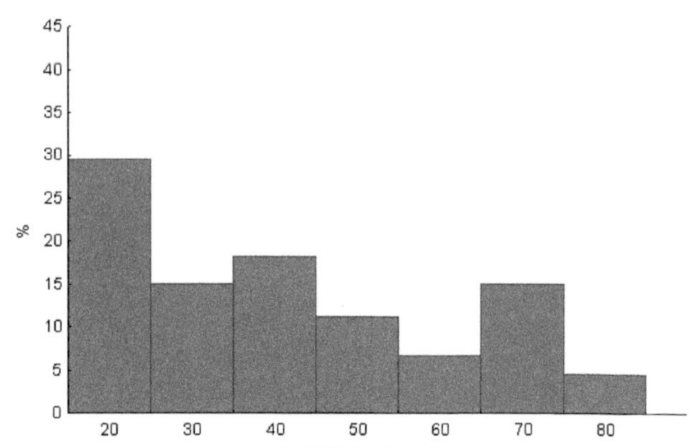

Figure 2.7e. Metal Period: Age Probability Profile Females, Ectocranial Suture Closure (n=18)

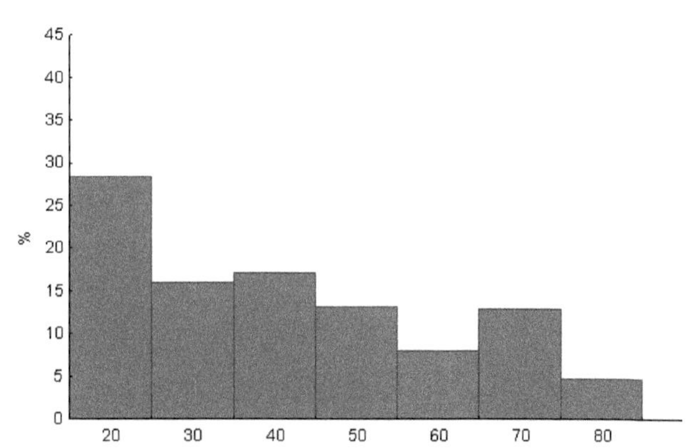

Figure 2.7f. Metal Period: Age Probability Profile Males and Females, Ectocranial Suture Closure (n=38)

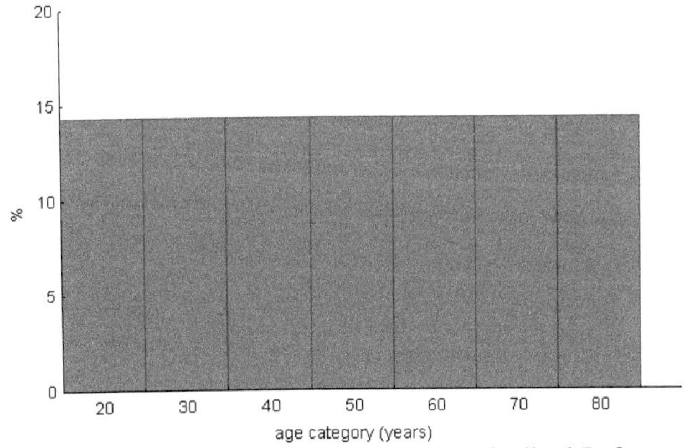

Figure 2.8. Age Probability Profile for Standardised Reference Population. Based on Cranial Suture Closure (all age classes equal)

(Kleinbaum & Kupper 1978). A residual is the error in prediction, or the difference between an actual value and its predicted value. Scatter-plots of age on each of the predictor variables are given in Figures 2.9a to 2.9d. All relationships are linear. There is less variation around the fitted line for the maxillary teeth (Figures 2.9a, 2.9b) than for mandibular teeth (Figures 2.9c, 2.9d). The following residual plots (Figures 2.10 to 2.12) all indicate that there are no significant departures from the chief assumptions of regression analysis. For instance, Figures 2.10a to 2.10d plot raw residuals against age predictors for each of the four predictors. These plots all indicate a random distribution of residuals, thus confirming the appropriateness of the linear model in each case. It can also be seen that the variance of the error terms are constant in all cases. Figures 2.11a to 2.11d plot standardised residuals on the predictors in determining if outliers are present. With few exceptions most residuals in each plot fall within 2 standard deviations of the fitted line and all points fall within 2.5 standard deviations. No outliers are evident. Figures 2.12a to 2.12d display normal probability plots, expected normal values against residuals, in determining if there are major departures from normality of the error terms. The linearity of the relationship in all cases indicates no major departures.

Table 2.10 summarises the results of the regression analysis in providing values of the correlation coefficient, mean of the sum of squared errors (residuals), sample size, slope for predictors, intercept of the *y* axis and the F statistic. Walker *et al.* (1991) suggest that predictive accuracy of a given regression model is best assessed by examining the sum of the squared errors (SSE), i.e. the sum of squared residuals. Walker *et al.* (1991) directly compared SSE values for each of their models, as sample size was constant across all of their models. For this analysis the mean of the squared errors (MSE) was used due to differing sample sizes for the models assessed. The smallest MSE indicates the best predictive accuracy for a given model.

Table 2.9. Intra-Observer Error in Assessing Wear Scores

	mean	sd	n	mean difference	t statistic	p	r
time 1	21.52	7.77	79	0.23	1.74	0.09	0.988
time 2	21.29	7.65					

When assessed individually, maxillary second and first molar wear, respectively, had the greatest predictive accuracy. The mandibular molars were also useful age predictors with the second molar being slightly better than the first.

Various combinations of molars were assessed to see if predictive accuracy could be increased. Only two multiple regression models were found to increase predictive accuracy. Both maxillary molars taken as the predictors gave the greatest predictive accuracy of all models with an adjusted r^2 value of 0.854 and an MSE of 0.309. The addition of the second molar to the two maxillary molars was nearly as good but the model is redundant in adding an extra predictor with a slight loss of predictive accuracy, albeit greater than the single variable models.

Walker *et al.* (1991:175) noted the potential role of sex differences in determining age. They added a dummy variable for sex in their analysis but found no improvement. To deal with indeterminate sex cases I have followed their example and coded for two dummy variables: 'Males' (coded 1 for males and 0 for others) and 'Females' (coded 1 for females and 0 for others). As can be seen from Table 2.10, the addition of sex into all models improved predictive accuracy. In applying these models it should be noted that due to differential preservation of molars, a range of these models needed to be applied to the sample to be aged.

The predictive accuracy of this modelling can be seen in examining the frequency scatter plot in Figure 2.13. Here true age minus estimated age is plotted against true age to examine errors in age prediction. The line drawn through zero indicates the position of no error in age prediction. As can be seen, the majority of cases fall along this line and all cases are within one age category of their true age. A negative score indicates over-aging and a positive score under-aging. It can be seen that those cases of over-aging tend to be in the younger to middle age categories while under-aging seems only to occur in the category 40-49 years. This discrepancy has been recently commented on by Masset (1989) and called attraction to the middle. Aykroyd *et al.* (1997) also comment on this apparent systematic bias and suggest the use of classical calibration over regression. However, the reduction in the efficiency of estimates with classical calibration (Aykroyd et al. 1997:263) would tend to indicate that regression is still the better option.

(c) Applying the Da But Period Sample Derived Equations to the Metal Period Sample

Sample specific models for predicting Metal period age-at-death could not be developed due to the small number of independently aged individuals in this sample. Nonetheless, the Da But period age prediction models

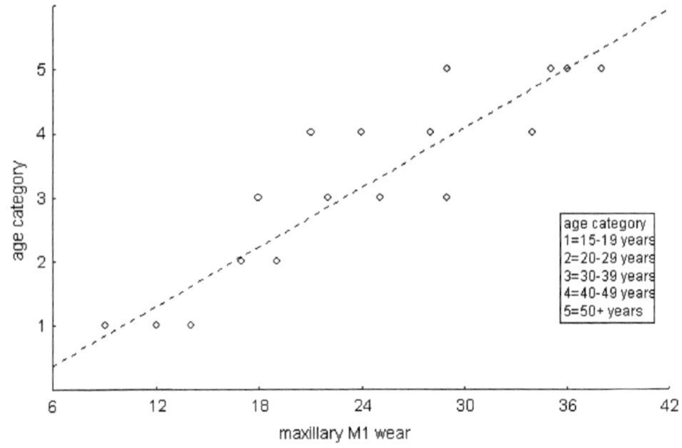

Figure 2.9a. Age on Maxillary M1 Wear Score (y=-0.556+0.155*x)

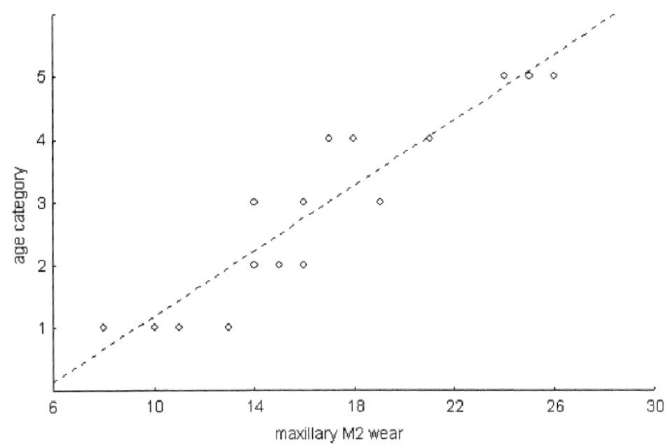

Figure 2.9b. Age on Maxillary M2 Wear Score (y=-1.428+0.261*x)

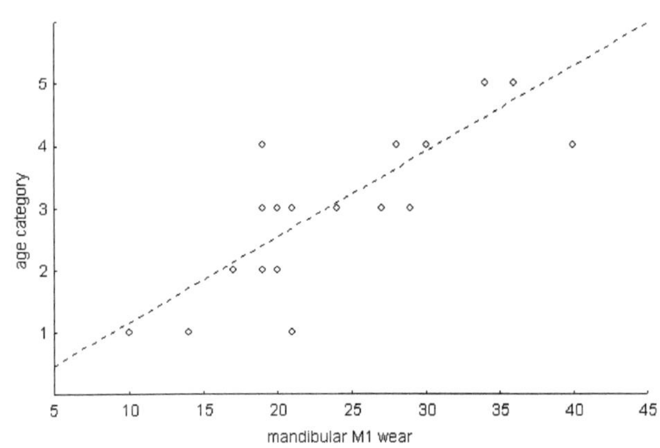

Figure 2.9c. Age on Mandibular M1 Wear Score (y=-0.237+0.138*x)

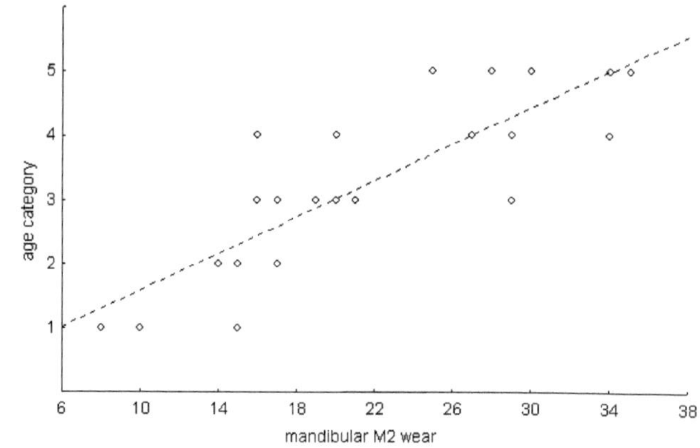

Figure 2.9d. Age on Mandibular M2 Wear Score (y=0.153+0.143*x)

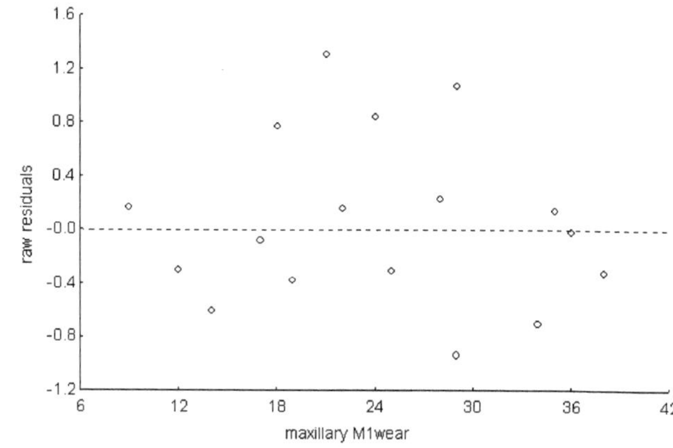

Figure 2.10a. Raw Residuals vs. Maxillary M1 Wear (r=0.000)

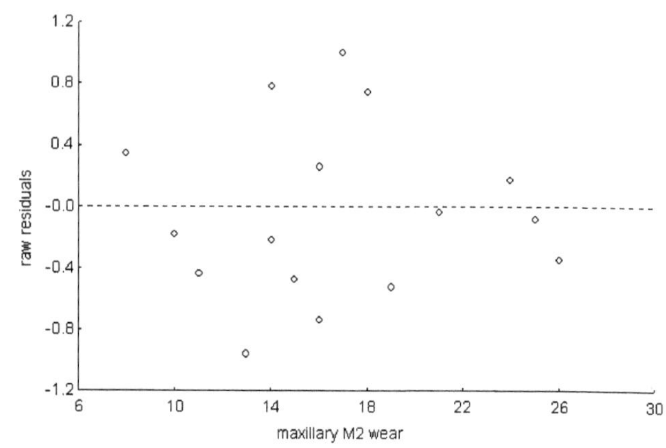

Figure 2.10b. Raw Residuals vs. Maxillary M2 Wear (r=0.000)

can be fairly confidently applied if two criteria are met, namely: (1) it needs to be shown that the Da But period models can accurately predict the age of individuals from the Metal period sample, for which there are independent age estimates; and (2) the rate of dental wear for both samples is similar, thus suggesting equivalence between degree of wear and age for each sample.

In addressing the first point, the Da But period models were applied to independently aged Metal period individuals. There are a total of 66 adult (≥15 years) Metal period individuals with dental remains that can be aged by way of the dental wear regression models. Of these there are 11/66 (16.7%) that can be both aged using dental remains and other independent age estimators. This series constitutes the control sample for assessing the accuracy of age assignment using the Da But period derived dental wear regression models. Of the control sample 7/11 were assigned to their respective correct categories. Four cases (identified by catalogue number) were misclassified by one age category and are as follows:

89MDM2a: originally assessed as age category 4 (40-49 years) based on pubic/auricular morphology. Assigned to age category 5 (50+ years) by the regression models.

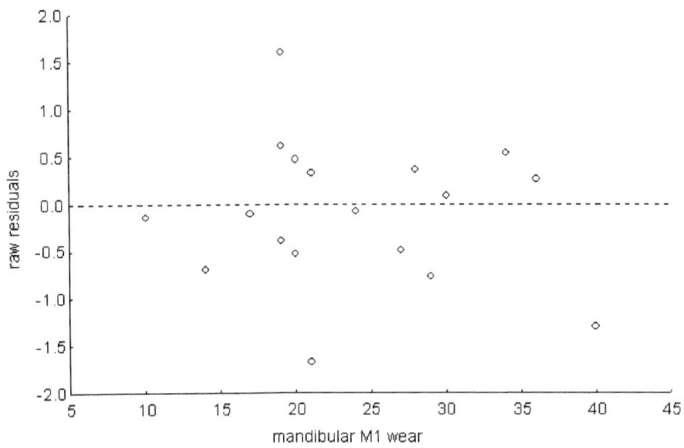

Figure 2.10c. Raw Residuals vs. Mandibular M1 (r=-0.000)

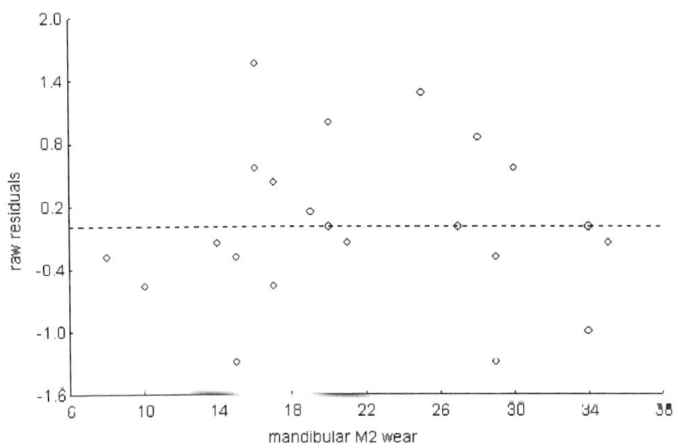

Figure 2.10d. Raw Residuals vs. Mandibular M2 (r=-0.000)

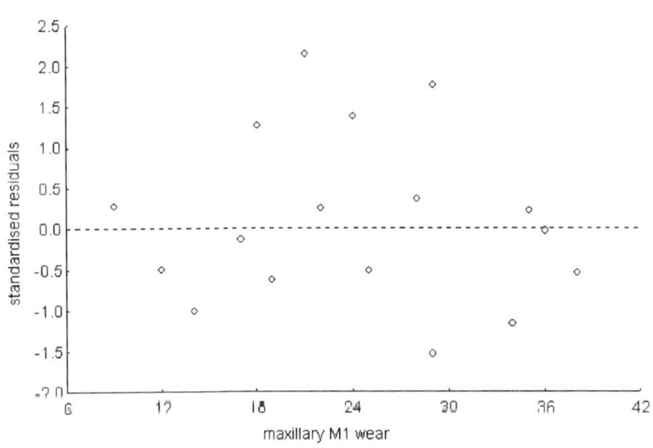

Figure 2.11a. Standardised Residuals vs. Maxillary M1 Wear (r=0.000)

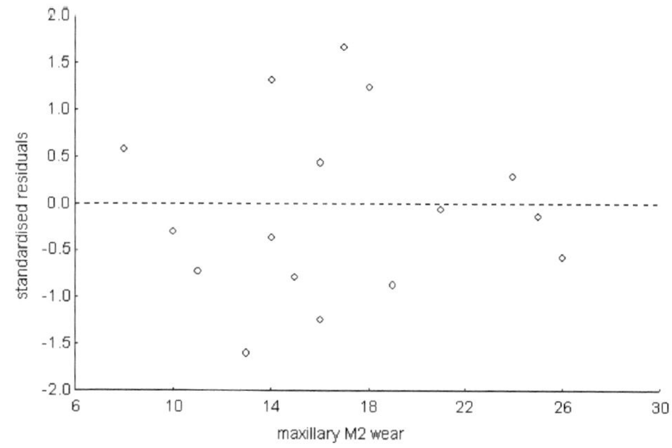

Figure 2.11b. Standardised Residuals vs. Maxillary M2 Wear (r=0.000)

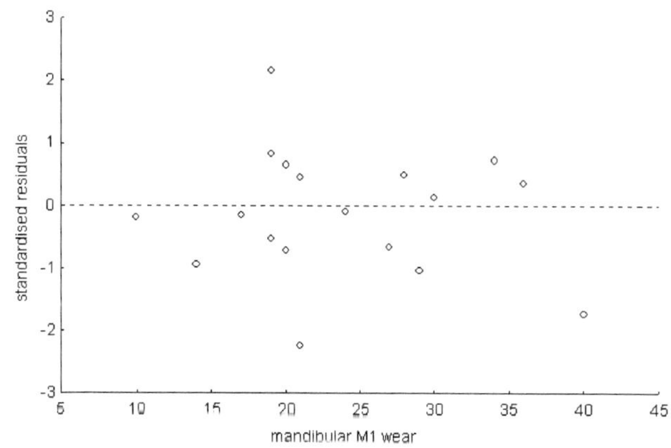

Figure 2.11c. Standardised Residuals vs. Mandibular M1 Wear (r=-0.000)

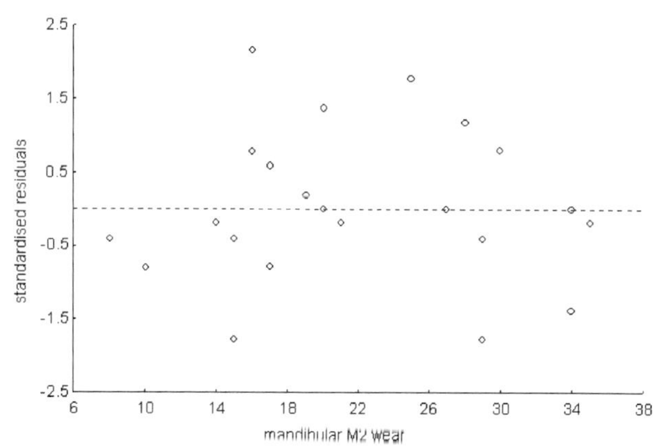

Figure 2.11d. Standardised Residuals vs. Mandibular M2 Wear (r=-0.000)

84DSM6a: originally assessed as age category 2 (20-29 years) based on auricular morphology, although epiphyseal age placed it at ≤25 years. Regression model assigned age was age category 1 (15-19 years).

95DXM1a: originally assessed as 30-34 based on auricular morphology and assigned to age category 3. The regression model assigned it to age category 2 (20-29 years).

78QCM19a: originally assessed as ≤22 years, based on epiphyseal fusion, and assigned to age category 2 (20-29 years). However, assignment could also as legitimately been made to age category 1 (15-19 years). Regression model assigned age was age category 1.

All of these cases are borderline with respect to initial age category assignment. Even with broad age categories there will always be cases that approach the category borders. The chief point, however, is that there is extremely good predictive accuracy, to within one age category, when applying the Da But period derived regression models to the Metal period data set. There is no clear age assignment bias within these four cases, with two being over-aged and two under-aged using the regression models. Moreover, none of the four cases represents a marked departure from expected error of assignment.

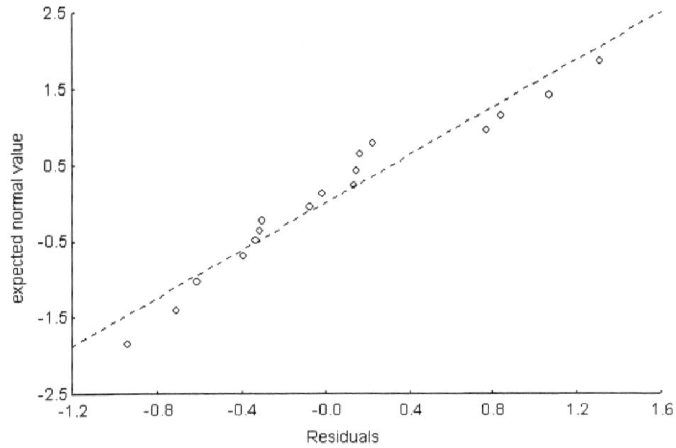

Figure 2.12a. Normal Probability Plot of Residuals: Age on Maxillary M1 Analysis

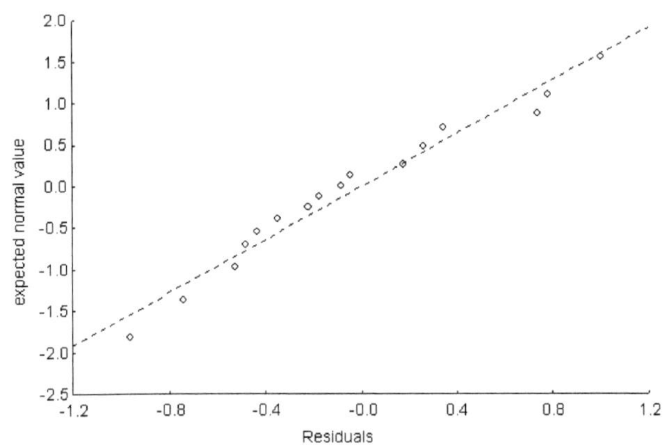

Figure 2.12b. Normal Probability Plot of Residuals: Age on Maxillary M2 Analysis

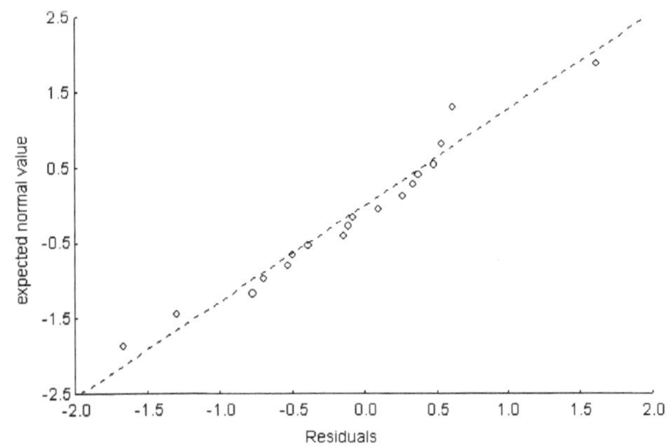

Figure 2.12c. Normal Probability Plot of Residuals: Age on Mandibular M1 Analysis

Examining the rate of dental wear for each sample revealed no significant differences in the age-independent rate of wear analysis. The complete results and discussion of this analysis can be found in Chapter three where the oral health of the samples is considered. It should be stressed that this particular method of determining rates of wear is age-independent, thus allowing a comparison between the two populations without knowing their respective age structures.

Age and Sex Estimation Summary

Table 2.11 summarises age category assignment based on morphology, epiphyseal union and tooth eruption as

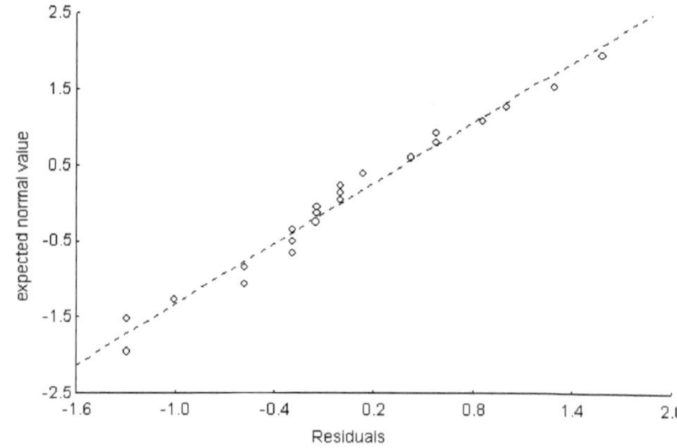

Figure 2.12d. Normal Probability Plot of Residuals: Age on Mandibular M2 Analysis

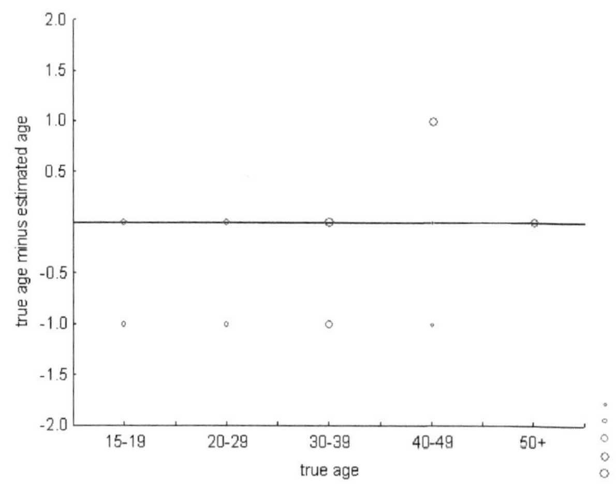

Figure 2.13. Age Assignment Error in Regression Modelling

well as age assignment based on these criteria and tooth wear age regression models. In the Metal period sample the proportion of cases assigned an age has increased from 29.17%, using morphological criteria alone, to 85.42% with the addition of the tooth wear aging models. Similarly, the Da But period aged sample increases from 44.68% to 89.36%.

Both this table and the derived age distribution histograms (Figures 2.14, 2.15) indicate differences in the adult age category make up. The Metal period is biased toward the 20-29 year category while the Da But period is biased to the 30-39 year category. Combining these categories indicates that 42.9% of the Da But period sample compared to 53.7% of the Metal period sample are aged between 20 and 39 years. Further, while relatively young adults (15 to 19 years) are similarly represented in both periods, the Da But period sample contains twice as many older adults (40+ years). This is in contrast to the cranial suture age probability profiles which indicated a younger Da But period population relative to the Metal period sample. However, the cranial suture age probability profiles were generated with much smaller sample sizes than the combined morphology and regression modelling method. The regression model aged sample represents a much larger proportion of the total sample. The good correlation

Table 2.10. Regression Models for Age Estimation from Tooth Wear Scores.

sample	n	r	r²	adjr²	MSE		slope	SE slope	intercept	SE int.	ANOVA F	p
age/max m1	19	0.908	0.825	0.816	0.369	max m1	0.155	0.016	-0.056	0.339	89.521	0.000
age/max m2	19	0.909	0.826	0.816	0.357	max m2	0.261	0.029	-1.423	0.506	80.757	0.000
age/man m1	22	0.841	0.708	0.693	0.553	man m2	0.138	0.020	-0.237	0.522	48.457	0.000
age/man m2	26	0.834	0.696	0.683	0.525	man m1	0.143	0.019	0.153	0.429	54.884	0.000
age/max m1, max m2	17	0.934	0.872	0.854	0.309	max m1	0.097	0.045	-1.233	0.487	47.689	0.000
						max m2	0.123	0.071				
age/max m1, male, female	21	0.92	0.847	0.820	0.360	max m1	0.149	0.017	-0.952	0.507	31.340	0.000
						female	0.742	0.491				
						male	0.494	0.475				
age/max m2, female, male	19	0.935	0.875	0.850	0.291	max m2	0 246	0.028	-2.068	0.560	35.032	0.000
						female	1.049	0.444				
						male	0.934	0.423				
age/man m1, female, male	22	0.897	0.805	0.772	0.411	man m1	0.121	0.019	-0.390	0.507	24.701	0.000
						female	1.218	0.468				
						male	1.280	0.435				
age/man m2, female, male	26	0.874	0.764	0.732	0.443	man m2	0.132	0.020	-0.517	0.475	23.797	0.000
						female	0.895	0.476				
						male	1.128	0.447				
age/max m1, max m2, female, male	17	0.955	0.912	0.883	0.247	max m1	0.077	0.042	-1.744	0.552	31.218	0.000
						max m2	0.136	0.066				
						female	1.005	0.429				
						male	0.708	0.408				

MSE=mean of the square errors (equivalent to the sum of squared residuals divided by appropriate df)
max m1: permanent first maxillary molar
man m1: permanent first mandibular molar
max m2: permanent second maxillary molar
man m2: permanent second mandibular molar

Table 2.11. Individual Age Assignment Summary

	Da But Period		Metal Period	
Status	non-metric aging	combined[1]	non-metric aging	combined[1]
0-11 months	1	1	2	2
1-14 years	6	6	11	11
15-19 years	6	9	3	9
20-29 years	9	14	5	27
30-39 years	7	22	5	17
40-49 years	7	14	2	9
50+ years	6	18	0	7
Indeterminate[2]	52	10	68	13
Indeterminate[3]	0	0	1	1
% aged	44.68%	89.36%	29.17%	85.42%
TOTAL	94	94	96	96

[1] combined non metric and tooth wear regression models

[2] Indeterminate= adult but could not be assigned to an age category

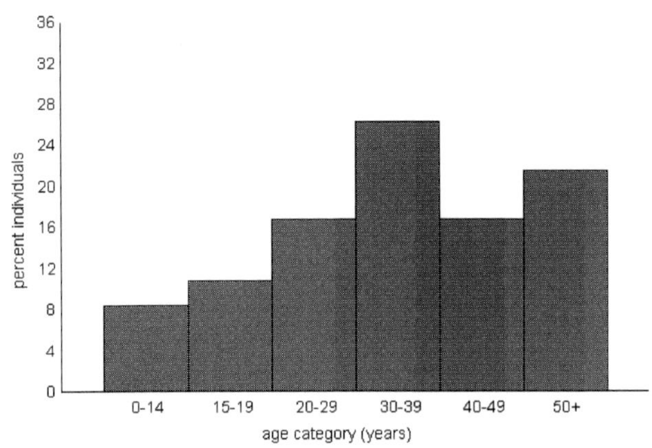

Figure 2.14. Da But Period Sample Age Distribution

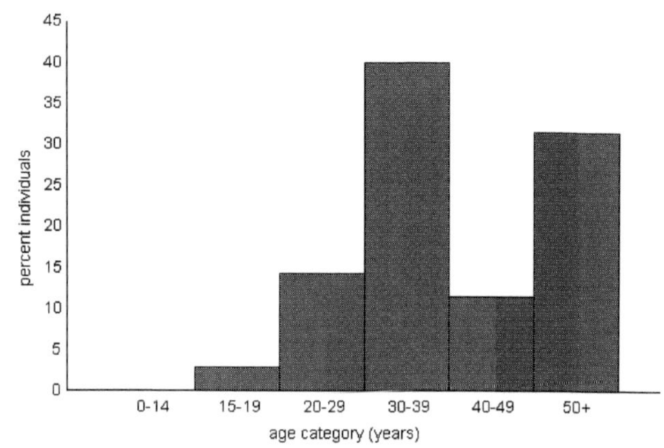

Figure 2.16a. Da But Period Age and Sex Distribution: Males (n=35 sexed and aged males)

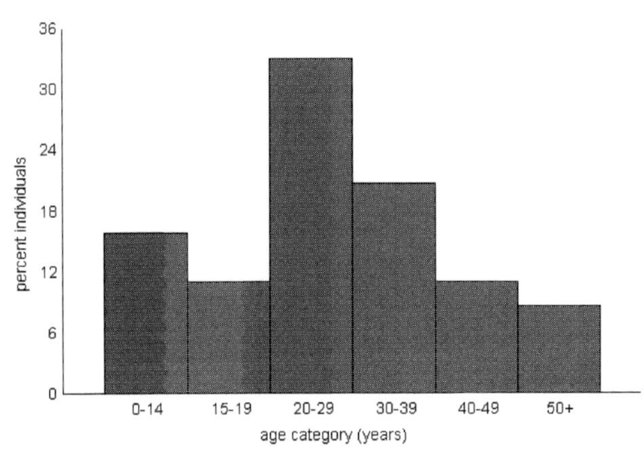

Figure 2.15. Metal Period Sample Age Distribution

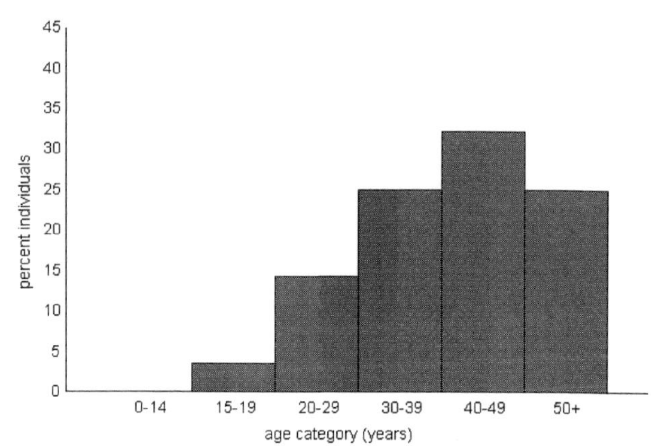

Figure 2.16b. Da But Period Age and Sex Distribution: Females (n=28 sexed and aged females)

[3] Indeterminate=too fragmentary to assign as adult or sub-adult

between age-at-death and tooth wear also means that the age distributions generated with the regression models are closer to the actual age composition of each sample. Finally, the regression models allow individual estimates of age-at-death, something that cannot be achieved using age probability profiles.

Given the greater representation of older adults in the Da But period assemblage, it is necessary to examine the age composition of any sub-samples of both populations used in later analyses. One important sub-sample includes those individuals with assessable gnathic remains used in examining oral health in Chapter three. It should be borne in mind that such sub-sampling may change the age structure of the new data sets in unknown ways. This is also true for the sex composition of these samples. Whenever

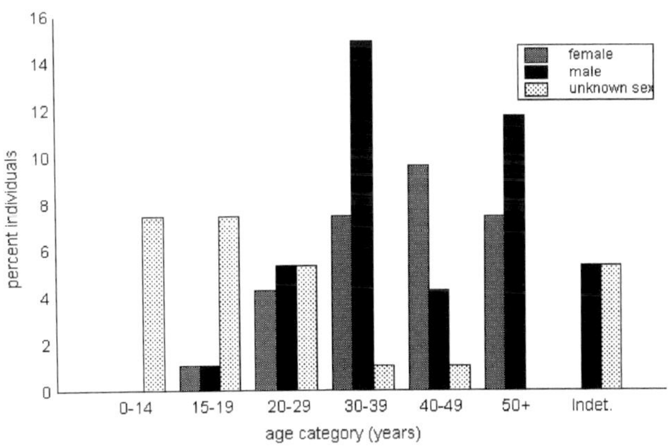

Figure 2.16c. Da But Period Age and Sex Distribution:Male, Female and Unknown Sex (n=94)

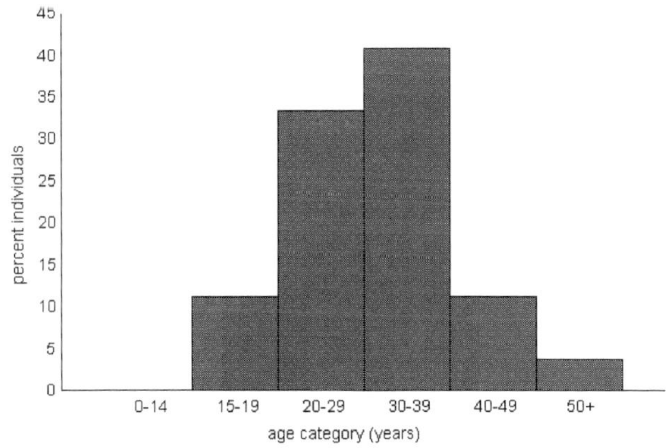

Figure 2.17b. Metal Period Age and Sex Distribution: Females (n=27 sexed and aged females)

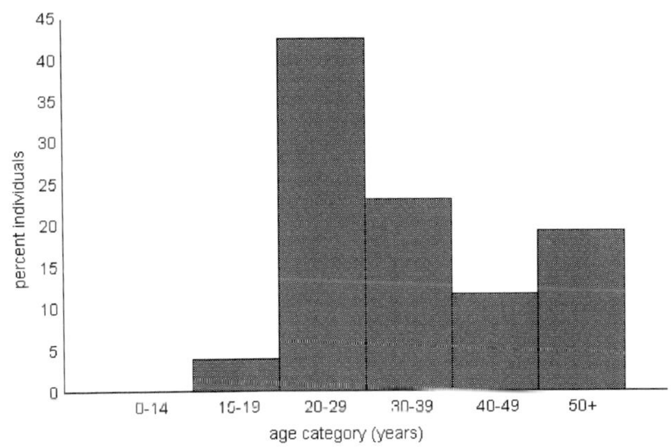

Figure 2.17a. Metal Period Age and Sex Distribution: Males (n=26 sexed and aged males)

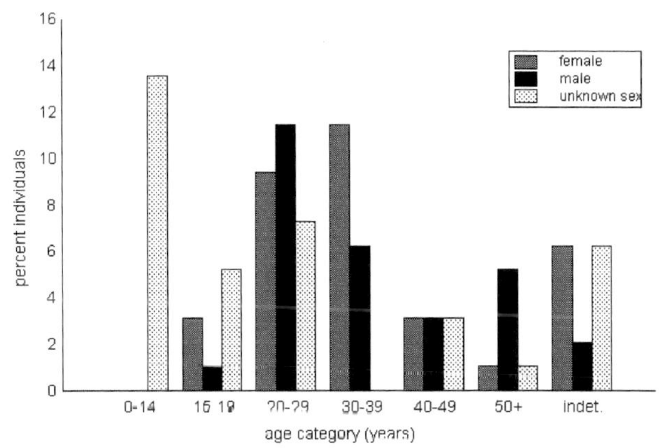

Figure 2.17c. Metal Period Age and Sex Distribution: Male, Female and Unknown Sex (n=96)

such sub-sampling occurs in later chapters the age and sex composition of each new sample is re-examined where practicable.

Figures 2.16 and 2.17 show the overall age and sex composition of each sample. When examining those Da But Period individuals for which sex and age have been estimated, males (Figure 2.16a) concentrate in the 30-39 and 50+ age categories while females (Figure 2.16b) are spread relatively evenly across the last three age categories. Figure 2.16c displays the proportion of sexed and unsexed individuals by age category. As would be expected, the greater proportion of unsexed individuals fell within the youngest two age classes (0 to 19 years) because sub-adult remains are extremely difficult to assess for sex. No cases of un-aged females occur in the Da But period sample, while unsexed cases and males make up the remaining 10% of un-aged cases.

When examining those Metal period individuals for which sex and age have been estimated, Figure 2.17a shows that the largest proportion of males fall within the 20 to 29 year age category. For females, it is the 30 to 39 year category that displays the greater percentage (Figure 2.17b). However, combining both categories reveals an almost equivalent proportion of males and females aged between 20 and 39 years. As was the case with the Da But period, the greatest proportion of unsexed individuals fell within the youngest two ages classes (0 to 19 years) in the Metal period sample (Figure 2.17c). Very few males were un-aged while females and unsexed cases make up the majority of un-aged individuals.

These distributions are presented here as descriptions of the sex and age composition of each assemblage only. The respective sample compositions may impact on later conclusions drawn regarding the past behaviour and health status of these people. Indeed, demonstrable age and sex composition differences in samples or sub-samples do not render comparison impossible. On the contrary, it simply means that these factors must be considered in explaining any differences found between the populations.

Stature

Stature is reported here for the Da But period and Metal period samples as part of the process of describing the biology of these people. Stature is affected by both genetic and socio-environmental factors. As such it is difficult to directly compare achieved adult stature between samples and make inferences with regard to diet, health and so forth. Generally, stature analysis is employed for making such inferences by way of growth studies using sub-adult material (Larsen 1997). The dearth of suitable sub-adult material in this study mitigated against attempting such work. Even simply looking at achieved adult stature is not without problems. Sample sizes for suitable intact post-cranial elements are quite small, especially for the Metal period. However, this problem can be somewhat ameliorated by examining a range of stature sensitive skeletal elements. So, how tall were the Da But period and Metal period people and were there any intra- and inter-sample differences?

Methods and Materials

Regardless of bone type, in all cases the left element was used in the following analyses when both sides were available for a given individual. This study looked at four separate skeletal elements involving five length variables. The elements examined include: femoral, tibial, humeral and calcaneal length (Table 2.1). The use of the calcaneus needs some explanation. Recently Holland (1995) devised regression models based on two length variables of the calcaneus and maximum length of the talus. He found a linear relationship between the lengths of these bones and stature that did not appear to be greatly affected by allometric, racial or sexual factors. The variable with the greatest correlation with stature was posterior calcaneal length (PCAL, r=0.817), followed by maximum talar length (MTAL, r=0.731) and maximum calcaneal length (MCAL, r=0.723) (Holland 1995:317). Holland (1995:319) goes on to point out that these models are based on 20th century American white and black samples and their "applicability to mongoloid remains is undocumented". However, it would be safe to assume that Asian populations will also display a positive correlation between these variables and stature. To circumvent the problem of a lack of Asian specific foot formulae, the Vietnamese data can be used to compare relative stature as expressed by the respective lengths of these bones. Only the calcaneal data are analysed here as the Metal period talar sample was not only limited, but it was heavily biased towards females.

Results

Intra Sample Differences by Element

Table 2.12 presents mean, standard deviation and range for these length variables by skeletal element. Male femoral length is longer in both samples relative to female length and these differences are statistically significant. Da But period tibial sex differences cannot be evaluated as the sample in biased in favour of males. However, for the Metal period there is a significant difference between male and female tibial length. Sample size for the humerus is quite a lot larger than for any other element in these samples. Da But period male humeral length is significantly longer than female length. However, this is not the case for the Metal period. Male calcaneal length, both MCAL and PCAL, is significantly greater for males in the Da But period but not in the Metal period. Regardless of element observed, the Da But period consistently exhibits evidence for sexual dimorphism in variables most sensitive to attained stature. The Metal period provides evidence for sexual dimorphism in the femur and tibia only. Nonetheless, these are the two most favoured elements used in stature estimation. Sample sizes are quite small when divided into separate sexes and as such any findings, significant or otherwise, need to be treated with some caution.

Inter Sample Differences

The only two elements to exhibit significant inter sample differences are the tibia and humerus (Table 2.12). It needs to be noted that there are few females in the Da But period sample and few males in the Metal period humeral samples.

Stature Estimates from Bone Lengths

Table 2.13 displays stature estimation figures derived from three separate sources. In the first instance stature has been estimated by way of the Trotter and Gleser (1952, 1958, cited in Bass 1987) formulae for Asian male femora and humeri. Unfortunately female formulae are not available. Further, the formula for the tibia was not used given recently discovered problems in the manner in which Trotter measured this bone to develop her regression formulae (Jantz *et al.* 1995).

In the next instance, sex and race non-specific formulae devised by Sjovold (1990) are used. Sjovold (1990) has revisited the problem of the population specific nature of stature estimation models. This point has already been noted in the previous discussion on the population specificity of discriminant function sexing equations. Sjovold (1990) has developed an alternative to population specific least squares models. Using a variation of the principal (major) axis technique, called reduced major axis, Sjovold develops a series of stature prediction models for a range of long bones. These models are not population specific. Sample sizes used in developing these models are quite large. For example, the weighted line of organic correlation (weighted reduced major axis) for the femur was based on 25 samples totalling 8247 individuals. Sjovold (1990) has tested these models on various ancestral groups with excellent results.

Feldesman & Fountain (1996) provide an Asian specific (broadly defined), sex non-specific regression formula for the femur. While Feldesman (see also Feldesman *et al.* 1990) has concentrated on deriving and proving the utility of the femur/stature ratio by ancestral group and generically, Feldesman & Fountain 1996 provide race specific regression formulae that perform better than his femur/stature ratios. This model has been used to provide a comparison with other femur stature estimates.

As can be seen from Table 2.13, there is not a great deal of discordance in estimated stature from the same type of bone, regardless of formula used. Further, differences in estimated stature by bone type are not particularly large either. Sample size must also be considered. While the femur is one of the favoured bones for stature estimation there is only a total sample size of 16 separate individuals. The humerus, on the other hand, gives a total sample size

Table 2.12. Skeletal Element Lengths Used in Stature Estimation

	Da But Period			Metal Period			Intra sample sex comparison		Inter sample sex comparison
	male	female	combined	male	female	combined	Da But	Metal	(combined sex)
Femur (cm)									
n	5	4	9	4	3	7	**t -2.963**	**t -4.300**	t 1.118
mean	45.98	41.40	43.94	44.35	39.23	42.16	**p 0.021**	**p 0.008**	p 0.282
sd	1.79	2.85	3.24	1.30	1.83	3.09			
min	43.10	38.90	38.90	43.40	38.00	38.00			
max	48.00	44.30	48.00	46.20	41.40	46.20			
Tibia (cm)									
n	8	1	10	4	3	7	na	**t -3.113**	**t 2.336**
mean	37.63	33.30	36.86	35.75	32.17	34.21		**p .026**	**p .034**
sd	1.77		2.26	0.90	2.11	2.36			
min	35.00		33.30	35.00	30.20	30.20			
max	39.40		39.40	36.80	34.40	36.80			
Humerus (cm)									
n	15	8	23	3	8	11	**t -3.555**	t -1.568	**t 2.250**
mean	32.01	29.56	31.16	30.77	29.23	29.65	**p .002**	p 0.151	**p .031**
sd	1.62	1.48	1.95	0.68	1.60	1.55			
min	29.20	28.00	28.00	30.00	27.40	27.40			
max	34.70	32.00	34.70	31.30	31.80	31.80			
Calcaneus (MCAL) mm									
n	16	8	29	3	5	8	**t -2.245**	t -1.997	t 1.126
mean	77.11	71.64	75.46	77.90	69.80	72.84	**p .035**	p .093	p 0.268
sd	5.00	6.81	5.62	2.50	3.57	6.63			
min	67.20	63.20	63.20	75.10	64.30	64.30			
max	85.80	82.50	85.80	79.90	81.20	81.20			
Calcaneus (PCAL) mm									
n	18	8	32	3	5	8	**t -2.410**	t -2.400	t 1.020
mean	54.92	50.96	54.09	57.27	49.34	52.31	**p 0.024**	p 0.053	p 0.314
sd	3.63	4.38	4.02	1.69	5.41	5.86			
min	47.8	45.3	45.3	56.1	45.7	45.7			
max	60.3	58.9	60.3	59.2	58.8	59.2			

Table 2.13. Stature Estimation from Long Bone Lengths

	Da But Period									Metal Period								
	male			female			combined sex			male			female			combined sex		
FEMUR																		
equation	a	b	c	a	b	c	a	b	c	a	b	c	a	b	c	a	b	c
n	5	5	5		4	4		9	9	4	4	4		3	3		7	7
mean height	171.43	170.47	170.83		158.05	157.82		164.95	165.05	167.92	166.05	166.20		152.18	151.66		160.11	159.97
sd	3.84	4.84	5.07		7.73	8.11		8.77	9.20	2.79	3.52	3.69		5.10	5.35		8.35	8.76
min	165.24	162.66	162.65		151.28	150.71		151.28	150.71	165.88	163.47	163.50		148.84	148.15		148.84	148.15
max	175.77	175.94	176.57		165.91	166.06		175.94	176.57	171.90	171.06	171.46		158.05	157.81		171.06	171.46
TIBIA																		
n		8			1			10			4			3			7	
mean height		171.13			156.90			168.61			164.96			153.17			159.91	
sd		5.83						7.43			2.96			6.95			7.76	
min		162.49						156.90			162.49			146.70			146.70	
max		176.97						176.97			168.41			160.52			168.41	
HUMERUS																		
n	15	15			8			23		3	3			8			11	
mean height	168.97	166.90			155.58			162.96		165.65	161.14			154.05			155.98	
sd	4.34	7.48			6.85			9.00		1.82	31.40			7.34			7.15	
min	161.45	153.90			148.36			148.36		163.59	157.60			145.59			145.59	
max	176.19	179.31			166.84			179.31		167.07	163.61			165.92			165.92	

a Sex specific Mongoloid stature formula (Bass 1987)
b Weighted line of organic correlation formula (Sjovold 1990)
c Asian specific (sex non-specific) regression formula (Feldesman & Fountain 1996)

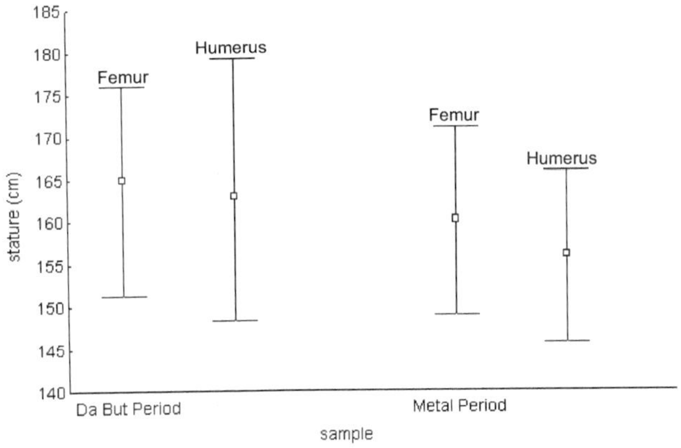

Figure 2.18. Range Plot of Da But Period and Metal Period Stature Estimates Based on the Humerus and Femur

of 34 individuals. However, estimated height using the femur (Sjovold's formula) is only 2 cm different to that when using the humerus for the combined Da But period sample. Although, this difference climbs to more than 5 cm for the combined sex Metal period sample.

In conclusion, regardless of bone or stature model used, Da But period height is greater in terms of both mean and maximum range than Metal period height both for combined sex samples and separately by sex. This pattern can be clearly seen in the range plots of stature estimated for combined sex samples using the femoral and humeral stature estimation models of Sjovold (Figure 2.18). If these samples are to a large extent genetically homogenous the implication is for a decline in health, as measured by reduced stature, in the Metal period.

3
ORAL HEALTH AND BEHAVIOUR

An examination of oral health is useful for issues relating to subsistence, behaviour and health in general. The principal aim of this chapter is to detail an oral health profile for the Da But period and Metal period samples. This profile is used to address two main questions: (1) What inferences can be made with respect to the diet and subsistence strategies of these people? (2) What behaviours can be inferred from the oral evidence preserved in these samples? Recent work on the oral health of samples from several sites in Thailand spanning the pre Metal-Metal transition is examined with the aim of determining if there are generalised Southeast Asian oral responses to particular subsistence strategies.

There are a wealth of variables that can be assessed in studies of gnathic remains (see contributions in Kelley & Larsen 1991; Hillson 1996). Given the expressed aims of this chapter, four broad types of variables were targeted. The first group includes the more commonly reported traits that have an important role with respect to inferring diet and oral behaviour generally: ante-mortem tooth loss, abscesses, caries, pulp chamber exposure, calculus deposits and tooth discolouration. The second group for the most part describes developmental and occlusal variation related traits: jaw robusticity, tooth impaction, rotation, malalignment or displacement and enamel anomalies such as pearls. The third group includes variables that reflect more specific oral behaviours not necessarily diet related: interproximal grooves, task-wear facets, tooth chipping and tooth ablation. The fourth group includes tooth wear, which is intimately correlated with many of the aforementioned traits. Specifically rates of tooth wear and the ratio of anterior to posterior tooth wear are examined.

General Diet and Behaviour Related Traits

Dental Caries

Larsen (1997:65) points out that "dental caries is a disease process" and the term does not refer to the lesions themselves. This condition is one of the most reported oral pathologies in the bio-archaeological literature (see Turner 1979; Larsen *et al.* 1991). Factors involved in this disease are myriad and complex. Larsen *et al.* (1991) provide a summary of the essential and modifying factors involved. Essential factors include:

> (1) teeth that have surfaces exposed to the oral environment; (2) presence of aggregates of complex indigenous oral bacterial flora (e.g. *Streptococcus mutans*, *Lactobacillus acidophilus*), salivary glycoproteins, and inorganic salts that adhere to the tooth surfaces (dental plaque); and (3) diet. Modifying factors include those that affect primarily site distribution and speed of carious lesion development ...These factors include, but are not limited to, tooth crown morphology and size, developmental enamel defects, occlusal surface attrition, food texture, certain systemic diseases, age, heredity, salivary composition and flow, nutrition, periodontal disease, enamel elemental composition, and presence of fluoride and other local geochemical factors... The genesis of dental caries is very closely tied to a complex interaction of both essential and modifying factors, making it important that they be understood in documenting the etiology of the disease process.
> (Larsen *et al.* 1991: 179)

With respect to local geochemical factors, an important contemporary study by Schamschula *et al.* (1978) demonstrated that while dental health in the Sepik River area of Papua New Guinea was independent of diet and food preparation techniques, it was heavily dependent upon local geochemical factors. These workers investigated caries prevalence, using an index of missing and carious teeth by tooth count, for sixteen villages in a very small geographical area and found a range in caries prevalence of between 0.0% to 8.6%. Diet and food preparation techniques for each village were the same. The differential presence of a host of major and trace elements were found to be highly correlated with the prevalence of dental caries. A decade later Hildebolt (Hildebolt *et al.* 1988; Hildebolt *et al.* 1989) also described a correlation between caries prevalence and different geochemical zones, in the North American State of Missouri, for both prehistoric and modern populations. A correlation between enamel calcification, essential element availability and subsistence has also been described by Schneider (1986) for prehistoric populations in the North American State of Ohio.

Perhaps the single most important element causally linked to caries prevalence is fluoride. Fluoride enhances "the ability of tooth structure to withstand acid dissolution" (Leverett 1982:26). The official position of the American Dietetic Association is that optimal water fluoride levels are between 0.7 and 1.2 ppm and this results "in a significant reduction in the prevalence of caries in permanent teeth in persons consuming fluoridated water from birth through adolescence" (Anon 1994:1428).

Siebert & Trautner (1985:54) note that:

> The fluoride concentrations in human foods are in most cases very low...with one major exception, when raw materials stem from the marine environment. Sea water throughout the world contains 1.3 mg F/l and thus provides a relatively fluoride-rich biotope for all marine life.

Nonetheless, the fluoride is concentrated in the bones rather than the flesh (Siebert & Trautner 1985). So to obtain the fluoride benefit from fish, the bones also need to be consumed. Walker & Erlandson (1986) and Sealy *et al.* (1992) have suggested that the fluoride benefits obtained from marine diets are possibly linked to low caries prevalence. Other studies of prehistoric populations (Kelley *et al.* 1991; Larsen *et al.* 1991) have also indicated the generally cariostatic nature of a marine diet. Even discounting fluoride, the protein rich marine diet has been associated with the likely prophylactic effect of increased levels of tooth abrasion (Sealy *et al.* 1992; Littleton & Frohlich 1993) as well as increased oral cavity alkalinity (Littleton & Frohlich 1993) inhibiting bacterial activity (Powell 1985).

Marine influences aside, caries prevalence has also been linked to local terrestrial fluoride levels. A pre-agricultural sample from Oakhurst, South Africa, that had a marine component in their diet albeit of unknown significance, exhibits one of the highest frequencies of carious lesions (17.7% tooth count) recorded in the literature (Sealy *et al.* 1991) for such a subsistence economy. Sealy *et al.* (1991) attribute this to extremely low fluoride levels in the local water table. Molnar & Molnar (1985a) in assessing samples spanning the Neolithic through Bronze Age in Hungary (5000 to 3200 years bp) suggest that fluoride intake is likely responsible for the relatively low prevalence of enamel lesions, specifically, experienced by these populations who essentially subsisted on domesticated grains.

Molnar & Molnar's (1985a) results are further interesting in that while the prevalence of lesions affecting the enamel is quite low for an agricultural population, overall lesion prevalence is somewhat high with a range of 3.8% to 14.7% by tooth count. This is despite the fact that high levels of fluorosis, between 30% and 60%, were observed. Molnar & Molnar (1985a) point out that 68% of all carious lesions initiated on the tooth root, rather than the enamel. In fact, the sample with the highest incidence of fluorosis also had the highest prevalence of carious lesions (14.7%), but with 82% of these lesions attacking the dentin of the root rather than the enamel. They also show that the generally expected effect of age on the frequency of root lesions was not evident in their samples. Moreover, there did not appear to be a correlation between gingival recession and the development of this type of lesion either. They suggest that an as yet unknown environmental factor was likely responsible.

Moore (1993) has demonstrated that from the Iron Age to the Medieval period in Britain, the vast majority of carious lesions were located on or near the enamel/cementum junction and very few occlusal or interstitial lesions were seen. Moore (1993) found a correlation between increasing coronal wear and an increasing prevalence of neck lesions during this period. There was also an observed negative correlation with a decreasing prevalence of occlusal lesions with increasing coronal wear. He noted that the principal diet during this period was a coarse rye or barley bread of a considerably abrasive nature. Moore (1993:55) hypothesised that with the destruction of the crown from this abrasive diet:

> the natural cleaning properties of the dentition would have been lost, so that food debris would have accumulated in stagnation areas around the necks of the teeth especially as oral hygiene was undoubtedly poor if non-existent. Under these conditions even the large molecule carbohydrates would have fermented sufficiently to produce enough acid to attack the dental tissues. In addition, food stagnation would have led to gum recession thus exposing further amounts of tooth root and so increasing the area open to caries' attack.

Moore (1993) goes on to say that with the improvement of milling and the introduction of sugar into Britain in the seventeenth century there was a concomitant decrease in tooth wear and an increase in the prevalence of occlusal caries. In fact occlusal sites now became the most common places for lesions. It is difficult to assess the global ubiquity of such a patterning of caries in the past as, until quite recently, very few authors have differentiated between dentin and enamel lesions. Of those who have, Lukacs (1992) reported that root and cervical lesions accounted for only 19.6% of all carious lesions in a Bronze Age agriculturalist sample from Harappa, Indus Valley. Reasonably high levels of tooth wear were evident in that 32.4% of all cases of pulp chamber exposure were attrition induced.

Littleton & Frohlich (1993) pointed out a high prevalence of cervical and interproximal lesions in a sample from Islamic Bahrain, Arabian Gulf (750-500 years bp). This population is described as displaying low to moderate amounts of tooth wear and an overall high prevalence of carious lesions, 14.0% by tooth count. Littleton & Frohlich (1993) suggest food impaction around the base of teeth, probably the result of a large grain component in the diet, was responsible for the high prevalence of cervical and interproximal lesions.

As for Southeast Asia, Douglas (1996) has examined dental material from two sites in Thailand: Ban Chiang dating from the late pre-metal period through to introduction of iron (5600 to 1700 years bp) and Non Nok Ta, pre-metal period through to Bronze Age (5500 to 3000 years bp). For each of these sites the archaeological and dental evidence suggest mixed economies rather than a reliance on domesticated grains. For Ban Chiang, Douglas (1996) found that while overall occlusal lesions were much more common than non-occlusal, this pattern was not consistent by sex. The vast majority of male lesions were non-occlusal while the opposite was the case for females. Moreover, the prevalence of cervical and interproximal lesions increased with increasing age. For Bronze Age Non Nok Ta, occlusal and non-occlusal lesions were equally distributed overall and within each sex. Both of Douglas' (1996) samples exhibited tooth wear to varying degrees.

Dental remains from the Iron Age site of Noen U-Loke, Thailand, have been examined by Nelsen (1999). Rice was found in abundance at this site (Higham & Thosarat 1998). Interestingly, caries prevalence is very low (3.0 % tooth count) even in the face of the likely agricultural subsistence base of this population. Of the recorded lesions 28.6% were interproximal, 42.9% were occlusal or smooth surface, while the remaining 28.6% were of unknown point of origin. The Noen U-Loke teeth evidence a reasonably high degree of wear.

Tayles (1992) has examined the dental remains from Khok Phanom Di (4000 to 3500 years bp) in Thailand. A high prevalence of carious lesions both by tooth count (10.8%) and individual count (58.2%) is reported and is consistent with the likely agricultural subsistence base. Moreover, 35.3% of all lesions are either interproximal or cervical.

The aetiology of dentin lesions has been both under-researched and its prehistoric occurrence under-reported. However, apart from the food impaction aetiology suggested by Moore (1993) and Littleton & Frohlich (199) there are other aetiological hints. It has been reported by van Houte (1994:672) that:

> At present, a few specific organisms in plaque are considered by many to play a special role in coronal development…Less detailed information suggests that a broader spectrum of bacteria is involved in root-caries development.

Specifically, he notes that "other studies suggest an etiologic role of organisms identified as *A.viscosis* in root-surface caries" (van Houte 1994:674).

This begs the question: is the higher prevalence of root lesions, compared to coronal lesions seen in some prehistoric samples, due to the species diversity, or even a single important vector, of the oral flora? This species diversity, or specific root lesion organism, could again be related to numerous dietary, environmental and cultural factors. Intriguingly, a recent study of calculus deposits in a range of samples between 500 and 12,000 years bp, including coastal gatherers, fishers, hunters, farmers, and participants in mixed economies, found:

> Hunters, fishermen and gatherers had a less diverse flora with bacillary and coccal morphotypes. Agricultural groups showed greater diversity with additional filamentous and spiral morphotypes.
> (Linossier *et al.* 1996:1005)

In a review of the literature on the aetiology of caries, Winter (1990) points out that changes in the virulence of bacteria could account for differences in the prevalence of caries, although this has not been demonstrated in modern populations. It is nonetheless plausible that changes in the virulence of cariogenic bacteria have occurred in prehistory.

It also appears that higher rates of dentin lesions are associated with both agricultural subsistence economies as well as moderate amounts of occlusal tooth wear. I propose a general hypothesis helping to explain this as follows: pre agricultural diets, particularly in marine/aquatic subsistence contexts, are generally relatively protein rich. Diets high in protein are believed to reduce oral acidity (ie. increase pH), thus neutralising the acids produced by cariogenic bacteria (Walker & Erlandson 1986). A change to a more agriculturally oriented diet will likely generally lead to a decrease in relative protein intake and a concomitant decrease in oral pH with an associated loss of protection against decalcifying bacterial acids. If the diet is still relatively abrasive, through grindstone introduced grit for example, the prophylactic benefits of coronal wear will keep down the frequency of coronal caries. However, the cemento-enamel junction, and root generally, will become increasingly prone to disease. This model should still apply where there is a high intake of natural fluoride. In such cases the enamel will benefit from the prophylactic effects of coronal attrition and the effects of fluoride strengthening the enamel. However, the dentin will still be unprotected. In such cases, even if tooth wear is minimal the fluoride will still protect the enamel. Added factors such as increased microfloral diversity and food impaction around tooth bases can easily be incorporated into this model. Indeed, such factors may play a dominant role in all or some cases.

It is generally believed that tooth wear has prophylactic properties with regard to carious lesion establishment (Jurmain 1990; Lukacs & Pal 1993). However, several studies have shown an increase in caries prevalence concomitant with an increase in tooth wear (Costa 1980; Hartnady & Rose 1991; Beckett & Lovell 1994). In Costa's (1980) study of three pre-contact Eskimo samples he found the highest prevalence of carious lesions among the Ipiutak (14.4% tooth count) and lowest among the

Jones Point sample (3.5%). The level of wear on the Jones Point dentitions was reported to be much less than that seen in the Ipiutak sample (Costa 1980). Hartnady & Rose (1991), looking at the dental health of the Archaic-period non-agricultural Pecos region people of Texas, found a high prevalence of lesions (14.0% tooth count) in association with extreme tooth wear. They noted that the "teeth are wearing, but not rapidly enough to out pace carious activity" (Hartnady & Rose 1991:271).

Beckett & Lovell (1994) found both a higher prevalence of caries and more cases of severe tooth wear in a later, agriculturally intensified, sample. They attribute higher levels of attrition in the agricultural population to the addition of grit to the diet added by way of airborne material, use of grinding stones, intentional addition of grit to assist with grinding and wood ash from fires.

Generally, it has been believed that foraging people experienced greater levels of tooth wear than did agriculturalists (Hinton 1981; Lukacs 1992). However the reverse situation with higher, or at least high, levels of tooth abrasion seen in agriculturally oriented populations have been recorded (Beckett & Lovell 1994). Further, Smith (1984:43) demonstrates that by "all evidence both hunter-gatherers and agriculturalists had an abrasive diet. Rate of wear alone (studied using chronological age estimates) will not differentiate them". Smith (1984) argues that it is the plane of wear that can be used to discriminate between agriculturalists and foragers, rather than rates of wear per se.

> Molar wear is more evenly distributed on the molars of hunter-gatherers, resulting in a relatively low wear plane angle in advanced wear. Agriculturalists show a more restricted pattern of wear and tend to develop oblique wear planes. The difference is attributed to a reduction in food toughness or fibrousness that is associated with the appearance of intensive collection of grains and intensive use of grinding stones and pottery in food preparation.
> (Smith 1984:54)

It is worth noting Molnar's (1972:514) comments in this context:

> In relating food and food-borne abrasives to attrition, a mistake is often made of taking only the staple items of the diet into account...In many populations the principal foods of the diet may be mainly soft and nonabrasive, but the constant addition of small quantities of supplemental food to the dietary intake may induce considerable wear.

This is a point not generally acknowledged in the debate over tooth wear and its relationship to subsistence.

Another factor potentially influencing the expression of caries is the habit of chewing betel nut (*Areca catechu*) quids. The chewing of betel nut on its own or in combination with pepper leaf (*Piper betel*) and/or a lime paste (calcium hydroxide) is a ubiquitous habit in Southeast Asia and the Pacific today. Several studies have reported on the cariostatic effects of chewing betel nut (Moller *et al.* 1977; Schamschula *et al.* 1977; Nigam & Srivastava 1990) while others have not found evidence for cariostatic properties (Williams *et al.* 1996; Reichart & Gehring 1984).

High lime levels in the oral environment are also known to reduce caries risk (Powell 1985). If betel nut chewers added lime paste, which would increase their oral pH, to the nut, this could explain a low prevalence of lesions on its own. Such a habit of adding lime would also lead to greater levels of plaque calcification. Powell (1985) notes the general correlation between high levels of calculus and low prevalence of caries.

In addition to the potentially cariostatic properties of both the nut and lime, is the increased salivary flow induced by constant chewing. It has been pointed out by van Houte (1994) that much evidence indicates an important role played by saliva in lesion development. In extreme cases of limited saliva flow, such as dryness of the mouth, or xerostomia, caries runs unchecked (van Houte 1994).

Not only is betel nut chewed for its narcotic effects, its cariostatic properties presumably being incidental, but a betel nut and lime slurry has also been reported to have been used to deliberately stain teeth for aesthetic purposes in a Mariana Island assemblage (Hocart & Fankhauser 1996). Also, in Vietnam today it is still common to observe people with black stained teeth (Phan Hai Linh 1998). Flynn (1977) has described the particular process as being a form of enamel etching that is still practised in modern dentistry. Flynn (1977) cites anecdotal evidence that the traditional procedure of tooth etching and blackening serves as a caries prophylaxis. Tooth blackening is seen in many Asian countries, Africa and South America (Phan Hai Linh 1998) and the process can involve the use of betel nut at various stages.

Other potentially very important factors affecting the expression of caries concern the oral health of the individual and population as a whole. Elvin-Lewis (1983:203) relates that globally "one hundred and seventy species [of plant] are known for teeth cleaning". A number of these plants specifically used for cleaning the teeth in Africa have also been shown to have antibacterial effects (Akpata & Akinrimisi 1977; Rotimi & Mosadomi 1987; Li *et al.* 1998). Other species have simply been noted for their utility in plaque removal (Olsson 1978; Hardie & Ahmed 1995; Hattab 1997). Elvin-Lewis (1983:208) has also related that:

> Genera in the Euphorbiales used for teeth cleaning in Africa are used there and in Southeast Asia to treat pain, fever, cough, and bleeding and for its astringency. Also, Alchornia are used to treat buccal ulcerations; Drypetes, toothache; and Jatropha, bleeding gums.

It is highly probable that a considerable pharmacopoeia existed for most populations in prehistoric Southeast Asia. Unfortunately, it is quite another matter to identify what may have been used, for what purpose and to what effect.

While it is clear from this discussion that the factors influencing the prevalence of caries are quite complex, there does appear to be an overall general correspondence between given subsistence economies and caries prevalence. Numerous studies (Turner 1979; Larsen 1983; Lukacs 1989; Larsen *et al.* 1991) document a higher caries prevalence in populations practising agriculture compared to foragers in most regions of the prehistoric world. Notwithstanding the range and complexity of caries modifying factors discussed above, the correlation between caries rates and subsistence economy would seem to be quite robust as a generalisation.

Calculus

Calculus has been mentioned above with respect to high pH oral environments. Hillson (1996) notes that the relationship between plaque mineralisation, or calculus formation, and saliva means that those teeth closest to the salivary ducts display the highest accumulations. Moreover, the differential siting of salivary glands of dissimilar composition may also play a role in the development of calculus deposits. This may also be a factor involved in the differences sometimes reported in pathological conditions between the upper and lower dentition.

With respect to diet, Littleton & Frohlich (1993) have noted that not only agricultural diets, high in starch, are conducive to the development of calculus. Diets high in proteins and fats, Eskimo diets for instance, are also known to favour calculus development. Lukacs & Pal (1993) have also suggested that extensive calculus development in a Mesolithic Indian sample was due to high meat consumption. An obvious question then, is to what extent are diet, calculus and caries related? Hillson (1996:260) addresses this by noting that:

> Dental caries involves progressive local demineralization of tooth surfaces, whereas calculus involves mineralization and the two conditions should therefore be mutually exclusive. But both conditions are frequently seen on the same tooth, where arrested caries lesions in crown fissures are often covered with calculus, and active caries may be seen in cement or dentine underneath calculus deposits. At a population level there may however be a slight inverse relationship between caries and calculus frequencies.

What would appear to be a crucial factor is the source of the bulk of the plaque substrate that has been mineralised into calculus. An essentially marine protein diet, with associated high oral pH levels, will have a presumably low associated caries prevalence. In the case of a diet high in agricultural products, any high calculus concentrations observed may well be predicted to be associated with a high prevalence of caries. Another potentially important factor influencing the degree of calculus build up is oral hygiene. However, evidence for oral hygiene practices would be extremely difficult to discern in the archaeological record and, contrary to Moore (1993), cannot be assumed for an archaeological sample.

Lieverse (1999) has recently surveyed the clinical and anthropological literature on calculus and points out, from the clinical literature, that the aetiology of calculus is quite complex. In reviewing a number of studies reporting both calculus and reconstructed diets in prehistoric populations, Lieverse (1999) reveals a great deal of variation with no clear pattern or relationship between diet and calculus levels. Lieverse (1999) confirms that both high protein consumption and conversely, high carbohydrate consumption can lead to large calculus deposits. As such, in the absence of supporting archaeological evidence for diet, it would be an extremely unwise exercise to infer diet from levels of calculus deposits.

Abscesses and Periodontal Disease

Hildebolt & Molnar (1991:225) stress that there "is a lack of agreement with regard to the cause, the treatment, and even the definition of this disease". They go onto say:

> A general working definition of periodontal disease as seen clinically is: the intermittent degeneration of the supporting tissues of the teeth, these tissues being the gingiva, cementum, periodontal ligament, and alveolar bone. The ultimate sequela of this process is the wasting of the periodontal structures to the point that teeth are lost due to inadequate surface attachment for the ligaments because of a reduction of bony support. By comparison, only the end result of periodontal disease is seen on dry skulls. This consists of loss of the alveolar bone and disruption of the smooth outer cortex of the alveolar bone.
> (Hildebolt & Molnar 1991:225)

and

> Perhaps the most important aspect of periodontal disease is loss of the alveolar bone, with the subsequent loss of teeth… Recently, periodontal disease as a cause of

> tooth loss has been vigorously challenged, and pupal necrosis with a sequela of periodontal ligament necrosis has been offered as an alternative explanation.
> (Hildebolt & Molnar 1991:227)

In an important literature review, Clarke & Hirsch (1991a) stress that one of the most common methods for assessing alveolar absorption, and by inference periodontal disease, is totally inappropriate. They point out the important effects of the continuous growth of the face and continuous tooth eruption as being the chief cause of much observed change in the relationship between the cemento-enamel junction and alveolar crest. Furthermore, they extensively review the literature on interpretations of alveolar defects ostensibly linked to periodontal disease and conclude that very seldom will such defects be due to this disease process. They note that, in general, discrete abscesses at root apices have been classified as being of pulpal origin while all others have tended to be seen as periodontitis. They stress however, that all "dental abscesses can affect all locations in the periodontal ligament and adjacent alveolar bone as well as extending in a coronal direction from the root apex" (Clarke & Hirsch 1991a:248). It would seem that it is the speed of any given infection that determines the location of bony destruction:

> Rapid spread would appear to limit the abscess location to the root apex, whereas slow spread provides the opportunity for internal/external interaction at any and all locations within the periodontium.
> (Clarke & Hirsch 1991:249-250)

It is important to emphasise here that obvious damage to a tooth, such as severe wear, caries or fracture, are not the only avenues open for bacterial invasion of the pulp.

> Developmental defects resulting in inadequate covering of root dentin by cementum in approximately 15-20% of teeth and direct invasion of cementum and dentin (exposed by continuous tooth eruption and/ or gingivitis) by plaque bacteria provide other pathways for bacteria to enter the endodontic cavity, thereby offering an explanation for the occurrence of severe localized defects in apparently sound teeth.
> (Clarke & Hirsch 1991a:253-254)

In practice, the only time an alveolar defect would be considered potentially connected with periodontitis is when specific diagnostic criteria regarding the morphology of surrounding alveolar bone are met. The morphological criteria for diagnosing periodontitis are as follows:

> The dense cortical plate is resorbed to reveal the spongy porous cancellous bone, which in turn results in the conversion of the smooth knife edged contour of the alveolar bone to be replaced by a rounded margin with obvious loss of form. In the absence of these changes, a diagnosis of periodontitis must be doubtful.
> (Clarke & Hirsch 1991a:256)

Clarke (1990; Clarke *et al.* 1986; Clarke & Hirsch 1991a) demonstrate that periodontitis was quite uncommon in past populations and it is a minimally destructive disease. It is bony destruction of a pulpal origin that is responsible for what has generally and erroneously been ascribed to periodontitis.

Given the most probable aetiology of these alveolar defects is invasion of the endodontic space by bacteria, the question is how inimical to human health are the processes that cause these lesions? Dias & Tayles (1997) have recently suggested that the term abscess cavity should be restricted to those alveolar defects that present fistulae, or drainage sinuses. They suggest most other defects be classified on size as either periapical granuloma (2-3mm in diameter) or its sequelae apical periodontal cysts (>3mm diameter). However, whether one choses to use the term granuloma or furcation defect is probably not important as long as its pulpal origin is recognised.

Dias & Tayles' (1997) chief point however, is that only the rarer abscesses, as they define such lesions, are potentially problematic while the cysts are essentially benign and thus without serious or immediate health consequences. Moreover, they point out the presence of such cysts or ostensibly benign lesions indicate the host's ability to combat these centres for infection over the long term. This, however, necessitates assumptions about these ostensibly benign lesions. Robert Hirsch (pers. comm.) points out that lesions in the alveolar bone, caused by such processes as pulpal pathosis, can occur during both chronic and acute phases of the disease process and manifest anywhere in the periodontium. Moreover, chronic abscesses commonly become acute while acute abscesses frequently become chronic. Regardless of the state of the infection, bone is lost and cavities formed. It would be extremely difficult to differentiate between cavities that were the result of chronic cysts and those that were due to acute abscesses. Hirsch (pers. comm.) says that the inflammatory response belongs to the soft tissues, but we are looking at the end result of those soft tissue reactions in the bones. We should be weary of falling into the trap of extrapolating from bones to soft tissues.

In this study of dental remains from northern Vietnam, alveolar defects of pulpal origin are classified as simply that: alveolar defects of pulpal origin (ADP). This classification does not include any assumptions regarding the status of the lesions as representing chronic or acute phases of pulpal disease, as such inferences simply cannot be made for the most part. The implications of this uncertainty, regarding the status of most of these lesions, clearly impact on the sort of statements that can be made with regard to the health of the host. Clearly however, it must be remembered, as Dias & Tayles (1997:554) point out, that "any lesion represents the potential for involvement of dental infection in an individual's death".

Tooth Discolouration

The most commonly reported tooth discolourations in the archaeological literature are opacities due to hypocalcification, ie. under-mineralisation (see Cook & Buikstra 1979; Blakey & Armelagos 1985), or fluorosis, ie. excessive fluoride intake (see Molnar & Molnar 1985a; Lukacs *et al.* 1985; Nayeem 1992). Such discolourations can be difficult to record in archaeological material with an array of diagenetic or post-depositional processes potentially complicating recognition (Hillson 1996). Other types of discolouration derive from either deliberate or unintentional staining of the teeth. In the previous discussion on caries it was pointed out that the dentition can be stained through the use of betel nut. Little is known with regard to the use, whether it be consumption or cosmetic, of betel nut during the distant past in Vietnam.

Developmental and Occlusal Variation Related Traits

Corruccini (1991:295) defines malocclusion as a:

> deviation from a rarely attained ideal arrangement among teeth; it does not necessarily cause a functional problem. Therefore, the developing trend has been to speak of *occlusal variation*, to avoid the connotation of the word malocclusion.

Occlusal variation can be assessed via a number of quantitative and qualitative variables. Hillson (1996) suggests that those variables based on fitting the upper and lower jaw are perhaps not appropriate in archaeological specimens due to preservation problems as well as the potential for warping of such remains. Such variables include overjet, overbite and posterior crossbite (Corruccini 1991). Other commonly used qualitative indicators, included as traits in the Federation Dentaire International (FDI) system for describing occlusion include: congenital absence (hypodontia); supernumerary (extra) teeth; malformed (e.g. peg molar) teeth; impaction; as well as transposed, missing and retained primary teeth (see Hillson 1996:109, Table 4.1). Other useful traits not covered by the FDI system include tooth rotation and displacement (Corruccini 1991).

Purely functional occlusal considerations aside, Calcagno & Gibson (1991:61-62) report that:

> Dental crowding is a critical problem because it can result in malocclusion and predispose dental impaction, dental caries, and periodontal disease. Malocclusion and occlusal stress, in turn, may predispose to dental chipping and fracture, and caries, impaction, and periodontal disease may also lead to tooth loss. Fractured teeth, caries, and partially erupted molars can result in dental abscesses and serious dental infections, which can lead to gangrene, septicemia, or osteomyelitis. Maxillary infections of odontogenic origin, for instance, may pass into the cranial cavity via the facial vein or the internal pterygoid plexus of veins, where they can produce cavernous sinus thrombosis or meningitis. Mandibular infections can result in Ludwig's angina, which may lead to asphyxiation, endocarditis, or pneumonia.

Perhaps the most important factors involved in occlusal variation are jaw size, tooth size and the relationship between these two factors. Brace *et al.* (1987) have suggested that tooth size has been reducing at a rate of about 1% per 1000 years during the Holocene. Brace *et al.* (1991) propose that that changes in food preparation techniques during the Holocene (increased levels of extra-oral food preparation) relaxed the evolutionary forces selecting for larger tooth size.

Other authors such as Calcagno & Gibson (1991) agree with the causal influence of changes in food preparation, but prefer an evolutionary model selecting for smaller teeth in opposition to Brace's model of disuse. Such a global reduction in tooth size would be contradictory in the face of the increasing frequency of occlusal problems, the closer populations are to the present, were it not for the concomitant trend to reduced orofacial size (Brown 1987). Whatever the role of complex evolutionary changes in jaw to tooth size relationships, it is perhaps the role of dietary softness or toughness that is critically involved in levels of occlusal variation. Hillson (1996) cites several studies that point to the role of dental wear, both approximal and occlusal, in maintaining good occlusal relationships. Further, Corruccini (1991:297) suggests that the role of masticatory stimulation in jaw development is a key variable in occlusion. Softer diets with their associated decreased levels of chewing stress will lead to a reduction in jaw size and thus occlusal problems. Hinton & Carlson (1979:330) also point out:

> There is a limited body of experimental evidence implicating induced changes in dietary consistency as a causative agent for increases in masticatory muscle attachment areas, mandibular corpus thickness, density of condylar bone and *condyle size* in rats.

A description of mandibular robusticity and assessment of tooth size would therefore be useful both with regard to an examination of diet and occlusal variation.

Non-Dietary Related Traits

> If attrition is a normal by-product of man's contact with certain facets of his environment, then a study of worn teeth should reveal some record of past activities. This record may indicate the type of diet, the way in which food was prepared, and the practice of crafts that involved the use of the teeth.
> (Molnar 1972:511)

The type of dental wear or modification of concern here is that which is not obviously or clearly associated with generally normal food mastication. Deliberate tooth mutilation, such as cross-hatched etching of the labial surface of the central maxillary incisors (Ikehara-Quebral & Douglas 1997:386-387, Figures 5-7) or task related deep occlusal grooving of anterior mandibular teeth (Larsen 1985:396, Figure 2), is relatively easy to discern and cannot readily be confused with normal masticatory function. In other cases, such as lingual abrasion of maxillary incisors (Turner & Machado 1983), differentiating between a purely masticatory and task orientated aetiology is not always so clear cut.

Molnar (1972) provides a literature survey of accidental tooth modification in the past. He relates a range of non-dietary chewing which can lead to specific wear patterns: habitual chewing of plants, twigs or animal skins; chewing of medicinal plants as part of the processing of this material; constant chewing of tough narcotic substances such as betel nut quids; and even the chewing of bones. In addition, Molnar (1972) points out that a range of task related activities can leave their own specific patterns of wear: retouching lithic artefacts; opening bivalves with the anterior dentition; processing material for basket making; using the teeth to clamp material; and processing various fibres and sinew. Milner & Larsen (1991) provide a recent survey of the literature concerning both deliberate and accidental tooth modification. As a clue to differentiating between deliberate and accidental modification Milner & Larsen (1991:358) note "intentionally modified teeth from archaeological contexts are only occasionally associated with the presumably subsequent development of periapical abscesses".

What is presumed to be task-related tooth wear is identified as task wear facets (TWF) in this study of Da But period and Metal period Vietnamese dental remains. A separate category for interproximal grooves (IG) is also recognised here. There was some controversy over the aetiology of IG which are essentially shallow grooves at or below the CEJ on the mesial or distal face of molars for the most part, although premolars and incisors have been noted to have these grooves on occasion. Lukacs & Pastor (1988) summarise the four main potential aetiologies as: tooth pick/probe induced; fibre pulling/processing; chemical erosion; and sucking grit-laden saliva between the teeth. Bermudez de Castro *et al.* (1997) point out the global occurrence of IG and also that IG have been described in hominids throughout the Pleistocene. Brown & Molnar (1990) suggested that sinew processing was the most likely aetiology with regard to IG in Australian Aboriginal samples. They based their argument on ethnographic film footage of sinew processing using the dentition. Frayer (1991:299) convincingly critiqued this hypothesis:

> Irrespective of geographic area or time period, interproximal grooves generally occur as single entities, seldom associated with occlusal channelling or buccal abrasion and are only occasionally found in the anterior teeth...most interproximal grooves do not fit the morphology nor arch location typical of documented cases of fibre processing.

Frayer (1991; Frayer & Russell 1987) and Lukacs & Pastor (1988) argue for habitual tooth picking or probing of the interstitial spaces as the most plausible explanation for IG.

Apart from specific wear patterns, teeth are susceptible to more immediate trauma manifesting along a gradient from small enamel chips through to major fractures exposing the pulp chamber. In reviewing the literature reporting on tooth trauma and chipping in particular, Milner and Larsen (1991) suggest more work needs to be done to clarify issues regarding the aetiology of such trauma. Separating damage due to particular subsistence and/or food preparation technologies on the one hand and extra dietary activities on the other is not straightforward.
Perhaps a more radical form of tooth modification is the deliberate removal of specific teeth. There is some evidence for the practice of tooth ablation in Southeast and East Asia (Tayles 1997), although it has not been reported for Vietnam. Tayles (1997) also points out that it is not necessarily an easy process identifying this particular behaviour. The Vietnamese assemblages were examined for any evidence of this behaviour.

Tooth Wear

An alternative to simply quantifying tooth wear is to describe the rate of wear between two samples of interest. Notwithstanding Smith's (1984) proposition that rates of wear alone cannot differentiate between agriculturalists and foragers, no detailed and comprehensive studies of this have been attempted to date. Tooth wear is intimately tied to most of the oral traits examined in this study and, as such, some attempt at describing wear is necessary. However, because estimations of the age-at-death of the Da But period and Metal period samples were heavily reliant on descriptions of tooth wear, an age independent method for describing rates of tooth wear is necessary. Such a description of tooth wear will serve two purposes: (1) provide an age-independent assessment of tooth wear that can be used to test the validity of using Da But period derived age estimation regression equations in estimating Metal period age-at-death; and (2) serve to describe rates of wear for each sample, thus potentially providing insights into diet and behaviour.

Principal axis analysis has been used (Scott 1979a; Richards 1984; Benfer & Edwards 1991) as an age-independent method for investigating the rate of dental wear in archaeological samples. What is of interest in the principal axis form of least squares analysis is the *y* intercept and the slope, both of which supply information about the rate of wear. The *y* intercept, where first molar (M1) wear is plotted against second molar wear (M2), measures the amount of wear on M1 when M2 wear is zero. Assuming an eruption time for M1 at six years and for M2 at 12 years, then the *y* intercept is measuring the amount of childhood wear within this six year period (Benfer & Edwards 1991). The slope cannot directly supply information about the rate of wear without information on the ages of the data points (Benfer & Edwards 1991). However, the slope will reduce in response to a reduction in the rate of M1 wear relative to M2 as M1 reaches the end on its functional life. Benfer & Edwards (1991:337) note:

> Smaller than unity slopes would be expected in populations experiencing very high wear and/or prolonged life expectancies. In those instances, childhood wear would also be anticipated to be high, and the intercept, wear at age 12 or 13, would probably be larger where b < 1.0 .

In effect, the slope can only impart information about the relative rates of M1 compared to M2 wear. Slopes less than 1.0 would indicate relatively faster M2 wear and, conversely, slopes greater than 1.0 would indicate relatively faster M1 wear. If the relationship described by the slope is linear, then the slope will describe a constant wear rate relationship between the two teeth. If there is a levelling out in the relative rate of wear between the two teeth the resultant asymptotic relationship will have to be transformed. Such a relationship is likely given the finite amount of enamel/dentin available on a tooth. It is because of this potential for data points from older individuals with highly worn teeth to affect the slope and intercept, that Benfer & Edwards (1991) suggest the removal of all heavily worn teeth. Doing this should ensure the age-independence of the method.

It is necessary to re-visit Benfer & Edwards' (1991) aforementioned assertion that high rates of wear will be reflected in a greater rate of M2 wear relative to M1, which will manifest as a lowering of the slope. They note that:

> The reverse is an unlikely event, but should it occur, M1 would be preferred for chewing after M2 erupts and the slope should exceed unity.
> (Benfer & Edwards 1991:337)

There are problems with this position however. For one, it is unlikely that a particular tooth class, especially adjacent and functionally equivalent teeth, will be preferred over another. The functional life of a tooth is not limited to the presence of its cusps, as Benfer & Edwards (1991) would suggest, but rather to a surface of the tooth being able to occlude and thus contribute to mastication. For example, Clarke & Hirsch (1991b) describe instances of dislocated molars where the tooth has tilted at right angles to the occlusal plane bringing the entire tooth length, including roots, into functional occlusion. Comuzzie and Steele (1989) describe an alternative scenario to dislocation in cases of extreme wear, whereby hypercementosis will lead to an increase in occlusal area once the crown has worn down below the CEJ and brought the enlarged roots into occlusion. Even without such radical outcomes, it is well known that teeth continue to migrate vertically throughout life and whether due to wear or not this effectively aids in bringing the tooth into functional occlusion (Clarke & Hirsch 1991b). Even with the likely most common scenario of the initial loss of occlusal cusps and enamel, along with continuous eruption, it could be quite plausibly hypothesised that the rate of M1 wear will exceed that of M2. This is because when both teeth are in occlusion, M2 with its greater level of occlusal enamel, will wear at a slower rate than M1, with its softer predominantly dentin occlusal surface.

Rather than use principal axis analysis, it can be argued that traditional least squares regression is more appropriate. Scott (1979a) originally proposed the use of principal axis because prediction was not an issue and only the overall trend, as expressed by the slope of the principal axis, was of interest. Scott (1979a) believed information on the rate of dental wear was available from the slope. This has been shown not to be the case. Further, Scott (1979a) was concerned with measurement error in the independent variable.

Traditional regression analysis is more appropriate in this study of the Vietnamese samples for several reasons. Prediction is a goal in this analysis, in that the wear of M1 when M2 is zero (the *y* intercept), is the single most important piece of information in the analysis. The *y* intercept is the estimate of childhood wear for the sample examined. The measurement error of the independent variable is minimal. Moreover, relatively accurate results are obtained in many studies where the measurement error of the independent variable is much larger. A good example is the prediction of age from macroscopically and/or microscopically assessed age indicators: osteon counts and cranial suture synostosis scores. Furthermore, perhaps the greatest advantage of traditional linear regression is the possibility of testing assumptions about the techniques used and the availability of significance testing.

Methods and Materials

Only permanent teeth are examined in this analysis.

Insufficient primary teeth from both sites mitigated against any sort of treatment of deciduous teeth. Preliminary to recording specific oral traits, gnathic remains were first described with respect to completeness. As the vast majority of traits assessed involved dental tissues, or closely associated structures such as alveoli, the following tooth status categorisations were made:

P	present (in situ)
PM	pre mortem loss
X	post mortem loss
I	isolated (not in situ) tooth
UP	unerupted (but present)
UPK	unerupted (and unknown if present)
UKE	not present (and unknown if ever erupted)
E	in the process of erupting
NP	not preserved (neither tooth nor alveolus)
R	roots only preserved
KA	cannot assess.

The reason for categories such as UPK is that radiographic facilities were not readily available and were only used very sparingly. Checking for unerupted molars was not a priority, but see the discussion on hypodontia. Any signs of trauma or lesions, other than alveolar defects of pulpal origin, found in these mandibular and maxillary remains are discussed in Chapter 6. Likewise, degenerative changes to the mandibular condyles are discussed in Chapter 4.

Trait Definitions and Recording Protocols

General Diet and Behaviour Related Traits

Ante-Mortem Tooth Loss (AMTL)

Ante-Mortem Tooth Loss was recognised after Lukacs (1989:271) as "progressive resorptive destruction of the alveolus".

Alveolar Defects of Pulpal Origin (ADP)

Clarke (1990; Clarke & Hirsch 1991a) outline the clinical classification of such bony defects as furcation, angular, lateral, apical and complex defects. This classification is for the most part site or locality specific but also furnishes information on the potential progression of pulpal pathosis. Nonetheless, in this study all ADP are recorded without reference to precise location or whether or not the defect was discrete or continuous with the alveolar crest. Moreover, as discussed previously, this classification avoids the difficulties involved in attempting to describe the condition as acute or chronic in the absence of soft tissue information.

Carious Lesions (CAR)

Carious lesions were recognised after Turner *et al.* (1991:27): a "tooth is considered carious if the lesion has an irregular border and discoloured, easily removed, necrotic dentin at the lesion site". Lesion location was recorded by principal crown face affected if this information was available (Lukacs 1992:139). Further, it was also noted if the lesion was located at the dental cervix (neck) or root of a tooth. Lesion size was recorded by way of four categories (Metress & Conway 1975 in Lukacs 1992): (1) small pit or fissure lesion; (2) lesion where less than half the tooth crown is destroyed; (3) lesion where greater than half the tooth crown is destroyed; (4) lesion where the entire tooth crown was obliterated.

Pulp chamber exposure (PEXP)

Pulp chamber exposure was categorised by aetiology after Lukacs (1992): attrition induced; caries induced; or unknown aetiology.

Calculus deposits (CAL)

When present, calculus was recorded for severity after Brothwell (1981) (1) slight: confined to crown and not reaching CEJ; (2) medium: confined to crown but extending to CEJ; (3) heavy: extending over the crown and overhanging the roots.

Tooth Discolouration

Two categories of enamel discolouration were defined. The first, tooth dying, was defined as discolouration due to what was provisionally believed to be betel nut. This discolouration may have been caused by deliberate staining of the teeth or have occurred incidentally as a side effect of chewing the nut. The second category, tooth staining, encompassed all other forms of discolouration the aetiology of which was not immediately apparent. For each category the location of the discolouration, that is tooth surfaces affected, were noted as was the distribution of the discolouration on the most severely affected surface. Stain and dying distribution was defined as: solid colour to 90%+ of face; solid colour to 50% to 90% of face; solid colour to <50% of face; colour confined to and/or concentrated on edges of face; diffuse/patchy colour distribution.

Hypocalcifications (opacities) were not recorded due to difficulty in discerning them in teeth where a high frequency were stained and/or dyed, deliberately or incidentally, during life or via post-depositional agents.

Developmental and Occlusal Variation Related Traits

Malaligned Teeth (MAL)

Following Corruccini (1991: 298), malaligned teeth were scored as: (1) displaying a slight (<2.0mm) buccal or lingual displacement (from normal or adjacent teeth); (2) displaying a large (>2.0mm) buccal or lingual displacement.

Rotated Teeth (ROT)

Rotated teeth were scored as: (1) having a rotation $<45^0$; (2) being rotated $>45^0$ (Corruccini 1991: 298).

Tooth Impaction (IM)

Hillson (1996:113) states that:

> Properly, impaction implies that the tooth remains inside the jaw and does not emerge into the mouth at all, but there are many

> variations and a tooth may erupt sideways into its neighbour, presenting one of its crown sides uppermost.

In this study a tooth is described as impacted if it is erupting into a neighbouring tooth, regardless of whether or not it would have emerged into the oral cavity during life.

Congenital Tooth Absence
Strictly speaking tooth agenesis, or hypodontia, was not purposefully examined in this study due to a lack of readily available radiographic equipment. However, the tooth status "UKE" refers to cases where the preserved alveolar bone shows no trace of remodelling, from a tooth lost ante-mortem, or any sign of an un-erupted tooth. Such a method has been used and defended by Bermudez de Castro (1989) in identifying third molar agenesis. This method has a conditional value in discussions regarding occlusal variations and jaw size.

Enamel Pearls (EP)
An enamel pearl is defined as:

> a separate nodule, or pearl, of enamel present on the root surface (actually a nodule of dentine covered by an enamel cap) varying from microscopic size to several millimetres in diameter.
> (Hillson 1996:98)

Enamel pearls are recorded in this study as little is known of the prevalence of these traits in Southeast Asia.

Mandibular Robusticity
Mandibular robusticity was assessed by way of measuring maximum mandibular body height at M1 and maximum body breadth perpendicular to this (Wood 1991). This measurement is easily obtained and this section of the mandible is frequently present even in fragmentary remains. Two indices are derived (after Chamberlain & Wood 1985), one being a measure of cross-sectional area treating this section of the mandible as an ellipse (height x breadth x $\pi/4$). The second is an index of gross size (height/breadth x 100). These measurements and indices do not describe the size of the dental arcade and therefore do not indicate available space for erupting teeth. Nonetheless, these measures do describe the robustness of the jaw and hence environmental/behavioural and perhaps also genetic influences on jaw robusticity.

Tooth Size
To measure tooth size the mesio-distal (MD) and bucco-lingual (BL) diameters of the left first and second molars of each jaw were taken. When the left tooth was missing the antimere was substituted. The measurement of these dimensions follows Buikstra & Ubelaker (1994:62). The MD diameter refers to the maximum diameter of the tooth crown in the mesio-distal plane. The BL diameter is the maximum diameter of the tooth perpendicular to the mesio-distal plane. Measurements were taken to the nearest 0.1mm using a Mitutoyo dial caliper.

Non-Dietary Related Traits
Interproximal Grooves (IG)
Identification of interproximal grooves followed the definition as outlined by Frayer & Russell (1987). The chief characteristics of such grooves are that they are artificial, not natural, grooves that run parallel to the CEJ and either cut through the CEJ or are just inferior to it. As the name suggests they are found within the interproximal spaces, and as noted by Frayer & Russell (1987) the grooves commonly exhibit striations within the groove parallel with the long axis of the groove.

Task Wear Facets (TWF)
Task wear facets are categorised as unintentional dental alterations. In this study they refer to teeth that were believed to have been altered through the use of the teeth as tools to perform particular tasks. TWF are recorded as distinct from IG as the aetiology of IG seems to suggest behaviour unrelated to use of teeth as tools.

Tooth Chipping (CP)
Milner & Larsen (1991:367) outline how pre-mortem and post-mortem chipping can be differentiated:

> A smoothing of the sharp edges of fractured enamel through continued attrition and the staining characteristics of adjacent broken and intact surfaces will help researchers identify when the damage occurred.

In order to meet these conditions it is likely that chips occurring shortly prior to death will be classed as post-mortem. Further, chips to the thin enamel edges of severely worn teeth, generally only molars, were disregarded as such damage is more representative of reduced tooth integrity than hard food stuffs or extra-dietary behaviour.

Tooth Ablation
There is no readily applicable method to test for tooth ablation. In the first instance only alveoli indicating ante-mortem tooth loss without evidence of alveolar defects of pulpal origin or periodontal disease were assessed. The patterning of the remaining ante-mortem tooth loss was then examined with respect to tooth type, age, sex and other associated oral traits, such as task wear facets for example, that might relate to observed ante-mortem tooth loss.

Tooth Wear
Molar wear was determined using Scott's (1979b) method. Incisor wear was determined using a modification of this system whereby the incisor was treated as a single quadrant. Scott's (1979b) system and the results of using wear scores to estimate age-at-death, and estimates of intra-observer error, have been discussed in Chapter 2. In this chapter, the data collected on tooth wear are used to examine: (1)

relative rates of childhood wear between each sample; and (2) the posterior tooth to anterior tooth wear gradient.

Oral Trait Reporting Method

Two reporting methods have been employed: the individual count method and tooth count method. Both methods can give different results for the same trait analysed. The individual count seems intuitively more meaningful, but intra- and inter-sample comparisons suffer from the effects of incomplete dentitions and small sample sizes when performing statistical tests. The chief advantage of the tooth count method is that it increases sample size. For ante-mortem tooth loss (AMTL) and alveolar defects of pulpal origin (ADP), the tooth count sample size refers to intact and assessable alveoli, thus addressing the problem of post-mortem tooth loss. For example, while there are 947 permanent Da But period teeth, there are 1430 assessable alveoli. The definition of an individual for defining the denominator when examining frequencies of ADP and AMTL is: an identified individual represented by bony maxillary and or mandibular remains. In the case of impacted teeth, the denominator is restricted to individuals represented by in situ third molars. In all other cases the denominator includes all identified individuals with gnathic remains.

Results

Age and Sex Structure of the Gnathic Samples

Preliminary to examining the frequency of various oral traits it is necessary to discuss the age and sex structure of each sample. Any observed difference, or conversely lack of difference, in trait occurrence may be due to the sex or age composition of the gnathic samples.

Figure 3.1 indicates that the age distribution of each adult sample represented by gnathic remains is virtually unchanged from the overall age distributions. This is not surprising given the role of the dentition in estimating age-at-death. Both five (Figure 3.1a) and three (Figure 3.1b) age class distributions are plotted. The five age class distributions document in more detail the age make up of each sample. However, the observed number of cases in each age class is too small for statistical testing. The age distributions are collapsed into three age class distributions to facilitate statistical testing. As previously noted, and still the case here, the Da But period sample has a greater proportion of individuals in the older age classes relative to the Metal period sample. Figures 3.2a and 3.2b look at the distribution of each sample with regard to sex and age. Again the distributions are almost identical with the overall sample distributions discussed in Chapter 2.

What is also of concern is that the distributions of assessable teeth and alveoli are similar by sex, jaw and age class. A useful way to examine this is by way of box plots indicating the median number of teeth or alveoli by sex or age class for each sample. Figure 3.3a is such a plot for the Da But period sample for assessable teeth, while Figure 3.3b plots assessable alveoli. The median number of both assessable teeth and alveoli is similar between the sexes. A Mann-Whitney test of medians indicates no significant differences between the male and female medians for teeth and alveoli (Table 3.1). Figures 3.4a and Figure 3.4b indicate that for the Metal period there is also a similar median number of teeth and alveoli by sex. A Mann-Whitney test of medians for this sample also indicated no significant differences between the male and female medians for teeth and alveoli (Table 3.1). It is, however, apparent that the Da But period sample has a lower number of assessable teeth per individual than is the case in the Metal period. This is also the case for assessable alveoli and so probably suggests poorer preservation/recovery of Da But period gnathic remains, than simply greater levels of post-mortem tooth loss.

The median number of assessable teeth and alveoli by jaw was also examined. Table 3.1 indicates that there was no significant difference in the median number of teeth nor alveoli for either the mandible or maxilla. This is the case for both periods.

As particular oral traits are also going to be discussed in relation to age, it is useful to explore the distribution of assessable teeth and alveoli by age class. Figures 3.5a and 3.5b are box plots of assessable teeth and alveoli, respectively, by age class for the Da But period sample. It can be seen that the median number of assessable teeth and alveoli are fairly constant across the five age classes. A Kruskal-Wallis test of medians (Table 3.1), used for 2 or more groups, indicates no significant differences in these median scores. Figures 3.6a and 3.6b plot the same information for the Metal period. For the teeth it can be seen that the youngest age class and 40-49 year class have somewhat lower levels of tooth preservation than the other age classes. However, there is no significant difference (Table 3.1) in the median scores for each age class. In the case of assessable alveoli there is a somewhat more pronounced difference in medians. The youngest two age classes exhibit significantly (Table 3.1) lower levels of alveolus preservation per individual than the older age groups.

One way to make the age classes more comparable for later statistical analysis is to combine them into a smaller number of classes. As will become apparent later, the frequency of most oral traits assessed is quite low for both samples. Combining age classes for both samples into young (15-29), middle age (30-39) and old (40+) will increase the number of observed traits for any age class and better facilitate statistical comparisons.

Figures 3.7a and 3.7b plot Da But period assessable teeth and alveoli, respectively, for the revised three age classes. Combining the age classes has made no appreciable difference to the medians of the new age classes (Table

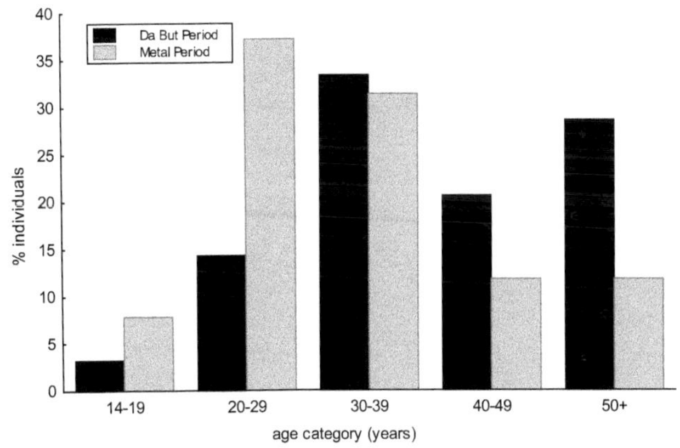

Figure 3.1a. Da But Period and Metal Period Age Distribution: Adults with Gnathic Remains (5 x age classes)

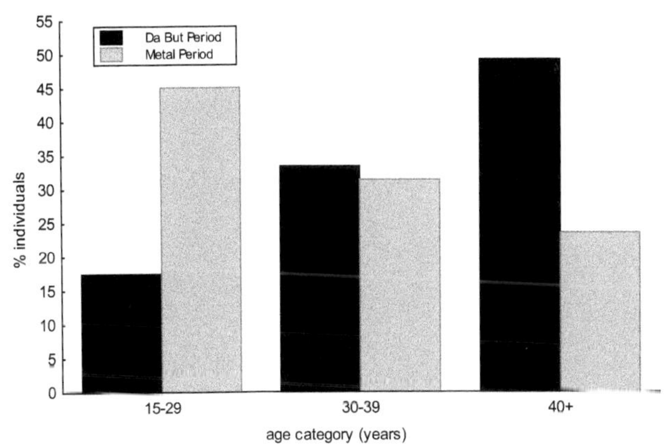

Figure 3.1b. Da But Period and Metal Period Age Distribution: Adults with Gnathic Remains (3 x age classes)

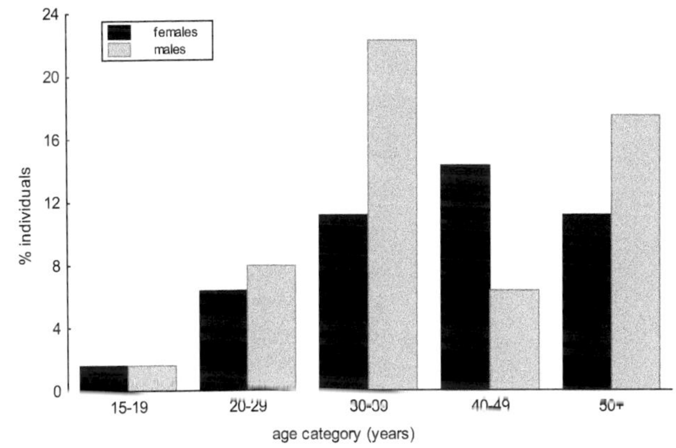

Figure 3.2a. Da But Period Age and Sex Distribution: Adults Represented by Gnathic Remains

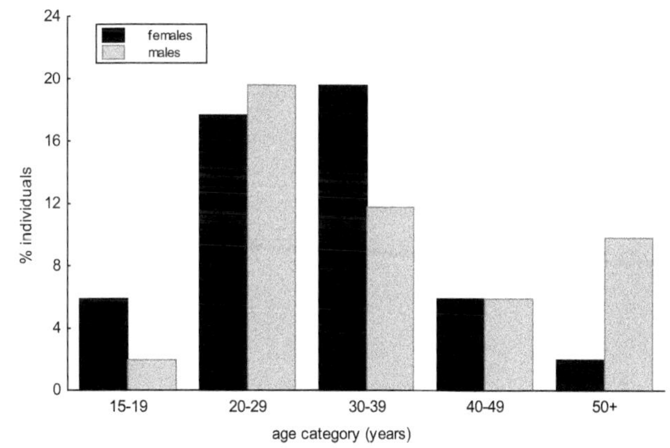

Figure 3.2b. Metal Period Sex and Age Distributions: Adults Represented by Gnathic Remains

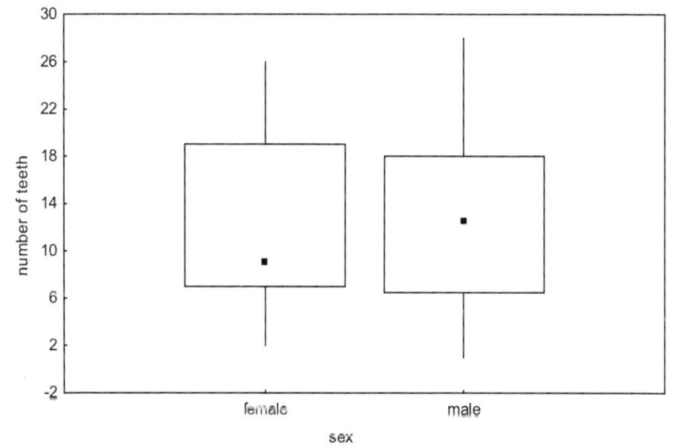

Figure 3.3a. Da But Period: Number of Assessable Adult Teeth by Sex per Individual (n=64)

Figure 3.3b. Da But Period: Number of Assessable Adult Alveoli by Sex per Individual (n=64)

3.1).

Figures 3.8a and 3.8b plot Metal period assessable teeth and alveoli, respectively, for the revised three age classes. While assessable teeth are now more evenly distributed within each age class (see also Table 3.1), the alveolus distribution is still uneven. The creation of three broad age classes has not served to lessen the impact of the lower number of assessable alveoli per individual in the lowest age class. This does not invalidate later comparisons, but this factor must be born in mind in any later interpretations of results where the alveolus is the unit of analysis. All subsequent analyses will use the revised three age class distributions when the frequency of oral traits is being examined in relation to age.

Oral Traits by Sex and Sample: An Overview

Before examining each oral trait individually it is useful to get an overview of the prevalence of these traits with respect to their distribution by sex, both within and between samples, and by individual and tooth count reporting methods.

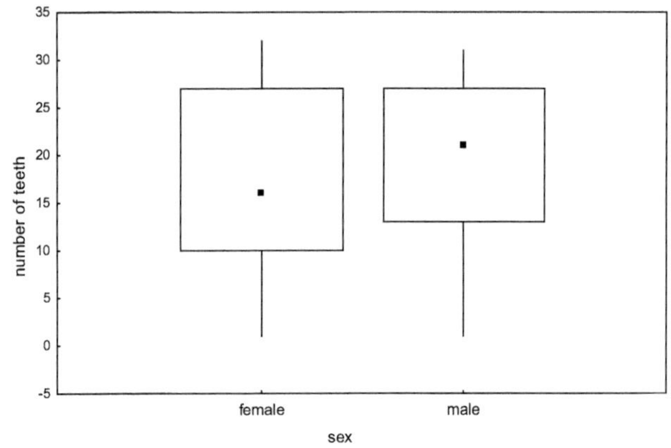

Figure 3.4a. Metal Period: Number of Assessable Adult Teeth by Sex per Individual (n=53)

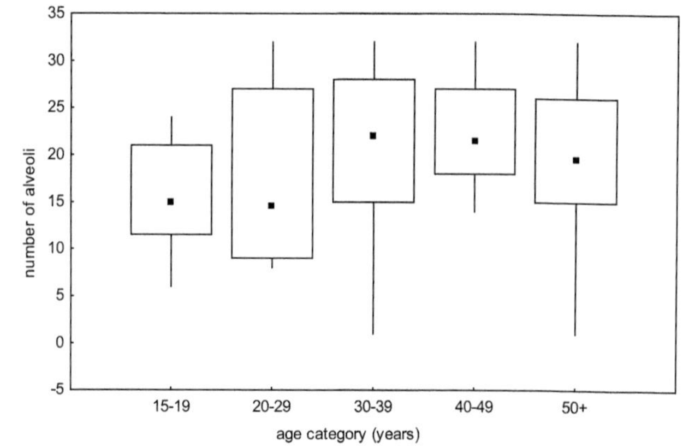

Figure 3.5b. Da But Period: Number of Assessable Adult Alveoli by Age per Individual (5 x age classes, n=75)

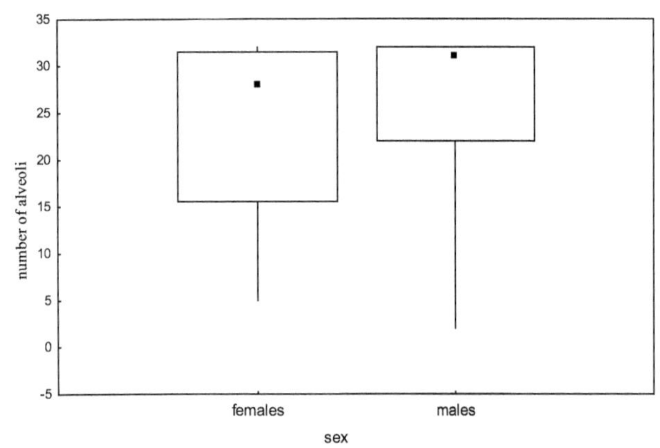

Figure 3.4b. Metal Period: Number of Assessable Adult Alveoli by Sex per Individual (n=53)

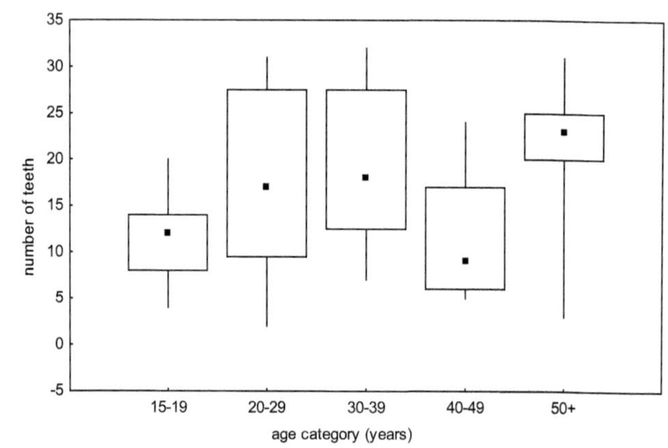

Figure 3.6a. Metal Period: Number of Assessable Adult Teeth by Age per Individual (5 x age classes, n=65)

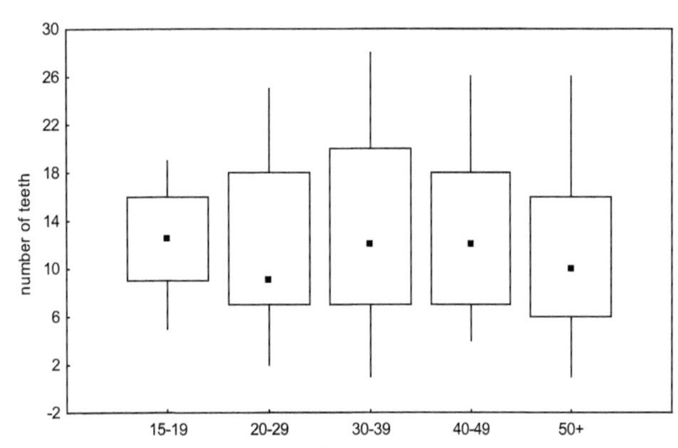

Figure 3.5a. Da But Period: Number of Assessable Adult Teeth by Age per Individual (5 x age classes, n=75)

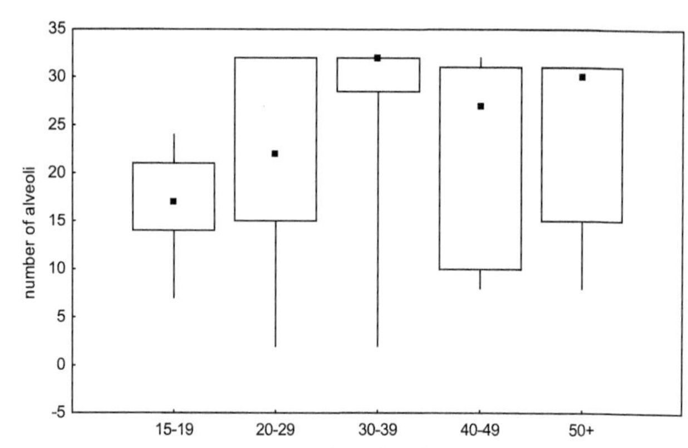

Figure 3.6b. Metal Period: Number of Assessable Adult Alveoli by Age per Individual (5 x age classes, n=66)

Individual Count Analysis

Intra-Sample Sex Differences

Table 3.2a lists sample sizes and oral traits frequencies for both periods by sex and total number of individuals with permanent dentitions. The sample size for all adults is slightly greater than the summed male and female sample due to the inclusion of unsexed young-adults (YA:15-19 year age class) with permanent dentitions and indeterminate sex cases. Figures 3.9a and 3.9b graphically display these results. Note that these figures only display medium and heavy instances of calculus.

Looking at the Da But Period first (Table 3.2a & 3.2b), there are few, and no statistically significant, differences between the sexes in manifestation of these oral traits. The largest differences between the sexes show a higher percentage of females exhibiting caries, malalignment and interproximal grooves, while a greater percentage of males displayed pulp exposure and alveolar defects of pulpal origin. By far the most frequently observed oral trait is calculus, which is present in over 90% of individuals if all three grades of calculus are included.

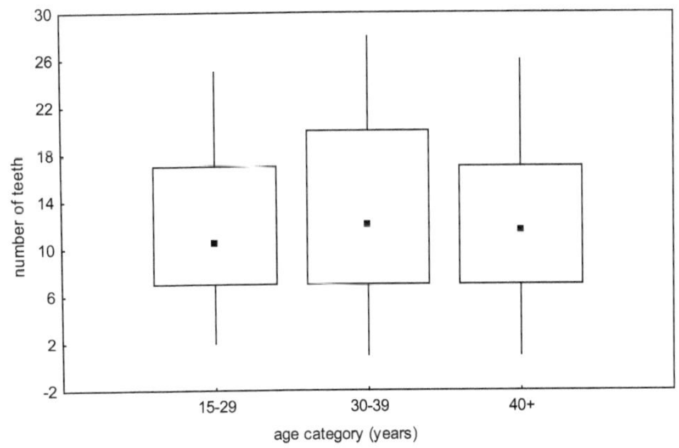

Figure 3.7a. Da But Period: Number of Assessable Adult Teeth by Age per Individual (3 x age classes, n=75)

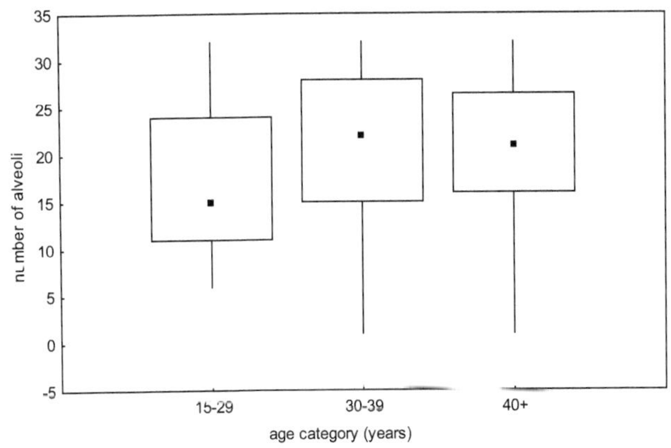

Figure 3.7b. Da But Period: Number of Assessable Adult Alveoli by Age per Individual (3 x age classes, n=75)

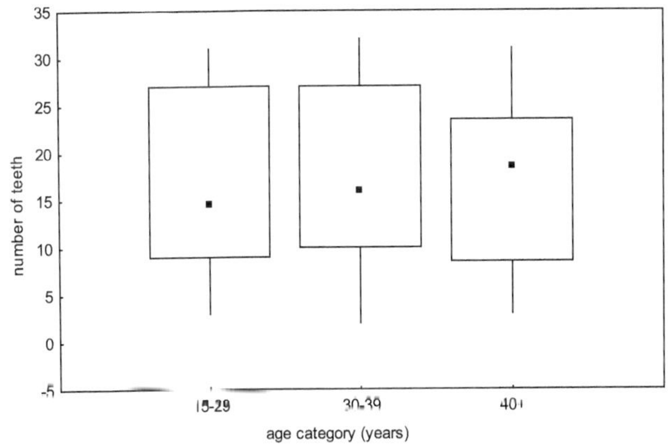

Figure 3.8a. Metal Period: Number of Assessable Adult Teeth by Age per Individual (3 x age classes, n=66)

With regard to the Metal period there are more striking sex differences in trait manifestation (Table 3.2a and 3.2b), but for the most part these differences are not significant. A greater proportion of males relative to females exhibit calculus, pulp exposure, tooth rotation, task wear facets, with only chipping being significantly greater. On the other hand, more females exhibit ante-mortem tooth loss, alveolar defects of pulpal origin, caries and impacted teeth. As with the Da But period, the most common trait is calculus with over 90% of all individuals exhibiting some degree of mineralised plaque. Tooth discolouration is almost evenly distributed between the sexes.

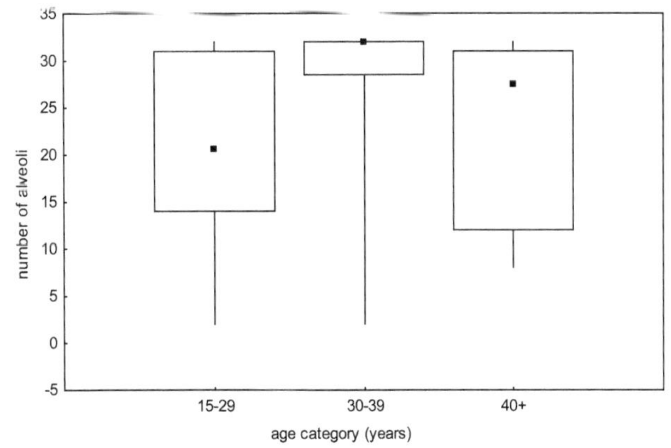

Figure 3.8b. Metal Period: Number of Assessable Adult Alveoli by Age per Individual (3 x age classes, n=62)

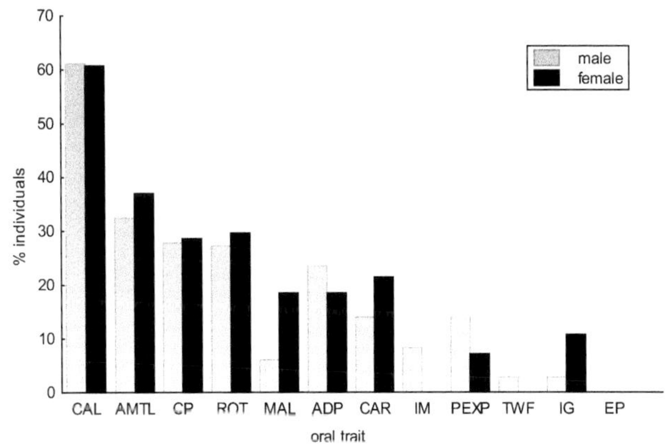

Figure 3.9a. Da But Period: Prevalence of Oral Traits by Individual

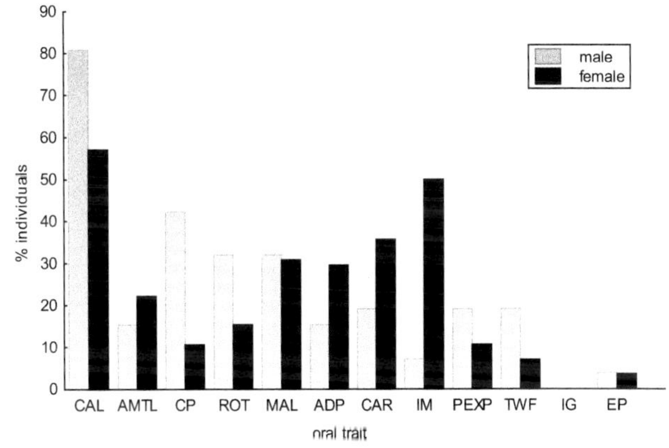

Figure 3.9b. Metal Period: Prevalence of Oral Traits by Individual

Inter-Sample Sex Differences

While there are differences in the patterning of dental disease between the Da But period and Metal periods (Figure 3.9c), the majority of individual differences are not significant. The Metal period displays a somewhat greater frequency of caries, tooth impaction, pulp exposure, enamel pearls, task wear facets and malaligned teeth, with only the last two traits displaying significant ($p<0.05$) differences. On the other hand, the Da But period exhibits greater, albeit non-significant, levels of calculus, ante-mortem tooth loss, chipped teeth, alveolar defects and interproximal grooves. Levels of calculus are somewhat

Table 3.1. Non-Parametric Tests for Sample Size Equivalence

Mann-Whitney U Statistic (for 2 independent groups)

period	group	variable	n	U	p
Da But	male:female	No of teeth/individual	64	486	0.806
	male:female	No alveoli/individual	64	496	0.914
	male:female	No teeth/jaw type	134	2097	0.564
	male:female	No alveoli/jaw type	136	1962	0.147
Metal	male:female	No of teeth/individual	53	297	0.337
	male:female	No alveoli/individual	55	284	0.114
	male:female	No teeth/jaw type	130	2009	0.629
	male:female	No alveoli/jaw type	137	2308	0.866

Kruskal-Wallis test (for 2 or more groups)

period	group	variable	n	H	p	Chi-sq.	p
Da But	age class (x5)	No teeth/individual	75	0.314	0.989	2.185	0.702
	age class (x5)	No alveoli/individual	75	3.971	0.410	2.503	0.644
	age class (x3)	No teeth/individual	75	0.009	0.995	0.216	0.898
	age class (x3)	No alveoli/individual	75	2.71	0.258	1.692	0.429
Metal	age class (x5)	No teeth/individual	61	8.42	0.077	6.624	0.157
	age class (x5)	No alveoli/individual	62	**10.257**	**0.036**	**12.675**	**0.013**
	age class (x3)	No teeth/individual	61	2.578	0.276	0.834	0.659
	age class (x3)	No alveoli/individual	62	**7.166**	**0.028**	**8.070**	**0.018**

high for both samples.

Tooth Count Analysis

Intra-Sample Sex Differences

Table 3.3a and Figures 3.10a to 3.10c show the patterning of dental disease by tooth count. For the most part the frequencies of these traits, for each sample by sex, remain unchanged from that seen in the analysis by individual. There is a slight reversal in the frequency of alveolar defects and tooth rotation by sex in the Da But period sample (Figure 3.10a). Within the Da But period sample, with the exception of significantly greater female mandibular AMTL there are no large sex differences (Table 3.3b).

Within the Metal period there is a slight reversal in malaligned teeth by tooth count. Most of the traits assessed in this sample (Figure 3.10b) are significantly different between the sexes (Table 3.3b). Males display significantly ($p<0.05$) greater frequencies of calculus, pulp exposure, chipped teeth, rotated teeth and task wear facets while females exhibit significantly ($p<0.05$) greater frequencies of ante-mortem tooth loss, caries and impacted teeth.

Inter-Sample Sex Differences

The patterns of dental disease when comparing the Da But period and Metal period are generally similar for the tooth count method as well (Figure 3.10c), with the exception of alveolar defects which have a greater frequency in the Metal period. The overall pattern is one of more a change in degree than kind. Using this reporting method there are few significant differences between each population (Table 3.3c). Alveolar defects ($p<0.05$) are significantly higher in the Metal period while ante-mortem tooth loss and calculus is significantly ($p<0.05$) greater in the Da But period sample.

Oral Traits by Tooth Class, Jaw Type and Age

General Diet and Behaviour Related Traits

Ante-Mortem Tooth Loss (AMTL)

In the Da But period sample AMTL is greatest amongst the incisor class with 43.5% of all teeth lost pre-mortem being incisors. This pattern is evident both by jaw type and sex (Table 3.4). However, significantly ($p<0.05$) more mandibular teeth exhibit AMTL than maxillary. Furthermore, significantly ($p<0.05$) more female mandibular teeth have AMTL than male mandibular teeth.

In the Metal period sample it is the molar class that exhibits the single greatest frequency of AMTL with 52.2% of all teeth lost pre-mortem being molars. This pattern is essentially the same by jaw type and sex although the distribution of AMTL in the maxillary dentition is less clear cut (Table 3.4). It is apparent that significantly ($p<0.05$) more female teeth, over all, are affected by AMTL than male. Moreover, this is also the case when comparing the maxillary dentition by sex. As with the Da But period, significantly ($p<0.05$) more mandibular teeth have been lost pre-mortem than maxillary.

When examining AMTL by age class in the Da But period sample (Table 3.5) it can be seen that there is a clear tendency for AMTL to be associated with older age, whether looked at by way of individual or tooth count reporting methods. This observation is confirmed from the results of the contingency table analysis. In fact, 72.2% of all individuals and 88.4% of all teeth affected by AMTL occur in the 40+ years age class.

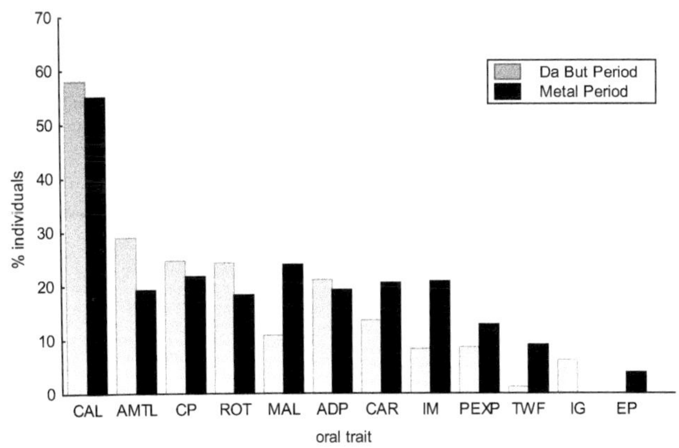

Figure 3.9c. Da But and Metal Periods: Prevalence of Oral Traits by Individual

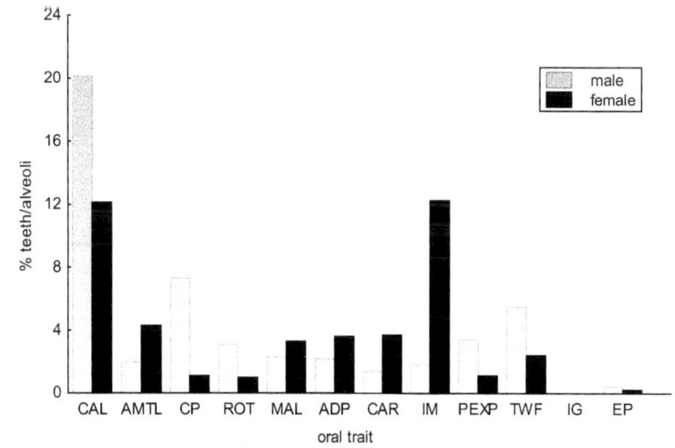

Figure 3.10b. Metal Period: Prevalence of Oral Traits by Tooth Count

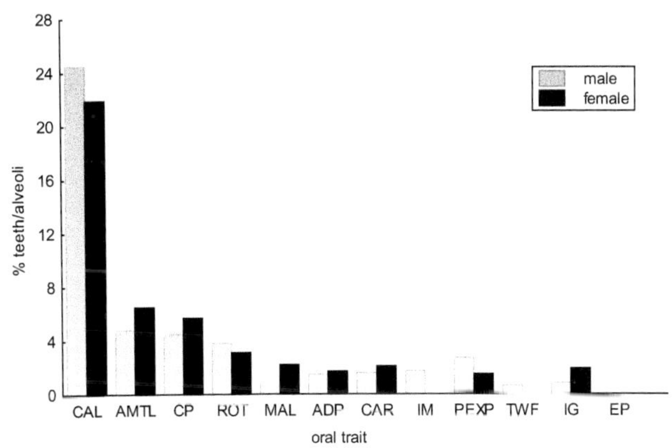

Figure 3.10a. Da But Period: Prevalence of Oral Traits by Tooth Count

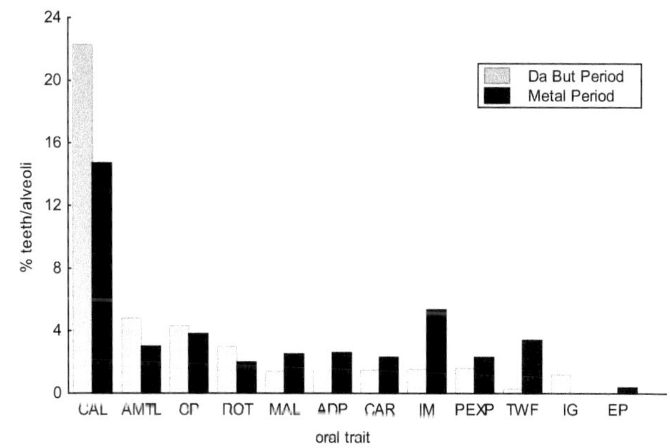

Figure 3.10c. Da But and Metal Periods: Prevalence of Oral Traits by Tooth Count

For the Metal period the same pattern is evident (Table 3.5) and this is also confirmed by the results of the contingency table analysis. However, AMTL in the younger age classes in the Metal period has brought down the frequency of AMTL in the 40+ year group to 46.2% of individuals and 63.0% teeth relative to the Da But period sample.

Alveolar Defects of Pulpal Origin (ADP)

When the Da But period sample is examined by way of tooth and jaw type (Table 3.6), mandibular teeth are seen to be slightly (1.7%) more affected than maxillary teeth (1.3%), but not to a significant degree. Moreover, there are no sex differences between jaw type either. Regardless of jaw type, by far the most afflicted tooth type is the molar, with 81.8% of all affected teeth belonging to this class.

In the Metal period (Table 3.6) significantly more mandibular teeth (3.7%) are affected than maxillary teeth at (1.4%). Unlike the Da But period, no one tooth class is particularly affected by ADP, although no cases of maxillary canine ADP were recorded. In fact overall, there appears to be an even distribution of ADP across all tooth classes.

When ADP is examined in relation to age in the Da But period sample, the vast majority of individuals with ADP are in the oldest age class (Table 3.7). This pattern is the same for the tooth count reporting method. While a slightly greater proportion of male individuals suffer from ADP, this is not significant. By tooth count, the percentages of female and male teeth affected by ADP are almost the same.

In the Metal period again, it is the older age class that is most affected by ADP, regardless of reporting method (Table 3.7). The youngest and middle age classes exhibit more ADP than their Da But period counterparts. The use of significance testing here is hampered due to the zero frequencies recorded in the middle age class of the Da But period.

Dental Caries (CAR)

While the overall prevalence of CAR, by tooth count, for both periods is quite low, there are differences in the patterning of caries between each sample. Caries prevalence is examined with regard to jaw type and tooth class in Table 3.8. During both the Da But period and Metal periods the vast majority of lesions manifested in the molars. There is a slight, but non-significant, bias toward the mandible in the Da But period sample. Only 1/14 carious lesions in the Da But period and 4/26 in the Metal period did not occur in molars. Neither period exhibited significant differences by jaw type for either sex. Further, as noted previously, neither period showed significant differences in caries frequency per se, although in the Metal period sample significantly ($p<0.05$) more

Table 3.2a. Prevalence of Dental Conditions by Individual

Da But Period

	Male			Female			YA (unsexed)			Indeterminate			Total		
	obs	n	%	obs	n	%	obs	n	%	obs	n	%	obs	n	%
ADP	8	33	24.2	5	27	18.5	1	7	14.3	2	7	28.6	15	74	20.3
AMTL	11	33	33.3	10	27	37.0	0	7	0.0	1	7	14.3	22	74	29.7
IM[1]	1	12	8.3	0	8	0.0	1	3	33.3	0	1	0.0	2	24	8.3
CAL[2]	22	36	61.1	17	28	60.7	4	9	44.4	4	7	57.1	47	80	58.8
CAL[3]	33	36	91.7	26	28	92.9	6	9	66.7	7	7	100.0	72	80	90.0
CAR	5	36	13.9	6	28	21.4	0	9	0.0	0	7	0.0	11	80	13.8
PEXP	5	36	13.9	2	28	7.1	0	9	0.0	0	7	0.0	7	80	8.8
IG	1	36	2.8	3	28	10.7	0	9	0.0	1	7	14.3	5	80	6.3
TWF	1	36	2.8	0	28	0.0	0	9	0.0	0	7	0.0	1	80	1.3
Areca	0	36	0.0	0	28	0.0	0	9	0.0	0	7	0.0	0	80	0.0
EP	0	36	0.0	0	28	0.0	0	9	0.0	0	7	0.0	0	80	0.0
CP	10	36	27.8	8	28	28.6	2	9	22.2	0	7	0.0	20	80	25.0
ROT	9	33	27.3	8	27	29.6	0	7	0.0	1	7	14.3	18	74	24.3
MAL	2	33	6.1	5	27	18.5	1	7	14.3	0	7	0.0	8	74	10.8

Metal Period

	Male			Female			YA (unsexed)			Indeterminate			Total		
	obs	n	%	obs	n	%	obs	n	%	obs	n	%	obs	n	%
ADP	4	26	15.4	8	28	28.6	0	9	0.0	2	13	15.4	14	76	18.4
AMTL	4	26	15.4	6	28	21.4	1	9	11.1	3	13	23.1	13	76	17.1
IM[1]	1	14	7.1	4	8	50.0	0	1	0.0	0	1	0.0	5	24	20.8
CAL[2]	21	26	80.8	16	27	59.3	1	10	10.0	5	14	35.7	42	77	54.6
CAL[3]	25	26	96.2	26	27	96.3	5	10	50.0	8	14	57.1	64	77	83.1
CAR	5	26	19.2	10	27	37.0	0	10	0.0	1	14	7.1	16	77	20.8
PEXP	5	26	19.2	3	27	11.1	0	10	0.0	2	14	14.3	10	77	13.0
IG	0	26	0.0	0	27	0.0	0	10	0.0	0	14	0.0	0	77	0.0
TWF	5	26	19.2	2	27	7.4	0	10	0.0	0	14	0.0	8	77	10.4
Areca	13	26	50.0	15	27	55.6	0	10	0.0	4	14	28.6	32	77	41.6
EP	1	26	3.85	1	27	3.7	1	10	10	0	14	0	3	77	3.9
CP	11	26	42.31	3	27	11.11	2	10	20	1	14	7.14	16	77	20.8
ROT	8	25	32.0	4	26	15.38	0	8	0.0	1	12	8.33	13	71	18.31
MAL	8	25	32.0	8	26	30.77	0	8	0.0	1	12	8.33	17	71	23.94

[1]sample size for IM based on preservation of insitu M3s.
[2]medium and heavy calculus concentrations
[3]slight, medium and heavy concentrations

ADP	alveolar defect of pulpal origin	TWF	task wear facet
AMTL	antemortem tooth loss	Areca	betel nut staining
IM	impacted tooth	EP	enamel pearl
CAL	calculus	CP	tooth chipping
CAR	dental caries	ROT	tooth rotation
PEXP	pulp chamber exposure	MAL	tooth malalignment
IG	interproximal groove		

females have lesions than males.

Caries was also investigated with regard to its distribution by sex, jaw and lesion size (Table 3.9). The majority of Da But period lesions are size 2, tend to manifest in the maxillary teeth and are most frequently occlusal, distal or mesial respectively. With regard to the Metal period, there is an increase in the frequency of size 1 lesions, with size 1 and 2 caries accounting for the vast majority of all lesions. More than half the lesions in the Metal period sample are occlusal, with distal lesions being the next most important category.

Carious lesions are classified as root, cervical or crown. Table 3.10 reveals a similar pattern of lesion distribution for both periods. Root caries are rare in both samples, cervical caries are only a little better represented. The vast majority of lesions are confined to the crown in both

Table 3.2b. Inter and Intra-Sample Chi-sq Results for Dental Conditions by Individual

	Da But Period: male/female			Metal Period: male/female			Da But/Metal Period		
	Chi-sq	p	p (Fisher exact 2 tailed)	Chi-sq	p	p (Fisher exact 2 tailed)	Chi-sq	p	p (Fisher exact 2 tailed)
ADP	0.29	0.592	0.755	1.36	0.244	0.332	0.24	0.624	0.686
AMTL	0.09	0.765	0.792	0.33	0.568	0.730	2.63	0.105	0.127
IM	na			2.85	0.091	0.148	1.51	0.220	0.416
CAL[1]	0.00	0.974	1.000	2.91	0.088	0.135	0.14	0.713	0.749
CAL[2]	0.03	0.860	1.000	0.00	0.978	1.000	1.60	0.205	0.245
CAR	0.63	0.428	0.513	2.07	0.150	0.224	1.36	0.243	0.293
PEXP	0.74	0.391	0.454	0.68	0.409	0.467	0.73	0.393	0.448
IG	1.69	0.193	0.311	na			na		
TWF	na			1.62	0.204	0.250	**4.99**	**0.026**	**0.032**
Areca	na			0.16	0.686	0.787	na		
EP	na			0.00	0.978	1.000	na		
CP	0.00	0.944	1.000	**6.63**	**0.010**	**0.014**	0.19	0.666	0.710
ROT	0.04	0.840	1.000	1.96	0.162	0.199	0.78	0.377	0.422
MAL	2.24	0.135	0.227	0.01	0.925	1.000	**4.68**	**0.031**	**0.047**

[1] medium and heavy calculus concentrations

[2] slight, medium and heavy calculus concentrations

samples. It should also be noted that cavities affecting the interstitial surfaces of the crown were rare. Only 2/9 crown lesions in the Metal period sample and 4/19 crown lesions in the Da But sample affected the interstitial surfaces.

Caries prevalence is examined by way of age for both the individual and tooth count reporting methods in Table 3.11. When all individuals with assessable permanent teeth are examined with respect to age in the Da But period sample, it is clear that the vast majority of lesions occur in older age (40+ years). Chi square and likelihood ratio tests of homogeneity of caries frequency across age classes confirm that lesion distribution is not even across age classes. The same pattern is apparent regardless of sex, although frequencies are too low for significance testing. Looking at caries by way of tooth count gives the same picture.

Quite a different pattern is evident in the Metal period sample however. Both the individual and tooth count methods indicate that most carious lesions occur in the younger age groups, especially between 20 and 39 years. This pattern is also evident for both sexes. However, significance tests fail to confirm this, with the null hypothesis that lesions are even across all age classes not being rejected.

Given the evidence for a relationship between betel nut use and reduced caries risk, the co-occurrence of each condition was explored for the Metal period sample. Of the 51 sexed individuals, 28 of these displayed evidence of betel staining. Of the 15 individuals with carious lesions, nine of these also displayed betel staining. By tooth count, 16/22 carious teeth were associated with betel staining. By sex, 4/5 male individuals with carious lesions also displayed betel staining, while 5/10 female individuals with carious lesions also had betel staining.

Pulp Chamber Exposure (PEXP)

In the Da But period sample there are no significant differences by sex or jaw type (Table 3.12). However, mandibular teeth were more affected by PEXP than maxillary and no female teeth were affected by PEXP.

In the Metal period there are several significant ($p<0.05$) differences (Table 3.12). Overall, male teeth are significantly more affected by PEXP than female and male maxillary teeth are significantly more affected than female maxillary teeth. With regard to jaw type, significantly more mandibular teeth are affected than maxillary. Further, male mandibular teeth were significantly more affected by PEXP than female mandibular teeth.

The greatest difference between the Da But period and Metal periods is in the type of teeth affected. In the Metal period only 4/26 of all affected teeth were molars with 11/26 affected teeth being incisors or canines and 11/26 being premolars. On the other hand, in the Da But period sample only premolars and molars were affected with the vast majority (15/17) being molars.

With regard to the distribution of PEXP by age, it can be seen that within the Da But period sample PEXP is concentrated in the older age classes (Table 3.13). This is the case when examined by both reporting methods, and is confirmed by contingency table results. This pattern is also evident when each sex is examined separately.

The Metal period sample is similar in PEXP being concentrated in the older age categories for both reporting methods (Table 3.13). Moreover, this is also confirmed by the contingency table results. However, two of the three females exhibiting PEXP are aged 20-29 years.

When examining PEXP by aetiology (Table 3.14) it can

Table 3.3a. Prevalence of Dental Conditions by Tooth Count

Da But Period	Male			Female			YA (unsexed)			Indeterminate			Total		
	obs	n	%	obs	n	%	obs	n	%	obs	n	%	obs	n	%
ADP	10	688	1.5	9	540	1.7	1	119	0.8	2	83	2.4	22	1430	1.5
AMTL	33	688	4.8	35	540	6.5	0	119	0.0	1	83	1.2	69	1430	4.8
CAL[1]	109	445	24.5	73	333	21.9	6	89	6.7	24	84	28.6	211	951	22.2
CAL[2]	280	445	62.9	199	333	59.8	31	89	34.8	67	84	79.8	576	951	60.6
CAR	7	445	1.6	7	333	2.1	0	89	0.0	0	84	0.0	14	951	1.5
PEXP	12	445	2.7	5	333	1.5	0	89	0.0	0	84	0.0	17	951	1.8
CP	20	445	4.5	19	333	5.7	2	89	2.2	0	84	0.0	41	951	4.3
TWF	3	445	0.7	0	333	0.0	0	89	0.0	0	84	0.0	3	951	0.3
EP	0	445	0.0	0	333	0.0	0	89	0.0	0	84	0.0	0	951	0.0
IM[3]	1	57	1.8	0	46	0.0	1	9	11.1	0	9	0.0	2	121	1.7
IG[4]	3	351	0.9	5	261	1.9	0	69	0.0	1	63	1.6	9	744	1.2
ROT	16	424	3.8	10	319	3.1	0	78	0.0	1	69	1.5	27	890	3.0
MAL	4	424	0.9	7	319	2.2	1	78	1.3	0	69	0.0	12	890	1.4

Metal Period	Male			Female			YA (unsexed)			Indeterminate			Total		
	obs	n	%	obs	n	%	obs	n	%	obs	n	%	obs	n	%
ADP	15	693	2.2	21	586	3.6	0	84	0.0	3	155	1.9	39	1518	2.6
AMTL	14	693	2.0	25	586	4.3	0	84	0.0	7	155	4.5	46	1518	3.0
CAL[1]	102	506	20.2	56	462	12.1	0	80	0.0	10	104	9.6	168	1152	14.6
CAL[2]	321	506	63.4	254	462	55.0	15	80	18.8	40	104	38.5	627	1152	54.4
CAR	7	506	1.4	17	462	3.7	0	80	0.0	2	104	1.9	26	1152	2.3
PEXP	17	506	3.4	5	462	1.1	0	80	0.0	4	104	3.8	26	1152	2.3
CP	37	506	7.3	5	462	1.1	1	80	1.3	1	104	1.0	44	1152	3.8
TWF	28	506	5.5	11	462	2.4	0	80	0.0	0	104	0.0	39	1152	3.4
EP	2	506	0.4	1	462	0.2	1	80	1.3	0	104	0.0	4	1152	0.4
IM[3]	1	55	1.8	6	49	12.2	0	1	0.0	0	9	0.0	7	114	6.1
IG[4]	0	355	0.0	0	328	0.0	0	49	0.0	0	83	0.0	0	815	0.0
ROT	15	486	3.1	4	390	1.0	0	61	0.0	1	83	1.2	20	1020	2.0
MAL	11	486	2.3	13	390	3.3	0	61	0.0	1	83	1.2	25	1020	2.5

[1]medium and heavy calculus
[2]slight, medium and heavy calculus
[3]n=number of in situ M3s
[4]n=number of assessable pre molars and molars

ADP	alveolar defect of pulpal origin	TWF	task wear facet
AMTL	antemortem tooth loss	IG	interproximal groove
IM	impacted tooth	EP	enamel pearl
CAL	calculus	CP	tooth chipping
CAR	dental caries	ROT	tooth rotation
PEXP	pulp chamber exposure	MAL	tooth malalignment

be seen that while PEXP is quite low overall, it is clear that attrition induced pulp chamber exposure is by far the dominant causal agent for both periods.

Calculus (CAL)

Because only a very slight deposit of calculus was needed for a positive score it was decided that it would be more meaningful to concentrate on examining only those individuals and teeth affected by moderate to large deposits. Looking at calculus distribution by tooth and jaw type in the Da But period sample, significantly ($p<0.01$) more mandibular teeth are affected by calculus than maxillary (Table 3.15). Similar frequencies of calculus are seen for each mandibular tooth class, whereas maxillary incisors are much less affected by calculus than other maxillary teeth.

In the Metal period there is no significant difference by jaw type, although it is the maxillary teeth that show a slightly higher frequency of calculus (Table 3.15). Overall, incisors, canines and premolars are similarly affected by calculus whereas the molars stand out as being the most affected. Interestingly, no mandibular incisors and very few canines exhibit calculus. On the other hand, maxillary incisors

Table 3.3b. Intra Sample Chi-sq Results for Dental Conditions by Tooth Count

		Metal Period			Da But Period	
		Chi-sq	p	p*	Chi-sq	p
ADP	m/f	2.34	0.126		0.09	0.764
	max/mand	**7.70**	**0.006**		0.41	0.522
	m/f max	1.17*	0.280		0.9*	0.342
	m/f mand	1.51	0.219		0.70	0.402
AMTL	m/f	**5.42**	**0.020**		1.64	0.200
	max/mand	**7.36**	**0.007**		**10.05**	**0.002**
	m/f max	**5.54***	**0.019**		2.17*	0.141
	m/f mand	1.51	0.219		**4.77**	**0.029**
CAL[1]	m/f	**11.42**	**0.001**		0.70	0.402
	max/mand	2.31	0.128		**17.77**	**0.000**
	m/f max	**3.72**	**0.054**		2.86	0.091
	m/f mand	**8.66**	**0.003**	**0.004**	0.02	0.893
CAL[2]	m/f	**7.17**	**0.007**		0.80	0.370
	max/mand	**27.83**	**0.000**		**18.25**	**0.000**
	m/f max	**2.76**	0.096		0.01	0.911
	m/f mand	**4.66**	**0.031**		1.39	0.238
CAR	m/f	**5.22**	**0.022**		0.28	0.595
	max/mand	0.46	0.499		3.44	0.064
	m/f max	2.4*	0.122		0.23	0.631
	m/f mand	2.86*	0.091		0.07	0.787
PEXP	m/f	**5.68**	**0.017**		1.31	0.252
	max/mand	**13.61***	**0.000**		2.12	0.145
	m/f max	0.01*	0.938		3.77	0.052
	m/f mand	**7.17***	**0.007**	**0.008**	0.01	0.905
CP	m/f	**22.71**	**0.000**		0.54	0.461
	max/mand	**4.00**	**0.046**		0.79	0.374
	m/f max	**19.92***	**0.000**	**0.000**	0.23	0.631
	m/f mand	3.96*	**0.047**	0.058	0.34	0.561
TWF	m/f	**6.27**	**0.012**			
	max/mand	**6.90**	**0.009**			
	m/f max	3.02	0.083			
	m/f mand	3.90	**0.048**	0.066		
IM	m/f	**4.49**	**0.034**	**0.050**	1.28	0.258
	max/mand	0.71	0.400		0.27	0.601
ROT	m/f	**4.33**	**0.037**		0.22	0.639
	max/mand	**5.34**	**0.021**		0.04	0.845
	m/f max	na			0.37	0.543
	m/f mand	2.16	0.142		1.40	0.237
MAL	m/f	0.93	0.335		1.95	0.162
	max/mand	0.07	0.793		0.1	0.746
	m/f max	1.26	0.262		0.72	0.398
	m/f mand	**5.44**	**0.020**	**0.024**	1.27	0.260

p* Fisher exact 2-tailed

[1]medium and heavy calculus

[2]slight, medium and heavy calculus

m=male, f=female, max=maxilla, mand=mandible

and canines exhibit the most calculus. Not only are male teeth significantly ($p<0.01$) more affected than female, this is true for both male and female maxillary ($p<0.05$) and mandibular ($p<0.01$) teeth examined separately.

Looking at calculus by age class in the Da But period it is

Table 3.3c. Inter-sample Chi-sq Results for Dental Conditions by Tooth Count

Da But Compared to Metal Period		
Condition	Chi-sq	p
ADP	**3.86**	**0.049**
AMTL	**6.33**	**0.012**
CAR	1.72	0.190
PEXP	0.57	0.449
CP	0.32	0.569
CAL[1]	**20.91**	**0.000**
CAL[2]	**7.63**	**0.006**
IM	**3.21**	0.073
ROT	2.28	0.131
MAL	**3.04**	0.081

[1]moderate and heavy
[2]slight, moderate and heavy

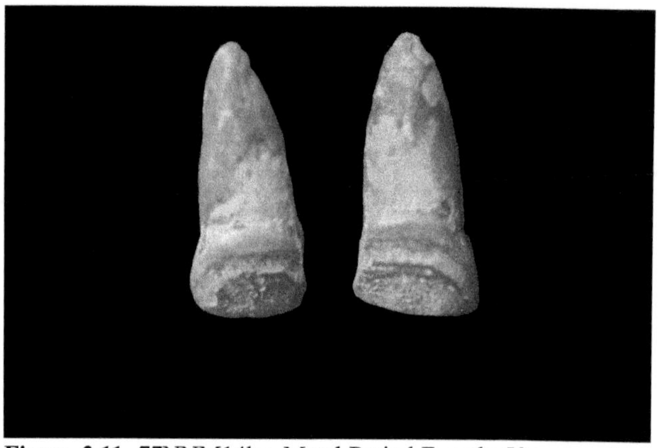

Figure 3.11. 77NNM14ka. Metal Period Female 50+ years (see also Figure 3.19). Etching of the enamel of the labial aspect of the labial aspect of both maxillary central incisors. Note also the heavily eroded enamel hypoplastic bands

apparent that significantly ($p<0.00$) more teeth are affected by calculus in the older age class (Table 3.16). However, this pattern is not evident when looked at by individual, where individuals displaying calculus are spread evenly across all age classes. Both males and females display a similar pattern by age class to each other.

In the Metal period there is a much clearer and significant ($p<0.00$) progression in the frequency of both individuals and teeth exhibiting calculus with increasing age (Table 3.16). As is the case in the Da But period, Metal period males and females display a similar pattern by age class to each other. Nonetheless, significantly more male teeth exhibit calculus than female.

Tooth Discolouration

The only form of tooth discolouration observed was that believed to be due to betel nut residues. Evidence for betel nut staining is absent in the Da But period sample but affects 39% of all individuals in the Metal period, males and females equally. The general pattern of staining manifests as a concentration on the labial aspect of the anterior teeth with a general marked lessening in density on the premolars and a general absence or only very slight staining of the molars. As such, the pattern is consistent with that described by Hocart & Fankhauser (1996), whereby the anterior teeth of female Mariana Islanders were believed to be deliberately stained with a betel nut and lime slurry for aesthetic purposes. It should be possible to differentiate between teeth incidentally stained as opposed to those deliberately stained by betel nut residues. Flynn (1977) outlines how the teeth are etched, or the enamel surface decalcified and roughened with citric acid preparatory to deliberate staining. Figure 3.11 is the labial view of two maxillary central incisors from a Metal period female aged 50+ years. The enamel surface apart from being darkly stained is also quite rough and extensively eroded. Moreover, the deep lateral grooves, likely due to enamel hypoplasia, have probably been exacerbated by the same agent that etched the rest of the tooth's exposed enamel surface. It is suggested that this is evidence of deliberate enamel etching preparatory to deliberate tooth staining. Further treatments throughout life presumably further degraded the enamel surface.

Developmental and Occlusal Variation Related Traits

Malaligned and Rotated Teeth (MAL & ROT)

When looking at both malaligned and rotated teeth ratios, the denominator is in situ assessable teeth. This differs from the denominator used for caries on the one hand and alveolar defects on the other.

In the Da But period 24.3% of individuals and 3.0% of assessable permanent teeth displayed some degree of rotation (Table 3.17). Of these rotated teeth, 44.4% were rotated in excess of 45^0. In the Metal period 18.3% of individuals and 2.0% of assessable teeth displayed rotation (Table 3.17). Of these teeth only 25% exhibited rotation in excess of 45^0. The differences between the two periods, for both reporting methods, were not significant.

When looked at with regard to jaw type, 3.1% of mandibular compared to 2.9% of maxillary Da But period teeth displayed rotation (Table 3.17). Again it is only in the Metal period that significant differences ($p<.05$) are seen with 3.0% mandibular compared to 1.0% maxillary teeth rotated. If tooth class is examined it is apparent that the canine is the single most commonly affected tooth for both periods and especially so in the Da But period sample where 15.2% of all assessable canines displayed some degree of rotation.

Examining malaligned teeth, 10.8% of individuals and 1.4% of assessable permanent teeth in the Da But period displayed some degree of malalignment or displacement (Table 3.18). Of these teeth, 50.0% were displaced in excess of 2.0mm. In the Metal period both significantly more ($p<0.05$) individuals (23.9%) and teeth (2.5%) displayed displacement. Of these teeth only 20% exhibited displacement in excess of 2.0mm.

Figure 3.12 represents a case of extreme displacement of the right mandibular incisors in a Da But period individual. The individual is a male, aged 50+ years. The

Table 3.4. Prevalence of Ante Mortem Tooth Loss by Tooth Count

Da But Period	Male			Female			YA (unsexed)			Indeterminate			Total		
Max.	obs	n	%	obs	n	%	obs	n	%	obs	n	%	obs	n	%
I	8	65	12.3	0	43	0.0	0	21	0.0	0	10	0.0	8	139	5.8
C	2	38	5.3	0	24	0.0	0	10	0.0	0	6	0.0	2	78	2.6
P	2	79	2.5	0	54	0.0	0	17	0.0	0	14	0.0	2	164	1.2
M	1	107	0.9	4	78	5.1	0	28	0.0	0	22	0.0	5	235	2.1
Total	13	289	4.5	4	199	2.0	0	76	0.0	0	52	0.0	17	616	2.8
Man.	obs	n	%	obs	n	%	obs	n	%	obs	n	%	obs	n	%
I	4	95	4.2	17	78	21.8	0	8	0.0	1	4	25.0	22	185	11.9
C	3	50	6.0	3	42	7.1	0	7	0.0	0	4	0.0	6	103	5.8
P	9	99	9.1	4	89	4.5	0	8	0.0	0	9	0.0	13	205	6.3
M	4	155	2.6	7	132	5.3	0	20	0.0	0	14	0.0	11	321	3.4
Total	20	399	5.0	31	341	9.1	0	43	0.0	1	31	3.2	52	814	6.4
Total	obs	n	%	obs	n	%	obs	n	%	obs	n	%	obs	n	%
I	12	160	7.5	17	121	14.0	0	29	0.0	1	14	7.1	30	324	9.3
C	5	88	5.7	3	66	4.5	0	17	0.0	0	10	0.0	8	181	4.4
P	11	178	6.2	4	143	2.8	0	25	0.0	0	23	0.0	15	369	4.1
M	5	262	1.9	11	210	5.2	0	48	0.0	0	36	0.0	16	556	2.9
Total	33	688	4.8	35	540	6.5	0	119	0.0	1	83	1.2	69	1430	4.8

Metal Period	Male			Female			YA (unsexed)			Indeterminate			Total		
Max.	obs	n	%	obs	n	%	obs	n	%	obs	n	%	obs	n	%
I	1	88	1.1	4	70	5.7	0	9	0.0	0	18	0.0	5	185	2.7
C	0	45	0.0	2	41	4.9	0	2	0.0	0	10	0.0	2	98	2.0
P	1	92	1.1	0	83	0.0	0	4	0.0	0	23	0.0	1	202	0.5
M	1	121	0.8	5	119	4.2	0	7	0.0	0	29	0.0	6	276	2.2
Total	3	346	0.9	11	313	3.5	0	22	0.0	0	80	0.0	14	761	1.8
Man.	obs	n	%	obs	n	%	obs	n	%	obs	n	%	obs	n	%
I	2	81	2.5	4	66	6.1	0	18	0.0	1	15	6.7	7	180	39.0
C	0	43	0.0	1	33	3.0	0	7	0.0	0	8	0.0	1	91	1.1
P	2	91	2.2	2	67	3.0	0	13	0.0	2	22	9.1	6	193	3.1
M	7	132	5.3	7	107	6.5	0	24	0.0	4	30	13.3	18	293	6.1
Total	11	347	3.2	14	273	5.1	0	62	0.0	7	75	9.3	32	757	4.2
Total	obs	n	%	obs	n	%	obs	n	%	obs	n	%	obs	n	%
I	3	169	1.8	8	136	5.9	0	27	0.0	1	33	3.0	12	365	3.3
C	0	88	0.0	3	74	4.1	0	9	0.0	0	18	0.0	3	189	1.6
P	3	183	1.6	2	150	1.3	0	17	0.0	2	45	4.4	7	395	1.8
M	8	253	3.2	12	226	5.3	0	31	0.0	4	59	6.8	24	569	4.2
Total	14	693	2.0	25	586	4.3	0	84	0.0	7	155	4.5	46	1518	3.0

max=maxilla, man=mandible

Intra-Sample Comparisons

Da But Period	Chi-sq.	p	Metal Period	Chi-sq.	p
male/female	1.64	0.40	male/female	**5.42**	**0.02**
maxilla/mandible	**10.05**	**0.00**	maxilla/mandible	**7.36**	**0.01**
male/female maxilla	2.17	0.14	male/female maxilla	**5.54**	**0.02**
male/female mandible	**0.02**	**0.03**	male/female mandible	1.51	0.22

mandibular body is preserved from RI1 to the alveolus of RM2. An in situ RI2 is lying on its labial face at 45^0 to correct orientation. Also, the morphology of the RI1 alveolus suggests it was also lying on its buccal side within the alveolar margin and pointing some 45^0 away from its correct orientation.

Figure 3.13 represents a case of either the extreme displacement of the permanent RC or a supernumerary tooth in a young adult aged 15-19 years from the Da But period. The extra tooth is erupting above the permanent RP3 and RP4. The RP3 is missing post mortem and so apparently is the RC, although this is probably the alveolus of the dc which was retained into adulthood. If this is the case, as appears likely, this is more a case of extreme displacement

Table 3.5. Ante Mortem Tooth Loss by Age (3 x age classes)

Da But Period

Individual Count

	male			female			total			tests for AMTL homogeneity by age class	
age	obs	n	%	obs	n	%	obs	n	%	Chi-sq.	p
<30	1	6	16.7	1	5	20.0	3	23	13.0	**12.248**	**0.002**
30-39	2	13	15.4	1	6	16.7	3	20	15.0		
40+	8	14	57.1	8	16	50.0	16	31	51.6	Likelihood ratio	p
indeterminate	0	0	0.0	0	0	0.0	0	0	0.0	**12.403**	**0.002**
TOTAL	11	33	33.3	10	27	37.0	22	74	29.7		

Tooth Count

	male			female			total				
age	obs	n	%	obs	n	%	obs	n	%	Chi-sq.	p
<30	1	106	0.9	1	90	1.1	3	377	0.8	**54.978**	**0.000**
30-39	2	296	0.7	3	107	2.8	5	408	1.2		
40+	30	286	10.5	31	343	9.0	61	645	9.5	Likelihood ratio	p
indeterminate	0	0	0.0	0	0	0.0	0	0	0.0	**60.242**	**0.000**
TOTAL	33	688	4.8	35	540	6.5	69	1430	4.8		

Metal Period

Individual Count

	male			female			total				
age	obs	n	%	obs	n	%	obs	n	%	Chi-sq.	p
<30	2	11	18.2	2	12	16.7	5	40	12.5	**6.750**	**0.034**
30-39	0	6	0.0	1	10	10.0	1	16	6.3		
40+	2	8	25.0	3	4	75.0	6	16	37.5	Likelihood ratio	p
indeterminate	0	1	0.0	0	2	0.0	1	4	25.0	**6.088**	**0.048**
TOTAL	4	26	15.4	6	28	21.4	13	76	17.1		

Tooth Count

	male			female			total				
age	obs	n	%	obs	n	%	obs	n	%	Chi-sq.	p
<30	4	273	1.5	5	233	2.2	13	695	1.9	**41.816**	**0.000**
30-39	0	187	0.0	3	244	1.2	3	431	0.7		
40+	10	228	4.4	17	93	18.3	29	362	8.0	Likelihood ratio	p
indeterminate	0	5	0.0	0	16	0.0	1	30	3.3	**36.476**	**0.000**
TOTAL	14	693	2.0	25	586	4.3	46	1518	3.0		

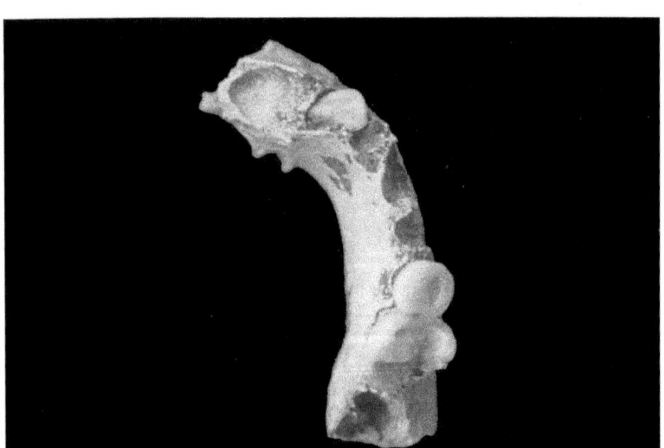

Figure 3.12. 80CCNM66. Da But Period Male 50+ years. Severe displacement of right first and second incisors

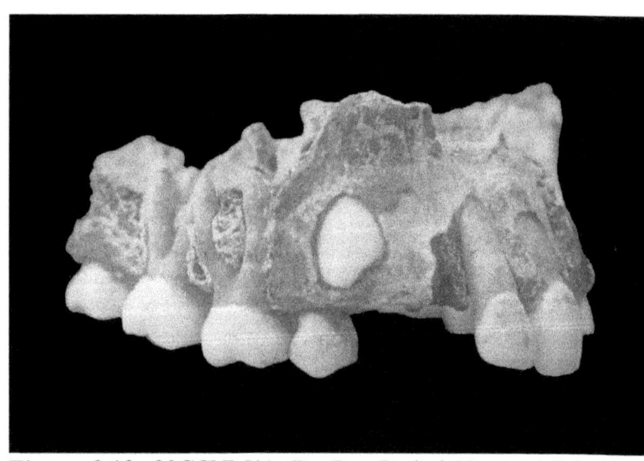

Figure 3.13. 80CCNM11. Da But Period Young Adult 15-19 years. Possible supernumerary tooth. However, it is more likely a severely displaced and impacted canine

and impaction of the permanent RC rather than a supernumerary tooth.

For the most part, teeth were either displaced or rotated but not both. In only two instances in the Da But period sample and two in the Metal period, were individual teeth both rotated and displaced. In the Da But period neither of the two individuals with impacted M3s suffered from rotated nor displaced teeth. However, in the Metal period three of the six individuals with impacted teeth also had displaced teeth. One of these had all three conditions.

Table 3.6. Prevalence of Alveolar Defects of Pulpal Origin by Tooth Count

Da But Period

	Male			Female			YA (unsexed)			Indeterminate			Total		
Max.	obs	n	%	obs	n	%	obs	n	%	obs	n	%	obs	n	%
I	0	65	0.0	0	43	0.0	0	21	0.0	1	10	10.0	1	139	0.7
C	0	38	0.0	0	24	0.0	1	10	10.0	0	6	0.0	1	78	1.3
P	0	79	0.0	0	54	0.0	0	17	0.0	0	14	0.0	0	164	0.0
M	4	107	3.7	1	78	1.3	0	28	0.0	1	22	4.5	6	235	2.6
Total	4	289	1.4	1	199	0.5	1	76	1.3	2	52	3.8	8	616	1.3
Man.	obs	n	%	obs	n	%	obs	n	%	obs	n	%	obs	n	%
I	0	95	0.0	0	78	0.0	0	8	0.0	0	4	0.0	0	185	0.0
C	0	50	0.0	0	42	0.0	0	7	0.0	0	4	0.0	0	103	0.0
P	1	99	1.0	1	89	1.1	0	8	0.0	0	9	0.0	2	205	1.0
M	5	155	3.2	7	132	5.3	0	20	0.0	0	14	0.0	12	321	3.7
Total	6	399	1.5	8	341	2.3	0	43	0.0	0	31	0.0	14	814	1.7
Total	obs	n	%	obs	n	%	obs	n	%	obs	n	%	obs	n	%
I	0	160	0.0	0	121	0.0	0	29	0.0	1	14	7.1	1	324	0.3
C	0	88	0.0	0	66	0.0	1	17	5.9	0	10	0.0	1	181	0.6
P	1	178	0.6	1	143	0.7	0	25	0.0	0	23	0.0	2	369	0.5
M	9	262	3.4	8	210	3.8	0	48	0.0	1	36	2.8	18	556	3.2
Total	10	688	1.5	9	540	1.7	1	119	0.8	2	83	2.4	22	1430	1.5

Metal Period

	Male			Female			YA (unsexed)			Indeterminate			Total		
Max.	obs	n	%	obs	n	%	obs	n	%	obs	n	%	obs	n	%
I	1	88	1.1	2	70	2.9	0	9	0.0	0	18	0.0	3	185	1.6
C	0	45	0.0	0	41	0.0	0	2	0.0	0	10	0.0	0	98	0.0
P	3	92	3.3	0	83	0.0	0	4	0.0	0	23	0.0	3	202	1.5
M	0	121	0.0	5	119	4.2	0	7	0.0	0	29	0.0	5	276	1.8
Total	4	346	1.2	7	313	2.2	0	22	0.0	0	80	0.0	11	761	1.4
Man.	obs	n	%	obs	n	%	obs	n	%	obs	n	%	obs	n	%
I	4	81	4.9	2	66	3.0	0	18	0.0	0	15	0.0	6	180	3.3
C	2	43	4.7	1	33	3.0	0	7	0.0	1	8	12.5	4	91	4.4
P	3	91	3.3	4	67	6.0	0	13	0.0	1	22	4.5	8	193	4.1
M	2	132	1.5	7	107	6.5	0	24	0.0	1	30	3.3	10	293	3.4
Total	11	347	3.2	14	273	5.1	0	62	0.0	3	75	4.0	28	757	3.7
Total	obs	n	%	obs	n	%	obs	n	%	obs	n	%	obs	n	%
I	5	169	3.0	4	136	2.9	0	27	0.0	0	33	0.0	9	365	2.5
C	2	88	2.3	1	74	1.4	0	9	0.0	1	18	5.6	4	189	2.1
P	6	183	3.3	4	150	2.7	0	17	0.0	1	45	2.2	11	395	2.8
M	2	253	0.8	12	226	5.3	0	31	0.0	1	59	1.7	15	569	2.6
Total	15	693	2.2	21	586	3.6	0	84	0.0	3	155	1.9	39	1518	2.6

max=maxilla, man=mandible

Intra-Sample Comparisons

Da But Period	Chi-sq.	p	Metal Period	Chi-sq.	p
male/female	0.09	0.76	male/female	2.34	0.13
maxilla/mandible	0.41	0.52	maxilla/mandible	**7.70**	**0.01**
male/female maxilla	0.9*	0.34	male/female maxilla	1.17*	0.28
male/female mandible	0.70	0.40	male/female mandible	1.51	0.22

Tooth Impaction (IM)

The vast majority of impacted teeth occurred with third molars. The denominator for assessing the frequency of IM is relatively small, for both reporting methods, as it is necessarily based on the number of assessable in situ third molars.

In the Da But period, a case in each of two individuals (2/24) was noted. One case was a left maxillary M3 in an unsexed individual aged 15-19 years. The only other assessable M3, for this individual, was the left mandibular tooth that was in the process of erupting without apparent problem. The other case was a right maxillary M3 in a male aged 50+ years. Interestingly, this tooth was in the process of erupting. The other three M3s, for this individual, had fully erupted, although wear on these M3s suggested

Table 3.7. Alveolar Defects of Pulpal Origin by Age (3 x age classes)

Da But Period

Individual Count

	male			female			total			tests for ADP homogeneity by age class	
age	obs	n	%	obs	n	%	obs	n	%	Chi-sq.	p
<30	0	6	0.0	0	5	0.0	2	23	8.7	***15.995**	**0.000**
30-39	1	13	7.7	0	6	0.0	0	20	0.0		
40+	7	14	50.0	5	16	31.3	13	31	41.9	Likelihood ratio	p
indeterminate	0	0	0.0	0	0	0.0	0	0	0.0	***18.855**	**0.000**
TOTAL	8	33	24.2	5	27	18.5	15	74	20.3		

Tooth Count

	male			female			total				
age	obs	n	%	obs	n	%	obs	n	%	Chi-sq.	p
<30	0	106	0.0	0	90	0.0	2	377	0.5	***19.297**	**0.000**
30-39	1	296	0.3	0	107	0.0	0	408	0.0		
40+	10	286	3.5	9	343	2.6	20	645	3.1	Likelihood ratio	p
indeterminate	0	0	0.0	0	0	0.0	0	0	0.0	***24.073**	**0.000**
TOTAL	11	688	1.6	9	540	1.7	22	1430	1.5		

Metal Period

Individual Count

	male			female			total				
age	obs	n	%	obs	n	%	obs	n	%	Chi-sq.	p
<30	0	11	0.0	3	12	25.0	4	40	10.0	**12.037**	**0.002**
30-39	0	6	0.0	2	10	20.0	2	16	12.5		
40+	4	8	50.0	3	4	75.0	8	16	50.0	Likelihood ratio	p
indeterminate	0	1	0.0	0	2	0.0	0	4	0.0	**10.691**	**0.005**
TOTAL	4	26	15.4	8	28	28.6	14	76	18.4		

Tooth Count

	male			female			total				
age	obs	n	%	obs	n	%	obs	n	%	Chi-sq.	p
<30	0	273	0.0	6	233	2.6	7	695	1.0	**39.152**	**0.000**
30-39	0	187	0.0	6	244	2.5	6	431	1.4		
40+	15	228	6.6	9	93	9.7	26	362	7.2	Likelihood ratio	p
indeterminate	0	5	0.0	0	16	0.0	0	30	0.0	**32.475**	**0.000**
TOTAL	15	693	2.2	21	586	3.6	39	1518	2.6		

* one category has no observations

considerable spacing between the eruption of these teeth.

In the Metal period, five individuals (5/24) and seven individual M3s (7/114) were impacted. Only one individual was male and five of the seven impacted teeth were mandibular. With the exception of one adult of unknown age, the rest were aged between 20 and 39 years.

Two individuals exhibited bilateral impaction of the mandibular M3s. One was the only male exhibiting IM and was aged 20-29 years. The left M3 had a large wear facet (75% of face) on the distal (now occlusal) face and the occlusal crown is now flat against the distal face of M2 (Figure 3.14). The same is true for the right M3, but the occlusal wear facet is much smaller (~25% of face). Further, a large (~25% of face and 4.4mm diameter) carious lesion of the neck occurs on the distal face of the right M2. The right M3 would have been impacting on this lesion and the impaction probably caused the lesion. In fact, the right M3 also has a small occlusal carious lesion.

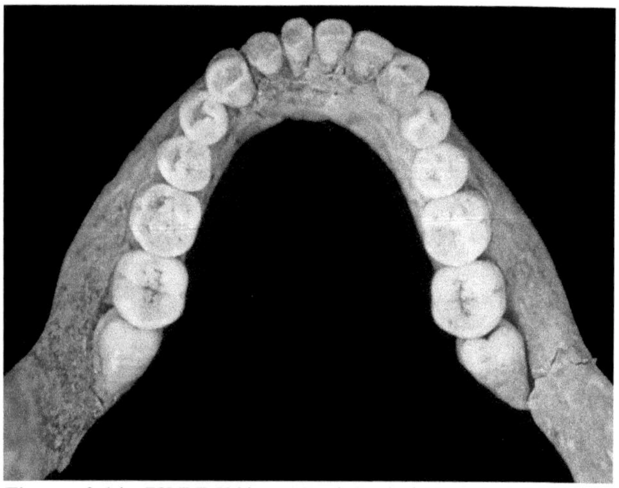

Figure 3.14. 78NNM23ka. Metal Period Male 20-29 years. Bilateral M3 impaction

The other case of bilateral IM was a female aged 20-29 years. Both molars were erupting into the distal surface

Table 3.8. Prevalence of Dental Caries by Tooth Count

Da But Period

	Male			Female			YA (unsexed)			Indeterminate			Total		
Max.	obs	n	%	obs	n	%	obs	n	%	obs	n	%	obs	n	%
I	0	21	0.0	0	21	0.0	0	7	0.0	0	9	0.0	0	58	0.0
C	0	22	0.0	0	15	0.0	0	7	0.0	0	8	0.0	0	52	0.0
P	1	54	1.9	0	37	0.0	0	13	0.0	0	14	0.0	1	118	0.9
M	4	99	4.0	5	71	7.0	0	25	0.0	0	23	0.0	9	218	4.1
Total	5	196	2.6	5	144	3.5	0	52	0.0	0	54	0.0	10	446	2.2
Man.	obs	n	%	obs	n	%	obs	n	%	obs	n	%	obs	n	%
I	0	25	0.0	0	18	0.0	0	4	0.0	0	2	0.0	0	49	0.0
C	0	26	0.0	0	18	0.0	0	2	0.0	0	2	0.0	0	48	0.0
P	0	68	0.0	0	48	0.0	0	5	0.0	0	7	0.0	0	128	0.0
M	2	130	1.5	2	105	1.9	0	26	0.0	0	19	0.0	4	280	1.4
Total	2	249	0.8	2	189	1.1	0	37	0.0	0	30	0.0	4	505	0.8
Total	obs	n	%	obs	n	%	obs	n	%	obs	n	%	obs	n	%
I	0	46	0.0	0	39	0.0	0	11	0.0	0	11	0.0	0	107	0.0
C	0	48	0.0	0	33	0.0	0	9	0.0	0	10	0.0	0	100	0.0
P	1	122	0.8	0	85	0.0	0	18	0.0	0	21	0.0	1	246	0.4
M	6	229	2.6	7	176	4.0	0	51	0.0	0	42	0.0	13	498	2.6
Total	7	445	1.6	7	333	2.1	0	89	0.0	0	84	0.0	14	951	1.5

Metal Period

	Male			Female			YA (unsexed)			Indeterminate			Total		
Max.	obs	n	%	obs	n	%	obs	n	%	obs	n	%	obs	n	%
I	0	49	0.0	0	40	0.0	0	13	0.0	0	6	0.0	0	108	0.0
C	0	30	0.0	0	29	0.0	0	3	0.0	0	9	0.0	0	71	0.0
P	0	75	0.0	1	57	1.8	0	4	0.0	2	22	9.1	3	158	1.9
M	4	101	4.0	8	110	7.3	0	11	0.0	0	30	0.0	12	252	4.8
Total	4	255	1.6	9	236	3.8	0	31	0.0	2	67	3.0	15	589	2.5
Man.	obs	n	%	obs	n	%	obs	n	%	obs	n	%	obs	n	%
I	0	43	0.0	1	37	2.7	0	12	0.0	0	4	0.0	1	96	1.0
C	0	29	0.0	0	28	0.0	0	3	0.0	0	2	0.0	0	62	0.0
P	0	67	0.0	0	62	0.0	0	13	0.0	0	9	0.0	0	151	0.0
M	3	112	2.7	7	99	7.1	0	21	0.0	0	22	0.0	10	254	3.9
Total	3	251	1.2	8	226	3.5	0	49	0.0	0	37	0.0	11	563	2.0
Total	obs	n	%	obs	n	%	obs	n	%	obs	n	%	obs	n	%
I	0	92	0.0	1	77	1.3	0	25	0.0	0	10	0.0	1	204	0.5
C	0	59	0.0	0	57	0.0	0	6	0.0	0	11	0.0	0	133	0.0
P	0	142	0.0	1	119	0.8	0	17	0.0	2	31	6.5	3	309	1.0
M	7	213	3.3	15	209	7.2	0	32	0.0	0	52	0.0	22	506	4.3
Total	7	506	1.4	17	462	3.7	0	80	0.0	2	104	1.9	26	1152	2.3

max=maxilla, man=mandible

Intra-Sample Comparisons

Da But Period	Chi-sq.	p	Metal Period	Chi-sq.	p
male/female	0.28	0.60	male/female	**5.22**	**0.02**
maxilla/mandible	3.44	0.06	maxilla/mandible	0.46	0.50
male/female maxilla	0.23	0.63	male/female maxilla	2.40	0.12
male/female mandible	0.07	0.79	male/female mandible	2.86	0.09

of their respective M2s at an angle of approximately 45°. Unlike the previous example, no other complicating factors were noted.

Of the other cases displaying IM one was a female, aged 30-39 years, with an IM left maxillary M3. None of her other M3s had yet erupted. The other three cases were all from the same site, Doi Son. The individual designated 84DSM1a, aged 30-39 years, had an IM left maxillary M3 with an associated abscess possibly caused by the impaction of M3 against M2. The abscess caused the destruction of most of buccal alveolar border at M3. The lingual root of M3 is still in situ, as are the buccal roots of M2, both of which have apical fenestrations. There was no evidence for IM of the other M3s.

Table 3.9. Size and Distribution of Carious Lesions

Da But Period: Lesion Size

	Lesion Size Categories								
	1		2		3		4		Total
	obs	%	obs	%	obs	%	obs	%	n
Maxilla	2	20.0	7	70.0	1	10.0	0	0.0	10
Mandible	2	50.0	1	25.0	0	0.0	1	25.0	4
Total	4	28.6	8	57.1	1	7.1	1	7.1	14
Male	0	0.0	5	71.4	1	14.3	1	14.3	7
Female	4	57.1	3	42.9	0	0.0	0	0.0	7

1=small pit or fissure lesion; 2=< half crown destroyed; 3=> half crown destroyed; 4=entire crown obliterated

Da But Period: Lesion Location

	Tooth Surface												
	Mesial		Distal		Buccal		Lingual		Occlusal		Unknown[1]		Total
	obs	%	obs	%	obs	%	obs	%	obs	%	obs	%	n
Maxilla	3	30.0	3	30.0	0	0.0	1	10.0	3	30.0	0	0.0	10
Mandible	0	0.0	1	25.0	0	0.0	0	0.0	2	50.0	1	25.0	4
Total	3	21.4	4	28.6	0	0.0	1	7.1	5	35.7	1	7.1	14
Male	2	28.6	2	28.6	0	0.0	0	0.0	2	28.6	1	14.3	7
Female	1	14.3	2	28.6	0	0.0	1	14.3	3	42.9	0	0	7

Metal Period: Lesion Size

	Lesion Size Ca tegories								
	1		2		3		4		Total
	obs	%	obs	%	obs	%	obs	%	n
Maxilla	10	66.7	3	20.0	2	13.3	0	0.0	15
Mandible	4	36.4	7	63.6	0	0.0	0	0.0	11
Total	14	53.8	10	38.5	2	7.7	0	0.0	26
Male	4	57.1	3	42.9	0	0.0	0	0.0	7
Female	8	47.1	7	41.2	2	11.8	0	0.0	17
indeterminate	2	100.0	0	0.0	0	0.0	0	0.0	2

Metal Period: Lesion Location

	Tooth Surface												
	Mesial		Distal		Buccal		Lingual		Occlusal		Unknown[1]		Total
	obs	%	obs	%	obs	%	obs	%	obs	%	obs	%	n
Maxilla	2	13.3	2	13.3	1	6.7	0	0.0	10	66.7	0	0.0	15
Mandible	0	0.0	7	63.6	0	0.0	0	0.0	4	36.4	0	0.0	11
Total	2	7.7	9	34.6	1	3.8	0	0.0	14	53.8	0	0.0	26
Male	1	14.3	1	14.3	1	14.3	0	0.0	4	57.1	0	0.0	7
Female	1	5.9	8	47.1	0	0.0	0	0.0	8	47.1	0	0.0	17
indeterminate	0	0.0	0	0.0	0	0.0	0	0.0	2	100.0	0	0.0	2

[1] caries point of origin unknown

Table 3.10. Cervical and Root Caries Frequency

Da But Period

	Cervical			Root			Crown		
	obs	n	%	obs	n	%	obs	n	%
	3	14	21.4	2	14	14.3	9	14	64.3

Metal Period

	Cervical			Root			Crown		
	obs	n	%	obs	n	%	obs	n	%
	6	26	23.1	1	26	3.7	19	26	73.1

Table 3.11. Dental Caries by Age (3 x age classes)

Da But Period

Individual Count

	male			female			total			tests for caries homogeneity by age class	
age	obs	n	%	obs	n	%	obs	n	%	Chi-sq.	p
<30	0	6	0.0	1	5	20.0	1	25	4.0	**8.813**	**0.011**
30-39	1	13	7.7	0	7	0.0	1	21	4.8		
40+	4	15	26.7	5	16	31.3	9	32	28.1	Likelihood ratio	p
indeterminate	0	2	0.0	0	0	0.0	0	2	0.0	**9.002**	**0.011**
TOTAL	5	36	13.9	6	28	21.4	11	80	13.8		

Tooth Count

	male			female			total				
age	obs	n	%	obs	n	%	obs	n	%	Chi-sq.	p
<30	0	73	0.0	2	60	3.3	2	289	0.7	**7.981**	**0.018**
30-39	1	187	0.5	0	71	0.0	1	262	0.4		
40+	6	181	3.3	5	202	2.5	11	396	2.8	Likelihood ratio	p
indeterminate	0	4	0.0	0	0	0.0	0	4	0.0	**8.250**	**0.016**
TOTAL	7	445	1.6	7	333	2.1	14	951	1.5		

Metal Period

Individual Count

	male			female			total				
age	obs	n	%	obs	n	%	obs	n	%	Chi-sq.	p
<30	4	10	40.0	6	12	50.0	11	40	27.5	1.631	0.442
30-39	0	6	0.0	3	10	30.0	3	16	18.8		
40+	1	8	12.5	1	4	25.0	2	16	12.5	Likelihood ratio	p
indeterminate	0	2	0.0	0	1	0.0	0	5	0.0	1.725	0.422
TOTAL	5	26	19.2	10	27	3.7	16	77	20.8		

Tooth Count

	male			female			total				
age	obs	n	%	obs	n	%	obs	n	%	Chi-sq.	p
<30	5	207	2.4	9	209	4.3	16	565	2.8	1.538	0.464
30-39	0	115	0.0	6	201	3.0	6	316	1.9		
40+	2	181	1.1	2	51	3.9	4	254	1.6	Likelihood ratio	p
indeterminate	0	3	0.0	0	1	0.0	0	17	0.0	1.571	0.456
TOTAL	7	506	1.4	17	462	3.7	26	1152	2.3		

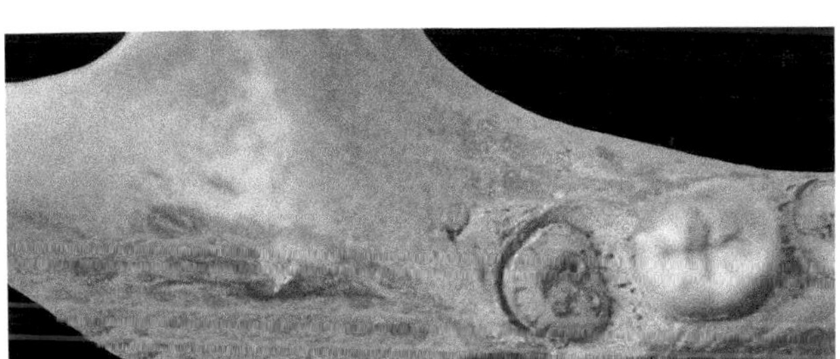

Figure 3.15. 84DSM8. Metal Period Female 20-29 years. Left M3 affected by what is most likely a dentigerous cyst

The individual 84DSM8a, aged 20-29 years, had an impacted right mandibular M3. There is loss of alveolar bone at the left M3 (Figure 3.15) in the form of a large shallow basin extending principally mesially. There are numerous smooth rimmed pits or perforations within this excavated basin. There is no intercrestal margin between the LM3 and LM2 as the aforementioned basin extends up to the face of the distal roots of the LM2. The mesio-angular orientation of what is left of this LM3 suggests that it was or would have impacted on the adjacent LM2. The roots up to the original cervix of the tooth are still firmly embedded in the alveolus and are not exposed. The buccal and lingual sides of the mandibular body are free of any evidence of pathology. A differential diagnosis is presented in the discussion section below.

The final case, 84DSM9a, age unknown, had a left mandibular M3 impacting mesially at a 45^0 angle. Only the left half of the mandible was preserved, so other teeth could not be assessed. No other complications were noted.

Congenital Tooth Absence

As the occurrence of hypodontia is inferred in this study, it is necessary to look at its distribution by age. Only post-teen adults are included to limit the potential for including cases where the M3s simply have not erupted before death. Table 3.19 presents the frequency of hypdontia by both tooth and individual count. In the Metal period it can be seen that M3 hypodontia occurs more frequently in the later age classes. This suggests that this method is measuring congenital tooth absence rather than late eruption. The reverse pattern is evident in the Da But period and suggests late eruption is being measured in

Table 3.12. Prevalence of Pulp Chamber Exposure by Tooth Count

Da But Period	Male			Female			YA (unsexed)			Indeterminate			Total		
Max.	obs	n	%	obs	n	%	obs	n	%	obs	n	%	obs	n	%
I	0	21	0.0	0	21	0.0	0	7	0.0	0	9	0.0	0	58	0.0
C	0	24	0.0	0	15	0.0	0	7	0.0	0	8	0.0	0	54	0.0
P	2	52	3.8	0	37	0.0	0	13	0.0	0	14	0.0	2	116	1.7
M	3	99	3.0	0	71	0.0	0	25	0.0	0	23	0.0	3	218	1.4
Total	5	196	2.6	0	144	0.0	0	52	0.0	0	54	0.0	5	446	1.1
Man.	obs	n	%	obs	n	%	obs	n	%	obs	n	%	obs	n	%
I	0	25	0.0	0	18	0.0	0	4	0.0	0	2	0.0	0	49	0.0
C	0	26	0.0	0	18	0.0	0	2	0.0	0	2	0.0	0	48	0.0
P	0	68	0.0	0	48	0.0	0	5	0.0	0	7	0.0	0	128	0.0
M	7	130	5.4	5	105	4.8	0	26	0.0	0	19	0.0	12	280	4.3
Total	7	249	2.8	5	189	2.6	0	37	0.0	0	30	0.0	12	505	2.4
Total	obs	n	%	obs	n	%	obs	n	%	obs	n	%	obs	n	%
I	0	46	0.0	0	39	0.0	0	11	0.0	0	11	0.0	0	107	0.0
C	0	48	0.0	0	33	0.0	0	9	0.0	0	10	0.0	0	100	0.0
P	2	122	1.6	0	85	0.0	0	18	0.0	0	21	0.0	2	246	0.8
M	10	229	4.4	5	176	2.8	0	51	0.0	0	42	0.0	15	498	3.0
Total	12	445	2.7	5	333	1.5	0	89	0.0	0	84	0.0	17	951	1.8
Metal Period	Male			Female			YA (unsexed)			Indeterminate			Total		
Max.	obs	n	%	obs	n	%	obs	n	%	obs	n	%	obs	n	%
I	0	49	0.0	0	40	0.0	0	13	0.0	0	6	0.0	0	108	0.0
C	0	30	0.0	0	29	0.0	0	3	0.0	0	9	0.0	0	71	0.0
P	2	75	2.7	0	57	0.0	0	4	0.0	0	22	0.0	2	158	1.3
M	0	101	0.0	2	110	1.8	0	11	0.0	0	30	0.0	2	252	0.8
Total	2	255	0.8	2	236	0.8	0	31	0.0	0	67	0.0	4	589	0.7
Man.	obs	n	%	obs	n	%	obs	n	%	obs	n	%	obs	n	%
I	6	43	14.0	0	37	0.0	0	12	0.0	1	4	25.0	7	96	7.3
C	3	29	10.3	0	28	0.0	0	3	0.0	1	2	50.0	4	62	6.5
P	6	67	9.0	2	62	3.2	0	13	0.0	1	9	11.1	9	151	6.0
M	0	112	0.0	1	99	1.0	0	21	0.0	1	22	4.5	2	254	0.8
Total	15	251	6.0	3	226	1.3	0	49	0.0	4	37	10.8	22	563	3.9
Total	obs	n	%	obs	n	%	obs	n	%	obs	n	%	obs	n	%
I	6	92	6.5	0	77	0.0	0	25	0.0	1	10	10.0	7	204	3.4
C	3	59	5.1	0	57	0.0	0	6	0.0	1	11	9.1	4	133	3.0
P	8	142	5.6	2	119	1.7	0	17	0.0	1	31	3.2	11	309	3.6
M	0	213	0.0	3	209	1.4	0	32	0.0	1	52	1.9	4	506	0.8
Total	17	506	3.4	5	462	1.1	0	80	0.0	4	104	3.8	26	1152	2.3

max=maxilla, man=mandible

Intra-Sample Comparisons

Da But Period	Chi-sq.	p	Metal Period	Chi-sq.	p	p*
male/female	1.27	0.26	male/female	**5.64**	**0.02**	
maxilla/mandible	2.12	0.15	maxilla/mandible	**13.61**	**0.00**	
male/female maxilla	n/a		male/female maxilla	0.01	0.94	
male/female mandible	0.01	0.91	male/female mandible	**7.17**	**0.01**	**0.01**

p* Fisher exact 2-tailed

this case. There are no apparent differences in expression of this condition within each sample by sex. While the condition is more prevalent in the Metal period sample, 18.5% of assessable individuals have congenitally missing M3s compared with 9.7% in the Da But period sample, this is not statistically significant.

There is also a probable case of P4 agenesis in a Da But period female aged 20-29 years. Figure 3.16 is an occlusal view of the mandible showing late retention of the right dm1 and absence of the left P4. The broken roots of all other permanent anterior teeth are present. It is difficult to assess whether the left P4 ever erupted or not. It is

Table 3.13. Pulp Chamber Exposure by Age (3 x age classes)

Da But Period

Individual Count

age	male obs	male n	male %	female obs	female n	female %	total obs	total n	total %	tests for PEXP homogeneity by age class		
											Chi-sq.	p
<30	0	6	0.0	0	5	0.0	0	25	0.0		***5.173**	**0.075**
30-39	1	13	7.7	0	7	0.0	1	21	4.8			
40+	3	15	20.0	2	16	12.5	5	32	15.6		Likelihood ratio	p
indeterminate	1	2	50.0	0	0	0.0	1	2	50.0		***6.527**	**0.038**
TOTAL	5	36	13.9	2	28	7.1	7	80	8.8			

Tooth Count

age	male obs	male n	male %	female obs	female n	female %	total obs	total n	total %		
										Chi-sq.	p
<30	0	73	0.0	0	60	0.0	0	289	0.0	***16.754**	**0.000**
30-39	1	187	0.5	0	71	0.0	1	262	0.4		
40+	9	181	5.0	5	202	2.5	14	396	3.5	Likelihood ratio	p
indeterminate	2	4	50.0	0	0	0.0	2	4	50.0	***19.901**	**0.000**
TOTAL	12	445	2.7	5	333	1.5	17	951	1.8		

Metal Period

Individual Count

age	male obs	male n	male %	female obs	female n	female %	total obs	total n	total %		
										Chi-sq.	p
<30	0	10	0.0	2	12	16.7	2	40	5.0	**11.771**	**0.003**
30-39	1	6	16.7	0	10	0.0	1	16	6.3		
40+	4	8	50.0	1	4	25.0	6	16	37.5	Likelihood ratio	p
indeterminate	0	2	0.0	0	1	0.0	1	5	20.0	**9.722**	**0.008**
TOTAL	5	26	19.2	3	27	11.1	10	77	13.0		

Tooth Count

age	male obs	male n	male %	female obs	female n	female %	total obs	total n	total %		
										Chi-sq.	p
<30	0	207	0.0	2	209	1.0	2	565	0.4	**56.361**	**0.000**
30-39	1	115	0.9	0	201	0.0	1	316	0.3		
40+	16	181	8.8	3	51	5.9	20	254	7.9	Likelihood ratio	p
indeterminate	0	3	0.0	0	1	0.0	3	17	17.7	**44.758**	**0.000**
TOTAL	17	506	3.4	5	462	1.1	26	1152	2.3		

*one category has no observations

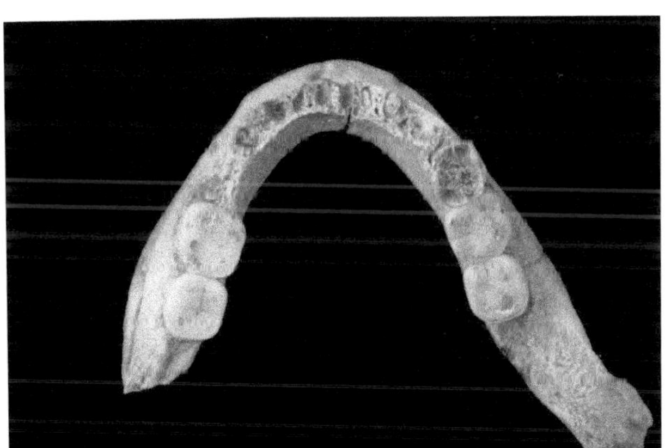

Figure 3.16. 80CCNM13. Da But Period Female 20-29 years. Probable bilateral P4 agenesis. Note retention of deciduous right first molar

probable that both P4s were congenitally absent. The maxilla displays an intact LP3 and the empty alveolus of the LP4. However, while there is a clear empty alveolus for the RP3, the normal location for the RP4 exhibits remodelled alveolar bone with no sign of the alveolus or missing RP4. This suggests that subsequent to the loss of the right maxillary dm1, it was not replaced by the RM1 due to congenital absence of this permanent tooth.

Enamel Pearls (EP)

Enamel pearls were not observed in the Da But period sample. In the Metal period sample three individuals (3/77) and a total of four teeth (4/1152) exhibited these enamel excrescences. All pearls occurred below the CEJ on either the buccal, mesial or distal tooth faces. In three cases the third molar was affected while in one individual both the right maxillary M2 and M3 had pearls.

Mandibular Robusticity

Comparing mandibular robusticity between samples (Table 3.20 and Figure 3.17) indicates that the Metal period had both greater levels of variation and more robust mandibles. The greater coefficient of variation (CV), of

Table 3.14. Pulp Chamber Exposure Aetiology

Da But Period

	Caries Induced			Attrition Induced			Unknown Aetiology		
	obs	n	%	obs	n	%	obs	n	%
I	0	0	0.0	0	0	0	0	0	0
C	0	0	0.0	0	0	0	0	0	0
P	1	2	50.0	1	2	50	0	2	0
M	2	15	13.3	11	15	73.3	2	15	13.3
Total	3	17	17.6	12	17	70.6	2	17	11.8

Metal Period

	Caries Induced			Attrition Induced			Unknown Aetiology		
	obs	n	%	obs	n	%	obs	n	%
I	0	7	0.0	7	7	100.0	0	7	0.0
C	0	4	0.0	4	4	100.0	0	4	0.0
P	0	11	0.0	10	11	90.9	1	11	9.1
M	2	4	50.0	2	4	50.0	0	4	0.0
Total	2	26	7.7	23	26	88.5	1	26	3.8

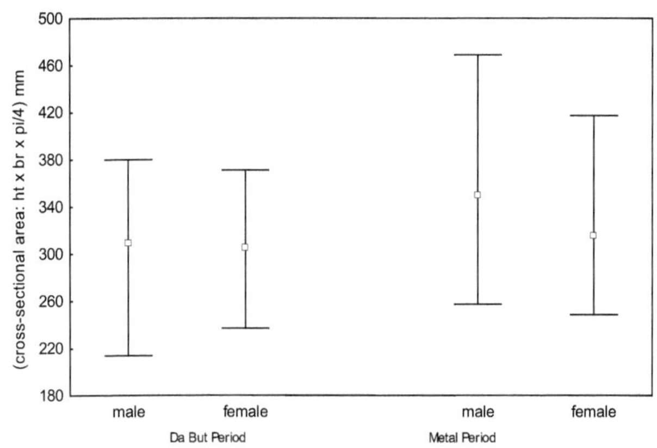

Figure 3.17. Da But Period and Metal Period Mandibular Robusticity.

11.18 compared to 9.73 for the Da But period, relates to the significant sex differences, or sexual dimorphism, within the Metal period sample. When comparing each sample it can be seen that while mandibular height is slightly, but not significantly, greater in the Da But period, breadth, size and area are all significantly ($p<0.01$) greater in the Metal period. Figure 3.17 displays this graphically for the cross-sectional area index. Interestingly, female robusticity in the Metal period is even greater than male robusticity in the Da But period. Moreover, the greater level of sexual dimorphism in the Metal period sample is clearly seen.

For the Da But period, while the male mean height and area was greater than that of the females, the females actually displayed greater mean breadth and size. None of these differences were significant (Table 3.20). With respect to the indices of robusticity, males displayed slightly greater levels of variability (cross-sectional area CV 10.52) than females (CV 9.19). For the Da But period the combined sex mandibular cross-sectional area CV was not high at 9.73.

For the Metal period, male mean height (significant at $p<0.01$) and area (significant at $p<0.05$) was greater than that of females. Female size and breath was greater than that for males, albeit not significantly so. Females (cross-sectional area CV 12.08) displayed higher levels of variation than males (CV 10.61), while the overall combined sex mandibular cross-sectional area CV was 11.18.

Molar Tooth Size

Table 3.21 reveals that with one exception, Da But period first and second molars are significantly ($p<0.01$) larger, for both jaws, than their Metal period counterparts. While Da But period maxillary M2 MD dimensions are larger than those in the Metal period, this is not a statistically significant difference.

Non-Dietary Related Traits

Interproximal Grooves (IG)

Interproximal grooves are absent from the Metal period sample but occur in five individuals, and nine teeth, from the Da But period. These grooves are found in four mandibular and five maxillary teeth, with five being found in molars and the remaining four in premolars. In two cases, representing two separate individuals, these grooves are associated with carious lesions.

One particular individual, a male aged 30-39 years, is worth examining in detail for the range of complicating and likely inter-correlated oral defects he displays. Figure 3.18 is an anterior right lateral view of the maxilla showing an association between an IG, a carious lesion and an alveolar defect of pulpal origin. The left maxillary canine has erupted buccally to the left maxillary P3. The P3 has rotated 90^0 causing its buccal face to point mesially. An interproximal cervical cavity formed on the original distal face of P3, the discomfort of which presumably led to palliative probing and the subsequent development of an IG on the distal face. There is also an IG on the mesial face of P4, probably also associated with the probing. Alternatively, the probing caused or contributed to the formation of the lesion. The P3 cavity broached the pulp chamber and is highly likely the cause of the ADP. This ADP is a complex defect that likely originated at the tooth apex and then spread coronally and in the process destroyed a considerable amount of bone. The canine was not preserved.

An IG is also present on the distal face of the RP4, while the RP3 has been lost post-mortem. Also associated with the RP4 is a large angular vertical defect and while the RP4 experienced more coronal wear, exposing a large area of secondary dentin, than the LP4, there is no clear breach of the pulp chamber. This individual is also the only one that displays clear evidence for TWF in the Da But period sample and is further described below.

Task Wear Facets (TWF)

Only one Da But period individual displays clear evidence of TWF and this case has been partially described above in relation to IG. The mesial-buccal-occlusal aspect of both maxillary M1s are worn into deep broad grooves (see arrow

Table 3.15. Prevalence of Moderate and Heavy Calculus by Tooth Count

Da But Period

	Male			Female			YA (unsexed)			Indeterminate			Total		
Max.	obs	n	%	obs	n	%	obs	n	%	obs	n	%	obs	n	%
I	1	21	4.8	0	21	0.0	0	7	0.0	3	9	33.3	4	58	6.9
C	3	22	13.6	3	15	20.0	0	7	0.0	4	8	50.0	10	52	19.2
P	9	54	16.7	4	37	8.1	0	13	0.0	2	14	14.3	15	118	12.7
M	25	99	25.3	11	71	15.5	4	25	16.0	3	23	13.0	43	218	19.7
Total	38	196	19.4	18	144	12.5	4	52	7.7	12	54	22.2	72	446	16.1
Man.	obs	n	%	obs	n	%	obs	n	%	obs	n	%	obs	n	%
I	8	25	32.0	4	18	22.2	0	4	0.0	2	2	100.0	14	49	28.6
C	9	26	34.6	5	18	27.8	0	2	0.0	2	2	100.0	16	48	33.3
P	20	68	29.4	15	48	31.3	0	5	0.0	2	7	28.6	37	128	28.9
M	34	130	26.2	31	105	29.5	1	26	3.9	6	19	31.6	72	280	25.7
Total	71	249	28.5	55	189	29.1	1	37	2.7	12	30	40.0	139	505	27.5
Total	obs	n	%	obs	n	%	obs	n	%	obs	n	%	obs	n	%
I	9	46	19.6	4	39	10.3	0	11	0.0	5	11	45.5	18	107	16.8
C	12	48	25.0	8	33	24.2	0	9	0.0	6	10	60.0	26	100	26.0
P	29	122	23.8	19	85	22.4	0	18	0.0	4	21	19.1	52	246	21.1
M	59	229	25.8	42	176	23.9	5	51	9.8	9	42	21.4	115	498	23.1
Total	109	445	24.5	73	333	21.9	5	89	5.6	24	84	28.6	211	951	22.2

Metal Period

	Male			Female			YA (unsexed)			Indeterminate			Total		
Max.	obs	n	%	obs	n	%	obs	n	%	obs	n	%	obs	n	%
I	12	49	24.5	8	40	20.0	0	13	0.0	0	6	0.0	20	108	18.5
C	6	30	20.0	3	29	10.3	0	3	0.0	0	9	0.0	9	71	12.7
P	8	75	10.7	6	57	10.5	0	4	0.0	0	22	0.0	14	158	8.9
M	29	101	28.7	18	110	16.4	0	11	0.0	5	30	16.7	52	252	20.6
Total	55	255	21.6	35	236	14.8	0	31	0.0	5	67	7.5	95	589	16.1
Man.	obs	n	%	obs	n	%	obs	n	%	obs	n	%	obs	n	%
I	0	43	0.0	0	37	0.0	0	12	0.0	0	4	0.0	0	96	0.0
C	3	29	10.3	1	28	3.6	0	3	0.0	0	2	0.0	4	62	6.5
P	8	67	11.9	5	62	8.1	0	13	0.0	1	9	11.1	14	151	9.3
M	36	112	32.1	15	99	15.2	0	21	0.0	4	22	18.2	55	254	21.7
Total	47	251	18.7	21	226	9.3	0	49	0.0	5	37	13.5	73	563	13.0
Total	obs	n	%	obs	n	%	obs	n	%	obs	n	%	obs	n	%
I	12	92	13.0	8	77	10.4	0	25	0.0	0	10	0.0	20	204	9.8
C	9	59	15.3	4	57	7.0	0	6	0.0	0	11	0.0	13	133	9.8
P	16	142	11.3	11	119	9.2	0	17	0.0	1	31	3.2	28	309	9.1
M	65	213	30.5	33	209	15.8	0	32	0.0	9	52	17.3	107	506	21.2
Total	102	506	20.2	50	402	12.1	0	80	0.0	10	104	9.6	168	1152	14.6

max=maxilla, man=mandible

Intra-Sample Comparisons

Da But Period	Chi-sq.	p	Metal Period	Chi-sq.	p	p*
male/female	0.70	0.40	male/female	**11.42**	**0.00**	
maxilla/mandible	**17.77**	**0.00**	maxilla/mandible	2.31	0.13	
male/female maxilla	2.86	0.09	male/female maxilla	**3.72**	**0.05**	0.06
male/female mandible	0.02	0.89	male/female mandible	**8.66**	**0.00**	**0.00**

p* Fisher exact 2-tailed

in Figure 3.18). These grooves correspond to a similar wear pattern on the lingual-distal-occlusal aspect of the mandibular M1s. The pattern is consistent with some sort of fibrous material being drawn, over some period of time, laterally through the teeth over the first molars. Further, the surviving right canine is severely worn and exhibits a sort of channel running labial-lingually along the occlusal surface. This may also be associated with fibre processing or clamping some type of reed-shaped object in the mouth.

Task wear facets are much more common in the Metal period with 10.4% of individuals and 3.4% of all permanent

Table 3.16. Moderate and Heavy Calculus by Age (3 x age classes)

Da But Period

Individual Count

	male			female			total			tests for caries homogeneity by age class	
age	obs	n	%	obs	n	%	obs	n	%	Chi-sq.	p
<30	3	6	50.0	3	5	60.0	14	25	56.0	0.659	0.719
30-39	9	13	69.2	3	7	42.9	12	21	57.1		
40+	10	15	66.7	11	16	68.8	21	32	65.6	Likelihood ratio	p
indeterminate	0	2	0.0	0	0	0.0	0	2	0.0	0.663	0.718
TOTAL	22	36	61.1	17	28	60.7	47	80	58.8		

Tooth Count

	male			female			total				
age	obs	n	%	obs	n	%	obs	n	%	Chi-sq.	p
<30	12	73	16.4	8	60	13.3	49	289	17.0	**27.047**	**0.000**
30-39	33	187	17.7	8	71	11.3	41	262	15.7		
40+	64	181	35.4	57	202	28.2	121	396	30.6	Likelihood ratio	p
indeterminate	0	4	0.0	0	0	0.0	0	4	0.0	**26.778**	**0.000**
TOTAL	109	445	24.5	73	333	21.9	211	951	22.2		

Metal Period

Individual Count

	male			female			total				
age	obs	n	%	obs	n	%	obs	n	%	Chi-sq.	p
<30	7	10	70.0	6	12	50.0	15	40	37.5	**14.389**	**0.001**
30-39	5	6	83.3	7	10	70.0	12	16	75.0		
40+	8	8	100.0	3	4	75.0	14	16	87.5	Likelihood ratio	p
indeterminate	1	2	50.0	0	1	0.0	1	5	20.0	**15.443**	**0.000**
TOTAL	21	26	80.8	16	27	59.3	42	77	54.6		

Tooth Count

	male			female			total				
age	obs	n	%	obs	n	%	obs	n	%	Chi-sq.	p
<30	30	207	14.5	13	209	6.2	46	565	8.1	**51.594**	**0.000**
30-39	21	115	18.3	31	201	15.4	52	316	16.5		
40+	50	181	27.6	12	51	23.5	69	254	27.2	Likelihood ratio	p
indeterminate	1	3	33.3	0	1	0.0	1	17	5.9	**49.573**	**0.000**
TOTAL	102	506	20.2	56	462	12.1	168	1152	14.6		

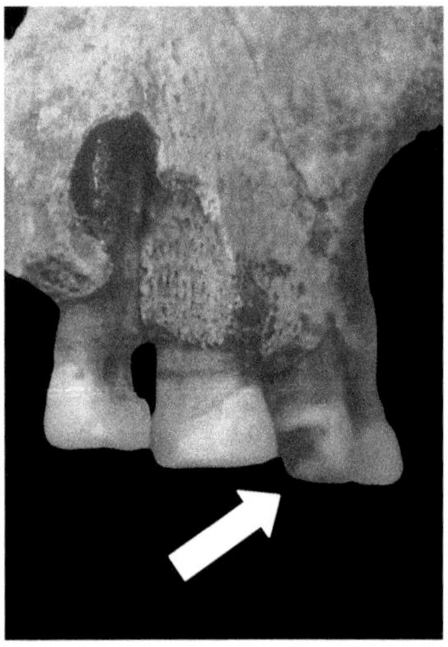

Figure 3.18. 80CCNM10a. Da But Period Male 30-39 years. Complex series of defects. Note also interproximal grooves to premolars and the task wear facet on the mesial-occlusal aspect of the first molar (arrow)

teeth exhibiting these facets. Task wear facets occurred significantly (p<0.01) more often in the maxillary dentition than the mandibular (Table 3.22). Moreover, significantly (p<0.01) more male maxillary teeth exhibit TWF than female maxillary teeth. If the facets are examined by tooth class it is apparent that, overall, they manifest in the anterior teeth with 61.5% of all observed facets occurring in the incisors and canines (Table 3.22).

When TWF are examined by way of tooth surface affected, it can be seen that the lingual aspect of the incisors is the single most commonly affected area (Table 3.23). The high frequency of occlusal TWF recorded for the canines, premolars and molars are all attributable to a single female individual aged 30-39 years.

Looking at the occurrence of these facets by age class is difficult by individual reporting method due to the relatively low number of individuals displaying the facet (Table 3.24). However, when examined by tooth count there is a clear increase in occurrence with increasing age, which is confirmed by the contingency table results.

Figure 3.19 is a particularly clear example of lingual

Figure 3.19. 77NNM14ka. Metal Period Female 50+ years. Deep task wear facets to the lingual face of both maxillary central incisors

facets from a single individual. A similar pattern of wear can be seen in Lukacs & Pastor (1988, Figure 12:396), the cause of which is suggested to be due to the use of the teeth as some sort of clamp (see also Turner & Machado 1983, Figure 2:127). The particular wear reported for the Vietnamese Metal period is believed to be TWF rather than deliberate mutilation, as their aesthetic, or otherwise, impact would have been unobservable.

It is possible that anterior tooth wear has masked further examples of TWF in the Metal period sample. While this cannot be tested for, an examination of the rate of anterior tooth wear compared to posterior tooth wear has the potential to shed light on this question. The results of this analysis are reported in the tooth wear section.

An additional somewhat unusual case of tooth mutilation occurred in a Metal period male aged 30-39 years. The maxillary RI2 and RC display a combination of labial enamel hypoplastic grooves and laterally directed filing. Adjacent teeth are not affected and neither are the corresponding mandibular teeth. The canine displays two clear wear or filing facets at the lower or coronal end of the crown on the labial face. The grooves superior to these facets are bands of enamel hypoplasia (Figure 3.20). The incisor displays three labial wear facets that appear to be associated with three hypoplastic grooves. It is not implausible that the filing was initiated in response to the presence of the enamel hypoplastic bands. It is also possible that these wear or filing facets may be incidental to using the teeth as tools, in which case they should be seen as TWF.

Tooth Chipping (CP)

The results from looking at teeth chipped pre-mortem are likely conservative in that only chips that appeared worn to some degree post-event were recorded. This was done to exclude the possibility of recording post-mortem chips, but would also have the effect of excluding chips formed shortly prior to death.

Examining chipping by jaw and tooth class for the Da But

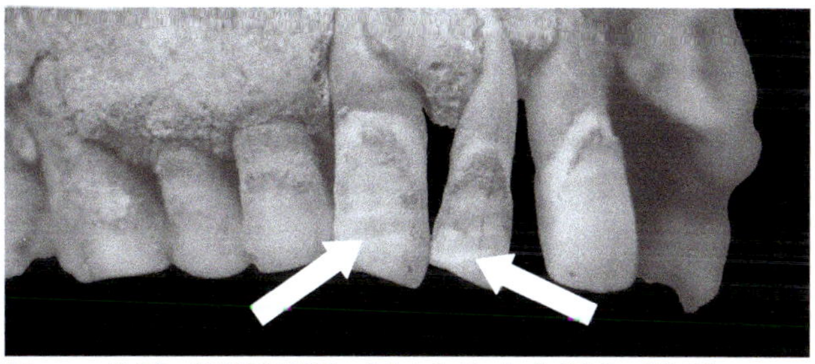

Figure 3.20. 78NNM6kb. Metal Period Male 30-39 years. Combination of enamel hypoplastic grooves and "filing" parallel and over the enamel defects. Arrows point to the filing

period, it can be seen that incisors are the least affected, only 0.9% of all assessable incisors, and molars the most affected at 5.4% of all assessable molars (Table 3.25). This pattern changes for each jaw and in fact no mandibular incisors or canines exhibited chipping. However, maxillary canines and premolars are the most frequently chipped teeth. The overall frequency of chipped maxillary and mandibular teeth is not significantly different.

In the Metal period, male teeth are significantly more affected than female teeth. With respect to tooth class, the most frequently chipped teeth are premolars, followed by molars (Table 3.25). While this pattern is the same for maxillary teeth, the premolar is the least affected mandibular tooth, with the molar class being the most affected. Maxillary teeth are significantly more affected than mandibular teeth overall and male maxillary teeth are significantly more affected than female. Female and male mandibular teeth are not significantly different in chipping frequency.

With regard to age class in the Da But period sample, while there appears to be a steady increase in affected individuals with increasing age, this is not statistically significant (Table 3.26). Male and female individuals display a similar pattern to each other with respect to age. Looking at the tooth count results indicates significant differences in the frequency of chipped teeth by age class. While few teeth in the youngest age class are chipped, most chipping is concentrated in the middle age group. This pattern is the same regardless of sex.

In the Metal period, chipped teeth are significantly unevenly distributed across age classes for both reporting methods. Again, it is the middle age class that exhibits the greatest frequency of individuals with chipped teeth and overall frequency of chipped teeth. This pattern is generally similar by sex as well, although the greatest frequency of male chipped teeth is in the oldest age class.

Tooth Ablation

No clear pattern of ante-mortem tooth loss that might indicate the practice of tooth ablation was discernible in the Metal period sample. However, the relatively high degree of anterior tooth loss in the Da But period sample needs to be examined in light of the possibility of deliberate removal. In all, 25 of the 30 incisors lost ante-mortem derive from nine people. These individuals were all 40+

Table 3.17. Prevalence of Tooth Rotation by Tooth Count

Da But Period	Male			Female			YA (unsexed)			Indeterminate			Total		
Max.	obs	n	%	obs	n	%	obs	n	%	obs	n	%	obs	n	%
I	0	17	0.0	1	18	5.6	0	6	0.0	0	6	0.0	1	47	2.1
C	3	21	14.3	1	14	7.1	0	5	0.0	0	6	0.0	4	46	8.7
P	3	52	5.8	1	35	2.9	0	13	0.0	0	13	0.0	4	113	3.5
M	0	96	0.0	3	65	4.6	0	24	0.0	0	21	0.0	3	206	1.5
Total	6	186	3.2	6	132	4.6	0	48	0.0	0	46	0.0	12	412	2.9
Man.	obs	n	%	obs	n	%	obs	n	%	obs	n	%	obs	n	%
I	0	24	0.0	1	17	5.9	0	4	0.0	0	2	0.0	1	47	2.1
C	7	24	29.2	2	18	11.1	0	2	0.0	1	2	50.0	10	46	21.7
P	2	63	3.2	0	48	0.0	0	4	0.0	0	6	0.0	2	121	1.7
M	1	127	0.8	1	104	1.0	0	20	0.0	0	13	0.0	2	264	0.8
Total	10	238	4.2	4	187	2.1	0	30	0.0	1	23	4.4	15	478	3.1
Total	obs	n	%	obs	n	%	obs	n	%	obs	n	%	obs	n	%
I	0	41	0.0	2	35	5.7	0	10	0.0	0	8	0.0	2	94	2.1
C	10	45	22.2	3	32	9.4	0	7	0.0	1	8	12.5	14	92	15.2
P	5	115	4.4	1	83	1.2	0	17	0.0	0	19	0.0	6	234	2.6
M	1	223	0.5	4	169	2.4	0	44	0.0	0	34	0.0	5	470	1.1
Total	16	424	3.8	10	319	3.1	0	78	0.0	1	69	1.5	27	890	3.0

Metal Period	Male			Female			YA (unsexed)			Indeterminate			Total		
Max.	obs	n	%	obs	n	%	obs	n	%	obs	n	%	obs	n	%
I	0	48	0.0	0	29	0.0	0	7	0.0	0	2	0.0	0	86	0.0
C	1	30	3.3	0	24	0.0	0	2	0.0	0	6	0.0	1	62	1.6
P	3	75	4.0	0	48	0.0	0	3	0.0	1	17	5.9	4	143	2.8
M	0	98	0.0	0	95	0.0	0	6	0.0	0	26	0.0	0	225	0.0
Total	4	251	1.6	0	196	0.0	0	18	0.0	1	51	2.0	5	516	1.0
Man.	obs	n	%	obs	n	%	obs	n	%	obs	n	%	obs	n	%
I	2	35	5.7	0	34	0.0	0	8	0.0	0	4	0.0	2	81	2.5
C	1	26	3.9	2	24	8.3	0	3	0.0	0	2	0.0	3	55	5.5
P	2	63	3.2	2	52	3.9	0	13	0.0	0	7	0.0	4	135	3.0
M	6	111	5.4	0	84	0.0	0	19	0.0	0	19	0.0	6	233	2.6
Total	11	235	4.7	4	194	2.1	0	43	0.0	0	32	0.0	15	504	3.0
Total	obs	n	%	obs	n	%	obs	n	%	obs	n	%	obs	n	%
I	2	83	2.4	0	63	0.0	0	15	0.0	0	6	0.0	2	167	1.2
C	2	56	3.6	2	48	4.2	0	5	0.0	0	8	0.0	4	117	3.4
P	5	138	3.6	2	100	2.0	0	16	0.0	1	24	4.2	8	278	2.9
M	6	209	2.9	0	179	0.0	0	25	0.0	0	45	0.0	6	458	1.3
Total	15	486	3.1	4	390	1.0	0	61	0.0	1	83	1.2	20	1020	2.0

max=maxilla, man=mandible

Intra-Sample Comparisons

Da But Period	Chi-sq.	p	Metal Period	Chi-sq.	p
male/female	0.22	0.64	male/female	**4.33**	**0.04**
maxilla/mandible	0.04	0.85	maxilla/mandible	**5.34**	**0.02**
male/female maxilla	0.37	0.54	male/female maxilla	na	
male/female mandible	1.40	0.24	male/female mandible	2.16	0.14

years old at death. Two patterns are evident, the first being that 28/30 of these teeth are mandibular incisors. The other pattern is that eight of the aforementioned individuals have a symmetrical pattern of the loss of two incisors or all four lower incisors. The ninth case displayed the loss of mandibular left I1 and left I2. Of these nine individuals, three had adjacent teeth also lost ante-mortem, while a total of four individuals had lost teeth other than incisors. In one case 11 teeth were lost ante-mortem including all the lower incisors.

Tooth Wear

Rates of Tooth Wear

In the following analysis M2 wear is regressed against M1 wear for the mandible and maxilla separately. Scatter plots were assessed for linearity in all cases. The same method

Table 3.18. Prevalence of Tooth Malalignment by Tooth Count

Da But Period

	Male			Female			YA (unsexed)			Indeterminate			Total		
Max.	obs	n	%	obs	n	%	obs	n	%	obs	n	%	obs	n	%
I	0	17	0.0	0	18	0.0	0	6	0.0	0	6	0.0	0	47	0.0
C	1	21	4.8	1	14	7.1	0	5	0.0	0	6	0.0	2	46	4.4
P	0	52	0.0	0	35	0.0	0	13	0.0	0	13	0.0	0	113	0.0
M	1	96	1.0	2	65	3.1	0	24	0.0	0	21	0.0	3	206	1.5
Total	2	186	1.1	3	132	2.3	0	48	0.0	0	46	0.0	5	412	1.2
Man.	obs	n	%	obs	n	%	obs	n	%	obs	n	%	obs	n	%
I	1	24	4.2	0	17	0.0	0	4	0.0	0	2	0.0	1	47	2.1
C	0	24	0.0	2	18	11.1	0	2	0.0	0	2	0.0	2	46	4.4
P	1	63	1.6	1	48	2.1	0	4	0.0	0	6	0.0	2	121	1.7
M	0	127	0.0	1	104	1.0	1	20	5.0	0	13	0.0	2	264	0.8
Total	2	238	0.8	4	187	2.1	1	30	3.3	0	23	0.0	7	478	1.5
Total	obs	n	%	obs	n	%	obs	n	%	obs	n	%	obs	n	%
I	1	41	2.4	0	35	0.0	0	10	0.0	0	8	0.0	1	94	1.1
C	1	45	2.2	3	32	9.4	0	7	0.0	0	8	0.0	4	92	4.4
P	1	115	0.9	1	83	1.2	0	17	0.0	0	19	0.0	2	234	0.9
M	1	223	0.5	3	169	1.8	1	44	2.3	0	34	0.0	5	470	1.1
Total	4	424	0.9	7	319	2.2	1	78	1.3	0	69	0.0	12	890	1.4

Metal Period

	Male			Female			YA (unsexed)			Indeterminate			Total		
Max.	obs	n	%	obs	n	%	obs	n	%	obs	n	%	obs	n	%
I	3	48	6.3	0	29	0.0	0	7	0.0	0	2	0.0	3	86	3.5
C	0	30	0.0	1	24	4.2	0	2	0.0	0	6	0.0	1	62	1.6
P	5	75	6.7	0	48	0.0	0	3	0.0	0	17	0.0	5	143	3.5
M	0	98	0.0	2	95	2.1	0	6	0.0	1	26	3.9	3	225	1.3
Total	8	251	3.2	3	196	1.5	0	18	0.0	1	51	2.0	12	516	2.3
Man.	obs	n	%	obs	n	%	obs	n	%	obs	n	%	obs	n	%
I	2	35	5.7	3	34	8.8	0	8	0.0	0	4	0.0	5	81	6.2
C	1	26	3.9	3	24	12.5	0	3	0.0	0	2	0.0	4	55	7.3
P	0	63	0.0	1	52	1.9	0	13	0.0	0	7	0.0	1	135	0.7
M	0	111	0.0	3	84	3.6	0	19	0.0	0	19	0.0	3	233	1.3
Total	3	235	1.3	10	194	5.2	0	43	0.0	0	32	0.0	13	504	2.6
	Male			Female			SA (unsexed)			Indeterminate			Total		
Total	obs	n	%	obs	n	%	obs	n	%	obs	n	%	obs	n	%
I	5	83	6.0	3	63	4.8	0	15	0.0	0	6	0.0	8	167	4.8
C	1	56	1.8	4	48	8.3	0	5	0.0	0	8	0.0	5	117	4.3
P	5	138	3.6	1	100	1.0	0	16	0.0	0	24	0.0	6	278	2.2
M	0	209	0.0	5	179	2.8	0	25	0.0	1	45	2.2	6	458	1.3
Total	11	486	2.3	13	390	3.3	0	61	0.0	1	83	1.2	25	1020	2.5

max=maxilla, man=mandible

Intra-Sample Comparisons

Da But Period	Chi-sq.	p	Metal Period	Chi-sq.	p
male/female	1.95	0.16	male/female	0.93	0.34
maxilla/mandible	0.10	0.75	maxilla/mandible	0.07	0.79
male/female maxilla	0.72	0.40	male/female maxilla	1.26	0.26
male/female mandible	1.27	0.26	male/female mandible	**5.44**	**0.02**

of residual analysis as performed in the age prediction section in Chapter 2 was carried out for each regression model. An ANOVA of regression was also carried out for each model and the assumptions of this test were also assessed. In particular, this includes tests of normality assessed both graphically and by way of the Kolmogorov-Smirnov test for goodness of fit. Further, the assumption of equality of variances was tested using Bartlett's test rather than Cochran's C, due to differences in sample size. Any violations of these tests are specifically noted. Assuming that there are no violations of the various assumptions and that the relationships are linear, intra and inter population comparisons are made of the *y* intercepts. Such comparisons provide information about the significance

Table 3.19. M3 Agenesis by Age

Da But Period

Individual Count

	male			female			total			Intra-Sample Comparisons males:females	
age	obs	n	%	obs	n	%	obs	n	%	Chi-sq.	p
20-29	0	5	0.0	1	4	25.0	2	14	14.3	0.070	0.797
30-39	2	12	16.7	1	5	20.0	3	18	16.7		
40+	1	13	7.7	0	16	0.0	1	30	3.3		
indeterminate	0	0	0.0	0	0	0.0	0	0	0.0		
TOTAL	3	30	10.0	2	25	8.0	6	62	9.7		

Tooth Count

	male			female			total				
age	obs	n	%	obs	n	%	obs	n	%	Chi-sq.	p
20-29	0	10	0.0	4	11	36.4	6	31	19.4	0.110	0.745
30-39	4	35	11.4	2	15	13.3	6	51	11.8		
40+	2	34	5.9	0	40	0.0	2	76	2.6		
indeterminate	0	0	0.0	0	0	0.0	0	0	0.0		
TOTAL	6	79	7.6	6	66	9.1	14	158	8.9		

Metal Period

Individual Count

	male			female			total				
age	obs	n	%	obs	n	%	obs	n	%	Chi-sq.	p
20-29	1	9	11.1	1	6	16.7	3	21	14.3	0.000	0.945
30-39	1	6	16.7	1	10	10.0	2	16	12.5		
40+	1	7	14.3	2	4	50.0	5	14	35.7		
indeterminate	1	1	100.0	0	2	0.0	0	3	0.0		
TOTAL	4	23	17.4	4	22	18.2	10	54	18.5		

Tooth Count

	male			female			total				
age	obs	n	%	obs	n	%	obs	n	%	Chi-sq.	p
20-29	3	30	10.0	1	18	5.6	5	60	8.3	0.030	0.860
30-39	4	23	17.4	1	29	3.4	5	52	9.6		
40+	3	25	12.0	5	11	45.5	9	40	22.5		
indeterminate	0	1	0.0	0	2	0.0	0	3	0.0		
TOTAL	10	79	12.7	7	60	11.7	19	155	12.3		

[1]a case <15 years old but with a permanent tooth (teeth)

note: n= assessable M3 tooth locations

inter-sample tests

Da But/ Metal tooth count		Da But/ Metal individual count	
Chi-sq.	p	Chi-sq.	p
1.90	0.168	0.96	0.328

of any differences in childhood wear between any given two samples under investigation. These comparisons are relatively straightforward by way of an analysis of covariance, ANCOVA, which is also advantageous in that it compares, in this case, M1 wear between samples while taking into consideration the effect of M2 wear in each sample. An ANCOVA assumes parallel slopes for each sample and this was assessed before proceeding in each case.

Table 3.27 presents information of the relationship between first and second molars of both the maxilla and mandible for each sample. Results for the entire assessable sample and also for samples that only included teeth with Scott wear scores less than 32 (ie. less worn) are included. As previously noted, the samples excluding severely worn teeth are to be preferred. In all cases the linear relationships are quite good with high Pearson correlation coefficients (0.88-0.92) and adjusted r^2 values. The regressions are significant in all cases as well. In all cases the slope approximates unity and unity falls within the 95% confidence intervals (CI) of the slope. Similar intercepts are seen when comparing Da But period and Metal period maxillary and mandibular teeth for the entire assessable sample. Looking at the sample, excluding extreme wear, it can be seen that both Metal period mandibular and maxillary M1 wear is greater than that for the Da But period. Nonetheless, the ANCOVA results testing the significance of these intercepts indicates that there are no significant inter sample differences in childhood wear

Table 3.20. Mandibular Robusticity

	n	mean	min	max	sd	CV (%)
Da But Period						
Females						
htm1	23	28.43	22.70	34.60	2.66	7.68
brm1	23	13.69	11.70	16.70	1.24	7.45
size	23	48.67	37.38	68.44	7.34	10.73
area	23	304.93	237.12	371.24	34.13	9.19
Males						
htm1	25	29.21	25.00	35.90	2.37	6.61
brm1	25	13.47	10.90	15.30	1.26	8.22
size	25	46.34	32.87	54.10	4.89	9.04
area	25	309.39	214.02	380.10	40.00	10.52
Combined						
htm1	48	28.83	22.70	35.90	2.52	7.01
brm1	48	13.58	10.90	16.70	1.24	7.44
size	48	47.46	32.87	68.44	6.23	9.11
area	48	307.25	214.02	380.10	36.98	9.73
Metal Period						
Females						
htm1	20	27.08	22.00	32.80	2.98	9.09
brm1	20	14.79	12.90	18.10	1.46	8.05
size	20	55.18	43.53	74.18	7.79	10.50
area	20	315.06	248.61	417.33	50.40	12.08
Males						
htm1	22	29.59	23.50	33.30	2.15	64.44
brm1	22	15.07	11.50	21.10	2.09	9.89
size	22	51.35	40.35	89.79	9.93	11.06
area	22	349.52	257.41	468.93	49.77	10.61
Combined						
htm1	42	28.39	22.00	33.30	2.84	8.53
brm1	42	14.94	11.50	21.10	1.80	8.52
size	42	53.17	40.35	89.79	9.07	10.11
area	42	333.11	248.61	468.93	52.43	11.18

Da But Period males:females

t-statistic	p
-1.08	0.29
0.59	0.56
1.31	0.20
-0.41	0.68

Metal Period males:females

t-statistic	p
-3.15	**0.00**
-0.50	0.62
1.38	0.18
-2.23	**0.03**

Da But Period: Metal period

t-statistic	p
0.78	0.44
-4.22	**0.00**
-3.52	**0.00**
-2.73	**0.01**

htm1= maximum mandibular height at M1
brm1=maximum mandibular breadth at M1
size=htm1/brm1 x 100
area=htm1 x brm1 x pi/4

(Table 3.28). Figure 3.21a graphically displays M1 wear (y intercept) and its 95% CI by jaw for each sample.

Intra sample sex comparisons are detailed in Tables 3.29-3.30. For the Da But period the slopes of both the complete samples and those excluding severely worn teeth include unity in their 95% CI (Table 3.29). Male maxillary teeth show greater wear than female, while female mandibular teeth are more worn than male. Nonetheless, these differences are not significant (Table 3.30). These results are displayed graphically in Figure 3.21b. The small differences in M1 wear are evident between the jaws and sexes.

For the Metal period only maxillary teeth can be examined (Table 3.31). The reduction in sample size caused by intra sample comparisons by sex has compromised the linearity of the female molar relationship. The resultant intercept (-4.937) is clearly at odds with expectations. An examination of the standardised residuals against the independent variable indicated a degree of variance heteroscedacity. Rather than apply a transformation to this rather small sample it was decided to simply forgo a comparison with the female mandibular sample. As for the maxilla, female wear is seen to be higher than male wear but not to a significant degree (Table 3.32). Figure 3.21c displays relative male and female wear for maxillary M1.

Anterior-Posterior Tooth Ratio of Tooth Wear

The rate of tooth wear between anterior and posterior

Table 3.21. Molar Tooth Size (Mesio-Distal and Bucco-Lingual Diameters)

	Mandibular M1			Mandibular M2		
MD	Da But	Metal	Student's t	Da But	Metal	Student's t
n[1]	30	46	**-3.015**	46	46	**-5.058**
mean	11.64	11.24		11.34	10.64	
SD	0.59	0.56	p	0.67	0.65	p
min	10.20	10.10	**0.004**	10.00	9.40	**0.000**
max	12.80	12.20		12.90	12.20	
BL						
n[1]	30	46	**-4.054**	46	46	**-4.026**
mean	11.16	10.65		10.77	10.30	
SD	0.52	0.53	p	0.57	0.55	p
min	10.2	9.80	**0.000**	9.30	9.00	**0.000**
max	12.4	12.00		12.10	11.40	

	Maxillary M1			Maxillary M2		
MD	Da But	Metal	Student's t	Da But	Metal	Student's t
n[1]	32	47	**-3.540**	40	49	-1.202
mean	10.78	10.30		9.91	9.77	
SD	0.57	0.61	p	0.74	0.64	p
min	9.60	8.90	**0.001**	8.40	8.30	0.232
max	12.10	11.90		11.50	11.90	
BL						
n[1]	32	47	**-3.588**	40	49	**-4.240**
mean	11.98	11.09		11.71	11.13	
SD	0.53	0.52	p	0.8	0.62	p
min	11.10	9.90	**0.000**	10.3	10.00	**0.000**
max	13.30	12.20		13.2	12.40	

[1] only teeth with a Scott wear score of <=28 used in analysis

permanent teeth is examined here. A sample of 20 Da But period individuals and 32 Metal period individuals were available that had assessable first maxillary molars and first maxillary incisors. The available sample of mandibular incisors was too small for statistical treatment and was therefore not assessed. The ratio of incisor wear to mean molar wear (molar wear score/4) was used as the index for assessing relative anterior to posterior wear. The higher the ratio the greater incisor wear relative to molar wear. The results (Table 3.33) indicated that there were no intra sample sex differences in the ratio of anterior to posterior wear (I1 wear/mean M1 wear). However, when comparing the two samples it was found the Metal period ratio of incisor to molar wear was significantly greater ($p<0.05$). The Metal period displays greater anterior tooth wear relative to the posterior teeth than is the case for the Da But period.

Prevalence of Dental Disease Across the Mouth

Having examined the various oral traits individually it is useful to summarise a number of these traits together by examining their relative distribution across the mouth. This is best approached by way of the tooth count reporting method. While the majority of oral traits have been individually assessed with regard to occurrence across the mouth there is an advantage in graphically summarising several traits together. Observing the patterning of several traits simultaneously, by tooth class, reveals differences in disease not readily apparent when concentrating on individual trait patterning.

With regard to alveolar defects of pulpal origin, it can be seen (Figure 3.22a) that in the Da But period this condition is concentrated on the posterior teeth, the molars in particular. In the Metal period, on the other hand, alveolar defects are more evenly distributed across the mouth (Figure 3.22b).

These results are consistent with the observation that pulp chamber exposure is also concentrated in the anterior of the mouth in the Metal period, while the reverse is true for the for the Da But period. Molar pulp exposure in the Metal period accounts for only 15.4% of all teeth so afflicted, while 88.2% of teeth with pulp exposure are molars in the Da But period sample.

The patterning of carious lesions across the mouth for both samples is quite similar, with the vast majority of caries concentrated in the molars. Ante-mortem tooth loss, on the other hand, is concentrated at the anterior of the mouth in the Da But period sample with the incisors recording the highest occurrence. The reverse trend is true for the Metal period, with the majority of cases affecting

Table 3.22. Prevalence of Task Wear Facets by Tooth Count

Metal Period	Male			Female			YA (unsexed)			Indeterminate			Total		
Max.	obs	n	%	obs	n	%	obs	n	%	obs	n	%	obs	n	%
I	8	49	16.3	6	40	15.0	0	13	0.0	0	6	0.0	14	108	13.0
C	2	30	6.7	2	29	6.9	0	3	0.0	0	9	0.0	4	71	5.6
P	4	75	5.3	1	57	1.8	0	4	0.0	0	22	0.0	5	158	3.2
M	5	101	5.0	0	110	0.0	0	11	0.0	0	30	0.0	5	252	2.0
Total	19	255	7.5	9	236	3.8	0	31	0.0	0	67	0.0	28	589	4.8
Man.	obs	n	%	obs	n	%	obs	n	%	obs	n	%	obs	n	%
I	2	43	4.7	0	37	0.0	0	12	0.0	0	4	0.0	2	96	2.1
C	2	29	6.9	2	28	7.1	0	3	0.0	0	2	0.0	4	62	6.5
P	3	67	4.5	0	62	0.0	0	13	0.0	0	9	0.0	3	151	2.0
M	2	112	1.8	0	99	0.0	0	21	0.0	0	22	0.0	2	254	0.8
Total	9	251	3.6	2	226	0.9	0	49	0.0	0	37	0.0	11	563	2.0
Total	obs	n	%	obs	n	%	obs	n	%	obs	n	%	obs	n	%
I	10	92	10.9	6	77	7.8	0	25	0.0	0	10	0.0	16	204	7.8
C	4	59	6.8	4	57	7.0	0	6	0.0	0	11	0.0	8	133	6.0
P	7	142	4.9	1	119	0.8	0	17	0.0	0	31	0.0	8	309	2.6
M	7	213	3.3	0	209	0.0	0	32	0.0	0	52	0.0	7	506	1.4
Total	28	506	5.5	11	462	2.4	0	80	0.0	0	104	0.0	39	1152	3.4

max=maxilla, man=mandible

Intra-Sample Comparison

Metal Period	Chi-sq.	p	p*
male/female	**6.21**	**0.01**	
maxilla/mandible	**6.9**	**0.01**	
male/female maxilla	3.02	0.08	
male/female mandible	3.9	0.05	0.07

p* Fisher exact 2-tailed

the posterior teeth. Although there is a noticeable peak in incisor ante-mortem tooth loss. For both samples there is a similar pattern with regard to tooth chipping, in that there is generally an increase in the incidence from the front to the back of the mouth.

The Vietnamese Samples Compared with other Southeast Asian Samples

While there are no other studies of the oral health of ancient Vietnamese remains, there are a number of recent studies of ancient dental assemblages from Thailand. The results from seven of these sites, including Noen U-Loke (Nelson 1999), Ban Na Di (Wiriyaromp 1984; Domett 1999), Non Nok Ta and Ban Chiang (Douglas 1996), Khok Phanom Di (Tayles 1992), Nong Nor (Tayles *et al.* 1998), and Ban Lum Khao (Domett 1999), are summarised for caries, ante mortem tooth loss, alveolar defects of pulpal origin and calculus (Table 3.34). The assessment of subsistence economy at each site is based on each author's assessment of the archaeological and environmental evidence. The use of the term 'mixed' refers to the inference that both agriculture, to a generally unknown degree, and foraging contributed to the subsistence economies at these sites. No attempt is made at testing the significance of observed differences among these assemblages due to differences in the age and sex structure of each sample. Further, these figures have been collated from the work of a number of authors and differences in the methods of data collection/recording cannot be controlled for. These comparisons are only meant as a gross analysis of these Southeast Asian dental series with respect to broad time and generalised subsistence considerations.

The ancient Southeast Asian oral response to subsistence/environment can be seen to be quite varied, even as assessed with the limited number of variables addressed in Table 3.34. This variation would be expected given the diversity and complexity of tropical regions themselves (Hutterer 1983). Further complications with generalising a Southeast Asian response centre on specifying in detail the subsistence strategies employed. The relative contributions of the components of the mixed subsistence economies, which are in the majority, are yet to be clearly elucidated for these sites from Thailand and Vietnam. Until the Southeast Asian Oral Profile database is substantially augmented, generalisations regarding Southeast Asian oral responses to subsistence/environment are of equivocal value. Assessing oral profiles against this limited data base thus also needs to proceed with some circumspection. Notwithstanding such cautions, one generalisation that seems fairly clear is that the frequency of caries, ante-mortem tooth loss and alveolar defects of pulpal origin is relatively low for all sites examined here. In only one instance does one of these oral health measures exceed 10%, and that is caries in agriculturally oriented Khok Phanom Di. However, the

Table 3.23. Task Wear Facets by Tooth Face Affected (Tooth Count)

Metal Period

	Labial				Lingual				Occlusal			
	I	C	P	M	I	C	P	M	I	C	P	M
obs	4	1	0	0	12	0	0	0	0	7	8	7
n	39	39	39	39	39	39	39	39	39	39	39	39
%	10.3	2.6	0.0	0.0	30.8	0.0	0.0	0.0	0.0	17.9	20.5	17.9

other two sites identified as agriculturally oriented, Ban Na Di and Noen U-Loke, display quite low frequencies of caries. It is possible that Southeast Asian oral responses to agriculture are somewhat different to those observed in other regions of the globe. Alternatively, the role of agriculture in the economies of these sites sampled in Southeast Asia to date is in need of revision.

Discussion

Ante-Mortem Tooth Loss (AMTL)

AMTL patterns are not obviously consistent with the patterning of alveolar defects, pulp exposure and task wear facets in the Metal period. However, the slight peaking of incisor AMTL at this time is consistent with the general pattern described for the Metal period proposed to be due to using the anterior teeth as tools. Also consistent with this hypothesis is the observation that AMTL is more common in the younger age classes in the Metal period compared to the Da But period, where the vast majority of AMTL occurs in individuals 40+ years of age.

With regard to the Da But period, the concentration of incisor AMTL is somewhat enigmatic. Given that the high frequency of incisor AMTL is unlikely to be due to attrition or caries induced pulp exposure and alveolar defects, which are concentrated in the posterior teeth, then some other causal factor must be sought. Deliberate tooth ablation could account for this pattern. Tooth ablation has been reported in prehistoric Thailand (Tayles 1997), but the clear patterning seen in that study was not apparent in the Da But period sample. The case for tooth ablation is explored below.

Alveolar Defects of Pulpal Origin (ADP)

In the Metal period ADP is greater by tooth count, affects a greater proportion of anterior teeth and more of the younger individuals than is the case in the Da But period sample. This is also consistent with the view forwarded for a greater use of the anterior teeth in non-masticatory tasks in the Metal period.

Dental Caries (CAR)

Lukacs (1995) has advocated the use of a caries correction factor that essentially assigns a percentage of those teeth lost ante-mortem to the carious tooth total. The percentage thus assigned is based upon the proportion of teeth exhibiting pulp exposure induced by caries. This correction factor has merit where a lot of teeth are lost ante-mortem and caries induced pulp exposure is the most common aetiology. Neither of these conditions are met in the Vietnamese samples and in fact the most common aetiology, by far, for PEXP is tooth wear. The caries correction factor is not employed in this analysis.

The low prevalence of caries reported here would at face value indicate a low fermentable carbohydrate component in the diet of each period. Such a conclusion is somewhat simplistic given the range and complexity of factors influencing caries prevalence previously discussed. One important factor that cannot be controlled for is that critically important geochemical, not the least of which includes ground water fluoride concentrations, information is not available for the study region. However, other factors

Table 3.24. Task Wear Facets by Age (3 x age classes)

Metal Period

Individual Count

	male			female			total			tests for TWF homogeneity by age class	
age	obs	n	%	obs	n	%	obs	n	%	Chi-sq.	p
<30	1	10	10.0	1	12	8.3	2	40	5.0	**4.669**	**0.097**
30-39	3	6	50.0	1	10	10.0	4	16	25.0		
40+	1	8	12.5	1	4	25.0	2	16	12.5	Likelihood	p
indeterminate	0	2	0.0	0	1	0.0	0	5	0.0	ratio	
TOTAL	5	26	19.2	3	27	11.1	8	77	10.4	4.299	0.117

Tooth Count

	male			female			total				
age	obs	n	%	obs	n	%	obs	n	%	Chi-sq.	p
<30	5	207	2.4	4	209	1.9	9	565	1.6	**20.932**	**0.000**
30-39	5	115	4.4	5	201	2.5	10	316	3.2		
40+	18	181	9.9	2	51	3.9	20	254	7.9	Likelihood	p
indeterminate	0	3	0.0	0	1	0.0	0	17	0.0	ratio	
TOTAL	28	506	5.5	11	462	2.4	39	1152	3.4	**18.411**	**0.000**

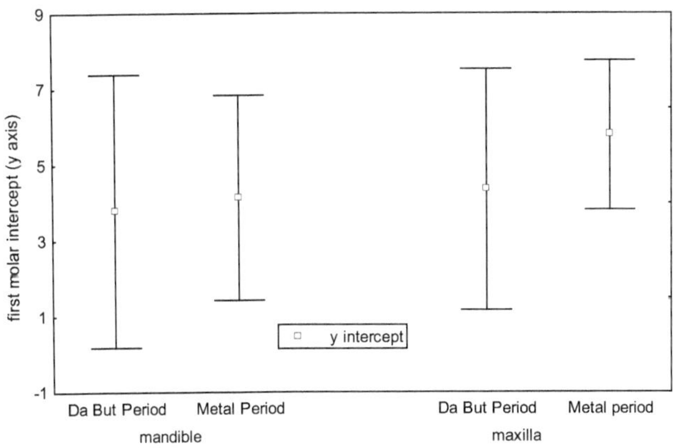

Figure 3.21a. Molar Rates of Wear: y Axis Intercepts and 95% C.I. Da But Period and Metal Period Inter-Sample Comparison

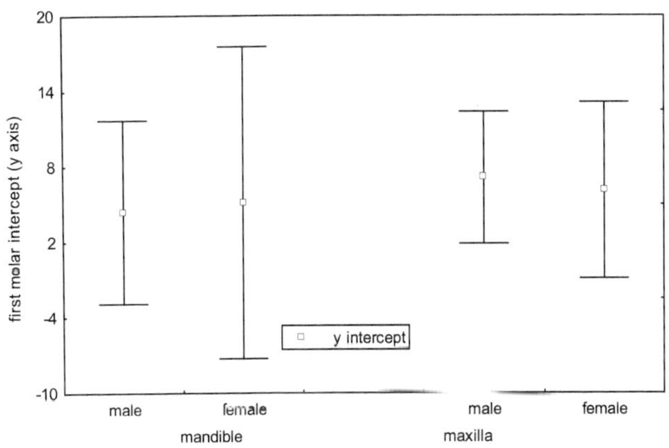

Figure 3.21b. Molar Rates of Wear: y Axis Intercepts and 95% C.I. Da But Period Intra-Sample Comparison

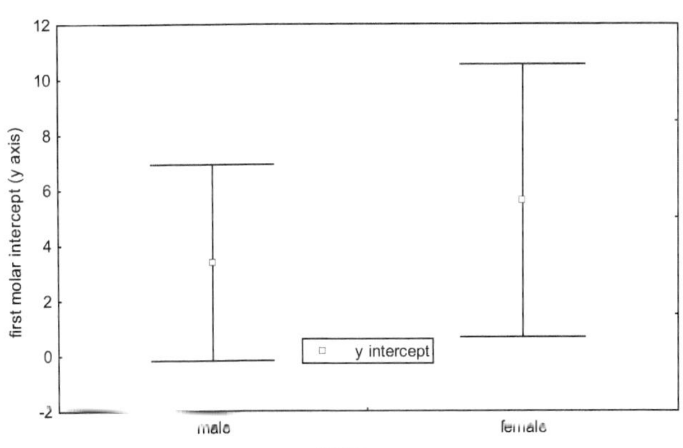

Figure 3.21c. Molar Rates of Wear: y Axis Intercepts and 95% C.I. Metal Period Intra-Sample Comparison

potentially influencing caries prevalence can be explored.

Given the archaeological evidence for a potentially significant rice component in the diet of the Metal period, the caries prevalence for this period might have been expected to have been greater than reported here, notwithstanding other complicating factors. However, rice consumption, as opposed to wheat, does not appear to be correlated with caries prevalence per se (Sreebny 1983), although high levels of rice consumption may be correlated with relatively higher caries prevalence. Krasse (1985) also points out that the caries-inducing potential of rice is very low. It would appear, then, that any marked

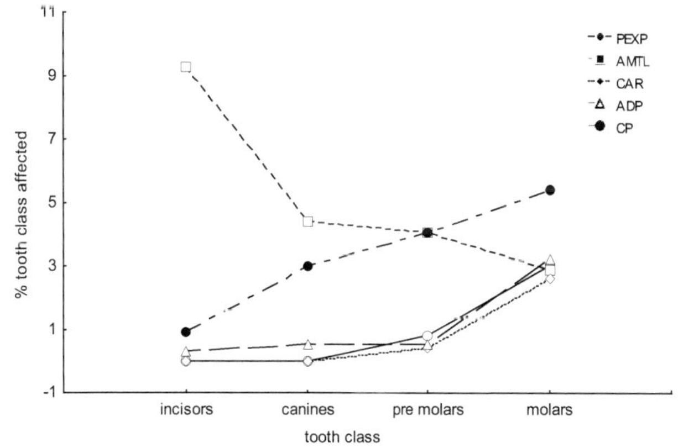

Figure 3.22a. Da But Period: Distribution of Traits Across the Mouth

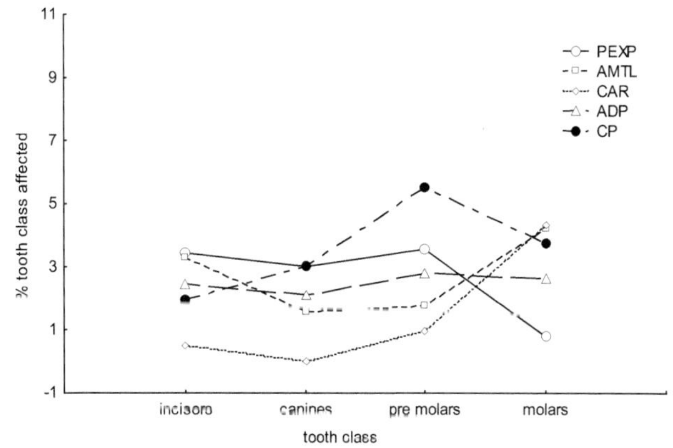

Figure 3.22b. Metal Period: Distribution of Traits Across the Mouth

increased reliance on rice may not leave the same general high caries rate signature that agricultural intensification has in general left in Europe and North America.

A further factor impacting on the nature of the diet in these samples is the very low prevalence of carious lesions affecting the dentin, as opposed to the enamel of the teeth. Earlier in this chapter, a model was developed whereby, regardless of total carious lesion frequency per se, it would be expected that a greater contribution to the total lesion count would come from dentin lesions if grains were a significant component of the diet. That this is not the case in either time period, in addition to the overall low caries prevalence, supports the view that rice or grains did not contribute to the diet of these people in any significant sense. That some quantity of rice, in the Metal period at least, contributed to the diet is not inconsistent with these results given the apparent low cariogenic status of this particular grain. These findings support the proposal developed in Chapter 1 that the historically recorded intensification of rice agriculture, shortly after increased Chinese colonising efforts in the region around AD 100, was essentially for non-subsistence reasons.

Archaeological evidence indicates a marine dietary component in both periods, although the relative contribution of marine resources to the diet is unknown. The cariostatic properties of such a diet have been

Table 3.25. Prevalence of Chipping by Tooth Count

Da But Period	Male			Female			YA (unsexed)			Indeterminate			Total		
Max.	obs	n	%	obs	n	%	obs	n	%	obs	n	%	obs	n	%
I	0	21	0.0	0	21	0.0	1	7	14.3	0	9	0.0	1	58	1.7
C	1	22	4.5	2	15	13.3	0	7	0.0	0	8	0.0	3	52	5.8
P	4	54	7.4	3	37	8.1	0	13	0.0	0	14	0.0	7	118	5.9
M	6	99	6.1	5	71	7.0	0	25	0.0	0	23	0.0	11	218	5.0
Total	11	196	5.6	10	144	6.9	1	52	1.9	0	54	0.0	22	446	4.9
Man.	obs	n	%	obs	n	%	obs	n	%	obs	n	%	obs	n	%
I	0	25	0.0	0	18	0.0	0	4	0.0	0	2	0.0	0	49	0.0
C	0	26	0.0	0	18	0.0	0	2	0.0	0	2	0.0	0	48	0.0
P	3	68	4.4	0	48	0.0	0	5	0.0	0	7	0.0	3	128	2.3
M	6	130	4.6	9	105	8.6	1	26	3.8	0	19	0.0	16	280	5.7
Total	9	249	3.6	9	189	4.8	1	37	2.7	0	30	0.0	19	505	3.8
Total	obs	n	%	obs	n	%	obs	n	%	obs	n	%	obs	n	%
I	0	46	0.0	0	39	0.0	1	11	9.1	0	11	0.0	1	107	0.9
C	1	48	2.1	2	33	6.1	0	9	0.0	0	10	0.0	3	100	3.0
P	7	122	5.7	3	85	3.5	0	18	0.0	0	21	0.0	10	246	4.1
M	12	229	5.2	14	176	8.0	1	51	2.0	0	42	0.0	27	498	5.4
Total	20	445	4.5	19	333	5.7	2	89	2.2	0	84	0.0	41	951	4.3

Metal Period	Male			Female			YA (unsexed)			Indeterminate			Total		
Max.	obs	n	%	obs	n	%	obs	n	%	obs	n	%	obs	n	%
I	3	49	6.1	0	40	0.0	0	13	0.0	0	6	0.0	3	108	2.8
C	2	30	6.7	0	29	0.0	0	3	0.0	0	9	0.0	2	71	2.8
P	14	75	18.7	2	57	0.5	0	4	0.0	0	22	0.0	16	158	10.1
M	7	101	6.9	0	110	0.0	0	11	0.0	1	30	3.3	8	252	3.2
Total	26	255	10.2	2	236	0.8	0	31	0.0	1	67	1.5	29	589	4.9
Man.	obs	n	%	obs	n	%	obs	n	%	obs	n	%	obs	n	%
I	0	43	0.0	1	37	2.7	0	12	0.0	0	4	0.0	1	96	1.0
C	2	29	6.9	0	28	0.0	0	3	0.0	0	2	0.0	2	62	3.2
P	1	67	1.5	0	62	0.0	0	13	0.0	0	9	0.0	1	151	0.7
M	8	112	7.1	2	99	2.0	1	21	4.8	0	22	0.0	11	254	4.3
Total	11	251	4.4	3	226	1.3	1	49	2.0	0	37	0.0	15	563	2.7
Total	obs	n	%	obs	n	%	obs	n	%	obs	n	%	obs	n	%
I	3	92	3.3	1	77	1.3	0	25	0.0	0	10	0.0	4	204	2.0
C	4	59	6.8	0	57	0.0	0	6	0.0	0	11	0.0	4	133	3.0
P	15	142	10.6	2	119	1.7	0	17	0.0	0	31	0.0	17	309	5.5
M	15	213	7.0	2	209	1.0	1	32	3.1	1	52	1.9	19	506	3.8
Total	37	506	7.3	5	462	1.1	1	80	1.3	1	104	1.0	44	1152	3.8

max=maxilla, man=mandible

Intra-Sample Comparisons

Da But Period	Chi-sq.	p	Metal Period	Chi-sq.	p
male/female	0.54	0.46	male/female	**22.71**	**0.00**
maxilla/mandible	0.79	0.37	maxilla/mandible	**4.00**	**0.05**
male/female maxilla	0.23	0.63	male/female maxilla	**19.92**	**0.00**
male/female mandible	0.34	0.56	male/female mandible	3.96	0.05

discussed. The Metal period is further distinctive in the large proportion of what is believed to be betel nut stained teeth. If betel nut use and/or associated factors, discussed previously, did significantly reduce caries disease in the Metal period population, this has implications for diet inference. However, the evidence from this study indicates quite the reverse. If anything, there appears to be a correlation between the use of betel nut and increased risk of caries. If the correlation is not simply fortuitous, a number of factors could be responsible: limited contact with betel nut residues if the teeth were purposely stained rather than stained through regular chewing of betel quids; betel nut does not in fact have cariostatic properties but rather cariogenic properties; possibly linked to this is evidence that betel nut chewing leads to reduced appetite (Schamschula *et*

Table 3.26. Tooth Chipping by Age (3 x age classes)

Da But Period

Individual Count

	male			female			total			tests for CP homogeneity by age class	
age	obs	n	%	obs	n	%	obs	n	%	Chi-sq.	p
<30	1	6	16.7	1	5	20.0	4	25	16.0	1.841	0.398
30-39	4	13	30.8	5	7	71.4	6	21	28.6		
40+	5	15	33.3	2	16	12.5	10	32	32.3	Likelihood	p
indeterminate	0	2	0.0	0	0	0.0	0	2	0.0	ratio	
TOTAL	10	36	27.8	8	28	28.6	20	80	25.0	1.946	0.378

Tooth Count

	male			female			total				
age	obs	n	%	obs	n	%	obs	n	%	Chi-sq.	p
<30	1	73	1.4	1	60	1.7	4	289	1.4	**11.500**	**0.003**
30-39	11	187	5.9	13	71	18.3	19	262	7.3		
40+	8	181	4.4	5	202	2.5	18	396	4.6	Likelihood	p
indeterminate	0	4	0.0	0	0	0.0	0	4	0.0	ratio	
TOTAL	20	445	4.5	19	333	5.7	41	951	4.3	**12.728**	**0.002**

Metal Period

Individual Count

	male			female			total				
	obs	n	%	obs	n	%	obs	n	%	Chi-sq.	p
<30	3	10	30.0	0	12	0.0	5	40	12.5	**9.442**	**0.009**
30-39	5	6	83.3	3	10	30.0	8	16	50.0		
40+	3	8	37.5	0	4	0.0	3	16	18.8	Likelihood	p
indeterminate	0	2	0.0	0	1	0.0	0	5	0.0	ratio	
TOTAL	11	26	42.3	3	27	11.1	16	77	20.8	**8.513**	**0.014**

Tooth Count

	male			female			total				
	obs	n	%	obs	n	%	obs	n	%	Chi-sq.	p
<30	9	207	4.4	0	209	0.0	11	565	2.0	**12.630**	**0.002**
30-39	16	115	5.2	5	201	2.5	21	316	6.7		
40+	12	181	6.6	0	51	0.0	12	254	4.7	Likelihood	p
indeterminate	0	3	0.0	0	1	0.0	0	17	0.0	ratio	
TOTAL	37	506	7.3	5	462	1.1	44	1152	3.8	**12.722**	**0.002**

al. 1978), which would suggest reduced health and hence a reduced ability to fight caries disease; staining is not due to betel nut but rather another plant extract with cariogenic properties. The important point however, is that there is no obvious or discernible caries reducing factor operating in the Metal period.

There are differences in the manifestation of carious lesions by age for each period. The frequency of caries in the Da But period sample increases with age while they are most prevalent in earlier age categories in the Metal period. This pattern is also consistent by sex within each sample. The difference in lesion distribution by age could be diet related or simply a change in the behaviour associated with eating the same foods. A change to more frequent snacking in the Metal period could explain the increase in lesions in younger age classes. Alternatively, a change in diet toward more cariogenic, high carbohydrate, foods in the Metal period could also account for this pattern. Another more prosaic reason could account for the patterning of caries by age in the Metal period. Schamschula *et al.* (1978) suggest that this pattern is not uncommon in studies in Papua New Guinea. They argue that the pattern is perhaps a "function of the slow progress in carious lesions combined with a high rate of attrition and complicated in older age groups by tooth loss through periodontal disease" (Schamschula *et al.* 1978:31). The low overall carious lesion count mitigates against the discovery of unequivocal reasons in this study.

In this context is also worth noting that there is a significant sex difference in caries for the Metal period. Larsen *et al.* (1991) point out that numerous studies have reported a greater prevalence of female caries, but also note that this is not a universally observed pattern. As with the explanation for more caries in the younger age classes, dietary differences and or snacking are commonly suggested as explanations for such differences (Larsen *et al.* 1991). Walker & Hewlett (1990), in an ethnographic study, suggested such a reason for the difference in male and female pygmy caries prevalence. This particular example is important in highlighting that differences in caries prevalence by sex can exist in otherwise non-hierarchical or egalitarian societies. Other possible sex

Table 3.27. Tooth Wear Rate Analysis (M1:M2)

Inter-Sample Comparisons

All Cases

	Da But Period				Metal Period			
	Mandible		Maxilla		Mandible		Maxilla	
n	47		40		44		44	
r	0.920		0.892		0.905		0.912	
r^2	0.846		0.797		0.819		0.832	
adjusted r^2	0.842		0.791		0.814		0.828	
slope	0.987		1.110		1.088		0.962	
y intercept	5.256		3.015		3.704		5.049	
95% CI for slope	0.860	1.113	0.926	1.294	0.929	1.248	0.827	1.096
95% CI for intercept	2.461	8.052	-0.672	6.702	1.021	6.388	2.983	7.115
ANOVA of regression								
F	246.498		148.768		189.521		208.148	
p	**0.000**		**0.000**		**0.000**		**0.000**	

Cases >32 Excluded

	Da But Period				Metal Period			
	Mandible		Maxilla		Mandible		Maxilla	
n	36		32		38		43	
r	0.886		0.896		0.878		0.907	
r^2	0.786		0.802		0.771		0.823	
adjusted r^2	0.779		0.795		0.765		0.819	
slope	1.061		0.977		1.019		0.899	
y intercept	3.790		4.354		4.137		5.791	
95% CI for slope	0.868	1.255	0.796	1.158	0.831	1.206	0.768	1.031
95% CI for intercept	0.192	7.388	1.177	7.530	1.433	6.841	3.818	7.765
ANOVA of regression								
F	124.671		121.529		121.389		191.06	
p	**0.000**		**0.000**		**0.000**		**0.000**	

factors such as the effects of pregnancy are both difficult to assess and not necessarily influential in affecting dental health (Larsen *et al.* 1991).

Pulp Exposure (PEXP)

Attrition is by far the chief cause of pulp exposure for both time periods. The trend toward posterior tooth PEXP in the Da But period and anterior tooth concentration in the Metal period suggests either quite different masticatory patterns between both periods or else one or both communities were using their teeth for activities other than food processing. Further, that PEXP occurs more often in younger ages in the Metal period, than the Da But period, suggests extra stresses on these teeth early in life. It is relevant then, that task wear facets are concentrated on the anterior teeth of the Metal period sample. This is also consistent with the evidence for relatively more rapid anterior tooth wear and high frequencies of anterior alveolar defects in the Metal period and suggests the causes of PEXP, alveolar defects and task wear facets are all related to using the teeth as tools.

The observation that more male Metal period teeth are affected by PEXP suggests that male teeth were under more stress than female teeth. Given that the chief cause of PEXP is attrition of the anterior teeth in the Metal period, this may indicate a bias toward males using their teeth in non-masticatory tasks.

With regard to the Da But period, PEXP and alveolar defects are both concentrated in the posterior teeth and are likely interrelated. Note also that for neither sample was dental caries a significant contributor to PEXP and hence probably not to alveolar defects either.

Calculus (CAL)

In discussing calculus deposits it is preferable to place more weight on the evidence from the individual count analysis when making comparisons. The reason is that it is virtually impossible to control for the loss of calculus deposits on individual teeth during the period of curation. Some of the remains in this study have been stored and handled for over 30 years.

For both periods there is a general pattern of increasing numbers of teeth affected by calculus and advancing age. Given the time dependent nature of plaque accumulation and calcification there is nothing particularly unusual in this observation. What is interesting however, is that males exhibit significantly more calculus, both by tooth count and individual count, in the Metal period. This pattern is not apparent in the Da But period where calculus frequency is similar by sex. The pattern seen in the Metal period may be due to under-representation of females in the oldest age class both by tooth and

Table 3.28. Tooth Wear Rate Analysis (M1:M2): ANCOVA Results

Inter-Sample Comparisons

Da But:Metal Period (maxillary M1: M2 all cases)

Test of Parallelism

	Sum of Squares	df	Mean Square	F	p-level
Effect	10.962	1	10.962	1.174	0.282
Error	737.881	79	9.340		

General MANOVA	df Effect	MS Effect	df Error	MS Error	F	p-level
	1	5.474	80	9.361	0.585	0.447

Da But:Metal Period (mandibular M1:M2 all cases)

Test of Parallelism

	Sum of Squares	df	Mean Square	F	p-level
Effect	11.130	1	11.130	1.025	0.314
Error	944.479	87	10.856		

General MANOVA	df Effect	MS Effect	df Error	MS Error	F	p-level
	1	1.283	88	10.859	0.118	0.732

Da But:Metal Period (maxillary M1:M2 cases<=32)

Test of Parallelism

	Sum of Squares	df	Mean Square	F	p-level
Effect	0.390	1	0.390	0.078	0.781
Error	351.457	70	5.021		

General MANOVA	df Effect	MS Effect	df Error	MS Error	F	p-level
	1	0.017	71	4.956	0.003	0.953

Da But:Metal Period (mandibular M1:M2 cases <=32)

Test of Parallelism

	Sum of Squares	df	Mean Square	F	p-level
Effect	0.785	1	0.785	0.102	0.750
Error	538.699	70	7.696		

General MANOVA	df Effect	MS Effect	df Error	MS Error	F	p-level
	1	1.554	71	7.598	0.205	0.652

individual. However, it may also be due to differences in diet and/or oral hygiene. Given that both samples exhibit very high frequencies of some degree of calculus, over 80% of individuals in each sample, levels of oral hygiene might reasonably be considered to have been quite low. If the difference between the sexes is dietary, it could indicate that the composition of the male diet was different to that of females. However, given the findings of Lieverse (1999) that diets high in carbohydrates or protein can lead to high calculus levels, it is impossible to suggest what the dietary difference, if real, consisted of. Alternatively, the oral environment of males may be differentially predisposed to calculus formation due to non-dietary factors. Such factors are unlikely to include behaviour associated with betel nut staining, as the habit appears to have been practised equally by both sexes. It is therefore not clear what factor(s) predisposed males to greater levels of calculus formation.

Tooth Discolouration (Betel Nut Staining)

A recent residue analysis of stained Metal period incisors indicates that the staining is consistent with betel nut use (Oxenham *et al.* in prep). Given the previous discussion on the probable prophylactic effect of betel nut residues with respect to caries disease, the results obtained here were not expected. If anything, betel nut use is associated with elevated levels of caries in the Metal period sample. One important factor associated with betel chewing in recent ethnographic accounts (see Schamschula *et al.* 1978; Moller *et al.* 1977) is the simultaneous use of lime (calcium hydroxide). As already noted the lime will have the effects of increasing the pH of the oral cavity and presumably neutralising the effects of acidogenic bacteria. It is possible that lime plays a crucial role in caries prevention and may not have been used with betel nut two millennia ago in northern Vietnam. Nonetheless, the role of betel nut itself in the prevention of caries disease is

Table 3.29. Da But Period Tooth Wear Rate Analysis (M1:M2)

Intra-Sample Comparisons

All Cases

	Male				Female			
	Mandible		Maxilla		Mandible		Maxilla	
n	21		18		17		14	
r	0.900		0.906		0.880		0.799	
r^2	0.809		0.820		0.774		0.639	
adjusted r^2	0.799		0.809		0.759		0.608	
slope	0.990		1.043		0.898		1.088	
y intercept	5.441		4.761		7.519		4.156	
95% CI for slope	0.760	1.221	0.784	1.301	0.631	1.165	0.573	1.603
95% CI for intercept	0.407	10.474	-1.006	10.527	0.899	14.139	-5.719	14.030
ANOVA of regression								
F	80.655		73.131		51.326		21.201	
p	**0.000**		**0.000**		**0.000**		**0.001**	

Cases <=32

	Male				Female			
	Mandible		Maxilla		Mandible		Maxilla	
n	17		12		10		12	
r	0.837		0.895		0.776		0.861	
r^2	0.701		0.801		0.603		0.742	
adjusted r^2	0.681		0.782		0.553		0.716	
slope	1.041		0.840		0.998		0.909	
y intercept	4.395		7.120		5.130		6.123	
95% CI for slope	0.667	1.415	0.545	1.135	0.338	1.659	0.531	1.287
95% CI for intercept	-2.892	11.681	1.868	12.372	-7.289	17.548	-0.881	13.127
ANOVA of regression								
F	35.207		40.347		12.142		28.741	
p	**0.000**		**0.000**		**0.008**		**0.000**	

still important. Various constituents of betel nut have been linked to both plaque prevention (Iwamoto *et al.* 1991) and bacterial suppression (De Miranda *et al.* 1996). It is also possible that the betel nut stains are the result of deliberate applications of a staining compound, comprising betel, to the teeth for purposes aesthetic or otherwise. If such is the case, betel may have been introduced into the oral cavity on an occasional to infrequent basis thus limiting its caries prophylactic properties. The recent study by Oxenham *et al.* (in prep.) indicates that these teeth may well have been deliberately stained. Nonetheless, even if teeth are demonstrated to have been deliberately stained, the owners may still have habitually chewed betel nut.

Given the large proportion of the adult population apparently using betel nut in the Metal period, it is worth examining other potential health risks associated with this habit. A wealth of recent literature has linked betel nut chewing with periodontal disease (Jeng *et al.* 1996; Chang *et al.* 1998), the pre-malignant condition oral squamous fibrosis and other subsequent malignant cancerous conditions (Canniff & Harvey 1981; Noor 1987; Sundqvist *et al.* 1989; Murti *et al.* 1990; Sinor *et al.* 1990; Sundqvist & Grafstrom 1992; Dave *et al.* 1992; van Wyk *et al.* 1993; Lu *et al.* 1993; Maher *et al.* 1994; van Wyk *et al.* 1994; Murti *et al.* 1995; Fang *et al.* 1997; Jeng *et al.* 1999). Differences in the level of genotoxicity, damage to the cell's DNA, of betel has also been observed. In its tender form, with attached husk, betel nut is less genotoxic than the ripe and dried form without the husk (Liu *et al.* 1996). Moreover, the calcium hydroxide contained in the lime paste commonly taken with betel nut, has also been shown to be genotoxic to buccal mucosa cells in betel chewers (Nair *et al.* 1990).

Even were the Metal period people experiencing pre-cancerous or cancerous conditions related to the use of betel nut, it would be unlikely to show in the skeletal record. In modern populations oral malignant conditions account for only 2% to 3% of all cancers (Sapp *et al.* 1997). Moreover, the majority of these conditions in modern populations occurs in individuals over 40 years old. The relative youth of the Metal period sample, the relative rarity of oral malignant conditions overall, the added infrequency of this neoplasm manifesting in bony tissue of the oral region and the vagaries of the skeletal preservation/excavation process, severely limits the likelihood of observing this condition. It is not improbable however, that oral submucous fibrosis at least, and perhaps advanced malignant cases, occurred in Metal period Vietnam, given the common use of betel nut at this time.

There are other potential side effects and benefits related to the use of betel nut. Boucher *et al.* (1994) suggest that the

Table 3.30. Da But Period Tooth Wear Rate Analysis (M1:M2): ANCOVA Results

Intra-Sample Comparisons

Maxillary M1/M2 Males:Females (all cases)

Test of Parallelism

	Sum of Squares	df	Mean Square	F	p-level
Effect	0.519	1	0.519	0.033	0.857
Error	442.031	28	15.787		

General MANOVA	df Effect	MS Effect	df Error	MS Error	F	p-level
	1	0.555	29	15.260	0.036	0.850

Mandibular M1/M2 Males:Females (all cases)

Test of Parallelism

	Sum of Squares	df	Mean Square	F	p-level
Effect	3.534	1	3.534	0.302	0.586
Error	397.409	34	11.689		

General MANOVA	df Effect	MS Effect	df Error	MS Error	F	p-level
	1	0.015	35	11.456	0.001	0.971

Maxillary M1/M2 Males:Females (cases <=32)

Test of Parallelism

	Sum of Squares	df	Mean Square	F	p-level
Effect	0.527	1	0.527	0.089	0.769
Error	118.692	20	5.935		

General MANOVA	df Effect	MS Effect	df Error	MS Error	F	p-level
	1	0.277	21	5.677	0.049	0.827

Mandibular M1/M2 Males:Females (cases <=32)

Test of Parallelism

	Sum of Squares	df	Mean Square	F	p-level
Effect	0.213	1	0.213	0.018	0.893
Error	266.835	23	11.602		

General MANOVA	df Effect	MS Effect	df Error	MS Error	F	p-level
	1	0.022	24	11.127	0.002	0.965

higher frequency of foregut cancers in habitual betel nut chewers indicates that the habit can cause diabetes. Perry (1980) cites Chinese sources, which place the use of betel nut as a specific treatment against tape worm infestations at least as far back as 1400 years. More recently, Chung & Ko (1976) have effectively treated human *Taenia saginata* (a tape worm) infestations with betel nut preparations. Inokuchi *et al.* (1986) have highlighted the antihypertensive properties of some constituents of betel nut. Perry (1980:302) lists a host of uses for the betel plant, both leaves and nut in various preparations, including the treatment of diarrhoea, dropsy, sunstroke, beri-beri, throat inflammations, oedema, lumbago, bronchial catarrh and urinary disorders.

Tooth Malalignment (MAL)

The greater level of tooth displacement or malalignment seen in the Metal period is likely related to the reduction in dental arch size discussed above. For both periods females displayed more, albeit not significantly so, tooth displacement. This is consistent with smaller female dental arch sizes. That Metal period females also had significantly more mandibular displaced teeth than males, is also consistent with this. In this context it is also relevant that three, two females and a male, of the six Metal period individuals with MAL also had impacted M3s. One of these, a female, suffered from tooth rotation, tooth displacement and impacted third molars.

Overall, the relatively low rates of tooth displacement and rotation in these samples may be due to the reduction in mesio-distal length of teeth due to heavy wear (see Calcagno & Gibson 1991). Nonetheless, although tooth size was smaller in the Metal period and jaw robusticity greater, an apparent confounding factor of reduced dental arch size seems to have contributed to greater occlusal

Table 3.31. Metal Period Tooth Wear Rate Analysis (M1:M2)

Intra-Sample Comparisons

All Cases

	Male				Female			
	Mandible		Maxilla		Mandible		Maxilla	
n	18		18		18		17	
r	0.921		0.939		0.885		0.819	
r^2	0.849		0.881		0.784		0.671	
adjusted r^2	0.840		0.873		0.770		0.649	
slope	0.916		1.077		1.691		0.912	
y intercept	6.243		3.380		-4.937		5.604	
95% CI for slope	0.712	1.121	0.867	1.287	1.220	2.161	0.561	1.263
95% CI for intercept	2.549	9.963	-0.177	6.938	-12.614	2.740	0.670	10.539
ANOVA of regression								
F	90.022		118.276		57.997		30.648	
p	**0.000**		**0.000**		**0.000**		**0.000**	

Cases <=32

	Male				Female			
	Mandible		Maxilla		Mandible		Maxilla	
n	16		17		15		17	
r	0.862		0.936		0.768		0.819	
r^2	0.742		0.876		0.589		0.671	
adjusted r^2	0.724		0.868		0.558		0.649	
slope	0.765		0.976		1.327		0.912	
y intercept	8.178		4.667		-0.248		5.604	
95% CI for slope	0.507	1.023	0.774	1.178	0.663	1.992	0.561	1.263
95% CI for intercept	4.125	12.231	1.383	7.952	-10.114	9.619	0.670	10.539
ANOVA of regression								
F	40.341		105.905		18.648		30.648	
p	**0.000**		**0.000**		**0.001**		**0.000**	

problems in this sample.

Tooth Rotation (ROT)

Similar numbers of teeth displayed various degrees of rotation, with the canine being the most commonly affected individual tooth, in both periods. It is possible that rotation is a variable not particularly sensitive to changes in the tooth size-dental arch size relationship. The reason for the significantly greater number of male Metal period teeth displaying rotation, as compared to female, is not readily apparent. It may be related to evidence for a greater proportion of males using their anterior teeth as tools during this period. It is also possible that tooth rotation is more under the control of genetic factors. If so, this would explain the poor correlation between tooth rotation and other variables expressing occlusal variation.

Tooth Impaction (IM)

The finding of a greater prevalence of IM in the Metal period, and a greater proportion of affected individuals being female, is likely related to a reduction in space for tooth eruption in this period. Using the frequency of IM as a measure of available space gives the somewhat paradoxical picture of increased jaw robusticity in the Metal period coupled with a decrease in the size of the dental arch. That this is a real decrease in the dental arch is borne out by the observation that tooth size is also reduced in the Metal period compared to the earlier Da But period period. However, while there are reductions in both tooth size and dental arch size in the Metal period, tooth size was apparently not reducing fast enough, for females at least, to compensate for the reduction in dental arch area.

A somewhat unusual dental pathology in association with an impacted left mandibular M3 was described previously (84DSM8: Figure 3.13). A differential diagnosis of this condition is presented here. The condition that caused these changes in the LM3 and surrounding bone are most consistent with the diagnosis of an odontogenic cyst.

> Cysts are pathological, fluid-filled cavities lined (in most cases) by epithelium. They are the most common cause of chronic swellings in the jaws...Cysts are more common in the jaws than in any other bone because of the many rests in the epithelium remaining in the tissues after dental development. Cysts formed from epithelium remaining after tooth formation (odontogenic cysts) account for all but a few cysts of the jaws.
> (Cawson 1991:225)

Cawson (1991:226) goes on to summarise the relative frequency of the various cysts as periodontal (65-

Table 3.32. Metal Period Tooth Wear Rate Analysis (M1:M2): ANCOVA Results

Intra-Sample Comparisons

Maxillary M1/M2 Males:Females (all cases)

Test of Parallelism		Sum of Squares	df	Mean Square	F	p-level
	Effect	4.751	1	4.751	0.763	0.389
	Error	192.962	31	6.225		

General MANOVA	df Effect	MS Effect	df Error	MS Error	F	p-level
	1	0.140	32	6.179	0.023	0.881

Mandibular M1/M2 Males:Females (all cases)

Test of Parallelism		Sum of Squares	df	Mean Square	F	p-level
	Effect	121.548	1	121.548	**11.742**	**0.002**
	Error	331.245	32	10.351		

General MANOVA	df Effect	MS Effect	df Error	MS Error	F	p-level
na						

Maxillary M1/M2 Males:Females (cases <=32)

Test of Parallelism		Sum of Squares	df	Mean Square	F	p-level
	Effect	0.665	1	0.665	0.121	0.731
	Error	165.448	30	5.515		

General MANOVA	df Effect	MS Effect	df Error	MS Error	F	p-level
	1	0.009	31	5.358	0.002	0.967

Mandibular M1/M2 Males:Females (cases <=32)

Test of Parallelism		Sum of Squares	df	Mean Square	F	p-level
	Effect	25.500	1	25.500	3.513	0.072
	Error	195.973	27	7.258		

General MANOVA	df Effect	MS Effect	df Error	MS Error	F	p-level
	1	0.129	28	7.910	0.016	0.899

70%), dentigerous (15-18%), keratocysts (3-5%) and nasopalatine. In this particular case the morphology of the cyst is not consistent with either an abscess or granuloma defect. The most likely odontogenic candidate is either a dentigerous cyst or a variant of this called an eruption cyst. The following discussion is based on Sapp *et al.* (1997:38-45). Dentigerous cysts generally affect unerupted, impacted maxillary and mandibular third molars. The cyst itself is derived from the enamel epithelium that surrounds the crown of the unerupted tooth. The cyst is distinct in that it only surrounds the crown leaving the roots outside the lumen created by the cyst. In this particular case it appears that the LM3 was in the process of erupting and therefore may be more likely an eruption cyst. These cysts form in an identical manner except that instead of forming within the bony tissues they develop in the soft alveolar tissues. In life this would be seen as a large swelling along the alveolar ridge. Chewing can lead to the haemorrhage of such a cyst. Apart from possible swelling and pain associated with both cysts and bleeding of a ruptured eruption cyst, these cysts are generally asymptomatic.

The only other possible candidate is a keratocyst. These cysts can envelop an unerupted tooth and thus give the appearance of a dentigerous cyst (Cawson 1991), however, they tend to be multilocular and would not confine themselves to the crown of the tooth, as appears to be the situation here.

Enamel Pearls (EP)

The only three individuals displaying EP all derived from the same Metal period site, Nui Nap. Clarke & Hirsch (1991a) point out the high incidence of this trait in some ancestral groups, so it is probable that genetic factors play a role in its expression. The site specific nature of its occurrence in this study would tend to support the genetic

Table 3.33. Relative Anterior to Posterior Tooth Wear Maxillary M1 and I1

	Total			Male			Female		
Da But Period	mean	sd	n	mean	sd	n	mean	sd	n
mean M1 wear	5.74	2.02	20	5.91	1.81	8	6.64	1.91	7
I1 wear	6.70	2.20	20	7.00	1.69	8	7.57	1.81	7
ratio (I1/M1)	1.20	0.33	20	1.25	0.36	8	1.17	0.23	7
Metal Period									
mean M1 wear	4.52	1.85	32	5.36	1.66	14	4.80	1.17	11
I1 wear	6.38	2.43	32	7.00	1.75	14	7.27	1.49	11
ratio (I1/M1)	1.42	0.33	32	1.35	0.30	14	1.55	0.30	11

t tests			
Da But M:F	t	df	p
mean M1 wear	-0.765	13	0.458
I1 wear	-0.632	13	0.539
ratio (I1/M1)	0.545	13	0.595
Metal M:F	t	df	p
mean M1 wear	0.170	23	0.352
I1 wear	-0.412	23	0.684
ratio (I1/M1)	-1.591	23	0.125
Metal: Da But	t	df	p
mean M1 wear	**2.221**	**50**	**0.031**
I1 wear	0.486	50	0.629
ratio (I1/M1)	**-2.407**	**50**	**0.020**

view.

Mandibular Robusticity

Both the greater level of mandibular sexual dimorphism and the overall greater level of robusticity in the Metal period sample was not expected given evidence for a global Holocene size reduction in the orofacial skeleton (Brown 1987). Aside from possible genetic differences between the two samples, the differences are likely dietary or oral task related. Hinton & Carlson (1979:330) point out that:

> there is a limited body of experimental evidence implicating induced changes in dietary consistency as a causative agent for increases in masticatory muscle attachment areas, mandibular corpus thickness, density of condylar bone and condyle size in rats.

It might be asked then: is the evidence for increased gnathic musculature and thus robust jaws in the Metal period due to the use of the anterior teeth as tools or to mastication of tougher food stuffs? That the Metal period people were masticating tougher food is not borne out by the equivalence of childhood wear rates of the permanent molars with the Da But period. Even if diet was essentially the same for both periods, the effects of the widespread Metal period inferred habit of betel quid chewing and/or the use of the teeth as tools, are the most likely contributors to this greater gnathic robusticity. While the Metal period gnathic apparatus may have been more robust, this did not apparently also translate into more space in the tooth rows for the permanent dentition as attested to by the results of tooth malalignment and impaction.

Molar Tooth Size

Often in such analyses of tooth size only the bucco-lingual (BL) dimension is assessed due to the effects of approximal wear of teeth in the mesio-distal (MD) direction. Nonetheless, Da But period tooth size was highly significantly greater than that for the Metal period in all cases except for M2 MD diameter. The reduction in tooth size from the Da But period through to the Metal period is consistent with an apparent global trend toward decreased tooth size from the later Pleistocene through Holocene (Brace *et al.* 1987; Brace *et al.* 1991; Calcagno & Gibson 1991). However, it is also possible that the difference in tooth size between these samples are due to ethnic differences (Harris & Rathbun 1991). Of the component of each sample that could be sexed, both sexes were equally represented, thus limiting the possibility that unbalanced samples were contributing to the significant differences. A possible explanation for greater Metal period mandibular robusticity, even in the face of smaller tooth size, has already been suggested above.

Interproximal Grooves (IG)

Where these grooves are confined to the molars and premolars, there are persuasive arguments to believe that the most likely aetiology is dental probing (Frayer 1991; Bermudez de Castro *et al.* 1997). Several IG

Table 3.34. The Prevalence of Several Oral Pathologies (Permanent Teeth) from Samples in Thailand and Vietnam

			Dental Pathology by Tooth Count % (rate in parentheses is individual count)				
Site	date*	subsistence	caries	AMTL	ADP	CAL	source
Vietnam (Da But)	~ 6.0	foraging	1.5 (13.8)	4.8 (29.7)	1.5 (20.3)	22.2 (58.8)	6
Non Nok Tha (early)	5.5 - 4.5	mixed	1.7	5.0	1.4	28.8	4
Ban Chiang (early)	5.6 - 3.4	mixed	6.2	6.6	5.5	32.8	4
Non Nok Tha (late)	4.5 - 3.0	mixed	4.1	10.4	2.7	35.7	4
Khok Phanom Di	4.0 - 3.5	agriculture	10.8 (58.2)	8.9 (50.7)	6.0 (55.2)	n/a	3
Ban Chiang (late)	3.6 - 1.7	mixed	7.7	6.9	6.6	20.7	4
Ban Na Di	2.6 - 2.4	mixed	4.7	5.4	2.1	n/a	5
Vietnam (Metal)	3.3 - 1.7	mixed	2.3 (20.8)	3.0 (17.1)	2.6 (18.4)	14.6 (54.6%)	6
Noen U-Loke	2.3 - 1.7	agriculture	3.0 (29.6)	8.4 (55.6)	1.8 (17.9)	54.7 (90.6)	2
Ban Lum Khao	1.0 - 1.5	mixed	4.5	5.1	1.3	n/a	6
Nong Nor	3.1 - 2.7	mixed	6.0 (49.3)	4.1 (27.1)	0.8	n/a	1, 5

* date: 1000 years bp
mixed (subsistence): foraging and agriculture of indeterminate ratio

1 Tayles *et al.* (1998)
2 Nelson (1999)
3 Tayles (1992)
4 Douglas (1996)
5 Domett (1999)
6 this study

AMTL ante-mortem tooth loss
ADP alveolar defect of pulpal origin
CAL calculus (moderate and severe expression)

were associated with a Da But period male who also had clear evidence of TWF on all molars. These IG are quite dissimilar to the molar TWF and it is highly improbable, given their position, that they could have been produced through some sort of industrial activity. It is probable that the molars were employed in fibre processing activities. It is therefore likely that the IG were caused by this individual probing and dislodging fibrous material that had been caught in the premolar interstitial spaces.

Task Wear Facets (TWF)

The only clear case of TWF in the Da But period is in the individual that has already received some attention with regard to alveolar defects and IG above. The TWFs described for this individual are consistent with the processing of fibrous or sinewy material. Brown & Molnar (1990) describe and illustrate sinew processing by an Australian Aboriginal male that is entirely consistent with the observed pattern of molar TWF seen in the Da But period individual. It should be noted that Brown & Molnar (1990) originally cited this example of sinew processing to illustrate how IG might have been formed. However, Frayer (1991) demonstrated that this particular example described and illustrated by Brown & Molnar (1990) could not have produced IG. This type of fibre/sinew processing will, it is suggested, produce the type of TWF seen in this Vietnamese example.

Tooth Chipping (CP)

The similar frequencies of chipped teeth and the lack of evidence for major tooth trauma in both samples, indicates no significant oral behavioural differences that traumatised teeth might be indicative of. Both samples reveal a tendency for posterior tooth chipping and neither sample indicates significant differences in chipping by sex. Milner & Larsen (1991:370) note that a number of studies indicate that the posterior permanent teeth are the typical focus of chipping. Even the Metal period sample, with evidence for the use of the anterior dentition in industrial activities, displays low levels of anterior tooth chipping. The lack of evidence for patterning by sex, the observation of an increase in chipping with age and the greater frequency of chipping of the posterior teeth are all consistent with a dietary aetiology for the chipping seen in both periods. Large diameter grit particles introduced into the oral cavity with food is a probable scenario.

Tooth Ablation

Tayles (1997) has surveyed the evidence for tooth ablation globally and notes evidence for the practice in Southeast and Eastern Asia. Tayles (1997:333) also stresses that the "distinction of deliberately removed teeth from those that have been accidentally or otherwise unintentionally lost is not straightforward".

Tooth ablation only appears to have occurred in the Da But period sample. One behaviour that can mimic ante-mortem tooth loss, otherwise due to tooth ablation, is use of the teeth in industrial activities. However, only one case of TWF was recorded for the Da But period sample and that occurred on the molars. Molnar (1972) cites ethnographic evidence for North American California Indians deliberately removing upper central incisors, in order to better facilitate the processing of materials used for basket making. However, task specific grooves were still present in certain remaining teeth, the upper canine particularly. It would appear unlikely that such behaviour was occurring in this Vietnamese sample given the dearth of evidence for TWF.

Further negative evidence that the teeth were not habitually used as tools in the Da But period sample comes from the analysis of tooth chipping. The frequency of tooth chipping was lowest on the incisors for this sample and in fact no cases of chipping were found on the mandibular incisors.

Task or diet related anterior tooth wear can also manifest in less dramatic forms than clear TWF. Nonetheless, it was found that anterior tooth wear relative to posterior wear was much lower in this sample compared with the Metal period sample. Further, only one incisor related abscess was found in the entire Da But period sample and that was maxillary and not one of the nine individuals displaying the symmetrical mandibular incisor ante-mortem tooth loss pattern. Also, no cases of incisor, or even canine, pulp exposure were found in the entire Da But period sample.

X-ray assessment of possible cases of permanent anterior tooth non-eruption due to congenital absence or impaction could not be carried out. However, the percentage of individuals with suspected tooth ablation, 12.2%, is much higher than frequencies of congenital absence surveyed by Tayles (1997).

It is potentially significant that all nine cases were at least 40 years old at death. It is arguable that these teeth were simply lost due to age related processes such as alveolar resorption. However, clear cases of alveolar resorption, over and above normal alveolar remodelling subsequent to tooth loss, were not observed. If these are examples of tooth ablation, the correlation with advanced age could be age dependent and/or status related. Alternatively, this could be an example of a secular trend in the cemetery population with ablation going out of fashion over the period of use of the cemetery. Unfortunately it is not possible to test statements about status against the distribution of grave goods, for the most part pottery and lithic artefacts, as such data have not been preserved in records of the original excavations. The few instances of ochre and stag-horns specifically documented as being associated with several burials (Bui Vinh 1980), were not connected with any of the nine individuals discussed here.

Tooth Wear

In Chapter 2, an analysis of tooth wear was carried out in order to assist in estimating the age-at-death of individuals in each sample. In this chapter, tooth wear was analysed in order to (1) describe age-independent rates of tooth wear, and (2) describe the ratio of anterior to posterior tooth wear in each sample. It was believed that a description of rates of tooth wear for each sample may throw some light on any possible dietary differences between the Da But period period and Metal period. The results of this analysis indicated that there were no significant differences in the rate of childhood tooth wear between the Da But period and Metal period samples. A best case interpretation of this would suggest little dietary difference between each sample. However the results, on their own, are equivocal with respect to diet. Nonetheless, the demonstration that the rate of childhood wear for the Da But period and Metal period samples is equivalent, supports the use of the Da But period derived age estimation regression equations developed in Chapter 2. Finally, The greater ratio of anterior to posterior tooth wear in the Metal period sample, relative to the Da But period sample, is consistent with evidence for the greater use of the anterior teeth in the Metal period for industrial activities.

Conclusions

Subsistence Strategies

Lukacs (1989; 1992) and Oxenham *et al.* (2002b) have used the frequencies of various dental traits in dental pathology profiles to infer subsistence strategies. However, I no longer believe this to be a greatly profitable procedure. A number of problems have been highlighted in this chapter with regard to inferences made from the frequencies or manifestations of particular traits. Lukacs (1989) sees high levels of attrition and low levels of calculus as hunter-gatherer traits. However, it has been pointed out that high levels of marine food or red meat consumption, generally hunter-gatherer traits, are also associated with high levels of calculus. Other traits, such as high caries prevalence and high prevalence of caries induced pulp exposure are so closely related as to be double counting when adding up those variables indicating one subsistence economy or the other. Also, caries prevalence itself can be contingent on a host of non-dietary factors, to the extent that simply inferring a change toward agriculture, based upon the observation of an increase in caries prevalence over time, would be incautious. Both the archaeological and bio-archaeological evidence must be considered in tandem if meaningful inferences regarding the subsistence economies of past groups are to be made.

Overall, the evidence from the oral health profiles detailed here, together with what little subsistence oriented archaeological evidence has been recovered, indicates that neither the Da But period nor the Metal period samples represent peoples engaged in intensive agriculture. The higher prevalence of caries, affecting more young people in the Metal period, may be indicative of an increased use of, or the beginnings of the utilisation of grain products. Evidence from early Chinese historical sources, discussed in Chapter 1, suggests some level of rice agriculture during the Dong Son phase shortly prior to Chinese colonising efforts. However, the oral health evidence from the Metal period does not indicate intensive agricultural subsistence.

Southeast Asian oral responses to agriculture, essentially rice cultivation, appear to be somewhat at odds with the pattern seen in the Americas and Europe. While the site Khok Phanom Di in Thailand has a relatively high prevalence of caries, alveolar defects of pulpal origin and ante-mortem tooth loss, the other ostensibly agricultural sites in Thailand (Noen U-Loke, Ban Na Di), display a low prevalence of caries and alveolar defects. That there is considerable variation in the expression of such oral traits in Southeast Asian sites, is probably a reflection of the very small number of sites that have been examined in comparison to the Americas and Europe. It is perhaps too early yet to generalise a Southeast Asian oral response to agricultural intensification.

There is no real need to posit the existence of agricultural intensification during the Metal period in order to explain the formation of complex semi-autonomous states (cf Higham 1989). Other workers have pointed out that complex societies can and did develop from aquatic foraging (Hayden *et al.* 1985) and marine-based economies (Yesner 1980). Moreover, Akazawa (1986a) has argued that the overwhelming concentration of Jomon period forager sites in eastern Japan, representing relatively high population densities and complex communities, can be explained in the context of the extremely highly productive combined forest-estuarine ecosystems present in the region during the period 6000 to 3000 years bp.

It is not necessary to accept that the availability of agricultural technology and know-how meant that agriculture was embraced in northern Vietnam some 2000 years ago. Akazawa (1986b) has convincingly argued that Jomon period communities in eastern Japan resisted moves toward food production economies, due to the existing efficiency of their highly developed foraging economies, and also owing to the perceived threat such subsistence changes would make to their socio-cultural systems. The oral health profile of Metal period northern Vietnam is consistent with work by Vietnamese archaeologists (Nguyen Kach Su 1997) outlining the development of the Dong Son directly from previous maritime/coastal cultures. The oral profile outlined here for the Metal period, while not consistent with an agricultural based economy, is certainly consistent with a transitional-mixed model. The degree to which foraging, including aquatic strategies, played a significant role in the subsistence economies of this period is difficult to assess, but it is possible that these strategies were vitally important.

Regardless of the presence of polished stone "hoes" and the appellation "Neolithic", there is neither archaeological nor bio-archaeological evidence indicting that the Da But period sample had an agricultural base. The archaeological (Bui Vinh 1980, 1991, 1994, 1996; Nguyen Khac Su 1997) and bio-archaeological evidence points to a sedentary forager society that was clearly highly dependent on rich coastal, estuarine and riverine resources. It is also highly probable, given its location close to the rich biota of steeply rising high country, that the hunting of terrestrial game as well as foraging for terrestrial plants played a significant role in this economy. It has been suggested that there is evidence for buffalo domestication from the earliest stages of the Da But period (Bui Vinh 1991). However, even if such was the case, it does not mean that these animals were used in pastoral or agricultural pursuits. For example, Ingold (1980) notes, in a northern European context, reindeer domestication within the context of a hunting subsistence economy.

Behaviour

The clearest oral evidence for behavioural differences between each period is that from betel nut use. Human use of betel nut appears to first occur in Northern Vietnam some 2500 years bp. An examination of the dentition (personal observation) of the remains of individuals excavated from Giong Co Vo in Southern Vietnam, indicate a similar time for the initial use of betel nut. The potential health effects, both positive and negative, of the use of betel nut have not left any readily apparent signature on the Metal period dental remains. One exception is the possibility that betel nut use augmented rather than protected against dental disease.

A further clear behavioural difference between periods is the use of the teeth, especially anterior, in the Metal period in industrial activities. Moreover, it seems that it is males who are engaged in such activities more so than females. Evidence complementing that from task wear facets comes from the age distribution and patterning of alveolar defects of pulpal origin and pulp exposure, as well as greater anterior tooth wear and mandibular robusticity in the Metal period sample. Inferences about the possible industrial activities involved would be equivocal at best. However, given the likely significance of a coastal/aquatic economy, the evidence suggests a division of labour, with males being more involved in orally manipulating certain subsistence related material. Such material could include nets, fishing line, binding material and so forth.

Given the evidence for much greater use of the dentition industrially in the Metal period, it is somewhat intriguing that interproximal grooves are only evident in the Da But period sample. That they are not exclusively associated with carious teeth suggests that their role was not simply palliative. Their presence may indicate the habitual use, manifested in extreme cases as IG in some individuals, of tooth picking for oral hygiene in the Da But period. However, no other evidence points to particularly high levels of oral hygiene in either period.

A further behavioural difference between each period is that tooth ablation possibly occurred in the Da But period. The evidence is somewhat equivocal, but if confirmed by later excavations of Da But period sites would prove an important insight into the culture of this period in Northern Vietnam.

Oral health profiles need to incorporate information on the distribution of oral traits by age, sex, jaw type and tooth class to be of any value in reconstructing the behaviour and oral responses of past populations to diet and environment. This chapter has revealed that it is these patterns, rather than overall frequencies of particular oral traits, that serve to delimit and reveal differences in health and cultural practices between each period.

4
ARTICULAR CHANGE

This chapter is concerned with changes to the articular regions of bones. It will become clear in the following discussion that not all the commonly reported changes to joints in the skeletal record are in fact strictly pathological. For this reason the term arthropathy is avoided. As most reported skeletally sensitive changes to joints are attributed to the rheumatic diseases, and osteoarthritis in particular, the first task of this chapter is a review of the clinical and epidemiological literature on osteoarthritis. The role of bio-archaeological studies of articular change is then examined in the context of the conclusions of that review.

Before looking at the skeletal manifestations of joint change in the archaeological record, it is first necessary to examine the pathogenesis, epidemiology, risk and predisposing factors, and clinical and skeletal manifestations of joint change. This is all the more important given the behaviour oriented nature of the vast majority of bio archaeological interest in joint change. Following a discussion of these factors, the results from an assessment of the Vietnamese skeletal samples are examined and discussed.

The two primary aims of this chapter are thus, to (1) review the literature on osteoarthritis and (2) document the nature, distribution and prevalence of joint changes in the Vietnamese samples. An additional aim is to compare, where possible and in a general manner, the nature, distribution and prevalence of articular change in the Vietnamese samples with other studies of past and contemporary populations.

A Review of the Literature

Osteoarthritis (OA) is the most common disease or condition affecting articular joints. OA has been documented in a wide range of vertebrates from *Iguanadon* ankles dating to 125 million years ago, to Pleistocene Kangaroos and on up to extant modern animals (Rothschild 1993). Moreover, joint changes are the most common and readily observed conditions in human skeletal material (Rogers & Waldron 1995).

While there is no consensus on a definition of what osteoarthritis is (Jurmain 1999), a useful starting point is that "pathologically it is a condition of synovial joints characterised by focal cartilage loss and an accompanying reparative bone response" (Jones & Doherty 1995:457). The umbrella classificatory scheme, of which OA is but one of a range of ostensibly distinct diseases, are the rheumatic diseases.

(1) Classification of Arthritis

The most widely accepted clinical classification of OA recognises a division into primary and secondary osteoarthritis (Altman *et al.* 1986). Primary, or idiopathic, OA covers that range of osteoarthritic conditions where the principal cause is unknown. In secondary OA a causal element can be identified. This fundamental classificatory division has been questioned however. Doherty *et al.* (1983), in stressing the multifactorial nature of OA, reported that the prior presence of primary OA significantly impacts on the likelihood of developing secondary OA, in the knee at least. The close relationship between primary OA susceptibility and likelihood of developing secondary OA, given an appropriate catalysing event, would appear to make a fundamental division between the two problematic. Ferguson (1987:439) has also suggested that "the use of the term secondary implies an understanding of the subsequent arthritic process which we do not possess". Notwithstanding this, however, a representative consensus classification of osteoarthritis is presented in Table 4.1. The major primary/secondary division is apparent with primary being further divided into localised and generalised forms. The secondary category includes six aetiologically recognised forms and one form of unknown cause.

(2) Pathogenesis of Osteoarthritis

There are currently models favouring either the bone or cartilage as the initiating site in the pathogenesis of osteoarthritis. While there is a consensus that it is the cartilage that is the principal initiating site (Mankin & Brandt 1997), histological and gross pathological changes can be observed in the earliest stages of arthritis in both joint cartilage and bone.

Cartilage

Changes in osteoarthritic cartilage can be grouped into biochemical, metabolic and morphological. Cartilage is a highly hydrated tissue with much of its dry weight being composed of type II collagen and proteoglycans. While the collagen is chiefly responsible for the architectural or structural integrity of the cartilage matrix, the proteoglycans are involved in packaging much of the water as well as playing a key role in the biochemistry of the matrix. In osteoarthritis the collagen fibres are reduced in size and they become architecturally more disorganised. Further, the frequency of proteoglycans markedly decreases

Table 4.1. Classification of Osteoarthritis (from Solomon 1997)

Primary	
A. Localised	
B. Generalised	
Secondary	
A. Dysplastic	1 Chondrodysplasias
	2 Epiphyseal dysplasias
	3 Congenital joint displacement
	4 Developmental disorders (Perthes' disease, epiphysiolysis)
B. Post-Traumatic	1 Acute
	2 Repetitive
	3 Postoperative
C.Structural Failures	1 Osteonecrosis
	2 Osteochondritis
D. Postinflammatory	1 Infection
	2 Inflammatory arthropathies
E. Endocrine and Metabolic	1 Acromegaly
	2 Ochronosis
	3 Hemochromatosis
	4 Crystal deposition disroders
F. Connective Tissue	1 Hypermobility syndromes
	2 Mucopolysaccaridoses
G. Etiology Obscure	1 Kashin-Beck disease

and the molecular structure of these macromolecules also changes (Mankin & Brandt 1997). Metabolically, the chondrocytes (cartilage cells embedded within the cartilage matrix) increase the synthesis of matrix degrading enzymes. Concomitantly, there is an increase in levels of the enzyme collagenase which is apparently involved in the reduction in size of collagen fibres and disruption of their normally tight weave, both factors of which lead to the commonly observed swelling of the cartilage matrix (Mankin & Brandt 1997). Specifically, it is thought that the proteoglycans, no longer inhibited by formerly tightly woven collagen bundles, are free to expand by becoming further saturated with water (Mankin & Brandt 1997).

Notwithstanding these changes, however, cartilage reacts to these degradative processes by way of increasing the synthesis of collagen, proteoglycans and other matrix compounds (Mankin & Brandt 1997). It would appear that degradative process outstripping reparative responses are responsible for the gross manifestations of osteoarthritis. Part of the reason for the inability of the reparative process to keep up may be related to the poorer quality of products, principally collagen and proteoglycans, synthesised in response to matrix degradation (Mankin & Brandt 1997). Finally, at a gross level, staining reveals discontinuities in the surface of cartilage in fibrillated regions where "the matrix has a fibrillary, disheveled appearance" (Hough 1993:1704). These areas are progressively eroded until the cartilage is totally destroyed, exposing the underlying bone.

Mankin & Brandt (1997) note that the consensus initiating factor in the pathogenesis of OA is damage, either via mechanical stress to the chondrocytes, thus initiating the release of degradative enzymes, or an initial damaging event, again mechanically induced, to the collagen network. Other models suggest that initiating predisposing events occur in the bone as a prelude to these established cartilaginous processes.

Bone

Hough (1993:1704) states that:

> new bone formation occurs in two separate locations in relation to the joint surface: in exophytic growths at the margins of the articular cartilage and in the immediately subadjacent bone marrow. Marginal osteophytes have two patterns of growth. One consists of a protuberance into the joint space [principally beneath areas of cartilage degradation], while the other develops within capsular and ligamentous attachments to the joint margins.

Neither the patho-physiological characteristics nor the relationship of osteophytosis to osteoarthritic events in cartilage are fully understood (Mankin & Brandt 1997).

Some of the possible aetiological pathways to osteophyte formation include vascularisation of the basal layers of damaged cartilage, hypertrophic responses to subchondral microfractures near the joint margins and changes to venous drainage patterns leading to venous congestion (Mankin & Brandt 1997).

A potentially very important recent study by Rogers *et al.* (1997) suggests that certain individuals may be predisposed, perhaps with an underlying genetic basis, to osteophytic development. They have termed these people bone-formers and suggest that "observed variation in bone formation could be due to differences in individual ability to form bone in response to stress rather than differences in stress" (Rogers *et al.* 1997:90). It is also worth noting that although osteophytes are seen as beneficial in increasing joint surface area and thus acting as a joint stabiliser by helping distribute loads over a greater area, it is unclear if this is a specific joint response to joint damage or simply a fortuitous side effect. What does seem clear is that osteophytes do not appear to be a primary, necessary nor definitively diagnostic feature of osteoarthritis.

The formation of osteophytes is not the only bony change to occur in osteoarthritis. As part of the normal process of bone turnover, endochondral ossification proceeds within the lowest layers of cartilage resting on the subchondral base. However, Hough (1993) notes that there is no evidence that this normal process of cartilage calcification contributes to the thinning of the articular cartilage. Further, there does not seem to be any evidence that endochondral ossification is involved in bone sclerosis, a trait commonly used to diagnose OA radiologically (Dequeker *et al.* 1997). Subchondral sclerosis is principally due to an increase in both cortical and trabecular bone mass (Dequeker *et al.* 1997). With the breakdown of cartilage, opposing bone surfaces are brought into contact, resulting in destruction of the subchondral surfaces in the form of eburnation. Cysts can also develop beneath eburnated surfaces, and can be caused by either infiltration of synovial fluid, passing through damaged cartilage, into the subchondral bone, or by way of necrotic defects within the subchondral bone (Mankin & Brandt 1997). The cumulative forces of cartilage and subchondral bone damage can eventually lead to collapse of the joint.

The most important model suggesting the primacy of subchondral changes in initiating the osteoarthritic process was proposed by Radin and colleagues three decades ago (Radin & Paul 1970; Radin *et al.* 1970). Assuming that OA was primarily due to mechanical stress, Radin & Paul (1970) suggested that cartilage was unable to attenuate peak impulsive loading of joints. Radin *et al.* (1970) went on to propose that the subchondral bone is principally responsible for attenuating peak impulse loads and any subsequent changes to this bone, reducing its energy absorption capacity, would then lead to degenerative changes in cartilage. The chief cause of a reduction in the attenuating ability of subchondral bone was seen to be bone stiffening due to callus formation from healed micro-fractures of subchondral trabeculae (Radin *et al.* 1973; Radin *et al.* 1978). While this model appealed to those looking for a relatively straightforward relationship between stress (or joint wear-and-tear) and arthritis, even Radin (Radin & Rose 1986) sees it as only one of a number of possible mechanisms leading to cartilage destruction.

Recently another important study has implicated subchondral changes in initiating osteoarthritis. Dequeker *et al.* (1997) while maintaining Radin's hypothesis that bone stiffening is responsible for subsequent degenerative changes to the cartilage, suggest that changes in bone mineralisation, rather than extensive trabecular micro-fracture repair, are responsible. While they note that weight, exercise, occupation and neuromuscular control all impact on subchondral bone remodelling, they propose that metabolic and systemic factors play an important role. Dequeker *et al.* (1997) ask why do osteoarthritis and osteoporosis generally present as mutually exclusive conditions if it is the cartilage that is the primary initiating site of OA? They suggest that if it is accepted that that force-attenuating properties of the subchondral bone are initially important, then the relationship between OA and osteoporosis can be better understood. In short, osteoporotic bone is too soft, lacks the mass, to stiffen to a sufficient degree to affect overlying cartilage.

In their own study Dequeker *et al.* (1997) examined biochemical, quantitative histomorphological and biomechanical properties of bone from the iliac crest area of patients assessed for OA using X-rays of the hand. The importance of using the iliac crest is that it would be expected to be free of local joint destruction. Findings for individuals with OA included: higher bone density; increased cortical and trabecular bone mass; higher bone stiffness; higher concentration of growth factors responsible for bone mass; and bone matrix composition was significantly different is several aspects between OA and non-OA cases. They went on to conclude that:

> The multiplicity of significant alterations found in patients with primary osteoarthritis and in osteoarthritic bone far away from the articular lesion indicates that osteoarthritis is part of a more generalised bone disease. (Dequeker *et al.* 1997:368)

It would be interesting to see if there is any correlation between this generalised, or systemic, bone disease proposed and the findings of Rogers *et al.* (1997) with regard to bone-forming individuals. The clear implications would appear to be that the underlying mechanisms leading to osteoarthritis may, in large part, be associated with a complex, systemic predisposition to the disease.

What is apparent, however, is that more research in the area of bone mineral density and its relationship to OA is required. It is still not entirely clear if changes to subchondral bone precede or are secondary to changes in cartilage tissue (Burr & Schaffler 1997; Matsui *et al.* 1997). Further, at least two bio-archaeological studies looking at OA and bone mineral density, have reported a negative relationship between the two (Burr *et al.* 1983; Brickley & Waldron 1998). Why the results of bio-archaeological studies should be inconsistent with the results of clinical studies needs further investigation. One potentially important difference is in the different criteria used by clinicians and bio-archaeologists to identify OA. Brickley & Waldron (1998), for example, used eburnation exclusively to diagnose osteoarthritis in their sample. While several bio-archaeologists (Rogers & Waldron 1995; Jurmain 1999) are pressing for eburnation to be exclusively used for the diagnosis of OA in skeletal samples it is, as they agree and others have pointed out (Rothschild 1997), a measure of severe joint degradation. While previous discussion has noted that high bone mineral density may be an initiating factor in the pathogenesis of OA, there is no research, to my knowledge, examining bone mineral density throughout the natural history of the osteoarthritic process. It may be that bone density decreases with advancement of the disease. If such is the case, it may help explain conflicting results between clinical and bioarchaeological studies of bone mineral density and osteoarthritis.

In defence of their own findings, Brickley & Waldron (1998) also stressed that it was unlikely that any systematic age allocation biases entered into their study. However, given that the males in their sample showed a clear positive relationship between higher bone mineral density and OA, while the females showed the reverse, age may be the important factor here. Interestingly, the same pattern was also found in the earlier Eskimo study by Burr *et al.* (1983). Again, if a model is hypothesised that sees high bone mineral density as important only as an initiating factor in OA, a potentially significant portion of females with OA may also have suffered from varying degrees of subsequent post-menopausal (and post-OA) osteoporosis. This would serve to lower their bone mineral density without removing the scars of OA.

(3) Risk Factors

Having examined models seeking to explain the underlying causes and processes of OA, this next section examines a range of factors that appear to increase the risk of either initiating or progressing OA. These risk factors can be divided into those that are intrinsic and those that are extrinsic to an individual or population. However, it must be noted that many of these factors probably act in complex relationships with one another, making the identification of any one element as primary extremely difficult.

Intrinsic Factors

Sex

Reviews of contemporary epidemiological studies have noted an overall OA prevalence of between just under a quarter to a third of the adult population assessed (Creamer & Hochberg 1997; Jurmain 1999). In modern population-based epidemiological studies OA is invariably assessed by way of radiographs. When examined in relation to sex, European studies have shown males to have a generally higher prevalence of OA up to about age fifty and thereafter it is women who are more at risk (Roberts & Burch 1966 cited in Moskowitz 1993). A similar pattern is seen in North America for hip and knee OA in particular (U.S. Department, H.E.W. 1979 cited in Jurmain 1999). Moreover, women appear to be at greater risk of severe OA (Jones & Doherty 1995). Interestingly, Bridges (1992), in a major survey of the American bio-archaeological literature, found that in general males tended to display a higher prevalence of OA than females. Given the attenuated mean age-at-death of many historic and prehistoric populations, the reasons for this pattern may be quite straightforward: females simply did not live long enough to present the pattern seen in current populations.

Ancestry

Several studies have suggested marked ancestral differences in OA. For example an indirect study of the prevalence of primary coxarthrosis (hip OA) by Hoaglund *et al.* (1995), in a San Francisco-based sample, found high rates in the white sample that were comparable with other American studies and slightly lower than several European studies. However, the white prevalence was some two and a half to three times that of the black sample and a massive 15 to 20 times that of the Asian sample. A further study using a Hawaii-based sample (Oishi *et al.* 1998), while displaying similar rates of coxarthrosis for Hawaii and San Francisco Asians, indicated however that Hawaii whites had a lower prevalence than their San Francisco counterparts. Lifestyle differences between the two white samples were posited as a possible explanation for these findings. An earlier study assessing OA of the knee, hip and hands radiologically in a Hong Kong sample also found a low prevalence of hip and knee OA in Chinese, especially when compared to an English sample (Hoaglund *et al.* 1973). Further, this particular study also found a markedly lower prevalence of proximal interphalangeal and first carpometacarpal (thumb base) OA in female Chinese, compared to their English counterparts (Hoaglund *et al.* 1973).

A large radiological study of OA in a total sample of 12,274 individuals including European whites, African and Jamaican blacks and American Indians has examined the prevalence and geographic distribution of OA (Lawrence & Sebo 1980). These researchers also report a very low prevalence of OA in the hip of black Africans, relative to their white and indigenous American samples. Relative to other groups, whites, especially from England, tended to display the highest prevalence of OA while African samples generally displayed the lowest levels. However, they note that there was no clear patterning of OA by latitude or longitude (Lawrence & Sebo 1980). As with Hoaglund *et al.* (1973), Lawrence & Sebo (1980) found a high prevalence of thumb base OA in the white samples and a low prevalence in the black and indigenous American samples. A further finding was a high prevalence of toe (first metatarsophalangeal joint) OA in white English samples, especially urban.

It has also been suggested that whites display a higher prevalence of hip OA as compared to blacks, while blacks tend to have a higher prevalence of shoulder OA than whites (Jurmain 1972 cited in Bridges 1992). Another recent study of the cervical spine in South African whites and blacks (Taitz 1999) has demonstrated a higher prevalence of osteophytosis in whites.

On the other hand, however, studies indicate that ancestral differences in the distribution and prevalence of OA are limited. For instance, the results from the large (6585 individuals) Dutch Zoetermeer survey population was compared with ten other samples totalling 22,629 individuals from England, North America, Bulgaria, South Africa and Japan (van Saase *et al.* 1989). The authors of this study concluded that although there were differences in the prevalence of OA for given joints between these populations, it was not clear if these were simply artefacts of varying methods or real differences based on differential risk factors. However, a particularly important observation is that:

> Joints with a low prevalence of osteoarthritis in one population are relatively spared in all populations, while frequently affected joints show signs of degeneration in all populations. It is therefore most likely that the aetiology of most osteoarthritis is the same in all populations.
> (van Saase *et al.* 1989:279)

It is difficult to reconcile the evidence for and against significant ancestral differences in the prevalence and distribution of OA. It is nonetheless potentially significant that lifestyle and behavioural variables have not been factored into these studies examining ancestral differences in OA prevalence and distribution. The potential for behaviour to seriously confound the issue cannot be underestimated.

Obesity

This factor is mediated by both intrinsic and extrinsic variables. Using the considerable Dutch Zoetermeer sample, van Saase *et al.* (1988) demonstrated a positive correlation between joints commonly affected by OA and obesity. In particular, obese females had a high prevalence of cervical disc, knee and distal interphalangeal joint OA. Men displayed a positive correlation between obesity and OA of the metatarsophalangeal joints of the feet and metacarpophalangeal joints of the hands. They went on to suggest that neither a straightforward wear-and-tear model nor a general metabolic model were consistent with their results. For instance they suggested that in a stress model the hip should not be protected at the expense of the knee. They also noted that:

> The pattern of OA and obesity also does not fully correspond to what one would expect if a general (metabolic) cause existed for OA. If a metabolically induced deterioration of cartilage were to take place, prevalence differences of OA between joints might be smaller than we found them.
> (van Saase *et al.* 1988:1156)

But this is not necessarily so. The metabolic effect, if it exists, may only predispose to OA (see above discussion of Dequeker's model). Other factors, such as joint geometry or age of onset of obesity, may then be crucial in any subsequent osteoarthritic development. In this context it is worth noting that in a longitudinal study spanning up to 36 years, Gelber *et al.* (1999:547) found "that cumulative exposure to greater weight is an important etiologic factor for knee osteoarthritis". They suggest that it is body weight in adolescence and youth that appears to be the greatest risk for the development of OA in later life, rather than mid-life obesity. Interestingly, however, the ankle is much less affected by OA despite it being subject to the same level of loading as the knee (Huch *et al.* 1997). The greater complexity of the knee joint, as well as histological differences in knee and ankle cartilage, may be mediating factors here (Huch *et al.* 1997).

New studies continue to find an association between OA of the hand and knee (Oliveria *et al.* 1999), although the correlation between hand OA and obesity is not universally confirmed (Hochberg *et al.* 1993). On the other hand, the association between knee OA and obesity would appear to be well established. For instance, a higher prevalence of knee OA in modern Holland, as compared to $18^{th}/19^{th}$ century Dutch forebears, has been attributed to greater levels of obesity in the modern population (Baetson *et al.* 1997).

Evolutionary Considerations

Evolutionary considerations here are principally those related to joint morphology and mechanics. Hutton (1987) suggests that rapid hominid evolutionary change over the last few million years has led to changes in both the morphology and use of certain joints. He hypothesises that this has meant some joints have become functionally over-designed while others are under-designed. Hutton (1987:1465) points out that a change to habitual bipedalism has meant that:

> In the foot the great toe is adducted and rotated. Push-off on the first metatarsophalangeal joint occurs. The leg is extended with changes at the hip and knee, but human and chimpanzee ankle bones are very similar. The arm is liberated and the hand becomes more dextrous. The spine now is a column and acts as a support, crane, and shock absorber. The changed loading suggests that the shoulders

> and metacarpals are now overdesigned. The first metatarsophalangeal and apophyseal joints and, to a lesser extent, the hip and knee are underdesigned. The dextrous hand increases loading on underdesigned distal interphalangeal and first carpometacarpal joints. OA should occur most commonly in the joints that are predicted to be underdesigned and be rare in the overdesigned joints. That is the joint distribution found in generalised OA.

This hypothesis has received some support from several recent studies. A comparison between a relatively small sample of human and macaque hands revealed OA of the thumb base was much higher in humans, while distal and proximal interphalangeal joints were generally similar in OA prevalence (Lim *et al.* 1995; Rogers *et al.* 1995; Lim *et al.* 1996). A later survey, with a larger sample size (170 humans, 134 apes and 108 monkeys), confirmed these findings (Mair *et al.* 1998). The human precision grip is likely responsible for this particular OA distribution. This later study also examined the prevalence of OA in the knee of humans, apes and monkeys. Humans, with their unique fully extendible knees, were the only primate to display patellofemoral OA, at a prevalence of 4.2% in this particular study (Mair *et al.* 1998).

While these results seem to confirm the evolutionary hypothesis, with respect to the thumb at least, the human and other primate samples need to be greatly expanded. Lim *et al.'s* (1995) findings for higher thumb base OA in humans compared to primates, was based exclusively on a small white human sample. However, Lawrence & Sebo (1980), utilising a very large sample of Native Americans, whites and blacks, reported a range of thumb base OA prevalence of between 2% and 5% for Nigerians and Liberians (n=440) and up to between 16% and 22% for whites (n=642). Using a white human comparative sample would have ensured a relatively high prevalence of human thumb base OA in Lim *et al.'s* (1995) study.

The evolutionary hypothesis also presents an alternative to another possible model for high levels of thumb base OA. Jonsson & Valtysdottir (1995) presented data supporting their view that joint hyper-mobility, or ligament laxity, was responsible for the high prevalence of thumb OA. However, their work has been criticised, particularly over quantification of hyper-mobility, an otherwise continuous variable, as well as potential sample selection problems (Snaith 1996).

An extremely useful study of the distribution of vertebral syndesmosal and apophyseal OA in a Medieval period English cemetery has been carried out by Knüsel *et al.* (1997). Their findings were more consistent with the mechanics of bipedalism than any occupational related factor. The findings of this study are all the more important due to the fairly tight controls Knüsel *et al.* (1997) were able to employ. For instance, the cemetery was divided into three sections of which socio-economic, if not occupational, status could be ascertained with some certainty. They were also able to demonstrate that the prevalence of other pathologies (spondylolysis, dental caries, DISH, tuberculosis and trauma) were consistent with these status divisions, while vertebral OA was not. They concluded by noting that the vertebral column is not a suitable place to assess skeletal populations for activity related behaviour. The upper limbs and pectoral girdle are better suited for this, given they are exempt from weight bearing functions.

It is worth noting in this context the work by Merbs (1996a, 1996b) on the occurrence of spondylolysis or separation of the neural arch of vertebrae. Merbs (1996a: 357-358) suggests that this condition:

> appears to be closely related to the development of the lumbar curve and the assumption of habitual erect posture by hominids, a relationship supported by the absence of spondylolysis in nonhominid primates and in adults who were never able to walk upright.

It would appear that the human back is victim to a number of design bugs.

While evolutionary factors are no doubt important in the prevalence of some arthropathies, it would be incautious to attribute all findings of knee, hand and back OA to the evolutionary vagaries of under and over-design. A recent study of osteophytosis, both syndesmosal and apophyseal, of the cervical spine of South African whites and blacks revealed some marked differences in distribution and severity (Taitz 1999). The whites not only had a greater prevalence of cervical osteophytosis, but a significant number of whites displayed "both cervical body and apophysial facet osteophytosis on the same vertebrae... In the white sample, the degree of facet osteophytosis is also more marked" (Taitz 1999:107). Because the black population was for the most part socio-economically disadvantaged and engaged in greater levels of physical labour than the whites, these findings appear somewhat unusual at face value. Taitz (1999) has suggested that evidence for a higher bone density in blacks might explain the relative protection of black vertebrae. In this model, increased bone density is seen to act as a joint stabiliser, thus reducing osteophyte development. This seems to be at odds with evidence for hypertrophic bone formation discussed previously however. To explain the distributional differences of apophyseal and cervical body osteophytes, Taitz (1999) suggests an increase in white cervical lordosis, spinal curvature, may be stressing the apophyseal joints. The specific cause of this increase in cervical lordosis is difficult to conjecture though. Nonetheless, recourse to a simple evolutionary explanation would seem inappropriate here.

It is also worth noting that a study by Waldron & Cox (1989), that controlled for occupation, found a greater prevalence of spine OA in non-manual workers. Given that these findings were not consistent with the wear-and-tear hypothesis, they were also at a loss to explain them.

Genetic factors

There is a vast and rapidly growing amount of literature linking an increasing number of genes with various forms of OA. It seems certain that some forms of OA can be inherited by way of mutations to portions of the type II collagen gene (Katzenstein *et al.* 1990; Pun *et al.* 1994). Type II collagen plays a major structural role in both cartilage and bone. Variations in the genotypes of the insulin-like growth factor I (IGF-1) gene have been found to be associated with OA of the knee, hip, spine and hand (Meulenbelt *et al.* 1998). Also, morphological variants of the oestrogen receptor gene have been associated with generalised OA. Further, a female-specific gene for idiopathic OA has also been tentatively recognised (Chapman *et al.* 1999).

Felson *et al.* (1998) suggest a significant polygenic component in the aetiology of generalised osteoarthritis . Spector *et al.* (1996) have also suggested a predisposition to OA at all sites. But importantly they also state, and this is applicable to all genetic studies of OA, that:

> These findings, in addition to current views that environmental risk factors act differently at different sites, such as the effects of obesity and occupation, point towards pronounced heterogeneity in environmental and genetic interactions. The strong genetic influence does not make osteoarthritis 'inevitable'. There are probably important genetic-environmental interactions such that changes in lifestyle in predisposed people may have a strong effect. The nature of the genetic effect is speculative and may entail either a structural defect (for instance, collagen) or alterations in cartilage or bone metabolism. It is therefore possible that the turnover or repair of cartilage and bone is under genetic control with different thresholds and responses in genetically distinct subjects.
> (Spector *et al.* 1996:946)

Extrinsic Factors

Activity and Occupation

There is an enormous clinical/epidemiological literature focusing on the relationship between occupation and risk of osteoarthritis. A recent critical examination of this literature base, that has developed over the past few decades, has highlighted a number of confounding systematic factors complicating the interpretations of many of these studies (Maetzel *et al.* 1997). Of 123 studies of risk factors associated with OA, Maetzel *et al.* (1997) could only find 17 that met reasonably high standards of methodology. A number of the more important study biases were discussed. Representativeness of the sample is very important and can be impacted by inclusion and exclusion protocols and movement of subjects between occupations prior to sampling. There tended to be a large variation in the manner in which ascertainment of exposure was recorded. The vast majority of studies recorded occupational exposure retrospectively, and a number did so in a particularly non-specific manner, by way of recording job titles for example. There was also inconsistency in that some studies informed subjects of the particular aims of their study while others did not. This is especially problematic when the subjects' occupational/ activity recall is used to ascertain exposure. Maetzel *et al.* (1997) note that in most cases OA assessment was by established radiological criteria. Nonetheless, they point out the potential bias in some studies of using more subjective patient reporting of symptoms. Another common problem was the lack of control for such confounding factors as age, sex and weight. The low participation rates in some studies could also have artificially affected results.

Of the 17 studies selected for inclusion in assessing the relationship between occupation and knee and hip OA, Maetzel *et al.* (1997) derived several conclusions. There is a positive correlation between work involving knee bending in males and OA. Such a relationship for females is not evident. There is also a positive correlation between occupation and hip OA in males, such as farming activities. Interestingly, none of the studies that met their selection criteria examined hip OA and occupation in females. Although they go on to note that their review only examined published studies assessing OA by way of radiological criteria, it is unlikely that this would have introduced any particular bias into this review. It would appear, on the face of it, that clear links between OA and specific activities and/or occupations are quite tenuous at present.

Activity and Sport

The link between the participation in sporting activities and OA is still not clear. For some specific sporting activities, such as running, an increase in hip OA has been reported (Marti *et al.* 1989), but most studies report no increased risk of knee and hip OA associated with running (e.g. Lane *et al.* 1986; Konradsen *et al.* 1990; Lane *et al.* 1998). In confirmation of such epidemiological studies, a recent long term experimental study looking at the effects of increased use and loading on canine knees found no association between these factors and cartilage deterioration (Newton *et al.* 1997).

Correlations between long-term exposure to strenuous sports activities in males, especially those involving fast acceleration and deceleration (track and field, racket sports) activities have been reported (Vingard *et al.* 1993). In a later study of hip OA in women, Vingard *et al.* (1998:80) even went as far as to suggest that there was "a dose-response relationship between physical load and osteoarthrosis of the hip". A study of former Finnish male elite athletes suggested a greater risk of developing lower limb OA for this group (Kujala *et al.* 1994).

In fact, correlations between everything from ballet dancing (van Dijk *et al.* 1995) to playing Australian Football (Deacon *et al.* 1997) have been implicated in increasing risks of developing OA. Nonetheless, the implications of these studies are not always straightforward. For instance, radiographic diagnosis of foot OA in the ballet dancers was asymptomatic in all cases, and the predicted increased risk of hip OA was not confirmed. The association between risk of knee OA in the Australian Football players was linked to knee injury. A recent review article of athletics and sports stresses the strong link between sporting injuries and subsequent OA (Buckwalter & Lane 1997). However, the evidence linking sporting activities to primary OA seems to be lacking. Buckwalter & Lane (1997:879) summarise the evidence for the correlation between sport and OA:

> The available evidence shows that moderate habitual exercise does not increase the risk of developing osteoarthritis....People who participated in competitive sports as adolescents or young adults, or who habitually engage in sports or vigorous exercise throughout life, appear to have an increased risk of developing marginal osteophytes that are not associated with cartilage degeneration...It seems likely that the osteoarthritis that develops in some people who participated in or continue to participate in athletics results from undiagnosed joint injuries.

While the vast majority of OA and activity-related literature examines overuse, some research has also suggested an aetiological role for under-use in the risk of OA. Alexander (1989) presented results that suggested that the risk of OA may be related to under-utilisation of the full range of available motion in a particular joint. Alexander (1989) suggested that a good test of this hypothesis would be comparing levels of OA in people who habitually wore shoes and people who habitually went barefoot. Wearing shoes limits the range of movement of the metatarsophalangeal joints and by this model would increase the risk of developing OA. I am unaware if this has been tested to date. Another intriguing recent study has suggested that under-use, as measured by quadriceps muscle weakness in this case, could be an aetiological factor in knee OA rather than an outcome of the disease (Slemenda *et al.* 1997). The link between regular moderate exercise and reducing the risk of developing OA, as well as moderating the effects of OA (Buckwalter & Lane 1997), is also indirect evidence for such a relationship.

(4) Identification of Osteoarthritis

The clinical identification of OA includes variables such as pain, joint mobility, joint stiffness, age, joint enlargement and synovial fluid composition (Moskowitz 1993; Jones & Doherty 1995; Creamer & Hochberg 1997). Specific radiographic signs include marginal osteophytosis, asymmetric joint space narrowing, subchondral cysts and sclerosis of subchondral bone beneath sites of cartilage damage (Solomon 1997). Clearly many of these variables generally only apply to living subjects. Confounding the issue is that often there is a high degree of discordance between gross morphological changes and an individual's experiences of discomfort (Creamer and Hochberg 1997). Moreover, other co-existing conditions (enthesopathy, bursitis, fibromyalgia and gout for example) can further complicate isolating OA as the chief morbid process responsible for an individual's symptoms (Jones & Doherty 1995).

In dry skeletal material a combination of clinical and other criteria believed to be associated with OA are recorded. These variables are osteophytic lipping of the joint margins, osteophytic development on the joint surface, porosity of the joint surface and eburnation of the joint surface. Eburnation is formed from the compaction, or sclerosis, and subsequent polishing (eburnation) of opposing subchondral bone surfaces after the loss of cartilage. Waldron & Rogers (1991), in a study of inter-observer variation in using such criteria to identify OA, found generally poor levels of consistency. By far the best or most consistently identified osteoarthritic trait was eburnation. In response to such recording problems, Buikstra and Ubelaker (1994) devised a set of standardised criteria for recording these variables. However, it is not at all clear what, even with high levels of consistency, the researcher is actually recording.

Cockburn *et al.* (1979) stressed that the presence of osteophytes is not necessarily indicative of OA. Recently, this has been endorsed by others (Dieppe & Rogers 1993; Rogers & Waldron 1995; Buckwalter & Lane 1997; Jurmain 1999). The aetiology of osteophytes and their relationship to osteoarthritis is discussed below.

The use of porosity also appears to be of dubious value in identifying OA in skeletal remains. Rothschild (1997) recently pointed out a lack of clinical acknowledgment for a correlation between porosity and OA. In his own study he did not find a significant correlation between porosity and OA in a sample of 400 individuals from the Hamman Todd collection. Rothschild (1997) suggests that porosity should not be used as a marker for OA. However, it is important to note that he did not use eburnation as a criterion for identifying OA in his sample, but rather subchondral sclerosis and marginal osteophytosis. As noted above, the use of marginal osteophytosis in identifying OA is problematic. Moreover, it is not clear that Rothschild's (1997) arguments (see below) for also excluding eburnation as a criterion for recognising OA would be widely accepted.

Jurmain (1999) has also commented that there does not appear to be a connection between porosity and eburnation. There are several aetiological pathways for the development of porosity including thinning of the articular plate, active avascular invasion of calcified cartilage and bony destruction as a prelude to eburnation (Jurmain 1999:33). Miles (1999) has suggested that apparent nutritional or vascular channels (i.e., porosity) in articular surfaces may be due to increased vascularity in stressed cartilage during the early phases of OA. If this is only a sign of the beginning of OA it may be a feature that does not continue as OA progresses. This would explain the poor correlation between porosity and other signs of OA such as osteophytosis and eburnation, which are variable stage and late stage signs respectively.

The only trait apparently pathognomonic for OA, in dry bones at least, is eburnation (Rogers & Waldron 1995; Jurmain 1999). Not uncommonly associated with eburnated subchondral bone are parallel ridges and grooves. The aetiology of these grooves is unknown, although Rogers & Dieppe (1993) have suggested that they may be the end product of the mechanism(s) producing porosity.

Rothschild (1997:530) argues against the use of eburnation as "the major marker for osteoarthritis". His position is that it is simply a gauge of the severity of OA. Moreover, in pointing out that eburnation is not unique to OA, the status of eburnation as pathognomonic for OA comes into question. Rothschild (1997:530) also goes on to state that eburnation "does not fulfill American College of Rheumatology established criteria for recognition of osteoarthritis". This comment is somewhat surprising given that this set of criteria was established for recognising OA in living subjects, where eburnation is radiographically invisible, though can be inferred from evidence of sclerosis.

Joint contour change is another macroscopic technique useful in identifying OA. Nonetheless, it is generally only the markedly changed joints, such as mushroom-like femoral heads and mandibular condyles, that are recorded as definitely osteoarthritic. The problem is one of quantification of such morphological changes. This issue has been addressed recently by Shepstone *et al.* (1999) who found a correlation between osteoarthritic femoral condyles and morphology of the condyles. Morphological assessment or quantification is relatively straightforward in this study and therefore adds another potentially useful quantifiable osteoarthritic trait.

(5) Common Sites for Osteoarthritis

In general, the most common osteoarthritic sites in contemporary populations are the hands, knee and hip in that order (Dieppe 1990; Cushnaghan & Dieppe 1991), although the spine is arguably the most common site for osteophytosis at least (Jurmain 1980). Further, the foot and in particular the first metatarsophalangeal joint, is prone to osteoarthritic changes (Kerr 1998). With the hand, the most common sites are the thumb base and the distal interphalangeal joints (Lim *et al.* 1996; Mair *et al.* 1998). With respect to the knee, it has been suggested that OA of the patellofemoral compartment is a fairly ancient form, while OA of the tibiofemoral compartment may be of relatively recent origin, in England at least (Rogers & Dieppe 1994).

Patterns of commonly co-occurring joints have prompted the recognition of so-called poly-articular subsets. One particularly common subset is knee and hand OA, especially prevalent in females (Cushnaghan & Dieppe 1991; Hirsch 1996). This sub-set has also been demonstrated to occur in 18th/19th century England, so seems not to be of recent origin (Waldron 1997). Another subset, prevalent in males, includes the elbow and 2nd/3rd metacarpophalangeal joints (Doherty & Preston 1989). Hand and hip OA have also been found to be correlated in elderly American women (Hochberg *et al.* 1995).

It is not clear that all these classificatory sub-sets are clinically meaningful. For example, OA of the hand in the hand-knee sub-set has been shown to be independent of both obesity (Hochberg *et al.* 1993) and bone mineral density (Sowers *et al.* 1999). Both of these factors are considered important in the aetiology of knee OA. Such findings suggest that the utility of poly-articular OA sub-sets must be questioned. Are such classifications measuring real conditions or simply describing fortuitous correlations?

(6) The Role of Bio-Archaeological Studies

The majority of bio-archaeological analyses of OA have tended to assume a primary role for mechanical stress in the aetiology of this disease. As such, much of the research has focused on inferring past activities and occupations from the distribution and prevalence of OA in skeletal samples. In a review of the literature Kennedy (1989) documented a bewildering array of occupational markers such as snow-mobiler's back; porter's neck; milker's neck; atlatl elbow; dog-walker's elbow; cowboy's thumb; seamstress' fingers; weaver's bottom; mountaineer's gait; miner's knee; rider's bone; and policeman's heel. However, Kennedy (1989:154) did note that "anthropological understanding of the causes of markers and their correct identification await further investigation".

Extremely specific and complex behaviours have been inferred from the severity and distribution of OA. Examples include Molleson's (1989, 1994) seed grinding females and the activities associated with fur trading in 18th/19th century Canada (Lovell & Dublenko 1999). However, a number of skeletal researchers have questioned the validity of such behavioural inferences from these types of studies. In a review of OA in pre-agricultural and agricultural period North Americans, Bridges (1991:387) concluded "there is no unitary response in level of arthritis to the advent or intensification of agriculture in the Eastern United States". She went on to note that the use of different hunting techniques, such as spear throwing as opposed to bow and arrow, could not be differentiated. Moreover, no clear evidence for OA associated with corn processing was found. Osteoarthritis did not appear to be sensitive to the broad behavioural changes involved in a change from a hunter-gatherer lifestyle to an agricultural one. These findings were corroborated in a further, broader review of the North American specific literature by Bridges (1992:86):

> Type of subsistence economy and level or patterning of arthritis cannot be coupled directly. Nor can arthritis be linked to any specific economic activity. Correlation of arthritis with particular activities in the past was often simplistic; it ignored the wide range of individuals' activities as well as the traumatic injuries they may have suffered.

Moreover, Larsen (1995:200), in reviewing the global prehistoric evidence for OA and subsistence, concluded that "no definitive pattern relating prevalence and subsistence mode emerges except with regard to specific geographic settings". Stirland (1991:40) has also questioned the use of all kinds of so-called palaeopathogical markers of occupation:

> No matter how good the preservation and recovery of artifacts from an archaeological site...the interpretation of their use and the activities involved relies heavily on assumptions based on ethnographic and historical parallels. Such interpretation is speculative.

Jurmain (1999:130), in his own extensive review of the literature, is even more scathing of much past and current bio-archaeological research:

> given the motivation, 'alluring prospects,' irresistible temptations, etc., anthropologists have devised a host of imaginative scenarios attempting to link osteoarthritic changes to a host of activities.

and

> It would thus seem that the accumulating literature (both clinical and anthropological) on OA has done little to dent enthusiasm regarding behavioral interpretations as derived from skeletal involvement. Quite the opposite trend, in fact, seems apparent. Indeed, in many cases, the assumptions regarding the theoretical bases of such conjectures are taken as established. While in more controlled epidemiological surveys, obvious controls regarding sampling and rigor of hypothesis testing are taken as basic, in some contemporary research, occupation (and other specific behavioral conclusions) are thought reasonable to infer from one (or a handful) of skeletons. What such an approach actually *proves*, is not at all clear. (Jurmain 1999:137)

It might be asked if there is any hope in pursuing such activity/occupation related research with skeletal remains. Stirland & Waldron (1997) have suggested that, at the very least, the specific criteria suggested by Merbs (1983) must be met if particular activities are to be inferred from skeletal remains. These criteria include: (1) reasonable sample size and completeness or preservation; (2) a known limited range of specific activities that the people engaged in; and (3) that the sample be temporally constrained. In their own study (Stirland 1991; Stirland & Waldron 1997) of the crew of the 16th century war ship the Mary Rose, they were able to meet all these criteria as well as the additional factor of a roughly contemporary late Medieval control sample. Nonetheless, they are very cautious in interpreting their results. While they suggest that skeletal changes in the crew of the Mary Rose were consistent with on-board life and activities, they avoid inferring specific shipboard activities from particular sets of skeletal changes.

Other studies have also examined specific occupational hypotheses using archaeological data. For instance, Waldron & Cox (1989) tested the view that hand OA was correlated with weaving in the 18th/19th century English Spitalfields sample. Some faith can be placed in their negative findings given that they met the criteria outlined above.

While not specifically addressed in this review, activity and occupational inferences from enthesial and syndesmosal development require adherence to these criteria as well, if they are to have any validity. Stirland's (1998) and Jurmain's (1999) recent reviews of the literature on these so-called occupational stress and musculoskeletal stress markers suggests an almost complete lack of clinical supporting literature for the occupational and activity-related claims being made. In a recent symposium titled *Activity Patterns and Musculoskeletal Stress Markers*, Stirland (1998) was the only voice of caution in a series of papers making extremely specific activity related inferences from entheses and syndesmoses. For example, some of the behaviours inferred included: increased agricultural production in post-contact New Mexico (Chapman 1997); male labour activity differences by biome (savanna compared to forest) in pre-European contact South Africa (Churchill & Morris 1998); use of atlatl or spearing throwing technology in late Pleistocene-early Holocene Levant (Peterson 1998); and arduous domestic duties in a female and the rigours of fur trading in two males from 18th/19th century Canada (Lovell & Dublenko 1999). None of these studies met the minimum criteria essential for inferring activity or occupation, described above, with any degree of plausibility.

As stressed by Jurmain (1999:253):

> Imaginatively constructed 'best fit' activity-based scenarios to explain skeletal variation can be used to accommodate and correlate *any* pattern... such reasoning becomes irresistibly circular.

What then is the role of bio-archaeological studies of OA, in particular, and joint change in general? The answer may appear somewhat disappointing at first but is nonetheless extremely important. The role of such studies of past populations is to add to the epidemiological database on osteoarthritis and other joint arthropathies and changes (Jurmain & Kilgore 1995; Jurmain 1999). Rogers & Dieppe (1994:612) have pointed out that:

> little is known about the relative frequency of OA at different sites in ancient skeletons. Any large variation in the relative prevalence of OA of different joints might imply that there has been a change in risk factor exposure, and could have implications in the development of strategies for disease prevention.

Clearly, this is a major and worthwhile reason to pursue any bio-archaeological study, whether it be one of dental health or the description of rare specific diseases. There is an enormous bio-archaeological literature (e.g. Kennedy 1989; Bridges 1992; Jurmain 1999) documenting primary osteoarthritis of the axial and peripheral joints in the past, although an assumption of behavioural or activity-related causation is generally implicitly or explicitly held. Nonetheless, on a reduced scale, rare subsets of primary and secondary osteoarthritis are being increasingly reported in the bio-archaeological and clinical literature. For instance, rheumatoid arthritis has been reported from Jomon period (Neolithic) Japan (Inoue *et al.* 1999), the Roman and Medieval periods of Normandy (Blondiaux *et al.* 1997), Medieval England (Hacking *et al.* 1994) and also North America some 5000 to 3000 years bp (Rothschild *et al.* 1988). Good evidence for a rare case of juvenile-onset rheumatoid arthritis comes from a 16th or 17th century individual in New Mexico (Hawkey 1998). Diffuse Idiopathic Skeletal Hyperostosis (DISH) has been reported from Medieval period Italy (Reale *et al.* 1999) and England (Waldron 1985). Spondyloarthropathy has recently been identified in North African skeletal samples dated to around 2000 years bp (Rothschild *et al.* 1999), as well as possibly Byzantine period Jerusalem (Nagar *et al.* 1999).

With regard to OA at least, this literature review has shown that the aetiology of OA is extremely complex and likely involves a heterogenous milieu of factors involved in manifold, and as yet largely unknown, interactions. Information on the distribution of OA, its prevalence and biases as regard to sex and age are epidemiologically important and can assist in the development of testable clinical models of osteoarthritis. Moreover, if there is a major mechanical stress component in the aetiology of OA, controlled bio-archaeological studies, meeting the Merbs-Stirland minimum criteria for samples, can add to an understanding of this aspect.

Articular Change in the Vietnamese Samples

Activity and occupational inferences cannot be made using the Vietnamese skeletal material as the Merbs-Stirland minimum for samples cannot be satisfied. This study documents the prevalence and distribution of specific joint changes in the Vietnamese samples. Very broad and tentative comparisons are made with other archaeological and contemporary findings.

Methods and Materials

The articular surfaces of all available material were assessed for the presence of pre-mortem changes to the joint surfaces. Porosity, eburnation (EB), marginal osteophytes (MO) and articular surface osteophytes (ASO) were all recorded as separate manifestations of bony joint change. Examples of these changes can be seen in Buikstra & Ubelaker (1994: Figures 111-112). In each case these bony changes were described, rather than scored on an ordinal scale, with respect to extent and severity. The reason for this is that for many joint surfaces the use of an ordinal scoring system would have been confounded by the frequent occurrence of joint surface post-mortem damage. Moreover, in the subsequent data summary severity and extent are disregarded in favour of presence or absence of any given trait.

Rogers & Waldron (1995) outline a number of conventions epidemiologists use for presenting data. One of these is that prevalence (cases with the disease/n) is given for whole individuals, an example being that if reporting OA for the hip only those individuals with two assessable hips can be included. They also point out that you could report by side, but this would be unusual epidemiological practice. Notwithstanding these conventions however, it would be extremely limiting to apply these constraints of analysis to skeletal remains from the majority of archaeological contexts. The chief problem is one of differential preservation of skeletal elements. Only to include those individuals with complete right and left elbows (the articular areas of six separate bones) would no doubt render attempts at gleaning information from many samples impossible.

In order to derive as much information as possible from the remains represented in these Vietnamese samples, the articular areas of individual bones are the unit of assessment. However, joint complexes form the focus of the discussion of results. For example, while the proximal ulna, proximal radius and distal humerus are assessed separately, the results are examined both by individual bone type and the elbow as a functional whole. While this is arguably not an ideal approach, it is the best possible compromise given the extreme paucity of complete joint complexes. Furthermore, the distribution and prevalence of joint surface changes for individual bones, much of it probably not strictly attributable to osteoarthritis specifically, is also of interest and the documentation of this is one of the aims of this chapter.

Statistical tests are not performed on the following results due to problems of sample size and differences in joint representation between samples. The problem is compounded when examining the biologically more meaningful joint complexes, rather than individual bone types, as these are composite structures with a varying contribution from their constituent parts. It is not practical to develop an age profile for each period in this analysis, as the age composition of each joint subset is a specific and different sub-sample. Where appropriate, specific reference is made to the age profile of any particular joint sub-set.

Results

Da But Period Sample

The Temporomandibular Joint

The temporomandibular joint (TMJ) consists of the mandibular condyle and the mandibular fossa. Of the 33 individuals with at least one assessable mandibular condyle, 21.2% displayed MO and 33.3% porosity (Table 4.2a). Neither ASO nor eburnation were seen in this sample. The cases of MO, five males and two females, were all individuals aged 30+ years. Porosity, four males and seven females, was likewise limited to those 30+ years old. The correlation matrix at the bottom of Table 4.2a illustrates the co-occurrence of the various traits on the same articular surface. It can be seen that porosity co-occurred with MO in five mandibular condyles.

Condyles were also assessed for gross morphological (contour) change. Only one individual displayed such gross changes to the contour of the condyle (Figure 4.1). This 40-49 year old female (80CCNM64a) also displayed MO and porosity. Neither mandibular fossa of this individual was preserved. While eburnation was not apparent, it is clear that this individual suffered from a severe degree of osteoarthritis of the temporomandibular joint.

The prevalence of a further feature, called condyle depressions (CD) for convenience, apparent in both the Da But Period and Metal period samples was also assessed (Table 4.2a). This feature manifests as a shallow depression

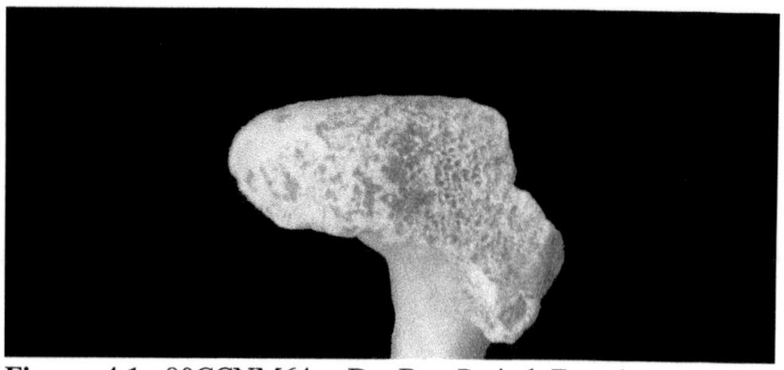

Figure 4.1. 80CCNM64a. Da But Period Female aged 50+ years. Dorsal aspect of a mushroom-like mandibular condyle. Appearance is consistent with osteoarthritis of the TMJ

ranging from several millimetres up to 15mm in diameter. The depression is always found on the posterior aspect of the mandibular condyle, within the area covered by articular tissues and the joint capsule in life (Figure 4.2). The depressions are usually roughly circular or oval with the long axis running medio-laterally. There is always some degree of pitting or porosity of the depressed basin. Such pitting associated with a clear depression was not scored as porosity separately. There were three individuals with such depressions, a male aged less than 30 years, a male aged 40+ years and a female aged 40+ years. While CD is always associated with porosity, no cases with porosity outside the depressed basin were recorded.

Of the 49 individuals with at least one assessable mandibular fossa, 12.2% displayed ASO and 34.7% porosity (Table 4.2b). No cases of MO (an unlikely event given the morphology of the fossa) or eburnation were recorded. An even number of males and females were affected by ASO and all were aged 40+ years. More females were affected by porosity and it was found in all age classes. An examination of the correlation matrix for Table 4.2b indicates of six mandibular fossae displaying ASO, porosity co-occurred with ASO in four instances.

Looking at the jaw as a whole, it is evident that of the 13 individuals exhibiting MO and/or ASO, seven had both assessable condyles and fossae (Table 4.2c). Moreover, none of these seven individuals displayed the condition

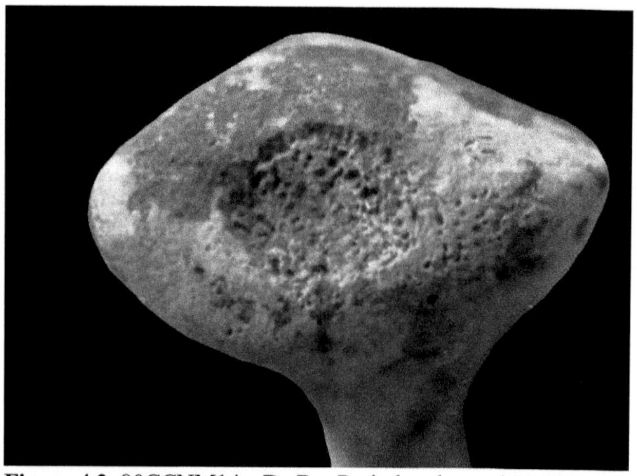

Figure 4.2. 80CCNM14a. Da But Period male aged 20-29 years. Dorso-posterior aspect of mandibular condyle displaying a well differentatied condylar depression. Note extensive porosity of the floor of the depression

Table 4.2a. Da But Period: Changes to the Articular Surface of the Mandibular Condyles (TMJ)

	Left Side									Right Side									Total Individuals*								
	male			female			total			male			female			total			male			female			total		
	obs	n	%	obs	n	%	obs	n	%	obs	n	%	obs	n	%	obs	n	%	obs	n	%	obs	n	%	obs	n	%
marginal osteophytosis																											
<30	0	2	0.0	0	2	0.0	0	5	0.0	0	1	0.0	0	1	0.0	0	5	0.0	0	2	0.0	0	3	0.0	0	9	0.0
30-39	2	6	33.3	0	1	0.0	2	7	28.6	1	4	25.0	0	1	0.0	1	5	20.0	3	8	37.5	0	1	0.0	3	9	33.3
40+	2	5	40.0	2	5	40.0	4	10	40.0	0	3	0.0	1	7	14.3	1	10	10.0	2	6	33.3	2	9	22.2	4	15	26.7
TOTAL	4	13	30.8	2	8	25.0	6	22	27.3	1	8	12.5	1	9	11.1	2	20	10.0	5	16	31.3	2	13	15.4	7	33	21.2
porosity																											
<30	0	2	0.0	0	2	0.0	0	5	0.0	0	1	0.0	0	1	0.0	0	5	0.0	0	2	0.0	0	3	0.0	0	9	0.0
30-39	2	6	33.3	1	1	100.0	3	7	42.9	0	4	0.0	1	1	100.0	1	5	20.0	2	8	25.0	1	1	100.0	3	9	33.3
40+	2	5	40.0	4	5	80.0	6	10	60.0	0	3	0.0	3	7	42.9	3	10	30.0	2	6	33.3	6	9	66.7	8	15	53.3
TOTAL	4	13	30.8	5	8	62.5	9	22	40.9	0	8	0.0	4	9	44.4	4	20	20.0	4	16	25.0	7	13	53.8	11	33	33.3
condyle depressions																											
<30	0	2	0.0	0	2	0.0	0	5	0.0	1	1	100.0	0	1	0.0	1	5	20.0	1	2	50.0	0	3	0.0	1	9	11.1
30-39	0	6	0.0	0	1	0.0	0	7	0.0	0	4	0.0	0	1	0.0	0	5	0.0	0	8	0.0	0	1	0.0	0	9	0.0
40+	1	5	20.0	0	5	0.0	1	10	10.0	0	3	0.0	1	7	14.3	1	10	10.0	1	6	16.7	1	9	11.1	2	15	13.3
TOTAL	1	13	7.7	0	8	0.0	1	22	4.5	1	8	12.5	1	9	11.1	2	20	10.0	2	16	12.5	1	13	7.7	3	33	9.1

Note: ASO and EB not observed for the condyles

Correlation Matrix (left and right total)

	MO	ASO	P	CD
MO	**8**			
ASO	0	**0**		
P	5	0	**13**	
CD	0	0	0	**3**

* Total Individuals (numerator and denominator include only one element side when both present)

Table 4.2b. Da But Period: Changes to the Articular Surface of the Mandibular Fossa (TMJ)

	Left Side									Right Side									Total Individuals*								
	male			female			total			male			female			total			male			female			total		
	obs	n	%	obs	n	%	obs	n	%	obs	n	%	obs	n	%	obs	n	%	obs	n	%	obs	n	%	obs	n	%
surface osteophytosis																											
<30	0	2	0.0	0	3	0.0	0	7	0.0	0	2	0.0	0	3	0.0	0	8	0.0	0	2	0.0	0	5	0.0	0	11	0.0
30-39	0	9	0.0	0	4	0.0	0	13	0.0	0	7	0.0	0	1	0.0	0	8	0.0	0	11	0.0	0	5	0.0	0	16	0.0
40+	2	7	28.6	3	7	42.9	5	15	33.3	3	8	37.5	2	9	22.2	5	17	29.4	3	11	27.3	3	10	30.0	6	22	27.3
TOTAL	2	18	11.1	3	14	21.4	5	35	14.3	3	17	17.6	2	13	15.4	5	33	15.2	3	24	12.5	3	20	15.0	6	49	12.2
porosity																											
<30	0	2	0.0	2	3	66.7	2	7	28.6	0	2	0.0	2	3	66.7	3	8	37.5	0	2	0.0	3	5	60.0	4	11	36.4
30-39	0	9	0.0	1	4	25.0	1	13	7.7	1	7	14.3	0	1	0.0	1	8	12.5	1	11	9.1	1	5	20.0	2	16	12.5
40+	0	7	0.0	2	7	28.6	3	15	20.0	3	8	37.5	6	9	66.7	9	17	52.9	3	11	27.3	7	10	70.0	11	22	50.0
TOTAL	0	18	0.0	5	14	35.7	6	35	17.1	4	17	23.5	8	13	61.5	13	33	39.4	4	24	16.7	11	20	55.0	17	49	34.7

Note: ASO and EB not observed for the mandibular fossa.

Correlation Matrix (left and right total)

	MO	ASO	P	EB
MO	**0**			
ASO	0	**6**		
P	0	4	**19**	
EB	0	0	0	**0**

* Total Individuals (numerator and denominator include only one element side when both present)

Table 4.2c. Da But Period: Individuals Affected by Articular Changes to the TMJ

			Condyle				Fossa	
case	sex	age	MO	P	CD	MC	ASO	P
80CCNM16b	M	5						Y
80CCNM38a	M	5	Y	Y				
80CCNM39a	M	5					Y	Y
80CCNM43a	M	5			Y		Y	
80CCNM61a	M	5						Y
80CCNM74a	M	5					Y	
80CCNM45a	F	5		Y				
80CCNM48a	F	5		Y				
80CCNM4a	F	5					Y	Y
80CCNM75a	F	5	Y	Y				Y
80CCNM33b	M	4	Y	Y				
80CCNM1y	F	4		Y				Y
80CCNM31a	F	4					Y	Y
80CCNM33c	F	4		Y				
80CCNM35b	F	4			Y			Y
80CCNM64a	F	4	Y	Y		Y		
80CCNM79a	F	4						Y
80CCNM83a	F	4					Y	Y
80CCNM22a	I	4						Y
80CCNM17a	M	3	Y					
80CCNM18a	M	3		Y				
80CCNM35a	M	3						Y
80CCNM8a	M	3	Y	Y				
80CCNM97a	M	3	Y					
80CCNM2a	F	3						Y
80CCNM53a	F	3		Y				
80CCNM14a	M	2			Y			
80CCNM33a	F	2						Y
80CCNM5a	F	2						Y
80CCNM71a	F	2						Y
80CCNM52a	I	1						Y

Y: articular change observed
(blank): articular change not observed
: articular area not preserved

M=male
F=female
I=indeterminate

5= 50+ years
4=40-49 years
3=30-39 years
2=20-29 years
1=15-19 years

both on the condyle and in the fossa. Of the three individuals with CD, one had associated ASO of the fossa, one had associated porosity of the fossa, while the third individual lacked an assessable mandibular fossa.

The Shoulder

The shoulder is defined here as the articular surfaces of the proximal humerus and glenoid fossa of the scapula. Of the 26 individuals with at least one assessable proximal humerus 15.4% displayed MO (Table 4.3a). Three of these individuals were male, one was female, all were aged 30+ years and the right side was more frequently affected than the left. Porosity was present in 15.4% of individuals, or one female and three males. The age distribution was fairly even and porosity affected the left side more so than the right. No cases of eburnation or ASO were recorded. Porosity co-occurred with MO in one individual.

For the glenoid fossa of the scapula, no cases of ASO or eburnation were recorded. MO affected 25% of individuals with a higher percentage of males affected than females (Table 4.3b). The left side was more commonly affected than the right and overall, most cases occurred in individuals 40+ years old. Porosity affected 39.3% of individuals with a fairly even distribution across age classes. A higher percentage of males were affected than females and a side bias is not apparent. Of the nine glenoid fossae displaying MO, porosity co-occurred in five of these.

Looking at the shoulder as a whole, of the four individuals displaying humeral MO two of these also displayed glenoid MO. The remaining two individuals with humeral MO did not have a corresponding scapula to assess. Of the four individuals displaying humeral porosity three of these also

Table 4.3a. Da But Period: Changes to the Articular Surfaces of the Proximal Humerus (Shoulder)

	Left Side									Right Side									Total Individuals*								
	male			female			total			male			female			total			male			female			total		
	obs	n	%	obs	n	%	obs	n	%	obs	n	%	obs	n	%	obs	n	%	obs	n	%	obs	n	%	obs	n	%
Marginal Osteophytosis																											
<30	0	2	0.0	0	2	0.0	0	4	0.0	0	1	0.0	0	2	0.0	0	4	0.0	0	3	0.0	0	2	0.0	0	6	0.0
30-39	1	6	16.7	0	1	0.0	1	7	14.3	2	5	40.0	0	2	0.0	2	7	28.6	2	9	22.2	0	2	0.0	2	10	20.0
40+	0	3	0.0	0	2	0.0	0	5	0.0	1	3	33.3	1	3	33.3	2	6	33.3	1	5	20.0	1	5	20.0	2	10	20.0
TOTAL	1	11	9.1	0	5	0.0	1	16	6.3	3	9	33.3	1	7	14.3	4	17	23.5	3	17	17.6	1	9	11.1	4	26	15.4
Porosity																											
<30	1	2	50.0	0	2	0.0	1	4	25.0	0	1	0.0	0	2	0.0	0	4	0.0	1	3	33.3	0	2	0.0	1	6	16.7
30-39	1	6	16.7	1	1	100.0	2	7	28.6	0	5	0.0	1	2	50.0	1	7	14.3	1	9	11.1	1	2	50.0	2	10	20.0
40+	1	3	33.3	0	2	0.0	1	5	20.0	0	3	0.0	0	3	0.0	0	6	0.0	1	5	20.0	0	5	0.0	1	10	10.0
TOTAL	3	11	27.3	1	5	20.0	4	16	25.0	0	9	0.0	1	7	14.3	1	17	5.9	3	17	17.6	1	9	11.1	4	26	15.4

Note: ASO and EB were not observed for the proximal humerus

Humerus Correlation Matrix (left and right total)

	MO	ASO	P	EB
MO	**5**			
ASO	0	**0**		
P	1	0	**5**	
EB	0	0	0	**0**

* Total Individuals (numerator and denominator include only one element side when both present)

Table 4.3b. Da But Period: Changes to the Articular Surfaces of the Scapula Glenoid (Shoulder)

	Left Side									Right Side									Total Individuals*								
	male			female			total			male			female			total			male			female			total		
	obs	n	%	obs	n	%	obs	n	%	obs	n	%	obs	n	%	obs	n	%	obs	n	%	obs	n	%	obs	n	%
Marginal Osteophytosis																											
<30	0	2	0.0	0	3	0.0	0	6	0.0	0	1	0.0	0	2	0.0	0	4	0.0	0	2	0.0	0	4	0.0	0	8	0.0
30-39	1	2	50.0	0	1	0.0	1	3	33.3	0	2	0.0	0	4	0.0	0	6	0.0	1	3	33.3	0	4	0.0	1	7	14.3
40+	2	3	66.7	3	8	37.5	5	11	45.5	2	3	66.7	1	7	14.3	3	10	30.0	3	4	75.0	3	9	33.3	6	13	46.2
TOTAL	3	7	42.9	3	12	25.0	6	20	30.0	2	6	33.3	1	13	7.7	3	20	15.0	4	9	44.4	3	17	17.6	7	28	25.0
Porosity																											
<30	1	2	50.0	0	3	0.0	1	6	16.7	1	1	100.0	1	2	50.0	3	4	75.0	1	2	50.0	1	4	25.0	3	8	37.5
30-39	1	2	50.0	1	1	100.0	2	3	66.7	1	2	50.0	1	4	25.0	2	6	33.3	2	3	66.7	1	4	25.0	3	7	42.9
40+	2	3	66.7	2	8	25.0	4	11	36.4	1	3	33.3	1	7	14.3	2	10	20.0	2	4	50.0	3	9	33.3	5	13	38.5
TOTAL	4	7	57.1	3	12	25.0	7	20	35.0	3	6	50.0	3	13	23.1	7	20	35.0	5	9	55.6	5	17	29.4	11	28	39.3

Note: ASO and EB were not observed for the scapula glenoid

Glenoid Correlation Matrix (left and right total)

	MO	ASO	P	EB
MO	**9**			
ASO	0	**0**		
P	5	0	**14**	
EB	0	0	0	**0**

* Total Individuals (numerator and denominator include only one element side when both present)

Table 4.4a. Da But Period: Changes to the Articular Surfaces of the Distal Humerus (Elbow)

Left Side										Right Side									Total Individuals*								
	male			female			total			male			female			total			male			female			total		
	obs	n	%	obs	n	%	obs	n	%	obs	n	%	obs	n	%	obs	n	%	obs	n	%	obs	n	%	obs	n	%
Marginal Osteophytosis																											
<30	0	3	0.0	0	3	0.0	0	10	0.0	0	2	0.0	0	3	0.0	0	9	0.0	0	4	0.0	0	4	0.0	0	13	0.0
30-39	0	9	0.0	1	5	20.0	1	14	7.1	0	8	0.0	1	3	33.3	1	11	9.1	0	10	0.0	1	5	20.0	1	15	6.7
40+	1	6	16.7	0	2	0.0	1	8	12.5	2	7	28.6	1	5	20.0	3	12	25.0	2	9	22.2	1	6	16.7	3	15	20.0
TOTAL	1	18	5.6	1	10	10.0	2	32	6.3	2	17	11.8	2	11	18.2	4	32	12.5	2	23	8.7	2	15	13.3	4	43	9.3
Articular Surface Osteophytosis																											
<30	1	3	33.3	0	3	0.0	1	10	10.0	0	2	0.0	0	3	0.0	0	9	0.0	1	4	25.0	0	4	0.0	1	13	7.7
30-39	2	9	22.2	0	5	0.0	2	14	14.3	2	8	25.0	1	3	33.3	3	11	27.3	2	10	20.0	1	5	20.0	3	15	20.0
40+	2	6	33.3	0	2	0.0	2	8	25.0	2	7	28.6	0	5	0.0	2	12	16.7	3	9	33.3	0	6	0.0	3	15	20.0
TOTAL	5	18	27.8	0	10	0.0	5	32	15.6	4	17	23.5	1	11	9.1	5	32	15.6	6	23	26.1	1	15	6.7	7	43	16.3
Porosity																											
<30	0	3	0.0	0	3	0.0	0	10	0.0	1	2	50.0	1	3	33.3	2	9	22.2	1	4	25.0	1	4	25.0	2	13	15.4
30-39	2	9	22.2	3	5	60.0	5	14	35.7	2	8	25.0	2	3	66.7	4	11	36.4	3	10	30.0	3	5	60.0	6	15	40.0
40+	2	6	33.3	1	2	50.0	3	8	37.5	2	7	28.6	1	5	20.0	3	12	25.0	2	9	22.2	2	6	33.3	4	15	26.7
TOTAL	4	18	22.2	4	10	40.0	8	32	25.0	5	17	29.4	4	11	36.4	9	32	28.1	6	23	26.1	6	15	40.0	12	43	27.9
Eburnation																											
<30	0	3	0.0	0	3	0.0	0	10	0.0	0	2	0.0	0	3	0.0	0	9	0.0	0	4	0.0	0	4	0.0	0	13	0.0
30-39	0	9	0.0	0	5	0.0	0	14	0.0	1	8	12.5	0	3	0.0	1	11	9.1	1	10	10.0	0	5	0.0	1	15	6.7
40+	0	6	0.0	0	2	0.0	0	8	0.0	0	7	0.0	0	5	0.0	0	12	0.0	0	9	0.0	0	6	0.0	0	15	0.0
TOTAL	0	18	0.0	0	10	0.0	0	32	0.0	1	17	5.9	0	11	0.0	1	32	3.1	1	23	4.3	0	15	0.0	1	43	2.3

Correlation Matrix (left and right total)

	MO	ASO	P	EB
MO	**6**			
ASO	3	**10**		
P	5	7	**17**	
EB	0	1	0	**1**

* Total Individuals (numerator and denominator include only one element side when both present)

displayed glenoid porosity. The remaining individual with humeral porosity did not have a corresponding scapula to assess.

The Elbow

The elbow is defined here as the articular surfaces of the distal humerus, proximal radius and proximal ulna. Of the 43 individuals with at least one assessable distal humerus, 9.3% displayed MO, 16.3% ASO, 27.9% porosity and 2.3% eburnation (Table 4.4a). The age distribution of both MO and ASO concentrated in the older age groups. The distribution of porosity was highest in the 30-39 year group followed by the oldest age class, but present in the youngest age group as well. More females displayed MO than males and MO overall tended to favour the right side. On the other hand, ASO was more common in males and overall there was no side preference. More females displayed porosity with occurrence by side being fairly even. The only case pathognomonic of osteoarthritis, that is displaying eburnation, occurred in the right humerus of a male aged 30-39 years. The left distal humerus of this individual also displayed porosity and ASO. Of the six humeri displaying MO three also displayed ASO, five porosity while none exhibited eburnation. Of the ten humeri displaying ASO seven also displayed porosity and one eburnation. None of the 17 humeri exhibiting porosity displayed eburnation.

Of the 39 individuals with at least one assessable proximal radial head only one case of MO and ASO, both occurring in the same female individual aged 40-49 years, was recorded (Table 4.4b). No cases of porosity or eburnation were recorded for the proximal radius.

Of the 47 individuals with at least one assessable proximal ulna head 14.9% displayed MO, 2.1% ASO, 10.6% porosity and 2.1% eburnation (Table 4.4c). The age distribution of MO concentrated in the older age groups and the only case of ASO also occurred in the oldest age class. The distribution of porosity was highest in the two oldest age groups, but present in the youngest age group as well. The only recorded case of eburnation occurred in the right ulna (left side not preserved) of a female aged 20-29 years. This ulna also displayed MO, ASO and porosity (Figure 4.3). Of the eight ulnae exhibiting MO only one also displayed ASO, four displayed porosity and one eburnation. The only ulna displaying ASO also exhibited porosity and eburnation.

Looking at the elbow as a whole, 14 individuals (26.4%) displayed either MO, ASO or both conditions (Table 4.4d). Two individuals displayed eburnation in addition to MO, ASO and also porosity. There is a clear pattern visible where ASO is more common on the distal humerus, while MO is more common on the corresponding proximal ulna. Moreover, of the seven individuals displaying MO of the proximal ulna, three also displayed ASO of the distal humerus, while three of the seven could not be assessed for the corresponding humerus. The proximal radius appears to be an uncommon site for the types of joint changes assessed in this study.

The Wrist

The wrist is defined here as the articular surfaces of the distal radius and distal ulna. Of the 32 individuals with at least one assessable distal radius 9.4% displayed MO, 3.1% ASO and 6.3% porosity while no cases of eburnation were recorded (Table 4.5a). The two individuals displaying porosity also displayed MO and one also exhibited ASO. Only one individual with MO did not also display ASO or porosity. Of these three individuals showing joint changes all were 30+ years old.

Of the 19 individuals with at least one assessable distal ulna the only recorded observation was porosity in one individual (5.3%), a male aged 40+ years (Table 4.5b).

Taking the wrist as a whole, 35 individuals have at least one assessable distal radius and or/distal ulna. Of these, four individuals (11.4%) display some form of joint change with three of these individuals (8.6%) having MO and/or ASO.

The Hand

The hand is defined here as the articular surfaces of all carpal, metacarpal and hand phalanges. Twenty four individuals are represented by a total of 56 assessable carpals (16 capitates, 9 hamates, 11 lunates, 11 naviculars, 4 trapezia, 4 trapezoids and 1 triquetral). None of these carpal elements exhibited articular changes (see Table 4.6a for sample sizes).

The fingers of the hand were assessed by joint complex. However, the carpometacarpal joint was defined here as the proximal articular surfaces of the metacarpals only. Of a total of 125 metacarpals, representing 43 individuals, only one joint (0.8%) displayed any form of articular change (Table 4.6b). This one instance occurred on the left first metacarpal of a male aged 50+ years. The left and right second metacarpals were also preserved, but did not display evidence for articular change.

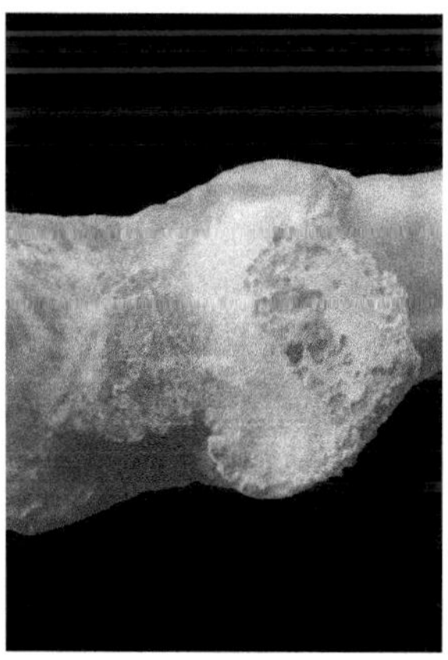

Figure 4.3. 80CCNM33c. Da But Period female aged 20-29 years. Dorsal aspect of left proximal ulna. Note extensive eburnation of the coronoid process

Table 4.4b. Da But Period: Changes to the Articular Surfaces of the Proximal Radius (Elbow)

	Left Side									Right Side									Total Individuals*								
	male			female			total			male			female			total			male			female			total		
	obs	n	%	obs	n	%	obs	n	%	obs	n	%	obs	n	%	obs	n	%	obs	n	%	obs	n	%	obs	n	%
Marginal Osteophytosis																											
<30	0	3	0.0	0	2	0.0	0	10	0.0	0	2	0.0	0	4	0.0	0	10	0.0	0	4	0.0	0	4	0.0	0	14	0.0
30-39	0	7	0.0	0	2	0.0	0	9	0.0	0	5	0.0	0	1	0.0	0	6	0.0	0	7	0.0	0	2	0.0	0	9	0.0
40+	0	7	0.0	1	6	16.7	1	13	7.7	0	5	0.0	0	5	0.0	0	10	0.0	0	7	0.0	1	9	11.1	0	16	0.0
TOTAL	0	17	0.0	1	10	10.0	1	32	3.1	0	12	0.0	0	10	0.0	0	26	0.0	0	18	0.0	1	15	6.7	1	39	2.6
Articular Surface Osteophytosis																											
<30	0	3	0.0	0	2	0.0	0	10	0.0	0	2	0.0	0	4	0.0	0	10	0.0	0	4	0.0	0	4	0.0	0	14	0.0
30-39	0	7	0.0	0	2	0.0	0	9	0.0	0	5	0.0	0	1	0.0	0	6	0.0	0	7	0.0	0	2	0.0	0	9	0.0
40+	0	7	0.0	1	6	16.7	1	13	7.7	0	5	0.0	0	5	0.0	0	10	0.0	0	7	0.0	1	9	11.1	1	16	6.3
TOTAL	0	17	0.0	1	10	10.0	1	32	3.1	0	12	0.0	0	10	0.0	0	26	0.0	0	18	0.0	1	15	6.7	1	39	2.6

Note: porosity and EB were not observed for the proximal radius

Correlation Matrix (left and right total)

	MO	ASO	P	EB
MO	**1**			
ASO	1	**1**		
P	0	0	**0**	
EB	0	0	0	**0**

* Total Individuals (numerator and denominator include only one element side when both present)

Table 4.4c. Da But Period: Changes to the Articular Surfaces of the Proximal Ulna (Elbow)

	Left Side									Right Side									Total Individuals*								
	male			female			total			male			female			total			male			female			total		
	obs	n	%	obs	n	%	obs	n	%	obs	n	%	obs	n	%	obs	n	%	obs	n	%	obs	n	%	obs	n	%
Marginal Osteophytosis																											
<30	0	3	0.0	0	3	0.0	0	11	0.0	0	2	0.0	0	3	0.0	0	8	0.0	0	4	0.0	0	4	0.0	0	13	0.0
30-39	0	9	0.0	0	3	0.0	0	12	0.0	0	9	0.0	1	4	25.0	1	13	7.7	0	10	0.0	1	4	25.0	1	14	7.1
40+	3	8	37.5	1	4	25.0	4	12	33.3	1	6	16.7	3	9	33.3	3	15	20.0	3	11	27.3	3	9	33.3	6	20	30.0
TOTAL	3	20	15.0	1	10	10.0	4	35	11.4	1	17	5.9	4	16	25.0	4	36	11.1	3	25	12.0	4	17	23.5	7	47	14.9
Articular Surface Osteophytosis																											
<30	0	3	0.0	0	3	0.0	0	11	0.0	0	2	0.0	0	3	0.0	0	8	0.0	0	4	0.0	0	4	0.0	0	13	0.0
30-39	0	9	0.0	0	3	0.0	0	12	0.0	0	9	0.0	0	4	0.0	0	13	0.0	0	10	0.0	0	4	0.0	0	14	0.0
40+	0	8	0.0	0	4	0.0	0	12	0.0	0	6	0.0	1	9	11.1	1	15	6.7	0	11	0.0	1	9	11.1	1	20	5.0
TOTAL	0	20	0.0	0	10	0.0	0	35	0.0	0	17	0.0	1	16	6.3	0	36	0.0	0	25	0.0	1	17	5.9	1	47	2.1
Porosity																											
<30	1	3	33.3	0	3	0.0	1	11	9.1	0	2	0.0	0	3	0.0	0	8	0.0	1	4	25.0	0	4	0.0	1	13	7.7
30-39	1	9	11.1	0	3	0.0	1	12	8.3	1	9	11.1	1	4	25.0	2	13	15.4	1	10	10.0	1	4	25.0	2	14	14.3
40+	1	8	12.5	0	4	0.0	1	12	8.3	1	6	16.7	1	9	11.1	2	15	13.3	1	11	9.1	1	9	11.1	2	20	10.0
TOTAL	3	20	15.0	0	10	0.0	3	35	8.6	2	17	11.8	2	16	12.5	4	36	11.1	3	25	12.0	2	17	11.8	5	47	10.6
Eburnation																											
<30	0	3	0.0	0	3	0.0	0	11	0.0	0	2	0.0	0	3	0.0	0	8	0.0	0	4	0.0	0	4	0.0	0	13	0.0
30-39	0	9	0.0	0	3	0.0	0	12	0.0	0	9	0.0	0	4	0.0	0	13	0.0	0	10	0.0	0	4	0.0	0	14	0.0
40+	0	8	0.0	0	4	0.0	0	12	0.0	0	6	0.0	1	9	11.1	1	15	6.7	0	11	0.0	1	9	11.1	1	20	5.0
TOTAL	0	20	0.0	0	10	0.0	0	35	0.0	0	17	0.0	1	16	6.3	1	36	2.8	0	25	0.0	1	17	5.9	1	47	2.1

Correlation Matrix (left and right total)

	MO	ASO	P	EB
MO	**8**			
ASO	1	**1**		
P	4	1	**7**	
EB	1	1	1	**1**

* Total Individuals (numerator and denominator include only one element side when both present)

Table 4.4d. Da But Period: Individuals Affected by Articular Changes to the Elbow

			Distal Humerus				Proximal Ulna				Proximal Radius			
case	sex	age	MO	ASO	P	EB	MO	ASO	P	EB	MO	ASO	P	EB
80CCNM16b	M	5	Y											
80CCNM43a	M	5		Y			Y							
80CCNM74a	M	5		Y	Y		Y		Y					
80CCNM45a	F	5	Y											
80CCNM50a	F	5					Y							
80CCNM75a	F	5					Y							
80CCNM15a	M	4		Y	Y									
80CCNM82a	M	4					Y							
80CCNM33c	F	4					Y	Y	Y	Y	Y	Y		
80CCNM8a	M	3	Y											
80CCNM10a	M	3	Y	Y	Y	Y			Y					
80CCNM18a	M	3		Y	Y									
80CCNM77a	F	3		Y	Y		Y		Y					
80CCNM65a	M	2		Y					Y					

Y: articular change observed
(blank): articular change not observed
(shaded): articular area not preserved

5= 50+ years
4=40-49 years
3=30-39 years
2=20-29 years
1=15-19 years

M=male
F=female
I=indeterminate

The metacarpophalangeal (MCP) joint is defined as the distal articular surfaces of the metacarpals and proximal articular surfaces of the proximal phalanges. Of the 213 assessable articular surfaces of the MCP joint, representing 47 individuals, one articular surface (0.5%) displayed MO and two (0.9%) displayed porosity (Table 4.6b).

The proximal interphalangeal joint (PIPJ) is defined as the distal articular surfaces of the proximal phalanges and the proximal articular surfaces of the middle phalanges. In the case of the thumb, the single interphalangeal joint comprising the distal articular surface of the proximal phalanx and proximal articular surface of the distal phalanx, is treated as a PIPJ. Of the 121 assessable articular surfaces of the PIPJ, representing 40 individuals, three articular surfaces (2.5%) displayed MO (Table 4.6c). No other articular changes for the PIPJ were observed.

The distal interphalangeal joint (DIPJ) is defined as the distal articular surfaces of the middle phalanges and the proximal articular surfaces of the distal phalanges. Note that the thumb does not have a DIPJ, only a PIPJ, in this analysis. Of the 34 assessable articular surfaces of the DIPJ, representing 20 individuals, three articular surfaces (8.8%) displayed MO (Table 4.6c). No other articular changes for the DIPJ were observed.

The two individuals displaying metacarpophalangeal porosity were both female and aged 30-39 years. All cases of carpopmetacarpal, metacarpophalangeal, PIPJ and DIPJ marginal osteophytosis occurred in the same individual. This was a male aged 50+ years. Looking at the hand as a whole, only one individual (out of 48, 2.1%) displayed evidence for MO. If porosity is included, 3/48 (6.3%) individuals displayed evidence for articular change.

The Hip

The hip is defined here as the articular surfaces of the proximal femoral head and the acetabulum of the os coxa. Of the 31 individuals with at least one assessable proximal femur 3.2% displayed MO, 6.5% ASO and 9.7% porosity, while no articular surfaces exhibited eburnation (Table 4.7a). The one individual with MO, affecting the left side (the right side was preserved but displayed porosity only) was male and aged 20-29 years. Of the individuals exhibiting ASO (around the rim of the fovea capitis) both observations occurred in the right side, one was male and the other female and both were aged 40+ years old. The left proximal femur in neither individual was preserved. The three observations of porosity included two males and one female, all less than 40 years old. None of these specific articular changes co-occurred among individuals.

Of the 12 individuals with at least one assessable acetabulum 50% displayed porosity (Table 4.7b). No other articular surface change was observed. Whilst each age class exhibited at least one case of porosity, it was most frequent in the oldest age class.

Looking at the hip as a whole, 33 individuals had at least an assessable proximal femur and/or acetabulum. Of these, 12 individuals (36.4%) display some form of joint change. The majority of cases, nine individuals (27.3%), displayed porosity and only three individuals (9.1%) displayed MO or ASO.

The Knee

The knee is defined here as the articular surfaces of the distal femoral condyles, proximal head of the tibia and the patella. The proximal fibula was too poorly preserved to be included in this analysis. Of the 22 individuals with at least one assessable distal femur 13.6% displayed MO and 9.1% porosity (Table 4.8a). No articular surfaces exhibited eburnation or ASO. All observations of MO occurred in males, in the case of the youngest the MO were bilateral while it was unilateral in the oldest (50+ years) individual.

Table 4.5a. Da But Period: Changes to the Articular Surfaces of the Distal Radius (Wrist)

	Left Side									Right Side									Total Individuals*								
	male			female			total			male			female			total			male			female			total		
	obs	n	%	obs	n	%	obs	n	%	obs	n	%	obs	n	%	obs	n	%	obs	n	%	obs	n	%	obs	n	%
Marginal Osteophytosis																											
<30	0	2	0.0	0	1	0.0	0	3	0.0	0	3	0.0	0	3	0.0	0	8	0.0	0	4	0.0	0	3	0.0	0	9	0.0
30-39	1	7	14.3	0	0	0.0	1	7	14.3	1	6	16.7	0	2	0.0	1	8	12.5	1	8	12.5	0	2	0.0	1	10	10.0
40+	0	4	0.0	1	2	50.0	1	6	16.7	1	6	16.7	0	4	0.0	1	10	10.0	1	8	12.5	1	5	20.0	2	13	15.4
TOTAL	1	13	7.7	1	3	33.3	2	16	12.5	2	15	13.3	0	9	0.0	1	26	3.8	2	20	10.0	1	10	10.0	3	32	9.4
Articular Surface Osteophytosis																											
<30	0	2	0.0	0	1	0.0	0	3	0.0	0	3	0.0	0	3	0.0	0	8	0.0	0	4	0.0	0	3	0.0	0	9	0.0
30-39	0	7	0.0	0	0	0.0	0	7	0.0	0	6	0.0	0	2	0.0	0	8	0.0	0	8	0.0	0	2	0.0	0	10	0.0
40+	0	4	0.0	0	2	0.0	0	6	0.0	1	6	16.7	0	4	0.0	0	10	0.0	1	8	12.5	0	5	0.0	1	13	7.7
TOTAL	0	13	0.0	0	3	0.0	0	16	0.0	1	15	6.7	0	9	0.0	0	26	0.0	1	20	5.0	0	10	0.0	1	32	3.1
Porosity																											
<30	0	2	0.0	0	1	0.0	0	3	0.0	0	3	0.0	0	3	0.0	0	8	0.0	0	4	0.0	0	3	0.0	0	9	0.0
30-39	0	7	0.0	0	0	0.0	0	7	0.0	0	6	0.0	0	2	0.0	0	8	0.0	0	8	0.0	0	2	0.0	0	10	0.0
40+	0	4	0.0	1	2	50.0	1	6	16.7	1	6	16.7	0	4	0.0	1	10	10.0	1	8	12.5	1	5	20.0	2	13	15.4
TOTAL	0	13	0.0	1	3	33.3	1	16	6.3	1	15	6.7	0	9	0.0	1	26	3.8	1	20	5.0	1	10	10.0	2	32	6.3

note: eburnation was not observed for the distal radius

Table 4.5b. Da But Period: Changes to the Articular Surfaces of the Distal Ulna (Wrist)

Porosity																											
<30	0	2	0.0	0	0	0.0	0	3	0.0	0	1	0.0	0	1	0.0	0	3	0.0	0	2	0.0	0	1	0.0	0	5	0.0
30-39	0	3	0.0	0	1	0.0	0	4	0.0	0	2	0.0	0	1	0.0	0	3	0.0	0	5	0.0	0	1	0.0	0	6	0.0
40+	0	2	0.0	0	2	0.0	0	4	0.0	1	4	25.0	0	2	0.0	1	6	16.7	1	5	20.0	0	3	0.0	1	8	12.5
TOTAL	0	7	0.0	0	3	0.0	0	11	0.0	1	7	14.3	0	4	0.0	1	12	8.3	1	12	8.3	0	5	0.0	1	19	5.3

Note: MO, ASO and EB was not observed for the distal ulna

Radius Correlation Matrix (left and right total)

	MO	ASO	P	EB
MO	**4**			
ASO	1	**1**		
P	2	1	**2**	
EB	0	0	0	**0**

Ulna Correlation Matrix (left and right total)

	MO	ASO	P	EB
MO	**0**			
ASO	0	**0**		
P	0	0	**1**	
EB	0	0	0	**0**

* Total Individuals (numerator and denominator include only one element side when both present)

Table 4.6a. Da But Period: Carpal Articular Surface Changes

Sample Sizes (no joint changes observed for any bone)						
Left and Right Side				Individuals*		
	male	female	total	male	female	total
	n	n	n	n	n	n
<30	8	9	24	1	3	7
30-39	15	2	17	6	2	8
40+	13	2	15	7	2	9
TOTAL	36	13	56	14	7	24

* individual: represented by at least one articular surface of a given joint

The remaining individual, aged 30-39 years, had only the right distal femur preserved. Porosity, not co-occurring with any of the above, manifested in a female aged 30-39 years and a male aged 50+years.

Of the 22 individuals with at least one assessable proximal tibia 4.5% displayed MO and 18.2% porosity (Table 4.8b). No articular surfaces exhibited eburnation or ASO. The observation of MO was in a female aged 40-49, who also displayed porosity. All observations of porosity occurred in individuals 30+ years old.

Of the 20 individuals with at least one assessable patella 5% displayed MO, 20% ASO and 10% porosity (Table 4.8c). No patella articular surfaces exhibited eburnation. The one observation of MO was in a male aged 30-39 years, who also displayed porosity. The right patella was not preserved in this case, and neither were the corresponding femoral and tibial articular areas. Of the four individuals with ASO of the patella, three were males aged 50+ years while the remaining individual was unsexed and aged 20-29 years. One male had ASO on both patellae, one male displayed ASO unilaterally, the remaining male only had one preserved patella to assess. The unsexed individual also only had one side preserved. The only two observations of patella porosity involved one male with MO and one male with ASO.

Looking at the knee as a whole, 36 individuals have at least an assessable distal femur, proximal tibia and/or patella (Table 4.8d). Of these, 10 individuals (27.7%) display some form of joint change. The patterning appears to indicate the femur is more at risk of developing MO, the tibia porosity and the patella ASO.

The Ankle

The ankle is defined here as the articular surfaces of the distal tibia, distal fibula and the trochlea of the talus. Of the 32 individuals with at least one assessable distal tibia 3.1% displayed MO (Table 4.9). No articular surfaces exhibited ASO, porosity or eburnation. The single observation of distal tibial MO occurred in a male aged 30-39 years, and manifested bilaterally.

Of the 22 individuals with at least one assessable distal fibula 4.5% displayed ASO (Table 4.9). No articular surfaces exhibited MO, porosity or eburnation. The only observation of ASO occurred in a female aged 50+ years. Only her left distal fibula was preserved.

Of the 37 individuals with at least one assessable talar trochlea 2.7% displayed MO and 13.5% porosity (Table 4.9). No talar trochlea exhibited ASO or eburnation. The single observation of MO occurred in the same male with distal tibial MO. This individual did not also display porosity. Of the five individuals displaying porosity two (a male and female) were aged 20-29 years, while the remaining three (two males and one female) were aged 50+ years.

Looking at the ankle as a whole, 44 individuals have at least one assessable distal tibia, distal fibula and/or talar trochlea. Of these, 7 individuals (15.9%) display some form of joint change. Porosity is the most common change accounting for 6/7 of all individuals with any change. MO only occurs in the ankle joints of one individual, or 4.5% of all assessable individuals.

The Foot

The foot is defined here as the articular surfaces of all tarsal (excluding the talar trochlea), metatarsal and foot phalanges. Very few articular surfaces in the various bones of the foot were affected by articular change. The following bones or articular surfaces, with the number of individuals represented in parentheses, did not display any changes: subtalar articular surface (37); first cuneiform (25); second cuneiform (19); and third cuneiform (9). Table 4.10a displays sample sizes for these bones without articular changes.

Of the 37 individuals represented by a talar head, 2.7% exhibited porosity and 2.7% eburnation. Both observations occurred in the same individual (Table 4.10b). This individual was male and aged 20-29 years. The right talar head was also preserved but unaffected. The corresponding naviculars were not preserved.

Of the 26 individuals represented by a calcaneal talar facet 15.4% exhibited MO (Table 4.10b). No calcaneal talar facets exhibited ASO, EB or porosity. Of those individuals displaying MO, three were female, belonging to each of the three age classes, and the fourth was male, in the oldest age class. The male and oldest female had both left and right facets preserved, but the condition manifested unilaterally. The other two females only had one side preserved.

Table 4.6b. Da But Period: Carpometacarpal and Metacarpophalangeal Joint Articular Changes (Hand)

	Articular Surface Count									Total Individuals								
	male			female			total			male			female			total		
	obs	n	%	obs	n	%	obs	n	%	obs	n	%	obs	n	%	obs	n	%
Carpometacarpal Joint[1] Articular Surfaces: Marginal Osteophytosis																		
<30	0	11	0.0	0	4	0.0	0	30	0.0	0	3	0.0	0	3	0.0	0	10	0.0
30-39	0	30	0.0	0	12	0.0	0	42	0.0	0	9	0.0	0	6	0.0	0	15	0.0
40+	1	30	3.3	0	23	0.0	1	53	1.9	1	9	11.1	0	9	0.0	1	18	5.6
TOTAL	1	71	1.4	0	39	0.0	1	125	0.8	1	21	4.8	0	18	0.0	1	43	2.3
Metacarpophalangeal Joint (MCP)[2] Articular Surfaces: Marginal Osteophytes																		
<30	0	23	0.0	0	14	0.0	0	55	0.0	0	3	0.0	0	3	0.0	0	11	0.0
30-39	0	44	0.0	0	20	0.0	0	64	0.0	0	11	0.0	0	5	0.0	0	16	0.0
40+	1	57	1.8	0	37	0.0	1	94	1.1	1	10	10.0	0	10	0.0	1	20	5.0
TOTAL	1	124	0.8	0	71	0.0	1	213	0.5	1	24	4.2	0	18	0.0	1	47	2.1
Metacarpophalangeal Joint (MCP)[2] Articular Surfaces: Porosity																		
<30	0	23	0.0	0	14	0.0	0	55	0.0	0	3	0.0	0	3	0.0	0	11	0.0
30-39	0	44	0.0	2	20	10.0	2	64	3.1	0	11	0.0	2	5	40.0	2	16	12.5
40+	0	57	0.0	0	37	0.0	0	94	0.0	0	10	0.0	0	10	0.0	0	20	0.0
TOTAL	0	124	0.0	2	71	2.8	2	213	0.9	0	24	0.0	2	18	11.1	2	47	4.3

Note: carpometacarpal joint ASO, EB and P not observed
metacarpophalangeal joint ASO and EB not observed
[1] Carpometacarpal Joint: defined as the articular surfaces of the proximal metacarpals only
[2] MCP: distal articular surfaces of metacarpals and proximal articular surfaces of proximal phalanges

MCP Correlation Matrix

	MO	ASO	P	EB
MO	**1**			
ASO	0	**0**		
P	0	0	**2**	
EB	0	0	0	**0**

* individual: represented by at least one articular surface of a given joint

Of the 28 individuals represented by a calcaneal cuboid facet, 7.1% displayed MO (Table 4.10b). No calcaneal cuboid facets displayed ASO, EB or porosity. Both individuals displaying MO were female, one aged 30-39 years and the other 50+ years. Moreover, the left calcaneal cuboid facet in each case was not preserved.

Of the 23 individuals represented by a navicular 8.7% displayed MO, 4.3% ASO and 8.7% porosity (Table 4.10b), No articular surfaces of the navicular displayed EB. The individuals with MO were the same two that displayed MO of the calcaneal cuboid, but in this case only the right sides were preserved. The younger female also displayed porosity while a further individual, male aged 30-39 years, also had porosity. The sole observation of navicular ASO was in a female, aged 30-39 years, who had a preserved left side only.

Of the 22 individuals represented by a cuboid, 4.5% displayed MO (Table 4.10b). No articular surfaces of the cuboid displayed ASO, EB or porosity. The sole individual displaying cuboid MO was a female, aged 30-39 years, and the right side was not preserved.

The toes of the foot were assessed by joint complex. However, the tarsometatarsal joint was defined here as the proximal articular surfaces of the metatarsals only. Of a total of 170 proximal metatarsals, representing 40 individuals, no observations of articular change were observed (Table 4.10c).

The metatarsophalangeal (MTP) joint is defined as the distal articular surfaces of the metatarsals and proximal articular surfaces of the proximal phalanges. Of the 182 assessable articular surfaces of the MTP joint, representing 38 individuals, 2.7% displayed MO and 3.3% porosity (Table 4.10c). The articular surfaces of the MTP joint did not exhibit ASO or EB. Articular surface changes were confined to those individuals aged 30+ years. Of the five observations of MO, porosity co-occurred in four instances. All observations of articular change occurred on the first MTP (big toe) joint.

The proximal interphalangeal joint (PIPJ) is defined as the distal articular surfaces of the proximal phalanges and the proximal articular surfaces of the middle phalanges. In the case of the big toe, the single

Table 4.6c. Da But Period: Interphalangeal Joint Articular Changes (Hand)

Articular Surface Count										Total Individuals								
	male			female			total			male			female			total		
	obs	n	%	obs	n	%	obs	n	%	obs	n	%	obs	n	%	obs	n	%
Proximal Interphalangeal Joint (PIPJ[1]) Articular Surfaces: Marginal Osteophytosis																		
<30	0	9	0.0	0	11	0.0	0	35	0.0	0	3	0.0	0	3	0.0	0	10	0.0
30-39	0	24	0.0	0	13	0.0	0	37	0.0	0	10	0.0	0	4	0.0	0	14	0.0
40+	3	29	10.3	0	20	0.0	3	49	6.1	1	10	10.0	0	6	0.0	1	16	6.3
TOTAL	3	62	4.8	0	44	0.0	3	121	2.5	1	23	4.3	0	13	0.0	1	40	2.5
Distal Interphalangeal Joint (DIPJ[2]) Articular Surfaces: Marginal Osteophytosis																		
<30	0	2	0.0	0	3	0.0	0	11	0.0	0	2	0.0	0	3	0.0	0	7	0.0
30-39	0	7	0.0	0	3	0.0	0	10	0.0	0	4	0.0	0	2	0.0	0	6	0.0
40+	3	9	33.3	0	4	0.0	3	13	23.1	1	4	25.0	0	3	0.0	1	7	14.3
TOTAL	3	18	16.7	0	10	0.0	3	34	8.8	1	10	10.0	0	8	0.0	1	20	5.0

Note: PIP joint ASO, EB and P not observed
DIP joint ASO, EB and P not observed
[1] PIPJ: distal articular surfaces of proximal phalanges and proximal articularsurfaces of middle phalanges
[2] DIPJ: distal articular surfaces of middle phalanges and proximal articular surfaces of distal phalanges

* individual: represented by at least one articular surface of a given joint

interphalangeal joint comprising the distal articular surface of the proximal phalanx and proximal articular surface of the distal phalanx, is treated as a PIPJ. Of the 88 assessable articular surfaces of the PIPJ, representing 28 individuals, 3.4% of the articular surfaces displayed MO, 1.1% ASO and 2.3% porosity (Table 4.10d). No observations of EB were made. All observations of articular change occurred in two males aged less than 30 years. Of the three observations of PIPJ articular surface MO, porosity co-occurred twice. Of the two individuals (7.1%) affected by articular changes, one individual manifested articular change unilaterally in the left first PIPJ (on both the proximal and distal phalanges), while the other individual had changes to the distal end of a proximal phalange of indeterminate position.

The distal interphalangeal joint (DIPJ) is defined as the distal articular surfaces of the middle phalange and the proximal articular surfaces of the distal phalanges. Note that the big toe does not have a DIPJ, only a PIPJ, in this analysis. Only 14 articular surfaces of the DIPJ, representing eight males, were preserved (Table 4.10e). No instances of DIPJ articular change were noted.

Looking at the foot as a whole, 14 individuals exhibited some form of osseous change to the articular surfaces of the foot (Table 4.10f). If the denominator is the number of individuals with at least one assessable element (n=52), then 26.9% of individuals display some form of articular change to the foot. Of the 14 individuals displaying some form of articular change, eight exhibited changes to the tarsals, with one case having no preserved tarsals to assess. If the denominator is taken as the number of individuals with at least an assessable talar head, calcaneal talar facet or cuboid facet (n=43), then 18.6% of these individuals displayed articular change to this region of the foot.

While the tarsometarsal joint is free from articular change, the metatarsophalangeal (MTP) joint displays a considerable degree of articular change. If individuals with an assessable first metatarsophalangeal joint (big toe) are taken as the denominator, 3/20 (15.0%) males, 2/12 (16.7%) females, or a total of 5/36 (13.9%) individuals (including unsexed cases) displayed marginal osteophytosis of the first MTP joint.

With regard to the PIPJ only two individuals displayed articular changes. One individual displayed changes to the first PIPJ (big toe). The other individual displayed articular changes to a PIPJ of indeterminate position.

The most common tarsal joint change is MO, especially of the calcaneal talar facet. The most common metatarsal/phalange joint change is MO followed by porosity. Of the 14 individuals exhibiting some form of joint change, 78.6% are aged over 30 years while 42.9% are aged over 40 years.

The Spine

In this analysis of osseous changes to the vertebrae, articular changes are classified as vertebral osteophytosis (VOP) of the intervertebral syndesmoses and zygapophyseal, or interfacetal, osteophytosis (ZOP). In many instances only one of the two end plates for a given vertebral body was available for assessment. Likewise, in most cases only two, or even one, articular facet was preserved.

Cervical Vertebrae:

The first cervical vertebra, the atlas, was too poorly preserved to assess here. However, the zygapophyses, albeit not the centra, of the axis could be assessed (Table 4.11a). Of the 18 individuals with assessable axis zygapophyses (including the dens facet), nine displayed

Table 4.7a. Da But Period: Changes to the Articular Surfaces of the Proximal Femur (Hip)

	Left Side									Right Side									Total Individuals*								
	male			female			total			male			female			total			male			female			total		
	obs	n	%	obs	n	%	obs	n	%	obs	n	%	obs	n	%	obs	n	%	obs	n	%	obs	n	%	obs	n	%
Marginal Osteophytosis																											
<30	1	3	33.3	0	1	0.0	1	7	14.3	0	2	0.0	0	2	0.0	0	6	0.0	1	3	33.3	0	2	0.0	1	8	12.5
30-39	0	7	0.0	0	4	0.0	0	11	0.0	0	4	0.0	0	2	0.0	0	6	0.0	0	7	0.0	0	4	0.0	0	11	0.0
40+	0	5	0.0	0	5	0.0	0	10	0.0	0	3	0.0	0	4	0.0	0	7	0.0	0	6	0.0	0	6	0.0	0	12	0.0
TOTAL	1	15	6.7	0	10	0.0	1	28	3.6	0	9	0.0	0	8	0.0	0	19	0.0	1	16	6.3	0	12	0.0	1	31	3.2
Articular Surface Osteophytosis																											
<30	0	3	0.0	0	1	0.0	0	7	0.0	0	2	0.0	0	2	0.0	0	6	0.0	0	3	0.0	0	2	0.0	0	8	0.0
30-39	0	7	0.0	0	4	0.0	0	11	0.0	0	4	0.0	0	2	0.0	0	6	0.0	0	7	0.0	0	4	0.0	0	11	0.0
40+	0	5	0.0	0	5	0.0	0	10	0.0	1	3	33.3	1	4	25.0	2	7	28.6	1	6	16.7	1	6	16.7	2	12	16.7
TOTAL	0	15	0.0	0	10	0.0	0	28	0.0	1	9	11.1	1	8	12.5	2	19	10.5	1	16	6.3	1	12	8.3	2	31	6.5
Porosity																											
<30	0	3	0.0	0	1	0.0	0	7	0.0	1	2	50.0	0	2	0.0	1	6	16.7	1	3	33.3	0	2	0.0	1	8	12.5
30-39	0	7	0.0	1	4	25.0	1	11	9.1	0	4	0.0	1	2	50.0	1	6	16.7	0	7	0.0	2	4	50.0	2	11	18.2
40+	0	5	0.0	0	5	0.0	0	10	0.0	0	3	0.0	0	4	0.0	0	7	0.0	0	6	0.0	0	6	0.0	0	12	0.0
TOTAL	0	15	0.0	1	10	10.0	1	28	3.6	1	9	11.1	1	8	12.5	2	19	10.5	1	16	6.3	2	12	16.7	3	31	9.7

note: eburnation was not observed for the proximal femur

Table 4.7b. Da But Period: Changes to the Articular Surfaces of the Acetabulum (Hip)

	male			female			total			male			female			total			male			female			total		
Porosity																											
<30							0	2	0.0							1	3	33.3	1	2	50.0	0	1	0.0	1	4	25.0
30-39							1	3	33.3							1	1	100.0	1	3	33.3	0	0	0.0	1	3	33.3
40+							1	2	50.0							3	4	75.0	2	3	66.7	2	2	100.0	4	5	80.0
TOTAL							2	7	28.6							5	8	62.5	4	8	50.0	2	3	66.7	6	12	50.0

Note: MO, ASO and EB were not observed for the acetabulum

Femur Correlation Matrix (left and rigt tota)

	MO	ASO	P	EB
MO	**1**			
ASO	0	**2**		
P	0	0	**3**	
EB	0	0	0	**0**

* Total Individuals (numerator and denominator include only one element side when both present)

Table 4.8a. Da But Period: Changes to the Articular Surfaces of the Distal Femur (Knee)

	Left Side									Right Side									Total Individuals*								
	male			female			total			male			female			total			male			female			total		
	obs	n	%	obs	n	%	obs	n	%	obs	n	%	obs	n	%	obs	n	%	obs	n	%	obs	n	%	obs	n	%
Marginal Osteophytosis																											
<30	1	3	33.3	0	1	0.0	1	5	20.0	1	1	100.0	0	2	0.0	1	4	25.0	1	3	33.3	0	2	0.0	1	6	16.7
30-39	0	3	0.0	0	2	0.0	0	5	0.0	1	4	25.0	0	3	0.0	1	7	14.3	1	4	25.0	0	3	0.0	1	7	14.3
40+	0	5	0.0	0	2	0.0	0	7	0.0	1	4	25.0	0	0	0.0	1	4	25.0	1	7	14.3	0	2	0.0	1	9	11.1
TOTAL	1	11	9.1	0	5	0.0	1	17	5.9	3	9	33.3	0	5	0.0	1	15	6.7	3	14	21.4	0	7	0.0	3	22	13.6
Porosity																											
<30	0	3	0.0	0	1	0.0	0	5	0.0	0	1	0.0	0	2	0.0	0	4	0.0	0	3	0.0	0	2	0.0	0	6	0.0
30-39	0	3	0.0	1	2	50.0	1	5	20.0	0	4	0.0	0	3	0.0	0	7	0.0	0	4	0.0	1	3	33.3	1	7	14.3
40+	0	5	0.0	0	2	0.0	0	7	0.0	1	4	25.0	0	0	0.0	1	4	25.0	1	7	14.3	0	2	0.0	1	9	11.1
TOTAL	0	11	0.0	1	5	20.0	1	17	5.9	1	9	11.1	0	5	0.0	1	15	6.7	1	14	7.1	1	7	14.3	2	22	9.1

note: articular surface osteophytes and eburnation were not observed for the distal femur

Table 4.8b. Da But Period: Changes to the Articular Surfaces of the Proximal Tibia (Knee)

Marginal Osteophytosis																											
<30	0	1	0.0	0	3	0.0	0	6	0.0	0	1	0.0	0	1	0.0	0	3	0.0	0	1	0.0	0	3	0.0	0	7	0.0
30-39	0	3	0.0	0	1	0.0	0	4	0.0	0	4	0.0	0	2	0.0	0	6	0.0	0	5	0.0	0	2	0.0	0	7	0.0
40+	0	6	0.0	0	1	0.0	0	7	0.0	0	5	0.0	1	1	100.0	1	6	16.7	0	7	0.0	1	1	100.0	1	8	12.5
TOTAL	0	10	0.0	0	5	0.0	0	17	0.0	0	10	0.0	1	4	25.0	1	15	6.7	0	13	0.0	1	6	16.7	1	22	4.5
Porosity																											
<30	0	1	0.0	0	3	0.0	0	6	0.0	0	1	0.0	0	1	0.0	0	3	0.0	0	1	0.0	0	3	0.0	0	7	0.0
30-39	0	3	0.0	0	1	0.0	0	4	0.0	0	4	0.0	1	2	50.0	1	6	16.7	0	5	0.0	1	2	50.0	1	7	14.3
40+	1	6	16.7	1	1	100.0	2	7	28.6	1	5	20.0	1	1	100.0	2	6	33.3	2	7	28.6	1	1	100.0	3	8	37.5
TOTAL	1	10	10.0	1	5	20.0	2	17	11.8	1	10	10.0	2	4	50.0	3	15	20.0	2	13	15.4	2	6	33.3	4	22	18.2

Note: ASO and EB were not observed for the proximal tibia

Femur Correlation Matrix (left and right total)

	MO	ASO	P	EB
MO	**4**			
ASO	0	**0**		
P	0	0	**2**	
EB	0	0	0	**0**

Tibia Correlation Matrix (left and right total)

	MO	ASO	P	EB
MO	**1**			
ASO	0	**0**		
P	1	0	**5**	
EB	0	0	0	**0**

* Total Individuals (numerator and denominator include only one element side when both present)

Table 4.8c. Da But Period: Changes the Articular Surfaces of the Patella (Knee)

	Left Side									Right Side									Total Individuals*								
	male			female			total			male			female			total			male			female			total		
	obs	n	%	obs	n	%	obs	n	%	obs	n	%	obs	n	%	obs	n	%	obs	n	%	obs	n	%	obs	n	%
Marginal Osteophytosis																											
<30	0	1	0.0	0	0	0.0	0	3	0.0	0	1	0.0	0	1	0.0	0	6	0.0	0	1	0.0	0	1	0.0	0	7	0.0
30-39	1	4	25.0	0	0	0.0	1	4	25.0	0	1	0.0	0	1	0.0	0	2	0.0	1	4	25.0	0	1	0.0	1	5	20.0
40+	0	2	0.0	0	4	0.0	0	6	0.0	0	4	0.0	0	2	0.0	0	6	0.0	0	4	0.0	0	4	0.0	0	8	0.0
TOTAL	1	7	14.3	0	4	0.0	1	13	7.7	0	6	0.0	0	4	0.0	0	14	0.0	1	9	11.1	0	6	0.0	1	20	5.0
Articular Surface Osteophytosis																											
<30	0	1	0.0	0	0	0.0	1	3	33.3	0	1	0.0	0	1	0.0	0	6	0.0	0	1	0.0	0	1	0.0	1	7	14.3
30-39	0	4	0.0	0	0	0.0	0	4	0.0	0	1	0.0	0	1	0.0	0	2	0.0	0	4	0.0	0	1	0.0	0	5	0.0
40+	2	2	100.0	0	4	0.0	2	6	33.3	2	4	50.0	0	2	0.0	2	6	33.3	3	4	75.0	0	4	0.0	3	8	37.5
TOTAL	2	7	28.6	0	4	0.0	3	13	23.1	2	6	33.3	0	4	0.0	2	14	14.3	3	9	33.3	0	6	0.0	4	20	20.0
Porosity																											
<30	0	1	0.0	0	0	0.0	0	3	0.0	0	1	0.0	0	1	0.0	0	6	0.0	0	1	0.0	0	1	0.0	0	7	0.0
30-39	1	4	25.0	0	0	0.0	1	4	25.0	0	1	0.0	0	1	0.0	0	2	0.0	1	4	25.0	0	1	0.0	1	5	20.0
40+	1	2	50.0	0	4	0.0	1	6	16.7	0	4	0.0	0	2	0.0	0	6	0.0	1	4	25.0	0	4	0.0	1	8	12.5
TOTAL	2	7	28.6	0	4	0.0	2	13	15.4	0	6	0.0	0	4	0.0	0	14	0.0	2	9	22.2	0	6	0.0	2	20	10.0

Note: EB was not observed for the patella

Correlation Matrix (left and right total)

	MO	ASO	P	EB
MO	**1**			
ASO	0	**5**		
P	1	1	**2**	
EB	0	0	0	**0**

* Total Individuals (numerator and denominator include only one element side when both present)

Table 4.8d. Da But Period: Individuals Affected by Articular Changes to the Knee

			Distal Femur				Proximal Tibia				Patella			
case	sex	age	MO	ASO	P	EB	MO	ASO	P	EB	MO	ASO	P	EB
80CCNM2a	F	3			Y									
80CCNM3a	M	5			Y							Y		
80CCNM14a	M	2	Y											
80CCNM16b	M	5							Y			Y		
80CCNM19a	I	2	▒	▒	▒	▒	▒	▒	▒	▒		Y		
80CCNM35a	M	3	Y								▒	▒	▒	▒
80CCNM42a	F	3							Y		▒	▒	▒	▒
80CCNM63a	M	3	▒	▒	▒	▒	▒	▒	▒	▒	Y		Y	
80CCNM74a	M	5	Y						Y			Y	Y	
80CCNM83a	F	4					Y		Y		▒	▒	▒	▒

Y: articular change observed
(blank): articular change not observed
▒: articular area not preserved

5= 50+ years
4=40-49 years
3=30-39 years
2=20-29 years
1=15-19 years

M=male
F=female
I=indeterminate

some degree of ZOP. Of the nine individuals exhibiting ZOP, seven were aged 30 years or more.

Only 28 cervical centra (C3 to C7), representing 18 individuals were available for assessing VOP (Table 4.11a). VOP was observed in five individuals (27.8%), three males and two females. Four of the five individuals displaying VOP were aged 40+ years.

Eleven cervical vertebrae (C3 to C7), representing seven individuals, had assessable zygapophyseal facets (Table 4.11a). ZOP was observed in one individual (14.3%), a female aged 40+ years.

Thoracic Vertebrae:
Of the 49 assessable thoracic centra, representing 16 individuals, 6.1% of these vertebrae (12.5% of individuals) exhibited VOP (Table 4.11b). The two individuals exhibiting VOP were both male and aged 30+ years. Of a total of 36 thoracic vertebrae, representing 11 individuals, 27.8% of these centra (36.4% of individuals) displayed ZOP. Two of these individuals exhibiting ZOP were male, two were female and they were all aged 30+ years. Only one of the two vertebrae displaying VOP could also be assessed for ZOP, and it did not also exhibit ZOP.

Lumbar Vertebrae:
Of the 38 assessable lumbar centra, representing 14 individuals, 7.9% of these vertebrae (14.3% of individuals) exhibited VOP (Table 4.11b). Of the two individuals displaying VOP, one was male and the other female and both were 30+ years old. The female did not also display ZOP, while the male could not be assessed for ZOP.

Of the 37 assessable thoracic vertebrae, representing 14 individuals, 29.7% of these vertebrae (35.7% of individuals) exhibited ZOP. Of the five individuals displaying ZOP, three were male and two females. All individuals with ZOP were aged 30+ years. Three of these five individuals also had assessable centra for which no instances of VOP co-occurred.

Sacrum:
The superior articular facets of the sacra in this sample were too poorly preserved to assess. However, 12 individuals had an assessable promontory of which one displayed VOP (Table 4.11b). This individual was a male aged 50+ years.

Looking at the spine as a whole (Table 4.11c), 15 individuals displayed osteophytic development to the zygapophyses and/or vertebral syndesmoses. If the denominator is taken as individuals with assessable vertebral syndesmoses and/or zygapophyses (n=34), then 44.1% of individuals had some form of osteophytic development. With the denominator being individuals with assessable zygapophyses (n=22), 54.5% displayed ZOP. When the denominator is individuals with assessable centra, including the sacral promontory (n=31), 22.6% exhibited VOP. Because too few cases had both assessable centra and zygapophyses in the same vertebral element, a meaningful assessment of correlation between VOP and ZOP cannot be made. Moreover, the observation that ZOP appears to be much more common than VOP is likely partly due to poor preservation of the centra in those cases displaying ZOP. Of the 15 individuals displaying vertebral joint change, 33.3% were 50+ years old and 80% were 30+ years old.

Metal Period Sample
The Temporomandibular Joint
Of the 41 individuals with at least one assessable mandibular condyle 4.9% displayed MO, 2.4% ASO and 26.8% porosity (Table 4.12a). No individuals displayed EB of the mandibular condyles. One individual exhibiting MO was a male aged 40+ years, while the other instance was in a female aged less than 30 years. The single individual with ASO was a male aged 40+ years. Porosity was fairly evenly distributed between the sexes and occurred in all age classes, but concentrated in the older age groups. MO and ASO did not co-occur, while porosity co-occurred with one of the cases of MO.

Table 4.9. Da But Period: Changes to the Articular Surfaces of the Ankle

	Left Side									Right Side									Total Individuals*								
	male			female			total			male			female			total			male			female			total		
	obs	n	%	obs	n	%	obs	n	%	obs	n	%	obs	n	%	obs	n	%	obs	n	%	obs	n	%	obs	n	%
Distal Tibia: Marginal Osteophytosis																											
<30	0	3	0.0	0	3	0.0	0	8	0.0	0	1	0.0	0	2	0.0	0	5	0.0	0	3	0.0	0	3	0.0	0	8	0.0
30-39	1	7	14.3	0	2	0.0	1	9	11.1	1	7	14.3	0	2	0.0	1	9	11.1	1	9	11.1	0	2	0.0	1	11	9.1
40+	0	9	0.0	0	4	0.0	0	13	0.0	0	6	0.0	0	2	0.0	0	8	0.0	0	9	0.0	0	4	0.0	0	13	0.0
TOTAL	1	19	5.3	0	9	0.0	1	30	3.3	1	14	7.1	0	6	0.0	1	22	4.5	1	21	4.8	0	9	0.0	1	32	3.1
Distal Fibula: Articular Surface Osteophytosis																											
<30	0	3	0.0	0	0	0.0	0	4	0.0	0	3	0.0	0	1	0.0	0	5	0.0	0	3	0.0	0	1	0.0	0	5	0.0
30-39	0	2	0.0	0	3	0.0	0	5	0.0	0	6	0.0	0	2	0.0	0	8	0.0	0	6	0.0	0	3	0.0	0	9	0.0
40+	0	1	0.0	1	3	33.3	1	4	25.0	0	1	0.0	0	3	0.0	0	4	0.0	0	2	0.0	1	6	16.7	1	8	12.5
TOTAL	0	6	0.0	1	6	16.7	1	13	7.7	0	10	0.0	0	6	0.0	0	17	0.0	0	11	0.0	1	10	10.0	1	22	4.5
Talar Trochlea: Marginal Osteophytosis																											
<30	0	2	0.0	0	3	0.0	0	9	0.0	0	4	0.0	0	2	0.0	0	11	0.0	0	4	0.0	0	3	0.0	0	13	0.0
30-39	1	4	25.0	0	2	0.0	1	6	16.7	1	9	11.1	0	2	0.0	1	11	9.1	1	9	11.1	1	3	33.3	1	12	8.3
40+	0	6	0.0	0	3	0.0	0	9	0.0	0	7	0.0	0	4	0.0	0	11	0.0	0	8	0.0	0	4	0.0	0	12	0.0
TOTAL	1	12	8.3	0	8	0.0	1	24	4.2	1	20	5.0	0	8	0.0	1	33	3.0	1	21	4.8	1	10	10.0	1	37	2.7
Talar Trochlea: Porosity																											
<30	1	2	50.0	1	3	33.3	2	9	22.2	0	4	0.0	1	2	50.0	1	11	9.1	1	4	25.0	1	3	33.3	2	13	15.4
30-39	0	4	0.0	0	2	0.0	0	6	0.0	0	9	0.0	0	2	0.0	0	11	0.0	0	9	0.0	0	3	0.0	0	12	0.0
40+	1	6	16.7	1	3	33.3	2	9	22.2	1	7	14.3	0	4	0.0	1	11	9.1	2	8	25.0	1	4	25.0	3	12	25.0
TOTAL	2	12	16.7	2	8	25.0	4	24	16.7	1	20	5.0	1	8	12.5	2	33	6.1	3	21	14.3	2	10	20.0	5	37	13.5

Note: ASO, EB and porosity not observed for the distal tibia; MO, EB and porosity not observed for the distal fibula; ASO and EB not observed for the talar trochlea.

Talar Trochlea Correlation Matrix (left and right total)

	MO	ASO	P	EB
MO	**2**			
ASO	0	**0**		
P	0	0	**6**	
EB	0	0	0	**0**

* Total Individuals (numerator and denominator include only one element side when both present)

Table 4.10a. Da But Period: Foot Bones Without Articular Changes (Sample Sizes)

	Left Side			Right Side			Individuals*		
	male	female	total	male	female	total	male	female	total
	n	n	n	n	n	n	n	n	n
First Cuneiform									
<30	2	3	8	3	0	3	4	3	10
30-39	3	1	4	5	1	6	7	1	8
40+	4	1	5	3	2	5	5	2	7
TOTAL	9	5	17	11	3	14	16	6	25
Second Cuneiform									
<30	2	1	4	1	0	1	2	1	4
30-39	5	0	5	3	0	3	7	0	7
40+	3	2	5	3	1	4	4	2	6
TOTAL	10	3	14	7	1	8	13	3	17
Third Cuneiform									
<30			2			1	1	0	2
30-39			2			1	2	1	3
40+			2			4	2	2	4
TOTAL			6			6	5	3	9
Subtalar Articular Surface									
<30	2	3	9	4	2	11	4	3	13
30-39	4	2	6	9	1	10	9	3	12
40+	6	3	9	7	4	11	8	4	12
TOTAL	12	8	24	20	7	32	21	10	37

* Total Individuals (numerator and denominator include only one element side when both present)

Only one individual, a male aged less than 30 years, displayed gross morphological changes in the contour of the mandibular condyle. The only other articular change noted in this individual's TMJ joint was porosity to the condyles.

With regard to condylar depressions (CD), 10/69 (14.5%) assessable condyles or nine individuals (22.0%) displayed this feature (Table 4.12a). Five of the nine individuals with CDs were female, three males and the remaining individual was of indeterminate sex. Seven of these nine individuals were aged less than 30 years old. The condition does not display a side bias. Further, in the five individuals where it could be assessed for bilateral expression, it expressed itself as such only once.

Of the 51 individuals with at least one assessable mandibular fossa, 7.8% displayed ASO and 25.5% porosity (Table 4.12b). No individuals were observed to express ASO or EB. An even number of males and females were affected by ASO and all were aged 30-39 years. With regard to porosity of the mandibular fossa the sexes were evenly affected. However, males tended to be in the older age class while females tended to be in the middle to younger age classes. Of the five observations of ASO, four co-occurred with porosity.

Looking at the jaw as a whole it can be seen that of the seven individuals exhibiting MO and/or ASO to at least one articular surface of the TMJ, six had both assessable condyles and fossae (Table 4.12c). For these particular six individuals, articular change only affected one of the corresponding articular surfaces, never the condyle and the fossa together.

Of the nine individuals exhibiting CD, one individual has associated ASO of the mandibular fossa, three display porosity of the fossa and one has porosity of the condyle other than that directly associated with the CD. The single individual where CD is associated with ASO is in a male aged 50+ years.

The Shoulder

Of the 12 individuals with at least one assessable proximal humerus one person (8.3%), a female aged 30-39 years with an assessable left element only, displayed MO (Table 4.13). Porosity was present in four (15.4%) individuals, three females and one male, all aged 30+ years. Bilateral expression of porosity manifested in only one of the three individuals it could be assessed for. No cases of eburnation or ASO were recorded. Porosity did not co-occur with MO.

Of the ten individuals with an assessable scapula glenoid cavity MO affected two individuals (Table 4.13). One individual was male, for which only one side was preserved, and one individual was female exhibiting the condition bilaterally. Both were aged 40+ years. Porosity of the glenoid affected six individuals, four females and two males, with a fairly even distribution across age classes. Bilateral expression of porosity manifested in

Table 4.10b. Da But Period: Changes to the Articular Surfaces of the Foot

	Left Side									Right Side									Total Individuals*								
	male			female			total			male			female			total			male			female			total		
	obs	n	%	obs	n	%	obs	n	%	obs	n	%	obs	n	%	obs	n	%	obs	n	%	obs	n	%	obs	n	%
Talar Head: Porosity																											
<30	1	1	100.0	0	0	0.0	1	3	33.3	0	6	0.0	0	2	0.0	0	14	0.0	1	4	25.0	0	3	0.0	1	14	7.1
30-39	0	4	0.0	0	3	0.0	0	8	0.0	0	6	0.0	0	3	0.0	0	9	0.0	0	9	0.0	0	3	0.0	0	12	0.0
40+	0	6	0.0	0	9	0.0	0	16	0.0	0	6	0.0	0	3	0.0	0	9	0.0	0	7	0.0	0	4	0.0	0	11	0.0
TOTAL	1	11	9.1	0	12	0.0	1	27	3.7	0	18	0.0	0	8	0.0	0	32	0.0	1	20	5.0	0	10	0.0	1	37	2.7
Talar Head: Eburnation																											
<30	0	1	0.0	0	0	0.0	1	3	33.3	0	6	0.0	0	2	0.0	0	14	0.0	0	4	0.0	0	3	0.0	1	14	7.1
30-39	0	4	0.0	0	3	0.0	0	8	0.0	0	6	0.0	0	3	0.0	0	9	0.0	0	9	0.0	0	3	0.0	0	12	0.0
40+	0	6	0.0	0	9	0.0	0	16	0.0	0	6	0.0	0	3	0.0	0	9	0.0	0	7	0.0	0	4	0.0	0	11	0.0
TOTAL	0	11	0.0	0	12	0.0	1	27	3.7	0	18	0.0	0	8	0.0	0	32	0.0	0	20	0.0	0	10	0.0	1	37	2.7
Calcaneal Talar Facet: Marginal Osteophytosis																											
<30	0	2	0.0	1	1	100.0	1	4	25.0	0	3	0.0	0	0	0.0	0	5	0.0	0	3	0.0	1	1	0.0	1	6	16.7
30-39	0	4	0.0	1	2	50.0	1	6	16.7	0	6	0.0	0	1	0.0	0	7	0.0	0	6	0.0	1	3	33.3	1	9	11.1
40+	1	6	16.7	0	4	0.0	1	10	10.0	0	2	0.0	1	3	33.3	1	5	20.0	1	6	16.7	1	5	20.0	2	11	18.2
TOTAL	1	12	8.3	2	7	28.6	3	20	15.0	0	11	0.0	1	4	25.0	1	17	5.9	1	15	6.7	3	9	33.3	4	26	15.4
Calcaneal Cuboid Facet: Marginal Osteophytosis																											
<30	0	2	0.0	0	2	0.0	0	5	0.0	0	3	0.0	0	0	0.0	0	4	0.0	0	3	0.0	0	2	0.0	0	7	0.0
30-39	0	1	0.0	0	2	0.0	0	3	0.0	0	6	0.0	1	1	100.0	1	7	14.3	0	6	0.0	1	3	33.3	1	9	11.1
40+	0	7	0.0	0	3	0.0	0	10	0.0	0	3	0.0	1	4	25.0	1	7	14.3	0	7	0.0	1	5	20.0	1	12	8.3
TOTAL	0	10	0.0	0	7	0.0	0	18	0.0	0	12	0.0	2	5	40.0	2	18	11.1	0	16	0.0	2	10	20.0	2	28	7.1

Note: MO and ASO not observed for the talar head; ASO, EB and porosity not observed for the calcaneal talar facet; ASO, EB and porosity not observed for the calcaneal cuboid facet

Talar Head Correlation Matrix (left and right total)

	MO	ASO	P	EB
MO	**0**			
ASO	0	**0**		
P	0	0	**1**	
EB	0	0	1	**1**

* Total Individuals (numerator and denominator includes only one element side when both present)

Table 4.10b continued. Da But Period: Changes to the Articular Surfaces of the Foot

Left Side										Right Side									Total Individuals*								
	male			female			total			male			female			total			male			female			total		
	obs	n	%	obs	n	%	obs	n	%	obs	n	%	obs	n	%	obs	n	%	obs	n	%	obs	n	%	obs	n	%
Navicular: Marginal Osteophytosis																											
<30	0	2	0.0	0	0	0.0	0	4	0.0	0	2	0.0	0	1	0.0	0	3	0.0	0	2	0.0	0	1	0.0	0	5	0.0
30-39	0	6	0.0	0	2	0.0	0	8	0.0	0	4	0.0	1	3	33.3	1	7	14.3	0	7	0.0	1	4	25.0	1	11	9.1
40+	0	2	0.0	0	2	0.0	0	4	0.0	0	3	0.0	1	3	33.3	1	6	16.7	0	3	0.0	1	4	25.0	1	7	14.3
TOTAL	0	10	0.0	0	4	0.0	0	16	0.0	0	9	0.0	2	7	28.6	2	16	12.5	0	12	0.0	2	9	22.2	2	23	8.7
Navicular: Articular Surface Osteophytosis																											
<30	0	2	0.0	0	0	0.0	0	4	0.0	0	2	0.0	0	1	0.0	0	3	0.0	0	2	0.0	0	1	0.0	0	5	0.0
30-39	0	6	0.0	0	2	0.0	0	8	0.0	0	4	0.0	1	3	33.3	1	7	14.3	0	7	0.0	1	4	25.0	1	11	9.1
40+	0	2	0.0	0	2	0.0	0	4	0.0	0	3	0.0	0	3	0.0	0	6	0.0	0	3	0.0	0	4	0.0	0	7	0.0
TOTAL	0	10	0.0	0	4	0.0	0	16	0.0	0	9	0.0	1	7	14.3	1	16	6.3	0	12	0.0	1	9	11.1	1	23	4.3
Navicular: Porosity																											
<30	0	2	0.0	0	0	0.0	0	4	0.0	0	2	0.0	0	1	0.0	0	3	0.0	0	2	0.0	0	1	0.0	0	5	0.0
30-39	1	6	16.7	0	2	0.0	1	8	12.5	0	4	0.0	1	3	33.3	1	7	14.3	1	7	14.3	1	4	25.0	2	11	18.2
40+	0	2	0.0	0	2	0.0	0	4	0.0	0	3	0.0	0	3	0.0	0	6	0.0	0	3	0.0	0	4	0.0	0	7	0.0
TOTAL	1	10	10.0	0	4	0.0	1	16	6.3	0	9	0.0	1	7	14.3	1	16	6.3	1	12	8.3	1	9	11.1	2	23	8.7
Cuboid: Marginal Osteophytosis																											
<30	0	2	0.0	0	0	0.0	0	3	0.0	0	3	0.0	0	1	0.0	0	4	0.0	0	3	0.0	0	1	0.0	0	5	0.0
30-39	0	5	0.0	1	3	33.3	1	8	12.5	0	5	0.0	0	0	0.0	0	5	0.0	0	7	0.0	1	3	33.3	1	10	10.0
40+	0	1	0.0	0	3	0.0	0	4	0.0	0	3	0.0	0	1	0.0	0	4	0.0	0	3	0.0	0	4	0.0	0	7	0.0
TOTAL	0	8	0.0	1	6	16.7	1	15	6.7	0	11	0.0	0	2	0.0	0	13	0.0	0	13	0.0	1	8	12.5	1	22	4.5

Note: EB was not observed for the navicular; ASO, EB and porosity was not observed for the cuboid.

Navicular: Correlation Matrix (left and right total)

	MO	ASO	P	EB
MO	**2**			
ASO	0	**1**		
P	1	0	**2**	
EB	0	0	0	**0**

* Total Individuals (numerator and denominator include only one element side when both present)

Table 4.10c. Da But Period: Tarsometatarsal and Metatarsophalangeal Articular Changes (Foot)

	Articular Surface Count									Total Individuals*								
	male			female			total			male			female			total		
	obs	n	%	obs	n	%	obs	n	%	obs	n	%	obs	n	%	obs	n	%
Tarsometatarsal Joint [1] Articular Surfaces (sample sizes)																		
<30		23			14			50			4			4			12	
30-39		43			22			65			10			5			15	
40+		32			24			55			7			6			13	
TOTAL		98			60			170			21			15			40	
Metatarsophalangeal (MTP)[2] Joint Articular Surfaces: Marginal Osteophytosis																		
<30	0	32	0.0	0	12	0.0	0	59	0.0	0	4	0.0	0	4	0.0	0	12	0.0
30-39	1	40	2.5	1	17	5.9	2	57	3.5	1	9	11.1	1	4	25.0	2	13	15.4
40+	2	39	5.1	1	27	3.7	3	66	4.5	2	8	25.0	1	5	20.0	3	13	23.1
TOTAL	3	111	2.7	2	56	3.6	5	182	2.7	3	21	14.3	2	13	15.4	5	38	13.2
Metatarsophalangeal (MTP)[2] Joint Articular Surfaces: Porosity																		
<30	0	32	0.0	0	12	0.0	0	59	0.0	0	4	0.0	0	4	0.0	0	12	0.0
30-39	2	40	5.0	0	17	0.0	2	57	3.5	1	9	11.1	0	4	0.0	1	13	7.7
40+	1	39	2.6	3	27	11.1	4	66	6.1	1	8	12.5	2	5	40.0	3	13	23.1
TOTAL	3	111	2.7	3	56	5.4	6	182	3.3	2	21	9.5	2	13	15.4	4	38	10.5

Note: Tarsometatarsal joint MO, ASO, EB and porosity was not observed
Metatarsophalangeal joint ASO and EB was not observed
[1] Tarsometatarsal Joint: defined as the articular surfaces of the proximal metatarsals only
[2] MTP: distal articular surfaces of metatarsals and proximal articular surfaces of proximal phalanges

MTP Correlation Matrix

	MO	ASO	P	EB
MO	**5**			
ASO	0	**0**		
P	4	0	**6**	
EB	0	0	0	**0**

* individual: represented by at least one articular surface of a given joint

two of the three individuals it could be assessed for. The two individuals displaying MO also showed evidence of porosity.

Looking at the shoulder as a whole, three individuals (20.0%) displayed MO. In only one of these three individuals were both elements, humerus and glenoid, assessable. In this particular individual MO only manifested in one element. Of the eight individuals displaying porosity, four could be assessed for both articulating elements and two of these displayed porosity on both joint surfaces.

The Elbow

Of the 14 individuals with at least one assessable distal humerus, one individual (7.1%) displayed MO, four (28.6%) ASO, five (35.7%) porosity and one (7.1%) eburnation (Table 4.14a). The single observation of MO was expressed unilaterally in the right side of a male, aged 40-49 years. ASO tended to be seen in the older age classes while porosity concentrated in the younger age classes. The only observation of eburnation occurred in the youngest age class. The majority of observations of ASO occurred in females and manifested bilaterally in one of the two individuals this could be assessed for. Most observations of porosity also occurred in females. Bilateral expression of porosity occurred in three of the five individuals this could be assessed for. Porosity co-occurred with two instances of ASO and also with the only instance of eburnation.

Of the 14 individuals with at least one assessable proximal radius, two individuals (14.3%) displayed MO, two (14.3%) ASO and three (21.4%) porosity (Table 4.14b). The four individuals displaying MO or ASO were all female and under 40 years old. No observations of eburnation were recorded for the proximal radius. MO manifested unilaterally in one of the two individuals that had both radii preserved. Neither individual expressing ASO could be assessed for bilateral expression. Two of the three individuals displaying porosity had both radii preserved. Porosity manifested bilaterally for both of these individuals. No observations of porosity co-occurred with MO. One observation of porosity did co-occur with ASO however.

Of the 14 individuals with at least one assessable proximal ulna, six (42.9%) displayed MO, two (14.3%) porosity and one (7.1%) eburnation (Table 4.14c). All observations of articular change were in individuals over 30 years

Table 4.10d. Da But Period: Proximal Interphalangeal Joint Articular Changes (Foot)

Articular Surface Count										Total Individuals*								
	male			female			total			male			female			total		
	obs	n	%	obs	n	%	obs	n	%	obs	n	%	obs	n	%	obs	n	%
Proximal Interphalangeal Joint (PIPJ)[1] Articular Surfaces: Marginal Osteophytosis																		
<30	3	15	20.0	0	3	0.0	3	29	10.3	2	4	50.0	0	3	0.0	2	11	18.2
30-39	0	24	0.0	0	0	na	0	24	0.0	0	7	0.0	0	0	na	0	7	0.0
40+	0	25	0.0	0	10	0.0	0	35	0.0	0	5	0.0	0	5	0.0	0	10	0.0
TOTAL	3	64	4.7	0	13	0.0	3	88	3.4	2	16	12.5	0	8	0.0	2	28	7.1
Proximal Interphalangeal Joint (PIPJ)[1] Articular Surfaces: Articular Surface Osteophytes																		
<30	1	15	6.7	0	3	0.0	1	29	3.4	1	4	25.0	0	3	0.0	1	10	10.0
30-39	0	24	0.0	0	0	na	0	24	0.0	0	7	0.0	0	0	na	0	7	0.0
40+	0	25	0.0	0	10	0.0	0	35	0.0	0	5	0.0	0	5	0.0	0	10	0.0
TOTAL	1	64	1.6	0	13	0.0	1	88	1.1	1	16	6.3	0	8	0.0	1	27	3.7
Proximal Interphalangeal Joint (PIPJ)[1] Articular Surfaces: Porosity																		
<30	2	15	13.3	0	3	0.0	2	29	6.9	1	4	25.0	0	3	0.0	1	10	10.0
30-39	0	24	0.0	0	0	na	0	24	0.0	0	7	0.0	0	0	na	0	7	0.0
40+	0	25	0.0	0	10	0.0	0	35	0.0	0	5	0.0	0	5	0.0	0	10	0.0
TOTAL	2	64	3.1	0	13	0.0	2	88	2.3	1	16	6.3	0	8	0.0	1	27	3.7

Note: no observations of PIPJ eburnation

[1] PIPJ: distal articular surfaces of proximal phalanges and proximal articularsurfaces of middle phalanges

PIPJ Correlation Matrix

	MO	ASO	P	EB
MO	**3**			
ASO	1	**1**		
P	2	0	**1**	
EB	0	0	0	**0**

* individual: represented by at least one articular surface of a given joint

old with an even distribution by sex. Three of the four individuals displaying MO that could be assessed for bilateral expression were so affected. Porosity manifested bilaterally in the one individual for which this could be assessed. Eburnation manifested unilaterally in the single individual displaying this articular change. Of the nine separate proximal ulnae displaying MO, only one also displayed porosity. The single case of eburnation co-occurred with porosity.

Looking at the elbow as a whole, of the 17 individuals with at least one assessable elbow element 11 displayed articular changes (Table 4.14d). Of the 11 individuals displaying some form of articular change, nine (52.9%) exhibited either MO, ASO, EB or a combination thereof. There is a clear pattern in that ASO is more common on the distal humerus, while MO is more common on the proximal ulna. Of the four individuals with ASO of the distal humerus three also displayed MO of the proximal ulna, while the remaining case could not be assessed for the corresponding ulna. Four individuals also displayed either MO or ASO of the proximal radius, which also corresponded to such changes in the humerus or ulna in three of these individuals, one person not having assessable corresponding joints.

The Wrist

Only one, female aged 40-49 years, of the 12 Metal period individuals with assessable distal ulnae and radii displayed any joint change (Table 4.15). This accounted for 8.3% of MO to the distal radius, 8.3% of MO and 8.3% ASO to the distal ulna. In both instances only one element side was preserved. A total of 13 individuals were represented by elements of the wrist. The aforementioned single individual displaying articular change to the wrist accounts for 7.7% of assessable individuals.

The Hand

Six individuals are represented by a total of 23 assessable carpals (4 capitates, 2 hamates, 4 lunates, 6 naviculars, 1 pisiform, 2 trapezia, 2 trapezoids and 2 triquetrals). None of these carpal elements exhibited articular changes (Table 4.16).

Results for the articular surfaces of the finger joints are given in Table 4.16. Of a total of 38 metacarpals, representing seven individuals, no cases of articular change were observed. Of a total of 62 MCP articular surfaces, representing nine individuals, no observations of articular change were observed. Of the 38 assessable articular surfaces of the PIPJ, representing 11 individuals,

Table 4.10e. Da But Period: Distal Interphalangeal Joint[1] Articular Changes to the Foot (Sample Sizes)

Articular Surface Count				Total Individuals		
	male	female	total	male	female	total
Age	n	n	n	n	n	n
<30	0	0	0	0	0	0
30-39	6	0	6	4	0	4
40+	8	0	8	4	0	4
TOTAL	14	0	14	8	0	8

Note: DIPJ MO, ASO, EB and porosity not observed

[1]DIPJ: distal articular surfaces of middle phalanges and proximal articular surfaces of the distal phalanges

* individual: represented by at least one articular surface of a given joint

no observations of articular change were observed. Of the 12 assessable articular surfaces of the DIPJ, representing seven individuals, one articular surface (8.3%) displayed MO. No other articular changes for the DIPJ were observed.

The Hip

Of the 11 individuals with at least one assessable proximal femur one individual (9.1%) displayed bilateral ASO and two (18.2%) porosity (Table 4.17a). No observations of MO or EB were made. One individual manifested porosity bilaterally, while the other individual only had one preserved element. The two individuals involved were both female and aged less than 40 years. The observations of ASO, around the rim of the fovea capitis, co-occurred with porosity.

Of the 10 individuals (Table 4.17b) with at least one assessable acetabulum two individuals displayed MO, one ASO, three porosity and one eburnation. One individual exhibited MO unilaterally and the other expressed it bilaterally. The individual with ASO manifested this bilaterally. The individual with eburnation expressed this unilaterally. Two of those with porosity expressed this bilaterally while the other individual had only one element side preserved. Porosity co-occurred with one observation of MO, both observations of ASO and the only observation of eburnation. While the MO and ASO were in individuals greater than 30 years old, porosity affected all age classes. The one instance of eburnation occurred in the youngest age class.

Looking at the hip as a whole, 15 individuals have at least an assessable proximal femur and/or acetabulum. Of these, four individuals (23.5%) displayed either MO, ASO and/or eburnation of one of the articular surfaces. The three individuals that could be assessed for changes to both joint surfaces, femur and acetabulum, did not display corresponding changes.

The Knee

Of the 10 individuals with at least one assessable distal femur two displayed MO and two porosity (Table 4.18a). No observations of ASO or EB occurred. MO could not be assessed for bilateral expression in any instances while both cases of porosity manifested bilaterally. MO and porosity did not co-occur. The two individuals displaying MO were aged 20-29 years and 30-39 years, while the individuals with porosity were both aged 30-39 years.

Of the 10 individuals with at least one assessable proximal tibia one displayed MO (Table 4.18a). No observations of ASO, EB or porosity were made. The individual with MO was a male aged 20-29 years and only one side could be assessed.

Of the eight individuals with at least one assessable proximal fibula one (12.5%) individual, female aged 30-39 years, displayed both porosity and eburnation (Table 4.18b). No observations of MO or ASO were made. For the female exhibiting articular change, the porosity expressed bilaterally while the eburnation manifested unilaterally.

Of the six individuals with at least one assessable patella one (16.7%) displayed ASO (Table 4.18b). No patellae exhibited MO, EB or porosity. The single individual displaying ASO was a male aged 40-49 years. No other knee articular surfaces were preserved for this individual.

Looking at the knee as a whole, of the 13 individuals with at least one assessable articular surface of the knee joint six (46.2%) displayed some form of articular change. Five individuals (38.5%) displayed some form of articular change other than porosity alone.

The Ankle

Of the 12 individuals with at least one assessable distal tibia, one (8.3%) displayed MO (Table 4.19). No articular surfaces of the ankle exhibited ASO, EB or porosity. The only individual to exhibit MO of the distal tibia was a female aged 40-49 years and only one side was assessable.

None of the seven individuals with an assessable distal fibula displayed articular changes (Table 4.19). Of the seven individuals with an assessable talar trochlea, one (14.3%) displayed porosity bilaterally (Table 4.19). This was the same individual that exhibited porosity and eburnation of the proximal fibula. No other articular changes to the talar trochlea were observed.

Looking at the ankle as a whole, 15 individuals have at least one assessable ankle articular surface. Of these, two

Table 4.10f. Da But Period: Individuals Affected by Articular Changes to the Foot

			Talar Head				Calcaneal Talar Facet				Calcaneal Cuboid Facet				Navicular				Cuboid			
case	sex	age	MO	ASO	P	EB	MO	ASO	P	EB	MO	ASO	P	EB	MO	ASO	P	EB	MO	ASO	P	EB
80CCNM66a	M	5																				
80CCNM45a	F	5					Y				Y				Y							
80CCNM56a	F	5																				
80CCNM15a	M	4					Y															
80CCNM85a	M	4																				
80CCNMX1a	F	4																				
80CCNM10c	M	3																				
80CCNM26a	M	3															Y					
80CCNM42a	F	3														Y						
80CCNM54a	F	3					Y												Y			
80CCNM77a	F	3									Y				Y		Y					
80CCNM60a	M	2																				
80CCNM14a	M	2			Y	Y																
80CCNM67a	F	1					Y															

			Tarsometatarsal Joint				MTP Joint				PIPJ				DIPJ			
case	sex	age	MO	ASO	P	EB	MO	ASO	P	EB	MO	ASO	P	EB	MO	ASO	P	EB
80CCNM66a	M	5					Y											
80CCNM45a	F	5																
80CCNM56a	F	5					Y		Y									
80CCNM15a	M	4																
80CCNM85a	M	4					Y		Y									
80CCNMX1a	F	4							Y									
80CCNM10c	M	3					Y		Y									
80CCNM26a	M	3																
80CCNM42a	F	3																
80CCNM54a	F	3					Y											
80CCNM77a	F	3																
80CCNM60a	M	2									Y							
80CCNM14a	M	2									Y	Y	Y					
80CCNM67a	F	1																

Y: articular change observed
(blank): articular change not observed
(shaded): articular area not preserved

5= 50+ years
4=40-49 years
3=30-39 years
2=20-29 years
1=15-19 years

M=male
F=female
I=indeterminate

Table 4.11a. Da But Period: Articular Changes to the Cervical Vertebrae

	Vertebral Count									Total Individuals*								
	male			female			total			male			female			total		
	obs	n	%	obs	n	%	obs	n	%	obs	n	%	obs	n	%	obs	n	%
ZOP: Axis (including dens)																		
<30										1	4	25.0	1	1	100.0	2	5	40.0
30-39										2	4	50.0	1	2	50.0	3	6	50.0
40+										3	5	60.0	1	2	50.0	4	7	57.1
TOTAL										6	13	46.2	3	5	60.0	9	18	50.0
VOP: Cervical C3 to C7																		
<30	0	7	0.0	0	3	0.0	0	11	0.0	0	3	0.0	0	2	0.0	0	6	0.0
30-39	1	11	9.1	0	0	na	1	11	9.1	1	6	16.7	0	0	na	1	6	16.7
40+	2	2	100.0	2	4	50.0	4	6	66.7	2	2	100.0	2	4	50.0	4	6	66.7
TOTAL	3	20	15.0	2	7	28.6	5	28	17.9	3	11	27.3	2	6	33.3	5	18	27.8
ZOP: Cervical C3-C7																		
<30	0	3	0.0	0	0	na	0	3	0.0	0	1	0.0	0	0	na	0	1	0.0
30-39	0	4	0.0	0	0	na	0	4	0.0	0	3	0.0	0	0	na	0	3	0.0
40+	0	2	0.0	1	2	50.0	1	4	25.0	0	1	0.0	1	2	50.0	1	3	33.3
TOTAL	0	9	0.0	1	2	50.0	1	11	9.1	0	5	0.0	1	2	50.0	1	7	14.3

*Individual: represented by at least one assessable vertebral element
ZOP: zygapophyseal osteophytosis
VOP: vertebral osteophytosis

individuals (13.3%) displayed some form of joint change, while only one individual (6.7%) displayed MO.

The Foot

Very few articular surfaces in the various bones of the foot were affected by articular change. The following tarsal or tarsal articular surfaces, with the number of individuals represented in parentheses, did not display any changes: subtalar articular surface (7); first cuneiform (5); second cuneiform (2); and third cuneiform (3). Table 4.20a displays sample sizes for those elements without articular changes.

Of the seven individuals represented by at least one talar head, one individual (14.3%) exhibited porosity and this manifested bilaterally (Table 4.20b). No other articular changes to the talar head were observed.

Of the eight individuals represented by at least one calcaneal talar facet, one individual (12.5%) exhibited MO and two displayed porosity (Table 4.20b). No other articular changes were observed. The single observation of MO, which co-occurred with porosity, manifested in a female aged 40-49 years. This female had only one preserved calcaneal talar facet.

Of the eight individuals represented by at least one calcaneal cuboid facet, one individual (12.5%) displayed MO bilaterally (Table 4.20a). No further articular changes to the calcaneal cuboid facet were observed. The individual displaying MO was a male aged 20-29 years. The corresponding articular facets on the cuboid bones also displayed MO.

Of the other tarsal bones examined only the cuboid and navicular displayed any articular changes (Table 4.20b). There were only seven preserved cuboids, representing one male and three females. The male, aged 20-29 years, exhibited bilateral MO. Likewise, there were only seven naviculars, representing three females and one male. The male, aged 30-39 years, displayed MO, ASO and porosity to his only preserved navicular bone (left side).

Results for the articular surfaces of the toe joints are given in Table 4.20a. Of a total of 32 proximal metatarsals (TMT), representing seven individuals, no observations of articular change were observed. Of the 23 assessable articular surfaces of the MTP joint, representing six individuals, no observations of articular change were observed. Only five assessable articular surfaces of the PIPJ, representing two individuals, were preserved and no articular changes were observed. Only one articular surface of the DIPJ, representing one individual, was preserved and no articular changes were observed.

Looking at the foot as a whole, 13 individuals were represented by foot elements. Five of these (38.5%) displayed some form of articular change, while four individuals (30.8%) exhibited changes other than porosity alone.

The Spine

Cervical Vertebrae:

Of the six individuals with assessable atlas facets two (33.3%) displayed ZOP to the inferior facets, a male and female both aged 40-49 years (Table 4.21a). The second cervical vertebra, the axis, was not represented by any

Table 4.11b. Da But Period: Articular Changes to the Thoracic, Lumbar and Sacral Vertebrae

	Vertebral Count									Total Individuals*								
	male			female			total			male			female			total		
	obs	n	%	obs	n	%	obs	n	%	obs	n	%	obs	n	%	obs	n	%
VOP: Thoracic T1 to T12																		
<30	0	3	0.0	0	5	0.0	0	10	0.0	0	2	0.0	0	1	0.0	0	4	0.0
30-39	2	21	9.5	0	11	0.0	2	32	6.3	1	5	20.0	1	1	100.0	1	6	16.7
40+	1	5	20.0	0	2	0.0	1	7	14.3	1	4	25.0	1	2	50.0	1	6	16.7
TOTAL	3	29	10.3	0	18	0.0	3	49	6.1	2	11	18.2	2	4	50.0	2	16	12.5
ZOP: Thoracic T1 to T12																		
<30	0	2	0.0	0	0	0.0	0	2	0.0	0	2	0.0	0	0	0.0	0	2	0.0
30-39	6	21	28.6	3	11	27.3	9	32	28.1	2	6	33.3	1	1	100.0	3	7	42.9
40+	0	1	0.0	1	1	100.0	1	2	50.0	0	1	0.0	1	1	100.0	1	2	50.0
TOTAL	6	24	25.0	4	12	33.3	10	36	27.8	2	9	22.2	2	2	100.0	4	11	36.4
VOP: Lumbar L1 to L5																		
<30	0	1	0.0	0	6	0.0	0	12	0.0	0	1	0.0	0	2	0.0	0	5	0.0
30-39	0	10	0.0	2	9	22.2	2	19	10.5	0	3	0.0	1	2	50.0	1	5	20.0
40+	1	6	16.7	0	1	0.0	1	7	14.3	1	3	33.3	0	1	0.0	1	4	25.0
TOTAL	1	17	5.9	2	16	12.5	3	38	7.9	1	7	14.3	1	5	20.0	2	14	14.3
ZOP: Lumbar L1 to L5																		
<30	0	1	0.0	0	1	0.0	0	6	0.0	0	1	0.0	0	1	0.0	0	4	0.0
30-39	5	12	41.7	2	9	22.2	7	21	33.3	2	3	66.7	1	2	50.0	3	5	60.0
40+	1	2	50.0	3	8	37.5	4	10	40.0	1	2	50.0	1	3	33.3	2	5	40.0
TOTAL	6	15	40.0	5	18	27.8	11	37	29.7	3	6	50.0	2	6	33.3	5	14	35.7
VOP: Sacral Promontory																		
<30										0	2	0.0	0	1	0.0	0	3	0.0
30-39										0	2	0.0	0	0	0.0	0	2	0.0
40+										1	5	20.0	0	2	0.0	1	7	14.3
TOTAL										1	9	11.1	0	3	0.0	1	12	8.3

*Individual: represented by at least one assessable vertebral element

ZOP: zygapophyseal osteophytosis

VOP: vertebral osteophytosis

Note: sacral articular facets only preserved in 3 cases (2 age class 2 and 1 age class 3). Slight marginal osteophytic lipping seen in one individual of age classes 2 and 3.

assessable centra. Of the 13 individuals with assessable articular surfaces of the zygapophyseal joints of the axis, five individuals (38.5%) displayed ZOP, one of these being the male with atlas ZOP. The other individuals included two males aged 30-39 years, a male aged 50+ years and a female aged 30-39 years.

None of the 12 individuals with assessable cervical (C3-C7) centra exhibited VOP (Table 4.21a). Of the nine individuals with assessable zygapophyseal facets, two individuals (22.2%), both males and aged 30-39 years, displayed ZOP.

Thoracic Vertebrae:
None of the seven individuals with assessable thoracic centra displayed VOP (Table 4.21a). However, four (57.1%) of these same seven individuals exhibited ZOP.

Lumbar Vertebrae:
Of the seven individuals with assessable lumbar centra, two individuals (28.6%) displayed VOP (Table 4.21a). Both individuals were male, one aged 20-29 years and the other 30-39 years. Of the six individuals with assessable lumbar zygapophyses, three displayed ZOP. One individual, aged 30-39 years also exhibited lumbar VOP. The remaining two individuals displaying ZOP included a male aged 50+ years and a female aged 40-49 years.

Sacrum:
The superior articular facets of the sacra in this sample were too poorly preserved to assess. However, five individuals had an assessable promontory of which two individuals displayed VOP. One was the 30-39 year old male with lumbar VOP and ZOP. The remaining individual was the aforementioned 50+ year old female with lumbar ZOP.

Looking at the spine as a whole, 17 individuals had at least one assessable vertebral element. Of these assessable cases, nine individuals displayed osseous changes to the zygapophyses and/or vertebral centra (Table 4.21b). If the denominator is individuals with assessable vertebral

Table 4.11c. Da But Period: Individuals Affected by Articular Changes to the Spine

			Axis		Cervical (C3 to C7)		Thoracic		Lumbar		Sacrum
case	sex	age	VOP	ZOP	VOP	ZOP	VOP	ZOP	VOP	ZOP	VOP
80CCNM14b	M	5		Y						Y	
80CCNM16b	M	5			Y						
80CCNM74a	M	5			Y		Y		Y		Y
80CCNM45a	F	5		Y	Y	Y		Y		Y	
80CCNM50a	F	5			Y						
80CCNM10a	M	3		Y				Y		Y	
80CCNM18a	M	3			Y						
80CCNM35a	M	3						Y		Y	
80CCNM62a	M	3		Y							
80CCNM42a	F	3						Y	Y		
80CCNM53a	F	3		Y							
80CCNM77a	F	3								Y	
80CCNM57a	M	2		Y							
80CCNM65a	M	2					Y				
80CCNM71a	F	2		Y							

Y : articular change observed
(blank) : articular change not observed
(shaded) : articular area not preserved

5= 50+ years
4=40-49 years
3=30-39 years
2=20-29 years
1=15-19 years

M=male
F=female
I=indeterminate

centra and/or zygapophyses (n=17), then 52.9% of the assessable sample displayed some form of articular change. With the denominator being individuals with assessable zygapophyses, 8/17 (47.1%) displayed ZOP. When the denominator is individuals with assessable centra, including the sacral promontory, 3/14 (21.4%) exhibited VOP. In most cases VOP and ZOP could be assessed for the same vertebra. By far, ZOP was the more common form of joint change.

Of the nine individuals displaying vertebral joint change, 22.2% were 50+ years, while 44.4% were aged 40+ years. Only one individual displaying articular changes to the spine was aged less than 30 years.

While few individuals displaying articular changes to the spine were excluded from this analysis, due to a lack of age-at-death information, the most extreme case of articular change to the spine was. This case is worth describing as it represents the only example so far, from both the Da But period and Metal period contexts of northern Vietnam, of the ankylosis of a joint.

The individual is from the site Vinh Quang and is represented by a relatively complete axis, C3, C4 and portions of a thoracic vertebra. The transverse processes of all cervical vertebrae are missing post-mortem. The ankylosis occurs between the axis and C3 at the right interarticular facets (Figure 4.4). The left and right superior and inferior facets of the axis display erosion of their latero-posterior aspects, but the fusion of the right articular processes is clear medially. The centra of the axis and C3 have not fused. The anterior aspect of the dens is porous and a high lip or sleeve has formed on the antero-superior aspect of the dens. This probably represents ossification of the apical ligament. The posterior aspect of the dens exhibits two large and confluent holes.

Figure 4.4. 66VQ1M8a. Metal Period individual of indeterminate sex and age. Right latero-posterior aspect of axis and C3. Note ankylosis of the interarticular facets of these two vertebrae

The superior interarticular facets of C4 are too badly eroded post-mortem to assess for articular changes. However, the inferior facets are preserved and display mild osteophytic lipping. The superior and inferior interarticular facets of the thoracic vertebra can also be assessed and display osteophytic lipping and porosity.

Summary of the Distribution of Articular Change by Sex and Age for Each Sample

The results of articular changes to the components of each joint complex have been detailed. In the following summary the distribution of articular change by particular joints (the elbow, wrist, knee and so forth) are examined with respect to the sex and age of the Da But period and Metal period samples.

Da But Period

A total of 73 individuals of known age had at least one

Table 4.12a. Metal Period: Changes to the Articular Surfaces of the Mandibular Condyles (TMJ)

Left Side										Right Side									Total Individuals*								
	male			female			total			male			female			total			male			female			total		
	obs	n	%	obs	n	%	obs	n	%	obs	n	%	obs	n	%	obs	n	%	obs	n	%	obs	n	%	obs	n	%
Condyle: Marginal Osteophytosis																											
<30	0	8	0.0	0	7	0.0	0	17	0.0	0	6	0.0	1	7	14.3	1	16	6.3	0	8	0.0	1	8	12.5	1	20	5.0
30-39	0	5	0.0	0	6	0.0	0	11	0.0	0	3	0.0	0	5	0.0	0	8	0.0	0	5	0.0	0	6	0.0	0	11	0.0
40+	0	7	0.0	0	2	0.0	0	9	0.0	1	6	16.7	0	2	0.0	1	8	12.5	1	8	12.5	0	2	0.0	1	10	10.0
TOTAL	0	20	0.0	0	15	0.0	0	37	0.0	1	15	6.7	1	14	7.1	2	32	6.3	1	21	4.8	1	16	6.3	2	41	4.9
Condyle: Surface Osteophytosis																											
<30	0	8	0.0	0	7	0.0	0	17	0.0	0	6	0.0	0	7	0.0	0	16	0.0	0	8	0.0	0	8	0.0	0	20	0.0
30-39	0	5	0.0	0	6	0.0	0	11	0.0	0	3	0.0	0	5	0.0	0	8	0.0	0	5	0.0	0	6	0.0	0	11	0.0
40+	1	7	14.3	0	2	0.0	1	9	11.1	1	6	16.7	0	2	0.0	1	8	12.5	1	8	12.5	0	2	0.0	1	10	10.0
TOTAL	1	20	5.0	0	15	0.0	1	37	2.7	1	15	6.7	0	14	0.0	1	32	3.1	1	21	4.8	0	16	0.0	1	41	2.4
Condyle: Porosity																											
<30	1	8	12.5	0	7	0.0	2	17	11.8	1	6	16.7	2	7	28.6	3	16	18.8	1	8	12.5	2	8	25.0	3	20	15.0
30-39	2	5	40.0	3	6	50.0	5	11	45.5	0	3	0.0	1	5	20.0	1	8	12.5	2	5	40.0	3	6	50.0	5	11	45.5
40+	1	7	14.3	0	2	0.0	1	9	11.1	2	6	33.3	0	2	0.0	2	8	25.0	3	8	37.5	0	2	0.0	3	10	30.0
TOTAL	4	20	20.0	3	15	20.0	8	37	21.6	3	15	20.0	3	14	21.4	6	32	18.8	6	21	28.6	5	16	31.3	11	41	26.8
Condyle: Depressions																											
<30	2	8	25.0	1	7	14.3	3	17	17.6	1	6	16.7	3	7	42.9	5	16	31.3	2	8	25.0	4	8	50.0	7	20	35.0
30-39	0	5	0.0	1	6	16.7	1	11	9.1	0	3	0.0	0	5	0.0	0	8	0.0	0	5	0.0	1	6	16.7	1	11	9.1
40+	1	7	14.3	0	2	0.0	1	9	11.1	0	6	0.0	0	2	0.0	0	8	0.0	1	8	12.5	0	2	0.0	1	10	10.0
TOTAL	3	20	15.0	2	15	13.3	5	37	13.5	1	15	6.7	3	14	21.4	5	32	15.6	3	21	14.3	5	16	31.3	9	41	22.0

Note: ASO was not observed for the mandibular condyles.

Correlation Matrix (left and right total)

	MO	ASO	P	EB
MO	**2**			
ASO	0	**2**		
P	1	0	**14**	
EB	0	0	0	**0**

* Total Individuals (numerator and denominator include only one element side when both present)

Table 4.12b. Metal Period: Changes to the Articular Surfaces of the Mandibular Fossa (TMJ)

Left Side

	male obs	male n	male %	female obs	female n	female %	total obs	total n	total %
Fossa: Articular Surface Osteophytosis									
<30	0	6	0.0	0	10	0.0	0	18	0.0
30-39	1	4	25.0	2	9	22.2	3	13	23.1
40+	0	7	0.0	0	4	0.0	0	12	0.0
TOTAL	1	17	5.9	2	23	8.7	3	43	7.0
Fossa: Porosity									
<30	1	6	16.7	2	10	20.0	3	18	16.7
30-39	1	4	25.0	4	9	44.4	5	13	38.5
40+	3	7	42.9	1	4	25.0	4	12	33.3
TOTAL	4	17	23.5	7	23	30.4	12	43	27.9

Right Side

	male obs	male n	male %	female obs	female n	female %	total obs	total n	total %
Fossa: Articular Surface Osteophytosis									
<30	0	6	0.0	0	6	0.0	0	16	0.0
30-39	1	3	33.3	1	9	11.1	2	12	16.7
40+	0	9	0.0	0	2	0.0	0	11	0.0
TOTAL	1	18	5.6	1	17	5.9	2	39	5.1
Fossa: Porosity									
<30	1	6	16.7	1	6	16.7	2	16	12.5
30-39	0	3	0.0	3	9	33.3	3	12	25.0
40+	4	9	44.4	0	2	0.0	4	11	36.4
TOTAL	5	18	27.8	4	17	23.5	9	39	23.1

Total Individuals*

	male obs	male n	male %	female obs	female n	female %	total obs	total n	total %
Fossa: Articular Surface Osteophytosis									
<30	0	8	0.0	0	11	0.0	0	23	0.0
30-39	2	4	50.0	2	10	20.0	4	14	28.6
40+	0	9	0.0	0	4	0.0	0	14	0.0
TOTAL	2	21	9.5	2	25	8.0	4	51	7.8
Fossa: Porosity									
<30	1	8	12.5	2	11	18.2	3	23	13.0
30-39	1	4	25.0	4	10	40.0	5	14	35.7
40+	4	9	44.4	1	4	25.0	5	14	35.7
TOTAL	6	21	28.6	7	25	28.0	13	51	25.5

Note: MO and EB not observed for the mandibular fossa.

Correlation Matrix (left and right total)

	MO	ASO	P	EB
MO	**0**			
ASO	0	**5**		
P	0	4	**21**	
EB	0	0	0	**0**

* Total Individuals (numerator and denominator include only one element side when both present)

Table 4.12c. Metal Period: Individuals Affected by Articular Changes to the TMJ

			Condyle					Glenoid	
case	sex	age	MO	ASO	P	CD	MC	ASO	P
77CHSM1a	M	5			Y				
77NNM22ka	M	5			Y				
78NNM16ka	M	5		Y		Y			Y
76NNM2c	M	4							Y
76NNM2d	M	4							Y
78QCM28a	M	4	Y		Y				Y
89MDM2a	F	4							Y
78NNM6kb	M	3			Y				
78QCM27a	M	3			Y			Y	Y
78QCM3a	M	3	(shaded)	(shaded)	(shaded)	(shaded)	(shaded)	Y	
77NNM5kb	F	3			Y			Y	Y
84DSM1a	F	3			Y				Y
92DMTS60-62M1a	F	3			Y				Y
95DXM1a	F	3				Y		Y	Y
66VQ1M1a	M	2			Y		Y		
78NNM23ka	M	2				Y		(shaded)	(shaded)
78QCM24a	M	2				Y			
78QCM25a	M	2							Y
77NNM21ka	F	2				Y		(shaded)	(shaded)
78NNM4kb	F	2				Y			
78QCM14a	F	2			Y	Y			Y
78QCM22a	F	2							Y
84DSM8a	F	2			Y				
89MDM1a	F	2	Y						
NNM3xa	I	2				Y		(shaded)	(shaded)
78QCM1a	F	1				Y			

5= 50+ years
4=40-49 years
3=30-39 years

2=20-29 years
1=15-19 years

Y: articular change observed
(blank): articular change not observed
(shaded): articular area not preserved

M=male
F=female
I=indeterminate

assessable joint complex. Of these, 56 individuals (76.7%) displayed some form of joint change while 45 individuals (61.6%) displayed changes other than porosity alone. Table 4.22 summarises the joints affected by osseous changes for those 56 individuals. Of these 56 individuals 55.4% are aged 40+ years, while 82.1% are aged 30+ years. One of the chief difficulties in dealing with fragmentary remains is illustrated here, in that rarely are individuals represented by all, or even most, joints.

Of the sexed and aged males with at least one assessable joint 29/35 (82.9%) displayed some form of joint change, while 27/35 (77.1%) displayed changes other than porosity alone. Of the sexed and aged females with at least one assessable joint 23/27 (85.2%) displayed some form of joint change, while only 16/27 59.3% displayed joint changes other than porosity alone. Of these sexed and aged individuals with at least one assessable joint 42.9% of the males and 59.3% of the females were aged 40 years or more.

Figure 4.5a graphically displays the prevalence of joint change for the combined sex sample. The denominator, in each instance, is all aged individuals with an assessable joint. The spine (15/34, 44.1%) exhibits the greatest frequency of articular change followed by the elbow (14/53, 26.4%), the shoulder (9/39, 23.1%) and foot (12/52, 23.1%), TMJ (13/58, 22.4%), knee (8/36, 22.2%), hip (3/33, 9.1%), wrist (3/35, 8.6%), ankle (2/44, 4.5%) and hand (2/47, 4.3%). As already noted, the changes in the spine principally affect the zygapophyses rather than syndesmoses.

For the most part, there is a clear positive correlation between increasing prevalence of joint change and increasing age (Table 4.23, Figure 4.5b). This is especially clear for the shoulder, elbow and TMJ. In joints such as the hand, hip and ankle only marginally affected by joint change, there are reversals to this pattern. The spine stands out in being both the most affected joint and in the middle age class showing the highest prevalence of joint change. Only the hand displays a greater prevalence of joint change in the youngest, relative to eldest, age class.

Looking at the distribution of joint change by sex, it can be seen that females display a greater prevalence of articular change in the spine, ankle and foot (Table 4.24). Males exhibit a much greater frequency of articular change to the shoulder and knee joints in particular, in addition to a greater frequency of change in the TMJ, elbow, wrist (slightly), hand (absent in females) and hip.

Metal Period

A total of 60 individuals of known age had at least one assessable joint complex. Of these, 29 individuals (48.3%) displayed some form of joint change while 20 individuals (33.3%) displayed changes other than porosity alone. Table 4.25 summarises the joints affected by osseous changes for those 29 individuals. Of these 29 individuals 27.6% are aged 40+ years, while 65.5% are aged 30+ years.

Of the sexed and aged males with at least one assessable joint 14/26 (53.8%) displayed some form of joint change, while 11/26 (42.3%) displayed changes other than porosity alone. Of the sexed and aged females with at least one assessable joint 15/26 (57.7%) displayed some form of joint change, while 9/26 (34.6%) displayed joint changes other than porosity alone. Of these sexed and aged individuals with at least one assessable joint 23.1% of the males and only 7.7% of the females were aged 40 years or more. However, 38.5% of the males and 34.6% of the females were aged 30+ years old.

Figure 4.6a graphically displays the prevalence of joint change. The denominator, in each instance, is all aged individuals with an assessable joint. The spine (9/17, 52.9%) and elbow (9/17, 52.9%) equally exhibit the greatest frequency of articular changes followed by the knee (5/13, 38.5%), foot (4/13, 30.8%), hip (4/15, 26.7%), shoulder (3/15, 20%), TMJ (8/57, 14.0%), wrist (1/13, 7.7%) and hand (1/13, 7.7% each) and finally the ankle (1/15, 6.7%). As already noted, the changes in the spine principally affect the zygapophyses rather than syndesmoses.

In general there is an overall positive correlation between prevalence of joint change and age (Table 4.23; Figure 4.6b). As was the situation for the Da But period sample, only the hand displays a greater prevalence of joint change in the youngest, relative to eldest, age class.

Looking at the distribution of joint change by sex, females display a higher frequency of articular changes in the majority of the various joints assessed (Table 4.24). Males exhibit a higher frequency of articular changes to the TMJ, shoulder and foot. The prevalence of joint change to the spine is about even for both sexes.

Vietnam Compared with other Southeast Asian Skeletal Series

Several studies have examined the prevalence and distribution of articular change in prehistoric sites in Thailand (Tayles 1992; Douglas 1996; Domett 1999).

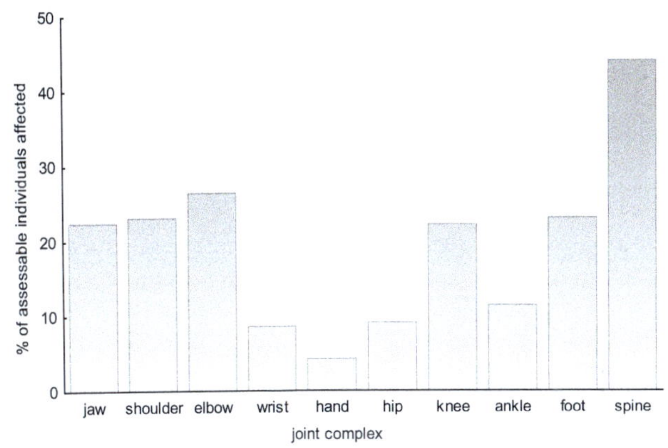

Figure 4.5a. Da But Period: Combined Sex Prevalence of Joint Change (joint change includes MO, ASO, MC and EB only)

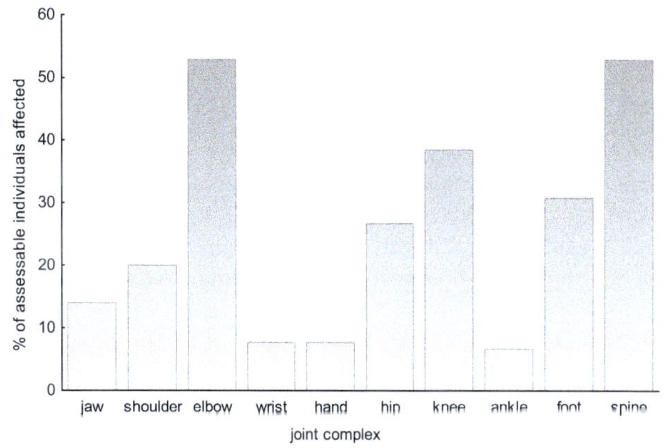

Figure 4.6a. Metal Period: Combined Sex Prevalence of Joint Change (joint change includes MO, ASO, MC and EB only)

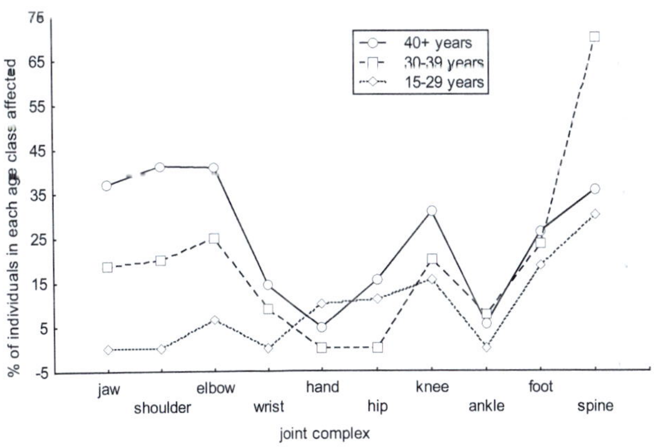

Figure 4.5b. Da But Period: Prevalence of Joint Change by Age Class (joint change includes MO, ASO, MC and EB only)

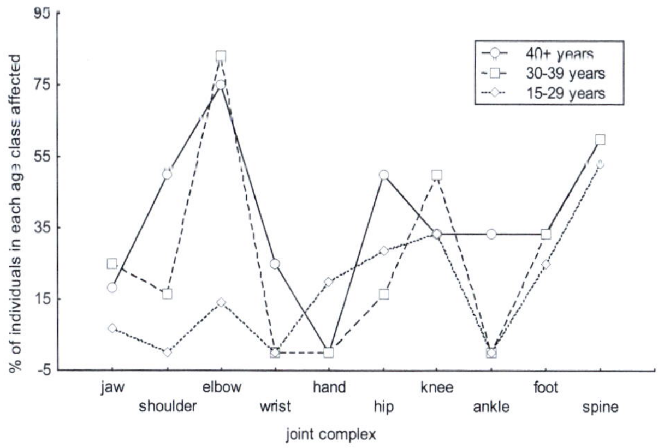

Figure 4.6b. Metal Period: Prevalence of Joint Change by Age Class (joint change includes MO, ASO, MC and EB only)

Table 4.13. Metal Period: Changes to the Articular Surfaces of the Shoulder

Left Side										Right Side									Total Individuals*								
	male			female			total			male			female			total			male			female			total		
	obs	n	%	obs	n	%	obs	n	%	obs	n	%	obs	n	%	obs	n	%	obs	n	%	obs	n	%	obs	n	%
Proximal Humerus: Marginal Osteophytosis																											
<30	0	1	0.0	0	0	na	0	1	0.0	0	2	0.0	0	1	0.0	0	3	0.0	0	2	0.0	0	1	0.0	0	4	0.0
30-39	0	1	0.0	1	4	25.0	1	5	20.0	0	1	0.0	0	3	0.0	0	4	0.0	0	1	0.0	1	4	25.0	1	5	20.0
40+	0	1	0.0	0	2	0.0	0	3	0.0	0	0	na	0	1	0.0	0	1	0.0	0	1	0.0	0	2	0.0	0	3	0.0
TOTAL	0	3	0.0	1	6	16.7	1	9	11.1	0	3	0.0	0	5	0.0	0	8	0.0	0	4	0.0	1	7	14.3	1	12	8.3
Proximal Humerus: Porosity																											
<30	0	1	0.0	0	0	na	0	1	0.0	0	2	0.0	0	1	0.0	0	3	0.0	0	2	0.0	0	1	0.0	0	4	0.0
30-39	1	1	100.0	2	4	50.0	3	5	60.0	0	1	0.0	1	3	33.3	1	4	25.0	1	1	100.0	2	4	50.0	3	5	60.0
40+	0	1	0.0	1	2	50.0	1	3	33.3	0	0	na	0	1	0.0	0	1	0.0	0	1	0.0	1	2	50.0	1	3	33.3
TOTAL	1	3	33.3	3	6	50.0	4	9	44.4	0	3	0.0	1	5	20.0	1	8	12.5	1	4	25.0	3	7	42.9	4	12	33.3
Scapula Glenoid: Marginal Osteophytosis																											
<30	0	0	na	0	0	na	0	1	0.0	0	1	0.0	0	1	0.0	0	3	0.0	0	1	0.0	0	1	0.0	0	3	0.0
30-39	0	1	0.0	0	1	0.0	0	2	0.0	0	1	0.0	0	2	0.0	0	3	0.0	0	1	0.0	0	2	0.0	0	3	0.0
40+	0	0	na	1	2	50.0	1	3	33.3	1	1	100.0	1	2	50.0	2	3	66.7	1	1	100.0	1	2	50.0	2	4	50.0
TOTAL	0	1	0.0	1	3	33.3	1	6	16.7	1	3	33.3	1	5	20.0	2	9	22.2	1	3	33.3	1	5	20.0	2	10	20.0
Scapula Glenoid: Porosity																											
<30	0	0	na	0	0	na	0	1	0.0	1	1	100.0	1	1	100.0	2	3	66.7	1	1	100.0	1	1	100.0	2	3	66.7
30-39	0	1	0.0	1	1	100.0	0	2	0.0	0	1	0.0	1	2	50.0	1	3	33.3	0	1	0.0	1	2	50.0	1	3	33.3
40+	0	0	na	2	2	100.0	3	3	100.0	1	1	100.0	1	2	50.0	2	3	66.7	1	1	100.0	2	2	100.0	3	4	75.0
TOTAL	0	1	0.0	3	3	100.0	3	6	50.0	2	3	66.7	3	5	60.0	5	9	55.6	2	3	66.7	4	5	80.0	6	10	60.0

Note: porosity and EB were ot observed for the proximal humerus; ASO and EB were not observed for the scapula glenoid.

Humerus Correlation Matrix (left and right total)

	MO	ASO	P	EB
MO	**1**			
ASO	0	**0**		
P	0	0	**5**	
EB	0	0	0	**0**

Glenoid Correlation Matrix (left and right total)

	MO	ASO	P	EB
MO	**3**			
ASO	0	**0**		
P	3	0	**8**	
EB	0	0	0	**0**

* Total Individuals (numerator and denominator include only one element side when both present)

Table 4.14a. Metal Period: Changes to the Articular Surfaces of the Distal Humerus (Elbow)

	Left Side									Right Side									Total Individuals*								
	male			female			total			male			female			total			male			female			total		
	obs	n	%	obs	n	%	obs	n	%	obs	n	%	obs	n	%	obs	n	%	obs	n	%	obs	n	%	obs	n	%
Marginal Osteophytosis																											
<30	0	2	0.0	0	0	na	0	3	0.0	0	2	0.0	0	1	0.0	0	3	0.0	0	2	0.0	0	1	0.0	0	4	0.0
30-39	0	1	0.0	0	5	0.0	0	6	0.0	0	1	0.0	0	4	0.0	0	5	0.0	0	1	0.0	0	5	0.0	0	6	0.0
40+	0	2	0.0	0	2	0.0	0	4	0.0	1	1	100.0	0	1	0.0	1	2	50.0	1	2	50.0	0	2	0.0	1	4	25.0
TOTAL	0	5	0.0	0	7	0.0	0	13	0.0	1	4	25.0	0	6	0.0	1	10	10.0	1	5	20.0	0	8	0.0	1	14	7.1
Articular Surface Osteophytosis																											
<30	0	2	0.0	0	0	na	0	3	0.0	0	2	0.0	1	1	100.0	1	3	33.3	0	2	0.0	1	1	100.0	1	4	25.0
30-39	0	1	0.0	1	5	20.0	1	6	16.7	0	1	0.0	1	4	25.0	1	5	20.0	0	1	0.0	1	5	20.0	1	6	16.7
40+	0	2	0.0	1	2	50.0	1	4	25.0	1	1	100.0	0	1	0.0	1	2	50.0	1	2	50.0	1	2	50.0	2	4	50.0
TOTAL	0	5	0.0	2	7	28.6	2	13	15.4	1	4	25.0	2	6	33.3	3	10	30.0	1	5	20.0	3	8	37.5	4	14	28.6
Porosity																											
<30	1	2	50.0	0	0	na	1	3	33.3	1	2	50.0	1	1	100.0	2	3	66.7	1	2	50.0	1	1	100.0	2	4	50.0
30-39	0	1	0.0	2	5	40.0	2	6	33.3	0	1	0.0	2	4	50.0	2	5	40.0	0	1	0.0	2	5	40.0	2	6	33.3
40+	0	2	0.0	1	2	50.0	1	4	25.0	0	1	0.0	0	1	0.0	0	2	0.0	0	2	0.0	1	2	50.0	1	4	25.0
TOTAL	1	5	20.0	3	7	42.9	4	13	30.8	1	4	25.0	3	6	50.0	4	10	40.0	1	5	20.0	4	8	50.0	5	14	35.7
Eburnation																											
<30	0	2	0.0	0	0	na	0	3	0.0	0	2	0.0	1	1	100.0	1	3	33.3	0	2	0.0	1	1	100.0	1	4	25.0
30-39	0	1	0.0	0	5	0.0	0	6	0.0	0	1	0.0	0	4	0.0	0	5	0.0	0	1	0.0	0	5	0.0	0	6	0.0
40+	0	2	0.0	0	2	0.0	0	4	0.0	0	1	0.0	0	1	0.0	0	2	0.0	0	2	0.0	0	2	0.0	0	4	0.0
TOTAL	0	5	0.0	0	7	0.0	0	13	0.0	0	4	0.0	1	6	16.7	1	10	10.0	0	5	0.0	1	8	12.5	1	14	7.1

Correlation Matrix (left and right total)

	MO	ASO	P	EB
MO	**1**			
ASO	1	**5**		
P	0	2	**8**	
EB	0	1	1	**1**

* Total Individuals (numerator and denominator include only one element side when both present)

Table 4.14b. Metal Period: Changes to the Articular Surfaces of the Proximal Radius (Elbow)

Left Side	male			female			total		
	obs	n	%	obs	n	%	obs	n	%
Marginal Osteophytosis									
<30	0	2	0.0	0	0	na	0	3	0.0
30-39	0	1	0.0	1	3	33.3	1	4	25.0
40+	0	1	0.0	0	1	0.0	0	2	0.0
TOTAL	0	4	0.0	1	4	25.0	1	9	11.1
Articular Surface Osteophytosis									
<30	0	2	0.0	0	0	na	0	3	0.0
30-39	0	1	0.0	0	3	0.0	0	4	0.0
40+	0	1	0.0	0	1	0.0	0	2	0.0
TOTAL	0	4	0.0	0	4	0.0	0	9	0.0
Porosity									
<30	0	2	0.0	0	0	na	0	3	0.0
30-39	1	1	100.0	1	3	33.3	2	4	50.0
40+	0	1	0.0	0	1	0.0	0	2	0.0
TOTAL	1	4	25.0	1	4	25.0	2	9	22.2

Right Side	male			female			total		
	obs	n	%	obs	n	%	obs	n	%
Marginal Osteophytosis									
<30	0	2	0.0	0	1	0.0	0	4	0.0
30-39	0	1	0.0	1	4	25.0	1	5	20.0
40+	0	1	0.0	0	2	0.0	0	3	0.0
TOTAL	0	4	0.0	1	7	14.3	1	12	8.3
Articular Surface Osteophytosis									
<30	0	2	0.0	1	1	100.0	1	4	25.0
30-39	0	1	0.0	1	4	25.0	1	5	20.0
40+	0	1	0.0	0	2	0.0	0	3	0.0
TOTAL	0	4	0.0	2	7	28.6	2	12	16.7
Porosity									
<30	0	2	0.0	0	1	0.0	0	4	0.0
30-39	1	1	100.0	1	4	25.0	2	5	40.0
40+	0	1	0.0	1	2	50.0	1	3	33.3
TOTAL	1	4	25.0	2	7	28.6	3	12	25.0

Total Individuals*	male			female			total		
	obs	n	%	obs	n	%	obs	n	%
Marginal Osteophytosis									
<30	0	3	0.0	0	1	0.0	0	5	0.0
30-39	0	1	0.0	2	4	50.0	2	5	40.0
40+	0	2	0.0	0	2	0.0	0	4	0.0
TOTAL	0	6	0.0	2	7	28.6	2	14	14.3
Articular Surface Osteophytosis									
<30	0	3	0.0	1	1	100.0	1	5	20.0
30-39	0	1	0.0	1	4	25.0	1	5	20.0
40+	0	2	0.0	0	2	0.0	0	4	0.0
TOTAL	0	6	0.0	2	7	28.6	2	14	14.3
Porosity									
<30	0	3	0.0	0	1	0.0	0	5	0.0
30-39	1	1	100.0	1	4	25.0	2	5	40.0
40+	0	2	0.0	1	2	50.0	1	4	25.0
TOTAL	1	6	16.7	2	7	28.6	3	14	21.4

Note: EB was not observed for the proximal radius.

Correlation Matrix (left and right total)

	MO	ASO	P	EB
MO	**2**			
ASO	0	**2**		
P	0	1	**5**	
EB	0	0	0	**0**

* Total Individuals (numerator and denominator include only one element side when both present)

Table 4.14c. Metal Period Changes to the Articular Surfaces of the Proximal Ulna (Elbow)

	Left Side									Right Side									Total Individuals*								
	male			female			total			male			female			total			male			female			total		
	obs	n	%	obs	n	%	obs	n	%	obs	n	%	obs	n	%	obs	n	%	obs	n	%	obs	n	%	obs	n	%
Marginal Osteophytosis																											
<30	0	5	0.0	0	0	na	0	6	0.0	0	2	0.0	0	0	na	0	2	0.0	0	5	0.0	0	0	na	0	6	0.0
30-39	1	1	100.0	1	2	50.0	2	3	66.7	0	1	0.0	2	3	66.7	2	4	50.0	1	1	100.0	2	3	66.7	3	4	75.0
40+	2	2	100.0	1	2	50.0	3	4	75.0	2	2	100.0	0	1	0.0	2	3	66.7	2	2	100.0	1	2	50.0	3	4	75.0
TOTAL	3	8	37.5	2	4	50.0	5	13	38.5	2	5	40.0	2	4	50.0	4	9	44.4	3	8	37.5	3	5	60.0	6	14	42.9
Porosity																											
<30	0	5	0.0	0	0	na	0	6	0.0	0	2	0.0	0	0	na	0	2	0.0	0	5	0.0	0	0	na	0	6	0.0
30-39	0	1	0.0	1	2	50.0	1	3	33.3	0	1	0.0	1	3	33.3	1	4	25.0	0	1	0.0	1	3	33.3	1	4	25.0
40+	0	2	0.0	1	2	50.0	1	4	25.0	0	2	0.0	0	1	0.0	0	3	0.0	0	2	0.0	1	2	50.0	1	4	25.0
TOTAL	0	8	0.0	2	4	50.0	2	13	15.4	0	5	0.0	1	4	25.0	1	9	11.1	0	8	0.0	2	5	40.0	2	14	14.3
Eburnation																											
<30	0	5	0.0	0	0	na	0	6	0.0	0	2	0.0	0	0	na	0	2	0.0	0	5	0.0	0	0	na	0	6	0.0
30-39	0	1	0.0	1	2	50.0	1	3	33.3	0	1	0.0	0	3	0.0	0	4	0.0	0	1	0.0	1	3	33.3	1	4	25.0
40+	0	2	0.0	0	2	0.0	0	4	0.0	0	2	0.0	0	1	0.0	0	3	0.0	0	2	0.0	0	2	0.0	0	4	0.0
TOTAL	0	8	0.0	1	4	25.0	1	13	7.7	0	5	0.0	0	4	0.0	0	9	0.0	0	8	0.0	1	5	20.0	1	14	7.1

Note: ASO was not observed for the proximal ulna.

Correlation Matrix (left and right total)

	MO	ASO	P	EB
MO	**9**			
ASO	0	**0**		
P	1	0	**3**	
EB	0	0	1	**1**

* Total Individuals (numerator and denominator include only one element side when both present)

Table 4.14d. Metal Period: Individuals Affected by Articular Changes to the Elbow

Metal Period			Distal Humerus				Proximal Ulna				Proximal Radius			
case	sex	age	MO	ASO	P	EB	MO	ASO	P	EB	MO	ASO	P	EB
77CHSM1a	M	5					Y							
76NNM2c	M	4	Y	Y			Y							
89MDM2a	F	4		Y	Y		Y		Y			Y	Y	
84DSM7a	M	3					Y						Y	
67VQ2M1a	F	3	(shaded)	(shaded)	(shaded)	(shaded)	Y				Y			
67VQ2M43a	F	3					(shaded)	(shaded)	(shaded)	(shaded)	Y			
84DSM1a	F	3		Y			Y							
89MDM3a	F	3			Y		(shaded)	(shaded)	(shaded)	(shaded)	(shaded)	(shaded)	(shaded)	(shaded)
95DXM1a	F	3			Y				Y	Y			Y	
66VQ1M1a	M	2			Y									
89MDM7a	F	2		Y	Y	Y	(shaded)	(shaded)	(shaded)	(shaded)		Y		

Y :articular change observed
(blank) :articular change not observed
(shaded) :articular area not preserved

5= 50+ years
4=40-49 years
3=30-39 years
2=20-29 years
1=15-19 years

M=male
F=female
I=indeterminate

Table 4.26 summarises articular change by ranking the prevalence of change to each joint for five skeletal assemblages in addition to the Vietnamese samples. Of the eight appendicular joint complexes examined, articular changes to the foot and hip are among the four most affected joints in 6/7 skeletal assemblages. The elbow is among the four most affected joints in five of these seven samples. If only the two most affected joints in each assemblage are examined, the foot, shoulder and elbow are the most commonly affected joints. Moreover, the elbow stands out as being the single most affected joint in three assemblages.

Discussion and Conclusions

Before discussing the results of the assessment of articular change, the presence and nature of the mandibular condyle depressions are examined. A range of aetiological factors could be responsible for these depressions that only occur on the posterior articular aspect of the mandibular condyle.

Mandibular Condyle Depressions

If the condition is pathologic, it may be some type of cyst such as an exposed subchondral bone cyst. However, the morphology and consistent positioning of the depressions suggests they are not subchondral bone cysts. They are possibly correlated with some type of trauma to the jaw. However, in lieu of any other specific evidence for trauma in any of the cases with this depression, trauma would seem an unlikely causal agent. This feature is possibly the result of unusual masticatory behaviours. This would be difficult to test using archaeological data and would also seem an unlikely cause given that the posterior region of the condyle is not greatly involved in load bearing. Another possibility is that these condylar depressions are some sort of epigenetic variant. If this were so they might have some value in reconstructing familial patterns.

Perhaps the most likely scenario is that this trait is a form of cortical defect, analogous to the rhomboid fossa of the clavicle or *teres major* cortical defect of the humerus. There is some clinical evidence for a developmental aetiology for this defect. The development of the temporomandibular joint is relatively late both during its embryonic development, in relation to diarthrosis of the extremities (Merida-Velasco *et al.* 1999), and in later post-subadult maturation (Luder 1997; 1998). In a microscopic study of condylar articular tissues, Luder (1998) points out that all zones, or levels, of mandibular condyle articular tissue undergo age-related changes up until about 30 years of age. By this (late) age endochondral ossification and the development of the compact subchondral bone plate has essentially finished. Moreover, there are fundamental differences in the nature or morphology of such tissues between the load bearing regions, anterior and central condylar region, and non-load bearing region, the posterior section, even after maturation at age 30 years (Luder 1997). Referring to the posterior section of the condyle, Luder (1997) suggests that this region resembles tissue organisation during active growth. The tissues of the anterior and central regions do not.

In the Da But sample only three individuals displaying condylar depressions (CD) were recorded, with two individuals being aged 40+ years and one 20-29 years. However, in the Metal period sample seven of the nine individuals displaying clear and well differentiated depressions were less than 30 years old. The only instances of CD in the Metal period sample, that were also associated with ASO or MO, were in the two older individuals. In the Da But period sample, only the oldest individual had both a CD and associated ASO. For the vast majority of people in these samples displaying CD, it is found in immature, by condyle developmental standards, individuals. The

Table 4.15. Metal Period: Changes to the Articular Surfaces of the Wrist

	Left Side									Right Side									Total Individuals*								
	male			female			total			male			female			total			male			female			total		
	obs	n	%	obs	n	%	obs	n	%	obs	n	%	obs	n	%	obs	n	%	obs	n	%	obs	n	%	obs	n	%
Distal Radius: Marginal Osteophytosis																											
<30	0	1	0.0	0	0	na	0	1	0.0	0	2	0.0	0	1	0.0	0	4	0.0	0	2	0.0	0	1	0.0	0	4	0.0
30-39	0	1	0.0	0	1	0.0	0	2	0.0	0	1	0.0	0	3	0.0	0	4	0.0	0	1	0.0	0	3	0.0	0	4	0.0
40+	0	2	0.0	0	1	0.0	0	3	0.0	0	0	na	1	2	50.0	1	2	50.0	0	2	0.0	1	2	50.0	1	4	25.0
TOTAL	0	4	0.0	0	2	0.0	0	6	0.0	0	3	0.0	1	6	16.7	1	10	10.0	0	5	0.0	1	6	16.7	1	12	8.3
Distal Ulna: Marginal Osteophytosis																											
<30	0	2	0.0	0	0	na	0	3	0.0	0	3	0.0	0	0	na	0	4	0.0	0	3	0.0	0	0	na	0	4	0.0
30-39	0	0	na	0	1	0.0	0	1	0.0	0	1	0.0	0	2	0.0	0	3	0.0	0	1	0.0	0	3	0.0	0	4	0.0
40+	0	1	0.0	1	1	100.0	1	2	50.0	0	2	0.0	0	1	0.0	0	3	0.0	0	2	0.0	1	2	50.0	1	4	25.0
TOTAL	0	3	0.0	1	2	50.0	1	6	16.7	0	6	0.0	0	3	0.0	0	10	0.0	0	6	0.0	1	5	20.0	1	12	8.3
Distal Ulna: Articular Surface Osteophytosis																											
<30	0	2	0.0	0	0	na	0	3	0.0	0	3	0.0	0	0	na	0	4	0.0	0	3	0.0	0	0	na	0	4	0.0
30-39	0	0	na	0	1	0.0	0	1	0.0	0	1	0.0	0	2	0.0	0	3	0.0	0	1	0.0	0	3	0.0	0	4	0.0
40+	0	1	0.0	1	1	100.0	1	2	50.0	0	2	0.0	0	1	0.0	0	3	0.0	0	2	0.0	1	2	50.0	1	4	25.0
TOTAL	0	3	0.0	1	2	50.0	1	6	16.7	0	6	0.0	0	3	0.0	0	10	0.0	0	6	0.0	1	5	20.0	1	12	8.3

Note: ASO, EB and porosity was not observed for the distal radius; EB and porosity was not observed for the distal ulna.

Ulna Correlation Matrix (left and right total)

	MO	ASO	P	EB
MO	**1**			
ASO	1	**1**		
P	0	0	**0**	
EB	0	0	0	**0**

* Total Individuals (numerator and denominator include only one element side when both present)

Table 4.16. Metal Period: Articular Changes in the Hand (sample sizes)

Carpals

Left and Right Side	male	female	total	Individuals[1] male	female	total
	n	n	n	n	n	n
<30	4	0	4	2	0	2
30-39	5	12	17	1	2	3
40+	2	0	2	1	0	1
TOTAL	11	12	23	4	2	6

Carpometacarpal Joint[2] Articular Surfaces

Articular Surface Count	male	female	total	Individuals male	female	total
	n	n	n	n	n	n
<30	0	0	5	0	0	1
30-39	10	16	26	1	3	4
40+	1	6	7	1	1	2
TOTAL	11	22	38	2	4	7

Metacarpophalangeal Joint (MCP)[3] Articular Surfaces

Articular Surface Count	male	female	total	Individuals male	female	total
	n	n	n	n	n	n
<30	2	0	10	1	0	2
30-39	15	22	37	2	2	4
40+	4	10	15	1	1	3
TOTAL	21	32	62	4	3	9

Proximal Interphalangeal Joint (PIPJ)[4] Articular Surfaces

Articular Surface Count	male	female	total	Individuals male	female	total
	n	n	n	n	n	n
<30	2	2	10	1	1	3
30-39	7	12	19	2	3	5
40+	5	3	9	1	1	3
TOTAL	14	17	38	4	5	11

Distal Interphalangeal Joint (DIPJ)[5] Articular Surfaces

Articular Surface Count	male	female	total	Individuals male	female	total
	n	n	n	n	n	n
<30	0	2*	3*	0	1*	2*
30-39	2	4	7	2	2	4
40+	3	0	2	1	0	1
TOTAL	5	6	12*	3	3	7*

[1] Individual: represented by at least one articular surface of one element
[2] Carpometacarpal Joint: defined here as the proximal articular surfaces of the metacarpals only
[3] MCP: distal articular surfaces of metacarpals and proximal articular surfaces of proximal phalanges
[4] PIPJ: distal articular surfaces of proximal phalanges and proximal articular surfaces of middle phalanges
[5] DIPJ: distal articular surfaces of middle phalanges and proximal articular surfaces of distal phalanges

*78QCM14a (Female, age 20-29 years) displayed MO to distal articular surface of one middle phalange

instances of articular changes in the older individuals with CD are most likely fortuitous. There does not appear to be a correlation between condylar depressions and osteophytic development.

The evidence tends to support the view, forwarded here, that CD may be a developmental defect or feature that remodels and disappears in later life, as the condyle matures. The depressions occur in an area of the condyle that is prone to tissue growth and remodelling up to at least 30 years of age. There is nothing to suggest a pathological or traumatic aetiology. An analogous situation would be the *teres major* and *pectoralis major* cortical defects commonly seen in subadults but rarely in adults (Mann and Murphy 1990). The rarity of these depressions in older adults would suggest the defects generally disappear with

Table 4.17a. Metal Period: Changes to the Articular Surfaces of the Proximal Femur (Hip)

	Left Side									Right Side									Total Individuals*								
	male			female			total			male			female			total			male			female			total		
	obs	n	%	obs	n	%	obs	n	%	obs	n	%	obs	n	%	obs	n	%	obs	n	%	obs	n	%	obs	n	%
Articular Surface Osteophytosis																											
<30	0	3	0.0	1	1	100.0	1	4	25.0	0	2	0.0	1	1	100.0	1	3	33.3	0	3	0.0	1	1	100.0	1	4	25.0
30-39	0	1	0.0	0	3	0.0	0	4	0.0	0	2	0.0	0	2	0.0	0	4	0.0	0	2	0.0	0	4	0.0	0	6	0.0
40+	0	0	na	0	1	0.0	0	1	0.0	0	0	na	0	1	0.0	0	1	0.0	0	0	na	0	1	0.0	0	1	0.0
TOTAL	0	4	0.0	1	5	20.0	1	9	11.1	0	4	0.0	1	4	25.0	1	8	12.5	0	5	0.0	1	6	16.7	1	11	9.1
Porosity																											
<30	0	3	0.0	1	1	100.0	1	4	25.0	0	2	0.0	1	1	100.0	1	3	33.3	0	3	0.0	1	1	100.0	1	4	25.0
30-39	0	1	0.0	0	3	0.0	0	4	0.0	0	2	0.0	1	2	50.0	1	4	25.0	0	2	0.0	1	4	25.0	1	6	16.7
40+	0	0	na	0	1	0.0	0	1	0.0	0	0	na	0	1	0.0	0	1	0.0	0	0	na	0	1	0.0	0	1	0.0
TOTAL	0	4	0.0	1	5	20.0	1	9	11.1	0	4	0.0	2	4	50.0	2	8	25.0	0	5	0.0	2	6	33.3	2	11	18.2

Note: MO and EB not observed for the proximal femur.

Femur Correlation Matrix (left and right total)

	MO	ASO	P	EB
MO	**0**			
ASO	0	**2**		
P	0	2	**3**	
EB	0	0	0	**0**

* Total Individuals (numerator and denominator include only one element side when both present)

Table 4.17b. Metal Period: Changes to the Articular Surfaces of the Acetabulum (Hip)

Left Side										Right Side									Total Individuals*								
	male			female			total			male			female			total			male			female			total		
	obs	n	%	obs	n	%	obs	n	%	obs	n	%	obs	n	%	obs	n	%	obs	n	%	obs	n	%	obs	n	%
Marginal Osteophytosis																											
<30	0	2	0.0	0	1	0.0	0	4	0.0	0	2	0.0	0	1	0.0	0	4	0.0	0	3	0.0	0	1	0.0	0	5	0.0
30-39	0	1	0.0	1	2	50.0	1	3	33.3	0	1	0.0	1	3	33.3	1	4	25.0	0	1	0.0	1	3	33.3	1	4	25.0
40+	0	0	na	0	1	0.0	0	1	0.0	0	0	na	1	1	100.0	1	1	100.0	0	0	na	1	1	100.0	1	1	100.0
TOTAL	0	3	0.0	1	4	25.0	1	8	12.5	0	3	0.0	2	5	40.0	2	9	22.2	0	4	0.0	2	5	40.0	2	10	20.0
Articular Surface Osteophytosis																											
<30	0	2	0.0	0	1	0.0	0	4	0.0	0	2	0.0	0	1	0.0	0	4	0.0	0	3	0.0	0	1	0.0	0	5	0.0
30-39	0	1	0.0	0	2	0.0	0	3	0.0	0	1	0.0	0	3	0.0	0	4	0.0	0	1	0.0	0	3	0.0	0	4	0.0
40+	0	0	na	1	1	100.0	1	1	100.0	0	0	na	1	1	100.0	1	1	100.0	0	0	na	1	1	100.0	1	1	100.0
TOTAL	0	3	0.0	1	4	25.0	1	8	12.5	0	3	0.0	1	5	20.0	1	9	11.1	0	4	0.0	1	5	20.0	1	10	10.0
Porosity																											
<30	1	2	50.0	0	1	0.0	1	4	25.0	0	2	0.0	0	1	0.0	0	4	0.0	1	3	33.3	0	1	0.0	1	5	20.0
30-39	0	1	0.0	1	2	50.0	1	3	33.3	0	1	0.0	1	3	33.3	1	4	25.0	0	1	0.0	1	3	33.3	1	4	25.0
40+	0	0	na	1	1	100.0	1	1	100.0	0	0	na	1	1	100.0	1	1	100.0	0	0	na	1	1	100.0	1	1	100.0
TOTAL	1	3	33.3	2	4	50.0	3	8	37.5	0	3	0.0	2	5	40.0	2	9	22.2	1	4	25.0	2	5	40.0	3	10	30.0
Eburnation																											
<30	1	2	50.0	0	1	0.0	1	4	25.0	0	2	0.0	0	1	0.0	0	4	0.0	1	3	33.3	0	1	0.0	1	5	20.0
30-39	0	1	0.0	0	2	0.0	0	3	0.0	0	1	0.0	0	3	0.0	0	4	0.0	0	1	0.0	0	3	0.0	0	4	0.0
40+	0	0	na	0	1	0.0	0	1	0.0	0	0	na	0	1	0.0	0	1	0.0	0	0	na	0	1	0.0	0	1	0.0
TOTAL	1	3	33.3	0	4	0.0	1	8	12.5	0	3	0.0	0	5	0.0	0	9	0.0	1	4	25.0	0	5	0.0	1	10	10.0

Correlation Matrix (left and right total)

	MO	ASO	P	EB
MO	**3**			
ASO	1	**2**		
P	1	2	**5**	
EB	0	0	1	**1**

* Total Individuals (numerator and denominator include only one element side when both present)

Table 4.18a. Metal Period Changes to the Articular Surfaces of the Knee

Left Side

	male obs	male n	male %	female obs	female n	female %	total obs	total n	total %
Distal Femur: Marginal Osteophytosis									
<30	0	1	0.0	0	0	na	0	2	0.0
30-39	0	1	0.0	0	2	0.0	0	3	0.0
40+	0	1	0.0	0	1	0.0	0	2	0.0
TOTAL	0	3	0.0	0	3	0.0	0	7	0.0
Distal Femur: Porosity									
<30	0	1	0.0	0	0	na	0	2	0.0
30-39	0	1	0.0	2	2	100.0	2	3	66.7
40+	0	1	0.0	0	1	0.0	0	2	0.0
TOTAL	0	3	0.0	2	3	66.7	2	7	28.6
Proximal Tibia: Marginal Osteophytosis									
<30	0	1	0.0	0	0	na	0	2	0.0
30-39	0	2	0.0	0	2	0.0	0	4	0.0
40+	0	1	0.0	0	1	0.0	0	2	0.0
TOTAL	0	4	0.0	0	3	0.0	0	8	0.0

Right Side

	male obs	male n	male %	female obs	female n	female %	total obs	total n	total %
Distal Femur: Marginal Osteophytosis									
<30	0	2	0.0	1	1	100.0	1	3	33.3
30-39	1	2	50.0	0	2	0.0	1	4	25.0
40+	0	0	na	0	1	0.0	0	1	0.0
TOTAL	1	4	25.0	1	4	25.0	2	8	25.0
Distal Femur: Porosity									
<30	0	2	0.0	0	1	0.0	0	3	0.0
30-39	0	2	0.0	2	2	100.0	2	4	50.0
40+	0	0	na	0	1	0.0	0	1	0.0
TOTAL	0	4	0.0	2	4	50.0	2	8	25.0
Proximal Tibia: Marginal Osteophytosis									
<30	1	3	33.3	0	0	na	1	4	25.0
30-39	0	1	0.0	0	2	0.0	0	3	0.0
40+	0	0	na	0	1	0.0	0	1	0.0
TOTAL	1	4	25.0	0	3	0.0	1	8	12.5

Total Individuals*

	male obs	male n	male %	female obs	female n	female %	total obs	total n	total %
Distal Femur: Marginal Osteophytosis									
<30	0	2	0.0	1	1	100.0	1	4	25.0
30-39	1	2	50.0	0	2	0.0	1	4	25.0
40+	0	1	0.0	0	1	0.0	0	2	0.0
TOTAL	1	5	20.0	1	4	25.0	2	10	20.0
Distal Femur: Porosity									
<30	0	2	0.0	0	1	0.0	0	4	0.0
30-39	0	2	0.0	2	2	100.0	2	4	50.0
40+	0	1	0.0	0	1	0.0	0	2	0.0
TOTAL	0	5	0.0	2	4	50.0	2	10	20.0
Proximal Tibia: Marginal Osteophytosis									
<30	1	3	33.3	0	0	na	1	4	25.0
30-39	0	2	0.0	0	2	0.0	0	4	0.0
40+	0	1	0.0	0	1	0.0	0	2	0.0
TOTAL	1	6	16.7	0	3	0.0	1	10	10.0

Note: ASO and porosity not observed for the distal femur; ASO, EB and porosity not observed for the proximal tibia.

Femur Correlation Matrix (left and right total)

	MO	ASO	P	EB
MO	**2**			
ASO	0	**0**		
P	0	0	**4**	
EB	0	0	0	**0**

* Total Individuals (numerator and denominator include only one element side when both present)

Table 4.18b. Metal Period: Changes to the Articular Surfaces of the Knee

Left Side										Right Side									Total Individuals*								
	male			female			total			male			female			total			male			female			total		
	obs	n	%	obs	n	%	obs	n	%	obs	n	%	obs	n	%	obs	n	%	obs	n	%	obs	n	%	obs	n	%
Proximal Fibula: Porosity																											
<30	0	1	0.0	0	0	na	0	1	0.0	0	2	0.0	0	1	0.0	0	4	0.0	0	2	0.0	0	1	0.0	0	4	0.0
30-39	0	1	0.0	1	2	50.0	1	3	33.3	0	1	0.0	1	2	50.0	1	3	33.3	0	1	0.0	1	2	50.0	1	3	33.3
40+	0	0	na	0	1	0.0	0	1	0.0	0	0	na	0	1	0.0	0	1	0.0	0	0	na	0	1	0.0	0	1	0.0
TOTAL	0	2	0.0	1	3	33.3	1	5	20.0	0	3	0.0	1	4	25.0	1	8	12.5	0	3	0.0	1	4	25.0	1	8	12.5
Proximal Fibula: Eburnation																											
<30	0	1	0.0	0	0	na	0	1	0.0	0	2	0.0	0	1	0.0	0	4	0.0	0	2	0.0	0	1	0.0	0	4	0.0
30-39	0	1	0.0	1	2	50.0	1	3	33.3	0	1	0.0	0	2	0.0	0	3	0.0	0	1	0.0	1	2	50.0	1	3	33.3
40+	0	0	na	0	1	0.0	0	1	0.0	0	0	na	0	1	0.0	0	1	0.0	0	0	na	0	1	0.0	0	1	0.0
TOTAL	0	2	0.0	1	3	33.3	1	5	20.0	0	3	0.0	0	4	0.0	0	8	0.0	0	3	0.0	1	4	25.0	1	8	12.5
Patella: Articular Surface Osteophytosis																											
<30	0	1	0.0	0	0	na	0	1	0.0	0	1	0.0	0	0	na	0	1	0.0	0	1	0.0	0	0	na	0	1	0.0
30-39	0	1	0.0	0	1	0.0	0	2	0.0	0	1	0.0	0	1	0.0	0	2	0.0	0	1	0.0	0	1	0.0	0	2	0.0
40+	1	2	na	0	1	na	1	3	na	0	0	na	0	1	na	0	1	na	1	2	na	0	1	na	1	3	na
TOTAL	1	4	25.0	0	2	0.0	1	6	16.7	0	2	0.0	0	2	0.0	0	4	0.0	1	4	25.0	0	2	0.0	1	6	16.7

Note: ASO and MO not observed for the proximal fibula; MO, EB and porosity not observed for the patella.

Fibula Correlation Matrix (left and right total)

	MO	ASO	P	EB
MO	**0**			
ASO	0	**0**		
P	0	0	**2**	
EB	0	0	1	**1**

* Total Individuals (numerator and denominator include only one element side when both present)

Table 4.19. Metal Period: Changes to the Articular Surfaces of the Ankle

	Left Side									Right Side									Total Individuals*								
	male			female			total			male			female			total			male			female			total		
	obs	n	%	obs	n	%	obs	n	%	obs	n	%	obs	n	%	obs	n	%	obs	n	%	obs	n	%	obs	n	%
Distal Tibia: Marginal Osteophytosis																											
<30	0	1	0.0	0	0	na	0	1	0.0	0	2	0.0	0	1	0.0	0	4	0.0	0	2	0.0	0	1	0.0	0	4	0.0
30-39	0	1	0.0	0	1	0.0	0	2	0.0	0	1	0.0	0	3	0.0	0	4	0.0	0	1	0.0	0	3	0.0	0	4	0.0
40+	0	2	0.0	0	1	0.0	0	3	0.0	0	0	na	1	2	50.0	1	2	50.0	0	2	0.0	1	2	50.0	1	4	25.0
TOTAL	0	4	0.0	0	2	0.0	0	6	0.0	0	3	0.0	1	6	16.7	1	10	10.0	0	5	0.0	1	6	16.7	1	12	8.3
Distal Fibula (Sample Sizes Only)																											
<30		1			1			2			2			1			3			2			1			3	
30-39		0			3			3			0			3			3			0			3			3	
40+		0			1			1			0			1			1			0			1			1	
TOTAL		1			5			6			2			5			7			2			5			7	
Talar Trochlea: Porosity																											
<30	0	1	0.0	0	0	na	0	2	0.0	0	2	0.0	0	0	na	0	3	0.0	0	2	0.0	0	0	na	0	3	0.0
30-39	0	0	na	1	3	33.3	1	3	33.3	0	0	na	1	3	33.3	1	3	33.3	0	0	na	1	3	33.3	1	3	33.3
40+	0	0	na	0	0	na	0	0	na	0	0	na	0	1	0.0	0	1	0.0	0	0	na	0	1	0.0	0	1	0.0
TOTAL	0	1	0.0	1	3	33.3	1	5	20.0	0	2	0.0	1	4	25.0	1	7	14.3	0	2	0.0	1	4	25.0	1	7	14.3

Note: ASO, EB and porosity were not observed for the distal tibia; no articular changes were oberved for the distal fibula; ASO, MO and EB were not observed for the talar trochlea

* Total Individuals (numerator and denominator include only one element side when both present)

Table 4.20a. Metal Period: Articular Surfaces of the Foot (Sample Sizes)

	Left and Right			Individuals[1]		
	M	F	T	M	F	T
	n	n	n	n	n	n
First Cuneiform						
<30	1	4	5	1	3	4
30-39	0	1	1	0	1	1
40+	1	0	1	1	0	1
TOTAL	2	5	7	2	4	6
Second Cuneiform						
<30	0	3	3	0	2	2
30-39	0	0	0	0	0	0
40+	0	0	0	0	0	0
TOTAL	0	3	3	0	2	2
Third Cuneiform						
<30	2	3	5	1	2	3
30-39	0	0	0	0	0	0
40+	0	0	0	0	0	0
TOTAL	2	3	5	1	2	3
Subtalar Articular Surface						
<30	3	6	11	2	3	6
30-39	0	1	1	0	1	1
40+	0	0	0	0	0	0
TOTAL	2	7	12	2	4	7

	Left and Right			Individuals[1]		
	M	F	T	M	F	T
	n	n	n	n	n	n
Calcaneal Cuboid Facet*						
<30	4	0	4	3	0	3
30-39	0	5	5	0	3	3
40+	0	3	3	0	2	2
TOTAL	4	8	12	3	5	8
Cuboid*						
<30	2	0	2	1	0	1
30-39	0	3	3	0	2	2
40+	0	2	2	0	1	1
TOTAL	2	5	7	1	3	4
Navicular*						
<30	0	0	0	0	0	0
30-39	1	4	5	1	2	3
40+	0	2	2	0	1	1
TOTAL	1	6	7	1	3	4

	Left and Right			Individuals[1]		
	M	F	T	M	F	T
	n	n	n	n	n	n
TMT Joint Articular Surfaces						
<30	8	0	8	1	0	1
30-39	1	16	17	1	4	5
40+	0	7	7	0	1	1
TOTAL	9	23	32	2	5	7
MTP Joint Articular Surfaces						
<30	3	0	3	1	0	1
30-39	1	12	13	1	3	4
40+	0	7	7	0	1	1
TOTAL	4	19	23	2	4	6
PIPJ Articular Surfaces						
<30	1	0	1	1	0	1
30-39	0	4	4	0	1	1
40+	0	0	0	0	0	0
TOTAL	1	4	5	1	1	2
DIPJ Articular Surfaces						
<30	1	0	1	1	0	1
30-39	0	0	0	0	0	0
40+	0	0	0	0	0	0
TOTAL	1	0	1	1	0	1

TMT: tarsometatarsal joint
MTP: metatarsophalangeal joint
PIPJ: proximal interphalangeal joint
DIPJ: distal interphalangeal joint
M: male
F: female
T:total

* articular change observed (see text)

[1]Total Individuals (numerator and denominator include only one element side when both present)

remodelling after the age of 30 years.

If these condylar defects are artefacts of the developmentally immature mandible, they may have some general use in estimating the age of severely damaged gnathic material. The presence of a CD would suggest an age equal to or less than 30 years. Unfortunately, the lack of a CD would be uninformative of age. The only other assemblages for which CDs have been observed are the Khok Phanom Di (George Dias pers. comm.) and Ban Chiang (Michele Douglas pers. comm.) samples from Thailand. An explanation for these depressions, other than that advanced here, has not been suggested to date.

Articular Change

If it is accepted that only eburnation is pathognomonic for OA, then the prevalence of OA is extremely low in these northern Vietnamese skeletal series. In the Da But period sample there are only three cases of eburnation, two individuals with OA of the elbow (2/53, 3.8%) and one with OA of the foot (1/52, 1.9%). Including the case of an individual with marked joint contour changes to the mandibular condyle brings the total number of Da But period individuals with changes pathognomonic of OA to 4/73 (5.5%). In the Metal period sample there were four instances of eburnation distributed among three individuals. Two individuals displayed OA of the elbow (2/17, 11.8%), one of the hip (1/17, 5.9%) and one of the individuals with OA of the elbow also had an affected knee (1/13, 7.7%). A total of 3/60 (5.0%) individuals of the Metal period sample displayed changes pathognomonic of OA. The frequency of OA is virtually identical for both samples when examined in this manner.

It is unlikely that OA was such an extremely uncommon condition in these skeletal samples from Vietnam, given the reported prevalence of OA in both contemporary and past populations. Perhaps the largest confounding factor for this discrepancy is the fairly restricted size of these samples. Are these samples representative of the actual state of affairs in ancient Vietnam or has restricted sample size lead to unknown, and unknowable, sampling biases? Two lines of evidence would suggest that these results are representative. The first is that both samples show low, and virtually equivalent, levels of OA as measured by eburnation alone. It would seem unlikely that both samples would be equally, and in the same direction, unrepresentative. The second reason is that while the Metal period sample is a composite, the Da But period assemblage samples the cemetery of one site most likely of a limited temporal span. The chances of sample representativeness are increased for this sample at least.

Table 4.20b. Metal Period: Changes to the Articular Surfaces of the Tarsals (Foot)

Left Side										Right Side									Total Individuals*								
	male			female			total			male			female			total			male			female			total		
	obs	n	%	obs	n	%	obs	n	%	obs	n	%	obs	n	%	obs	n	%	obs	n	%	obs	n	%	obs	n	%
Talar Head: Porosity																											
<30	0	1	0.0	0	0	na	0	2	0.0	0	2	0.0	0	0	na	0	3	0.0	0	2	0.0	0	0	na	0	3	0.0
30-39	0	0	na	1	3	33.3	1	3	33.3	0	0	na	1	3	33.3	1	3	33.3	0	0	na	1	3	33.3	1	3	33.3
40+	0	0	na	0	0	na	0	0	na	0	0	na	0	1	0.0	0	1	0.0	0	0	na	0	1	0.0	0	1	0.0
TOTAL	0	1	0.0	1	3	33.3	1	5	20.0	0	2	0.0	1	4	25.0	0	7	0.0	0	2	0.0	1	4	25.0	1	7	14.3
Calcaneal Talar Facet: Marginal Osteophytosis																											
<30	0	1	0.0	0	0	na	0	1	0.0	0	2	0.0	0	0	na	0	3	0.0	0	3	0.0	0	0	na	0	3	0.0
30-39	0	0	na	0	2	0.0	0	2	0.0	0	0	na	0	3	0.0	0	3	0.0	0	0	na	0	3	0.0	0	3	0.0
40+	0	0	na	1	2	50.0	1	2	50.0	0	0	na	0	1	0.0	0	1	0.0	0	0	na	1	2	50.0	1	2	50.0
TOTAL	0	1	0.0	1	4	25.0	1	5	20.0	0	2	0.0	0	4	0.0	0	7	0.0	0	3	0.0	1	5	20.0	1	8	12.5
Calcaneal Talar Facet: Porosity																											
<30	0	1	0.0	0	0	na	0	1	0.0	1	2	50.0	0	0	na	1	3	33.3	1	3	33.3	0	0	na	1	3	33.3
30-39	0	0	na	1	2	50.0	1	2	50.0	0	0	na	1	3	33.3	1	3	33.3	0	0	na	1	3	33.3	1	3	33.3
40+	0	0	na	0	2	0.0	0	2	0.0	0	0	na	0	1	0.0	0	1	0.0	0	0	na	0	2	0.0	0	2	0.0
TOTAL	0	1	0.0	1	4	25.0	1	5	20.0	1	2	50.0	1	4	25.0	2	7	28.6	1	3	33.3	1	5	20.0	2	8	25.0

Note: MO, ASO and EB were not observed for the talar head; ASO and EB were not observed for the calcaneal talar facet.

Calcaneal Correlation Matrix (left and right total)

	MO	ASO	P	EB
MO	**1**			
ASO	0	**0**		
P	1	0	**3**	
EB	0	0	0	**0**

* Total Individuals (numerator and denominator include only one element side when both present)

Table 4.21a. Metal Period: Articular Changes to the Vertebrae

Cervical										Total Individuals*								
	male			female			total			male			female			total		
	obs	n	%	obs	n	%	obs	n	%	obs	n	%	obs	n	%	obs	n	%
Atlas: ZOP[1]																		
<30										0	2	0.0	0	0	na	0	2	0.0
30-39										0	1	0.0	0	1	0.0	0	2	0.0
40+										1	1	100.0	1	1	100.0	2	2	100.0
TOTAL										1	4	25.0	1	2	50.0	2	6	33.3
Axis (including dens): ZOP																		
<30										0	1	0.0	0	0	na	0	4	0.0
30-39										2	3	66.7	1	2	50.0	3	5	60.0
40+										2	3	66.7	0	1	0.0	2	4	50.0
TOTAL										4	7	57.1	1	3	33.3	5	13	38.5
Cervical C3 to C7: VOP[2]																		
<30	0	10	0.0	0	0	na	0	13	0.0	0	4	0.0	0	0	na	0	5	0.0
30-39	0	8	0.0	0	5	0.0	0	13	0.0	0	3	0.0	0	1	0.0	0	4	0.0
40+	0	4	0.0	0	5	0.0	0	9	0.0	0	2	0.0	0	1	0.0	0	3	0.0
TOTAL	0	22	0.0	0	10	0.0	0	35	0.0	0	9	0.0	0	2	0.0	0	12	0.0
Cervical C3 to C7: ZOP																		
<30	0	2	0.0	0	0	na	0	5	0.0	0	2	0.0	0	0	na	0	3	0.0
30-39	4	6	66.7	0	5	0.0	4	11	36.4	2	3	66.7	0	1	0.0	2	4	50.0
40+	0	1	0.0	0	5	0.0	0	6	0.0	0	1	0.0	0	1	0.0	0	2	0.0
TOTAL	4	9	44.4	0	10	0.0	4	22	18.2	2	6	33.3	0	2	0.0	2	9	22.2
Thoracic T1 to T12: VOP																		
<30	0	1	0.0	0	0	na	0	2	0.0	0	1	0.0	0	0	na	0	2	0.0
30-39	0	10	0.0	0	12	0.0	0	22	0.0	0	1	0.0	0	1	0.0	0	2	0.0
40+	0	10	0.0	0	15	0.0	0	25	0.0	0	1	0.0	0	2	0.0	0	3	0.0
TOTAL	0	21	0.0	0	27	0.0	0	49	0.0	0	3	0.0	0	3	0.0	0	7	0.0
Thoracic T1 to T12: ZOP																		
<30	0	1	0.0	0	0	na	0	2	0.0	0	1	0.0	0	0	na	0	2	0.0
30-39	1	10	10.0	1	12	8.3	2	22	9.1	1	1	100.0	1	1	100.0	2	2	100.0
40+	1	10	10.0	1	15	6.7	2	25	8.0	1	1	100.0	1	2	50.0	2	3	66.7
TOTAL	2	21	9.5	2	27	7.4	4	49	8.2	2	3	66.7	2	3	66.7	4	7	57.1
Lumbar L1 to L5: VOP																		
<30	1	1	100.0	0	0	na	1	1	100.0	1	1	100.0	0	0	na	1	1	100.0
30-39	1	5	20.0	0	6	0.0	1	11	9.1	1	1	100.0	0	2	0.0	1	3	33.3
40+	0	5	0.0	0	6	0.0	0	11	0.0	0	1	0.0	0	2	0.0	0	3	0.0
TOTAL	2	11	18.2	0	12	0.0	2	23	8.7	2	3	66.7	0	4	0.0	2	7	28.6
Lumbar L1 to L5: ZOP																		
<30	0	1	0.0	0	0	na	0	1	0.0	0	1	0.0	0	0	na	0	1	0.0
30-39	1	5	20.0	0	5	0.0	1	10	10.0	1	1	100.0	0	1	0.0	1	2	50.0
40+	1	5	20.0	1	6	16.7	2	11	18.2	1	1	100.0	1	2	50.0	2	3	66.7
TOTAL	2	11	18.2	1	11	9.1	3	22	13.6	2	3	66.7	1	3	33.3	3	6	50.0
Sacral Promontory: VOP																		
<30										0	0	na	0	1	0.0	0	1	0.0
30-39										1	1	100.0	0	2	0.0	1	3	33.3
40+										0	0	na	1	1	100.0	1	1	100.0
TOTAL										1	1	100.0	1	4	25.0	2	5	40.0

* Individual: represented by at least one assessable vertebral element

[1] ZOP: zygapophyseal osteophytosis [2] VOP: vertebral osteophytosis

Table 4.21b. Metal Period: Individuals Affected by Articular Changes to the Spine

Metal Period			Atlas	Axis		Cervical (C3 to C7)		Thoracic		Lumbar		Sacrum
case	sex	age	ZOP	VOP	ZOP	VOP	ZOP	VOP	ZOP	VOP	ZOP	VOP
77CHSM1a	M	5							Y		Y	
77NNM22ka	M	5			Y							
76NNM2c	M	4	Y		Y							
89MDM2a	F	4	Y						Y		Y	Y
78QCM27a	M	3			Y		Y					
84DSM7a	M	3			Y		Y		Y	Y	Y	Y
77NNM5kb	F	3			Y							
95DXM1a	F	3							Y			
67VQ2M44a	M	2								Y		

Y: articular change observed
(blank): articular change not observed
(shaded): articular area not preserved

5=50+ years
4=40-49 years
3=30-39 years
2=20-29 years

M= male
F= female
I= indeterminate

It is highly probable that quite a number of the individuals sampled here, that did not display eburnation, would have met clinical and radiological criteria for OA during life. The reason for this is that although hypertrophic bone formation (marginal and joint surface osteophytes) is not pathognomonic of OA, it is however positively correlated with OA. It is therefore probable that the prevalence of osteophytc development in these samples reflects, to a degree, the approximate prevalence and distribution of OA within these samples.

There is however, yet another confounding factor with respect to bone-former activity. Rogers *et al.'s* (1997) study noted a positive correlation between osteophyte development, age and being male. For their sample they calculated the prevalence of bone-formers to be about 22%. A majority of this portion would also be male. This is only one study and it is not clear that this proportion of bone-formers is biologically universal. A whole range of confounding genetic, behavioural and environmental factors could be operating to modify the proportion of bone-formers within any population. The bone-former proportion of any sample probably accounts for much of the residual variation unaccounted for in the correlation between OA and osteophytic development. In lieu of any evidence suggesting that bone-formers will markedly skew the correlation between OA and osteophytc development, it is still useful to employ the prevalence and distribution of osteophytc development as an approximate surrogate for the prevalence of OA.

Distribution of Joint Changes

In the following discussion, articular change or joint change refers to changes other than porosity as the only evidence of articular change. While there are differences in the prevalence of joint change by joint class for both samples, it is very difficult to draw any conclusions from this. The greatest problem is one of poor joint representation in the Metal period sample. In all cases it should be noted that any apparent substantive differences between samples may be artefacts of sampling. Nonetheless, the relative or proportional intra-sample prevalence of joint change between the various joint classes is generally similar for both time periods. From an inter-sample perspective, the elbow, knee and hip joint changes show a somewhat higher prevalence in the Metal period and this is despite the attenuated age structure of this sample.

When all evidence for joint change, with the exception of porosity, is examined, the high prevalence in the spine is consistent with both bio-archaeological (Bridges 1992) and contemporary studies (van Saase *e al.* 1989). In fact, osteophytic development of the spine has been promoted as a surrogate age-at-death estimation method (Stewart 1958). The poor preservation of vertebral remains in these samples severely limits an examination by anatomical region. Nevertheless, it is worth noting that the results for both samples indicate that thoracic bodies are the least affected by osteophytosis. This is consistent with the findings of other bio-archaeological studies (e.g. Bridges 1992).

The elbow, in both time periods, is the next most commonly affected joint. Why the elbow appears to display a greater prevalence of joint change in archaeological populations is not clear. It is also apparent that the elbow is a commonly affected joint in other Southeast Asian skeletal assemblages as well (see Table 4.26). Bridges (1991) refers to the commonly held view, by bio-archaeologists at least, that this is related to increased physical demands on the elbow in many pre-industrial populations. That hypothesis cannot be tested here. However, given that most of the osteophytic development observed in archaeological samples would likely be radiologically invisible in the living subject, the reported differences in prevalence are possibly artefacts of different recording protocols. Reviews of the distribution of OA in archaeological samples (e.g. Bridges 1991; Jurmain 1999) also note that inter-observer differences in scoring criteria for OA make inter-sample comparisons difficult. The majority of bio-archaeological studies have used both porosity and osteophytic development as diagnostic of OA. It may well be that the elbow is particularly prone

Table 4.22. Da But Period: Individuals Affected by Articular Changes

case	sex	age	jaw	spine	shoulder	elbow	wrist	hand	hip	knee	ankle	foot
80CCNM3a	M	5								Y		
80CCNM14b	M	5		Y								
80CCNM16b	M	5	P	Y	Y	Y			P	Y	P	
80CCNM38a	M	5	Y									
80CCNM39a	M	5	Y									
80CCNM43a	M	5	Y			Y		Y				
80CCNM58a	M	5			Y		P					
80CCNM61a	M	5	P						Y			
80CCNM66a	M	5									P	Y
80CCNM74a	M	5	Y	Y	Y	Y	Y			Y		
80CCNMX2	M	5			Y							
80CCNM4a	F	5	Y									
80CCNM45a	F	5	P	Y	Y	Y						Y
80CCNM48a	F	5	P									
80CCNM50a	F	5		Y		Y					Y	
80CCNM56a	F	5			P						P	Y
80CCNM75a	F	5	Y		Y	Y						
80CCNM15a	M	4				Y			P			Y
80CCNM33b	M	4	Y									
80CCNM82a	M	4				Y						
80CCNM85a	M	4										Y
80CCNM31a	F	4	Y						P			
80CCNM33a	F	4	P									
80CCNM33c	F	4	P			Y						
80CCNM35b	F	4	P									
80CCNM64a	F	4	Y		P		Y					
80CCNM79a	F	4	P									
80CCNM83a	F	4	Y		Y				P	Y		
80CCNMX1	F	4							Y			P
80CCNM1y	F	4	P									
80CCNM22a	I	4	P									
80CCNM8a	M	3	Y		Y	Y						
80CCNM10a	M	3		Y	Y	Y						
80CCNM10c	M	3			P							Y
80CCNM17a	M	3	Y				Y					
80CCNM18a	M	3	P	Y		Y						
80CCNM26a	M	3										P
80CCNM35a	M	3	P	Y					P	Y		
80CCNM62a	M	3		Y								
80CCNM63a	M	3								Y		
80CCNM97a	M	3	Y								Y	
80CCNM2a	F	3	P						P	P		
80CCNM42a	F	3		Y	P			P	P	P		Y
80CCNM53a	F	3	P	Y								
80CCNM54a	F	3										Y
80CCNM77a	F	3		Y		Y						Y
80CCNM14a	M	2	P						Y	Y		Y
80CCNM57a	M	2		Y	P						P	
80CCNM60a	M	2										Y
80CCNM65a	M	2		Y	P	Y		Y	P			
80CCNM5a	F	2	P									
80CCNM71a	F	2	P	Y	P						P	
80CCNM19a	I	2								Y		
80CCNM67a	F	1										Y
80CCNM27a	I	1			P							
80CCNM52a	I	1	P									

Y: articular change including one or a combination of joint contour change, eburnation, marginal and articular surface osteophytosis
P: porosity only
(blank): articular change not observed
(grey shading): articular area not preserved

5=50+ years; 4=40-49 years; 3=30-39 years; 2=20-29 years; 1=15-19 years

M= male; F= female; I= indeterminate

Table 4.23. Prevalence of Articular Change* by Age

Da But Period

	40+ years			30-39 years			15-29 years		
	obs	n[1]	%	obs	n[1]	%	obs	n[1]	%
TMJ	10	27	37.0	3	16	18.8	0	15	0.0
spine	5	14	35.7	7	10	70.0	3	10	30.0
shoulder	7	17	41.2	2	10	20.0	0	12	0.0
elbow	9	22	40.9	4	16	25.0	1	15	6.7
wrist	2	14	14.3	1	11	9.1	0	10	0.0
hand	1	21	4.8	0	16	0.0	1	10	10.0
hip	2	13	15.4	0	11	0.0	1	9	11.1
knee	4	13	30.8	2	10	20.0	2	13	15.4
ankle	1	18	5.6	1	13	7.7	0	13	0.0
foot	5	19	26.3	4	17	23.5	3	16	18.8

Metal Period

	40+ years			30-39 years			15-29 years		
	obs	n[1]	%	obs	n[1]	%	obs	n[1]	%
TMJ	2	11	18.2	4	16	25.0	2	30	6.7
spine	4	5	80.0	4	6	66.7	1	6	16.7
shoulder	2	4	50.0	1	6	16.7	0	5	0.0
elbow	3	4	75.0	5	6	83.3	1	7	14.3
wrist	1	4	25.0	0	4	0.0	0	4	0.0
hand	0	3	0.0	0	5	0.0	1	5	20.0
hip	1	2	50.0	1	6	16.7	2	7	28.6
knee	1	3	33.3	2	4	50.0	2	6	33.3
ankle	1	3	33.3	0	7	0.0	0	5	0.0
foot	1	3	33.3	2	6	33.3	1	4	25.0

* articular changes only include marginal osteophytosis, articular surface osteophytosis, eburnation and morphological changes to the joint contour.

[1] n for each joint refers to the number of assessable individuals with that particular joint

to bone-forming processes, as well as those processes leading to porosity, and so artificially inflating the apparent level of elbow OA. Differential predisposition of joints to bone-forming activities is an area that requires further investigation.

It would appear, however, that the bone-former hypothesis cannot solely account for the elevated levels of elbow joint change in these samples. Only 23.5% of the Metal period sample assessable for joint changes to the elbow were 40+ years old, compared to 55.0% in the Da But period sample. However, 26.4% of the assessable Da But period sample exhibited elbow joint changes compared to 52.9% of the Metal period sample. A simple one to one relationship with age, and also a bone-forming predisposition, does not appear to explain these differences. The older Da But period sample would have been expected to display the greater prevalence of hypertrophic bone development to the elbow. These unexpected results would tend to suggest a difference in one or a combination of risk factors, other than age and bone-forming, between the two samples targeting elbow changes specifically. What such factors are can only be speculative.

The prevalence of knee and hip joint changes is also higher in the Metal period. The frequency of changes to these joints, for either temporal period, is not however unusual given the wide range of reported variation in OA for these joints (van Saase *et al.* 1989). In the Metal period sample the knee was the third most commonly affected joint, after the back and elbow. Reviewing 25 bio-archaeological studies of Native American Indian assemblages, Bridges (1992) noted that for 17 of these assemblages the knee was the most commonly affected joint. In the Southeast Asian assemblages examined to date, the knee and hip were among the top four most affected joints in 4/7 and 6/7 assemblages respectively, including the Vietnamese samples. As such, the prevalence of hip and knee changes for the Metal period does not appear unusual on the world or even Southeast Asian stage. However, the prevalence of hip and knee joint change does appear elevated in light of studies that indicate modern Asians display a very low prevalence of hip OA (Oishi *et al.* 1998) and also apparently knee OA (Hoaglund *et al.* 1973). It is possible that in this study, of the Metal period assemblage, the evidence for articular changes to the hip, at least, are not correlated with OA. The recording protocol in this study scored all articular surface osteophytic activity as evidence for articular change. In the case of the hip joint, the common finding of ossification of the femoral insertion of the *ligamentum teres* has probably elevated the recorded prevalence of hip OA. Ossification of *ligamentum teres*

Table 4.24. Prevalence of Articular Change* by Sex

Da But Period

	Male			Female			Total		
	obs	n[1]	%	obs	n[1]	%	obs	n[1]	%
TMJ	8	28	28.6	5	24	20.8	13	58	22.4
spine	9	20	45.0	6	11	54.5	16	34	44.1
shoulder	6	20	54.5	3	11	27.3	9	39	23.1
elbow	9	26	34.6	5	20	25.0	14	53	26.4
wrist	2	21	9.5	1	11	9.1	3	35	8.6
hand	2	21	9.5	0	11	0.0	2	47	4.3
hip	2	16	12.5	1	13	7.7	3	33	9.1
knee	6	18	33.3	1	11	9.1	8	36	22.2
ankle	1	24	4.2	1	14	7.1	2	44	4.5
foot	6	27	22.2	6	17	35.3	12	52	23.1

Metal Period

	Male			Female			Total		
	obs	n[1]	%	obs	n[1]	%	obs	n[1]	%
TMJ	5	24	20.8	3	25	12.0	8	57	14.0
spine	6	10	60.0	3	5	60.0	9	17	52.9
shoulder	1	4	25.0	2	10	20.0	3	15	20.0
elbow	3	8	37.5	6	8	75.0	9	17	52.9
wrist	0	6	0.0	1	6	16.7	1	13	7.7
hand	0	6	0.0	1	6	16.7	1	13	7.7
hip	1	7	14.3	3	7	42.9	4	15	26.7
knee	3	8	37.5	2	4	50.0	5	13	38.5
ankle	0	6	0.0	1	8	12.5	1	15	6.7
foot	2	5	40.0	2	7	28.6	4	13	30.8

* articular changes only include marginal osteophytosis, articular surface osteophytosis, eburnation and morphological changes to the joint contour.

[1] n for each joint refers to the number of assessable individuals with that particular joint.

may not strictly be correlated with OA. However, the only instance of hip eburnation in the combined Vietnamese samples was from the Metal period.

The low prevalence of changes to the toes and fingers in both samples is likely more an artefact of extremely poor preservation of these joints than anything else. Toe joints only made up a small proportion of the preserved joints in the foot joint. It is changes to the larger tarsal bones that account for most of the observed foot changes. The somewhat low prevalence of bony changes to the joints of the ankle and wrist are also consistent with global evidence, past and contemporary, of these joint sites not being overly affected by OA. The ankle and wrist also displayed the least degree of articular change in other Southeast Asian sites examined. However, the foot was one of the top three joint complexes affected by articular change in 6/7 Southeast Asian sites, including the Vietnamese, summarised in Table 4.26.

Because of the good preservation and sample size of both Da But period and Metal period mandibular condyles, the observed greater prevalence of Da But period TMJ changes may be more real than artefactual. Nonetheless, the greater proportion of older individuals in the Da But period sample is the most likely factor responsible for the elevated prevalence of mandibular joint changes in this sample.

Age Distribution

The correlation between age and articular change (Rogers and Waldron 1995) in both temporal periods is consistent with expectations. This pattern is particularly clear in the better represented joints of the Da But period. The same pattern, albeit somewhat complex due to problems of representativeness, is also evident is the Metal period sample. The pattern is entirely consistent with skeletal (Jurmain 1980) and living contemporary (van Saase *et al.* 1989) studies of OA that demonstrate a clear tendency for the prevalence of OA to increase with increasing age. However, the underlying aetiological reasons remain extremely difficult to disentangle.

Sex Distribution

The sex differences in joint change between the Da But period and Metal period samples are somewhat intriguing. Females exhibit a greater frequency of joint change in the Metal period, while males display a greater prevalence of joint change in the Da But period sample. Nonetheless, as pointed out, it is likely that joint representation problems within the Metal period sample have distorted the picture considerably. The Metal period sample itself, was biased

Table 4.25. Metal Period: Individuals Affected by Articular Changes

case	sex	age	jaw	spine	shoulder	elbow	wrist	hand	hip	knee	ankle	foot
77CHSM1a	M	5	P	Y	Y	Y						
78NNM16ka	M	5	Y									
77NNM22ka	M	5	P	Y								
76NNM2c	M	4	P	Y		Y				Y		
76NNM2d	M	4	P									
78QCM28a	M	4	Y									
89MDM2a	F	4	P	Y	Y	Y	Y		Y		Y	Y
89MDM8a	F	4			P							
78NNM6kb	M	3	P									
78QCM3a	M	3	Y							Y		
78QCM27a	M	3	Y	Y								
84DSM7a	M	3		Y	P	Y						Y
67VQ2M1a	F	3				Y						
67VQ2M43a	F	3			Y	Y				P		
77NNM5kb	F	3	Y	Y								
92DMTS60/62	F	3	P									
84DSM1a	F	3	P			Y			Y			
89MDM3a	F	3			P	P			P			
95DXM1a	F	3	Y	Y	P	Y			P	Y	P	Y
66VQ1M1a	M	2	Y		P	P						
67VQ2M44a	M	2		Y					Y			P
78QCM25a	M	2	P									
NNM1x	M	2								Y		Y
84DSM8a	F	2	P									
89MDM1a	F	2	Y									
89MDM7a	F	2				Y			Y	Y		
78QCM14a	F	2	P					Y				
78QCM22a	F	2	P									
78QCM1a	F	1			P							

Y: articular change including one or a combination of joint contour change, eburnation, marginal and articular surface osteophytosis

P: porosity only

(blank): articular change not observed

: articular area not preserved

5=50+ years
4=40-49 years
3=30-39 years
2=20-29 years
1=15-19 years

M= male
F= female
I= indeterminate

toward older males, and it might have been predicted that this would have led to an increase in the prevalence of male articular change relative to female change. As this is not the case these results may indicate a real higher prevalence of female articular change in the Metal period.

In Bridges' (1991) extensive survey of the bio-archaeological literature on OA, the prevalence of male OA was found to be generally higher than female OA. This finding is in contrast to the results of contemporary studies of OA prevalence (Jones & Doherty 1995), and the possible confounding reasons have been discussed previously. The greater prevalence of joint change, for most joints in the Da But period sample, is consistent with the findings of other bio-archaeological studies. However, the reason for a higher prevalence of joint change in the Da But period is no more clear in this study than those. This is particularly so in light of the observation that while the Da But sample displays an older age structure, relative to the Metal period sample, the older component of this sample is biased toward females. It would have been expected, given these circumstances, that a greater frequency of articular change would be seen in Da But period females. This would have been in line with the results of contemporary studies that suggest females have a higher prevalence of OA in later middle age (Roberts & Burch cited in Moskowitz 1993). The observation that males, even though under-represented relative to females in the older ages classes, display a greater prevalence of joint change may be indicating aetiological factors other than simply those related to age. It would be difficult to support the view that the observed heightened levels of male articular change is simply related to bone-former factors. While it has been noted that evidence suggests that older males are preferentially affected by bone-forming, the lower proportion of older males relative to older females in the Da But sample, would suggest that male bone-formers were not overtly contributing to observed levels of joint change. It is unclear why Da But period males display a greater prevalence of joint change.

In conclusion, there is an apparent greater prevalence of articular change in the Da But period sample relative to the Metal period. This is most likely related to the older age structure of the Da But period sample. On the other hand, it would appear that the prevalence and distribution

Table 4.26. Appendicular Articular Change in Southeast Asian Skeletal Series (ranked by prevalence)

	Vietnam Da But	Ban Chiang[1]	Non Nok Tha[1]	Khok Phanom Di[2]	Vietnam Metal	Ban Lum Khao[3]	Ban Na Di[3]
Joint	~6.0kya	5.6-1.7kya	4.5-3.0kya	4.0-3.5kya	3.3-1.7kya	1.0-1.5kya	0.6-0.4kya
shoulder	**2**	6	7	**2**	5	**2**	**3**
elbow	**1**	**4**	**3**	6	**1**	**1**	6
wrist	5	5	**4**	7	6	6	**1**
hand	7	7	8	5	6	8	6
hip	**4**	**1**	**2**	9	**4**	**4**	**3**
knee	**3**	8	6	**3**	**2**	5	**4**
ankle	6	**2**	**1**	**4**	7	7	5
foot	**2**	**3**	5	**1**	**3**	**3**	**2**

bold type = top four ranked joints 1 = most prevalent, 8 = least prevalent

[1] Douglas (1996)
[2] Tayles (1992)
[3] Domett (1999)

of joint change in the Da But period and Metal periods of Vietnam is not particularly unusual when compared to contemporary and other bio-archaeological studies of OA. The somewhat elevated levels of elbow, hip and knee joint change seen in the Metal period sample is potentially artefactual and needs to be re-examined at some future time when an enhanced sample size is available. The same is also true of the sex differences.

There is no compelling evidence in either the distribution of articular change by sex, joint complex or by age that would suggest recourse to the mechanical stress model, as a result of particular behaviours or activities, to explain the Vietnamese findings. This is not discounting the role of such factors, but even if they have played a role, significant or otherwise, in the distribution and prevalence of articular change in these samples it must of necessity remain hidden.

The distribution of articular changes by joint and by age in past communities from northern Vietnam is generally consistent with the findings from modern and bio-archaeological studies. Previous to this study, nothing was known about the prevalence and distribution of articular change from this part of the world during the mid-Holocene and Bronze periods. This study adds another unique spatio-temporally defined study to the world database on articular change.

5
MARKERS OF PHYSIOLOGICAL STRESS

Bio-archaeological studies of enamel hypoplasia (EH), a defect developing during tooth formation, and porotic hyperostosis, localised skeletal scarring caused by increased red blood cell production, have been extensively used in assessing levels of physiological or metabolic stress in past populations. Enamel hypoplasia is commonly used as a marker of non-specific stress (Goodman *et al.* 1984; Goodman & Rose 1991; Mittler *et al.* 1992; Duray 1996; Malville 1997). Some evidence has also been forwarded suggesting a causal link between the prevalence of EH and marginal living conditions (Goodman *et al.* 1987; Goodman *et al.* 1992). A more specific link between the occurrence of EH and nutritional status has also been reported (Goodman & Rose 1991; Zhou & Corruccini 1998). Other studies have examined the prevalence of EH in relation to social stratification (Wood 1996; Cucina & Iscan 1997) and also changes in subsistence-economic circumstance (Cook 1984; Goodman *et al.* 1984; Rose *et al.* 1984; Yamamoto 1992; Wood 1996).

Porotic hyperostosis, manifesting in the cranial vault and orbital roofs, has been linked with a number of vitamin deficiency conditions (Schultz 1993; Carli-Thiele & Schultz 1997). However, it is generally believed to be indicative of anaemia and iron-deficiency anaemia specifically (Stuart-Macadam 1987a, 1987b, 1992). While nutritional deficiencies have been favoured as causing iron-deficiency anaemia (Mittler & Van Gerven 1994), other studies (e.g. Walker 1986) have suggested that parasitic loads and associated diarrhoeal diseases can play a greater role in the aetiology of porotic hyperostosis than nutritional deficiencies alone. In tropical Southeast Asia genetic haemoglobinopathies have been suggested as an important aetiological factor in the prevalence of porotic hyperostosis (Tayles 1996).

Clearly, an examination of the prevalence and distribution of enamel hypoplasia and porotic hyperostosis is useful in shedding light on the health and behaviour of past communities. The following investigation of enamel hypoplasia and porotic hyperostosis first reviews the literature on the aetiology and identification of each condition. The frequency and distribution of these markers of physiological stress are then explored for the Vietnamese skeletal assemblages. The health and behavioural implications of these results are discussed.

Enamel Hypoplasia

EH has been reported in most fossil hominid species from the Australopithecines (Bombin 1990) through to Neandertals (Molnar & Molnar 1985b; Skinner 1996; Hutchinson & Larsen 1997). In northern Vietnam the earliest hominoids with evidence of EH are the remains of Pleistocene Pongids (personal observation).

Most bio-archaeological interest has focused on the role of EH in exploring variations in human adaptation and behaviour. For instance, Larsen (1997) points out a general globally evident, albeit not blanket, trend of increased enamel defect prevalence with subsistence shifts from foraging to agriculture and agricultural intensification. The prevalence of EH has also been used to explore the presence and/or effect of culturally imposed stress buffers, differential status for example. Some studies addressing status and EH have found probable relationships between the two (e.g. Goodman & Armelagos 1988), while others have failed to find such correlations (e.g. Powell 1988).

An implicit assumption of nutritional or dietary stress underlies many bio-archaeological analyses of enamel hypoplasia. One of the aims of this chapter is to review the literature specifically examining the ostensible relationship between nutritional status and enamel hypoplasia. This review is extended to cover recent work examining these enamel defects with respect to developmental chronologies. In particular, the so-called weaning-hypothesis frequently cited to explain intra-population peaks in enamel defect occurrence is examined. The second aim of this study of enamel hypoplasia is to document the nature, distribution and prevalence of enamel hypoplasia in the Da But period and Metal period dental assemblages. The implications of these results with regard to the nutritional status and physiological health of these samples are then discussed.

Aetiopathogenesis of Enamel Hypoplasia

The pathogenesis of enamel hypoplastic defects is fairly well understood. Suga (1992:287-288) states that functional disturbances to the secretory ameloblasts (enamel forming cells) lead to changes in matrix histological morphology (the shape and arrangement of the enamel rods), enamel thickness and occasional hypomineralisation. A broad range of contributing factors have been implicated in the aetiology of enamel defects (Neiberger 1990; Hillson 1996). Cutress & Suckling (1982:117) are often quoted: "Almost 100 factors have been held responsible for inducing tooth defects, many of them involving enamel".

For the most part EH is treated as a general marker of systemic metabolic stress. Goodman & Rose (1991) categorise hypoplasiae aetiologically as 1) hereditary, 2) due to localised trauma and 3) those due to systemic metabolic stress. The vast majority of hypoplasiae are suggested to be due to the latter agent with trauma-induced defects being rare and hereditary defects being extremely rare, less than 1% of defects in modern populations (Goodman & Rose 1991).

Enamel Hypoplasia and Nutritional Stress

Metabolic stress induced defects have been specifically linked to socio-economic conditions and even more specifically to variations in levels of nutrition. In particular, work has suggested a link between the prevalence of EH and living in marginal environments where the population is mildly to moderately undernourished (Goodman *et al.* 1987; Goodman *et al.* 1992).

An important recent retrospective study has specifically tested the nutritional stress hypothesis on a sample of more than 3000 urban and rural individuals in China (Zhou & Corruccini 1998). The tooth development times of these individuals spanned the periods before, during and after one of the world's worst famines in China from the late 1950s to early 1960s. This study strongly supports the link between levels of nutrition and the prevalence of EH. However, their results must be appreciated in the same context as Goodman's (Goodman & Rose 1991) own reflections on the results of his studies of nutrition and EH. Goodman & Rose (1991) note that in their studies of the link between nutrition and EH in Mexican children, it was clear that better nourished individuals displayed less sick time (respiratory and diarrhoeal illness). They go on to suggest that:

> it is possible that the enamel defects are more related to illness than to nutritional intake, or to the synergistic interaction between undernutrition and infectious disease. Poor nutritional status decreases immune function and predisposes one to infection, while infectious diseases can further diminish scarce nutritional resources.
> (Goodman & Rose 1991:284)

The difficulty of finding a causal relationship between EH and nutritional status, in modern epidemiological studies, only underlines the impossibility of finding specific causal factors for EH frequencies in archaeological populations.

Enamel Hypoplasia and Weaning

Another area of EH research that has attracted some attention is the ostensible link between weaning and the timing of EH. In a review article of approaches to identifying weaning in the archaeological record, Katzenberg *et al.* (1996) suggest that most positive correlations between peaks in the frequency of EH and supposed weaning age are fortuitous. Moreover, most of this correlation is forced in that it is assumed that weaning occurs at these times without any direct evidence whatsoever. They state that weaning is a process involving:

> the introduction of other foods and reduction of dependence on breast milk, and differentiate this from complete weaning which is the cessation of breast feeding. Much of the literature on enamel hypoplasia tends to regard weaning as an event, the cessation of breast-feeding.
> (Katzenberg *et al.* 1996:179)

Katzenberg *et al.* (1996) cite studies by Blakey *et al.* (1994) and Wood (1996) that appear to reject convincingly a correlation between EH timing and weaning. Such retrospective studies are important for their insights. Nonetheless, when such studies purport to test specific hypotheses they bear closer inspection.

Blakey *et al.* (1994) cite historical evidence suggesting that weaning generally occurred between nine and 12 months in African American slave women. They also cited evidence for a modal birth spacing average of two years, with a range of about one to three years. This birth spacing pattern supports the weaning age timing of nine to 12 months due to the anovulatory effects of lactation and subsequent delayed resumption of normal ovulation post-weaning. The hypothesis to be tested was that the peak timing of enamel hypoplasia would cluster around the time of weaning. The results indicated an average clustering of EH between two and three years. This resulted in a rejection of the hypothesis that this was due to weaning, which occurred at least a year previously based on historical data.

There are several problems with the reasoning in this study however. The first is quite straightforward in that the sample tested was an archaeological assemblage, without any documentation on the timing of weaning for any individual in the sample. The only way to test such a hypothesis is to examine a sample of teeth from individuals where the specific timing and duration of weaning is known. The lack of such data, unavoidable with bio-archaeological assemblages, renders the argument circular.

There are further problems with this study, especially with respect to interpretations of the historical evidence for weaning. It was not stipulated exactly what was meant by the term weaning. As already noted, weaning is a process. Infants could quite conceivably, and likely, have continued to be suckled (morning and night feeds) for some considerable period of time after nine to 12 months. Such feedings would not have interfered with the duties of mothers in the field. Unless complete cessation of suckling was vigorously enforced, night and morning feeding was probably common. As Blakey *et al.* (1994) themselves point out, ovulation will also resume under much reduced nursing. The modal birthing interval of two years is therefore not incompatible with continued, albeit reduced, nursing for an extended period of time. It could very reasonably be argued that either pregnancy or childbirth was the chief factor contributing to the timing of weaning. This would occur every two to three years on average and would be, coincidentally, occurring at the same time that EH is reported to be clustering.

While Saunders & Keenleyside (1999) argue that weaning should not be seen as the most important single cause in the development of enamel hypoplasia, it may be premature to reject the hypothesis that peaks in the age of EH formation are not related to weaning in all studies. As Katzenberg *et al.* (1996) note, weaning is a complex and sometimes protracted process. It involves the introduction of a variable range of solid foods (as well as non-food stuffs that infants universally seem to enjoy putting in their mouths) into the infant's diet well before the final cessation of breast milk. Further, the impact of the final cessation of breast feeding, regardless of the length of the weaning process, may still have a significant and largely unknown psycho-physiological impact on the infant. This, combined with the eventual loss of breast milk introduced immunity and the development of the child's own independent immune repertoire, will also have an important and variable impact on the infant's health.

There are some data on contemporary breast feeding and weaning behaviour in ethnic groups living in highland (the Ede of Dac Lac province) and lowland (the Viet of Hau Giang province) southern Vietnam (Dang Phuong Kiet & Bui Vu Huy 1995). Among the Ede most infants appear to be weaned, here meaning complete cessation of suckling, at about 16 months with a range of between 10 and 36 months. There also appears to be no sex bias in the timing of weaning. Most infants in this survey were given additional foods at about seven and a half months. With regard to the Viet, the average age of weaning was about 10 months with a range of six to 12 months. Foods, additional to breast milk, were introduced at around six months with a range between two and 15 months.

Weaning is a process that can be quite variable in duration as the above Ede example suggests. Nonetheless, certain key periods in the process will leave the infant more prone to psycho-physiological stress. It is during these periods, likely also to coincide with the period of maximum tooth sensitivity to physiological disruption, that stress is potentially being recorded. Nonetheless, the real problem is perhaps not in isolating weaning, or some other factor, as the critical specific aetiological agent in the development of certain enamel defects. What may be much more important is the question of what role tooth morphology, architecture and developmental processes play in the timing and prevalence of defects.

Enamel Hypoplasia Chronologies

With readily available data on tooth development (Goodman *et al.* 1980) and labour saving computer programs (Murray & Murray 1989), the last two decades have seen a wealth of studies investigating the timing of EH on individual teeth. However, the meaning of much of this work is now being questioned.

Skinner & Goodman (1992) review, in some detail, work centred on the elucidation of enamel hypoplastic defect chronologies. They note that tooth development is highly variable and the formation of EH defects is heavily dependent upon the periods that a tooth is actively forming enamel. The development time of a tooth will affect the time available for the formation of EH defects. Prolonged growth times will enable a tooth to record more episodes of stress. It is thus important to have accurate knowledge of the development times of the teeth under study. Moreover, there is not a one to one relationship between the external surface of a tooth and the timing of its formation. Skinner & Goodman (1992) note that the external surface of the maxillary canine, for example, represents growth between two and five years. The earliest EH defect that this tooth can record is thus from the second year of the individual's life and not from the first six months as previously believed.

Skinner & Goodman (1992) point out that much research attributing peak periods of EH formation to the timing of weaning is methodologically flawed. They convincingly argue that what is being measured in such instances are not peak periods of stress associated with weaning, or whatever, but rather the time of peak enamel formation. Inter-individual and inter-tooth variation in periods of active enamel formation are distributed normally about the first two to four years of an individual's life. This period of time encompasses the formation of the permanent incisors, canines, first molar and most of the development period of the first premolar. Even with the future availability of more accurate crown formation data, for a range of ancestral groups, it would seem that the available stress sensitive windows in studies of EH are quite limited. Hillson (1996), for reasons outlined above, has questioned the usefulness of just recording the position of EH defects. Although he proposes that buccal tooth impressions are potentially useful for matching EH events with tooth growth sequences.

In order to address the problem of unrealistic tooth development chronologies, Malville (1997) has rendered Skinner & Goodman's (1992) published tooth development data into a readily useable chart for the anterior teeth. However, this still does not deal with the problem of inter and intra-population variability in tooth development rates. Moreover, even the use of more realistic tooth development chronologies does not tackle the issue of the timing of enamel defects being more a function of differential enamel sensitivity to stress than periods of peak stress in an individual's life. Further, Malville's own case study, using anterior teeth, revealed an even narrower window of enamel sensitivity to defect development than previously thought: "For the combined sample the average length of time during which individuals show evidence of enamel disruption is approximately one year" (Malville 1997:357). This should not be surprising since Goodman & Armelagos (1985a) pointed out, some time ago, that up to 65% of enamel defects appear in the middle third of the crown of anterior teeth. This bias to macroscopically visible defects in this region turns out to be a function of the geometry of tooth development (Hillson 1996; Hillson & Bond 1997) rather than systemic stress.

A further problem in studies of enamel defects, whether simply scoring prevalence or developing chronologies, is the identification of EH. With respect to linear enamel hypoplasia (LEH) or furrow-form defects, Hillson & Bond (1997) demonstrate that these defects form a continuum from microscopic lines up to easily seen furrows. The point being that studies of EH that only record defects visible to the naked eye are assessing a variable proportion of the total number of defects present.

Methods and Materials

The left maxillary and mandibular canines, or when absent their antimere substitution, form the focus of analysis in this study. While it has been noted that the incisors tend to be more sensitive to EH development than the canines (Goodman & Armelagos 1985a, 1985b), this is only marginally so. The advantage of using the maxillary and mandibular canines in this study is the much greater preservation, compared to incisors, in both the Da But period and Metal period samples. Condon & Rose (1992) note that studies using single tooth types are necessarily restricted to recording stress during the period of time the tooth is sensitive to enamel disruption. In the case of the maxillary and mandibular canines this period roughly corresponds to between 1.5 and 4.5 years. All severely worn canines were excluded from this study. An extremely worn canine is defined as a tooth where more than 25% of the occlusal surface displays exposed dentin.

The presence of all linear enamel hypoplastic defects corresponding to the DDE index type 4 (Federation Dentaire International 1982) on the labial face of the left maxillary and mandibular canines, or their antimere substitutions, were recorded. The DDE type 4 defect corresponds to what is referred to as linear (horizontal) enamel hypoplasia or LEH. Because the DDE index does not allow for coding of severity, the severity definitions of Duray (1996) were adopted. Duray (1996:278) classified DDE type 4 defects as 1) severe (grooves >0.5mm in width), 2) moderate (grooves <0.5 mm in width) and 3) mild (questionable cases of only slightly broadened perikymata). LEH of the third, or mild, category of groove severity is not addressed in this study. LEH defects were recorded under a combination of natural and incandescent light without the aid of magnifying devices. DDE type 4 defects of moderate to severe degree can be readily recorded using the naked eye. The use of a magnifying lens may have led to the accidental recording of smaller defects than the ones targeted. As such, less visible and certainly all microscopic defects, are necessarily ignored.

It is reasonable to suppose that these data are still meaningful in that they follow the same method as a great many other studies of EH, thus allowing some level of inter-study comparison. Furthermore, this method is also pursued because it also seems reasonable to assume a fairly good correlation between the prevalence of LEH as measured in this manner and the prevalence of all defects from one perikyma in width upwards. In other words, the prevalence of macroscopically scored LEH should still be a good indicator of stress during the 1.5 to 4.5 year window, even if it does not record all stress events. The demonstrated correlation between metabolic stress and macroscopically scored EH in studies of modern populations (e.g. Goodman *et al.* 1987; Goodman *et al.* 1992; Zhou & Corruccini 1998) would tend to reinforce the continued utility of this method.

The unit of analysis is the individual, as represented by maxillary and/or mandibular canines. The distribution of defects is examined for each individual by each tooth separately and with the maxillary and mandibular data combined. For the purposes of analysis, moderate and severe defects are combined. Tests of LEH prevalence homogeneity by age class are performed on the three age category data only. The prevalence of LEH by sex and age as well as by sample is assessed. While LEH develops in childhood, it is necessary to determine if there is a correlation between the age-at-death of adults in these samples displaying evidence of LEH acquired in childhood.

Results

Mandibular Canines

Da But Period Sample

Of the 37 individuals with assessable mandibular canines, 56.8% displayed at least one LEH event (Table 5.1a). LEH affected more males (65.0%) than females (53.8%), although not to a significant degree. Similarly, although LEH affected a lower percentage of the oldest age category, relative to the two younger age classes, the null hypothesis of homogeneity of EH prevalence by age class was not rejected (Table 5.1a).

Metal Period Sample

Of the 40 individuals with assessable mandibular canines, 60.0% displayed LEH (Table 5.1b). LEH affected more males (66.7%) than females (55.6%), although not to a significant degree. Similar to the pattern in the Da But period sample, the oldest age class was the least affected by LEH, although again not to a significant degree (Table 5.1b).

While the Metal period sample displayed a slightly higher prevalence of LEH than the Da But period sample, the difference was not significant (Chi-sq. 0.083, $p=0.773$). However, the Metal period sample mean number of LEH events per individual (2.50) was significantly ($p<0.05$) greater than the mean (1.62) for the Da But period sample (Table 5.1c).

Table 5.1a. Da But Period. Mandibular Canine LEH[1] by Age

Five Age Classes

	male			female			total individuals		
	obs	n	%	obs	n	%	obs	n	%
<15	0	0	na	0	0	na	0	0	na
15-19	1	1	100.0	0	0	na	1	3	33.3
20-29	3	4	75.0	2	2	100.0	6	8	75.0
30-39	4	7	57.1	2	2	100.0	6	9	66.7
40-49	1	3	33.3	1	6	16.7	2	9	22.2
50+	4	5	80.0	2	3	66.7	6	8	75.0
TOTAL	13	20	65.0	7	13	53.8	21	37	56.8

Chi-sq. tests male:female	
Chi-sq.	p
0.441	0.522

Three Age Classes

	male			female			total individuals		
	obs	n	%	obs	n	%	obs	n	%
<30	4	5	80.0	2	2	100.0	7	11	63.6
30-39	4	7	57.1	2	2	100.0	6	9	66.7
40+	5	8	62.5	3	9	33.3	8	17	47.1
TOTAL	13	20	65.0	7	13	53.8	21	37	56.8

tests for LEH homogneity by age class	
Chi-sq.	p
1.224	0.542
Likelihood ratio	p
1.229	0.541

Table 5.1b. Metal Period. Mandibular Canine LEH[1] by Age

Five Age Classes

	male			female			total individuals		
	obs	n	%	obs	n	%	obs	n	%
<15	0	0	na	0	0	na	0	0	na
15-19	1	1	100.0	0	0	na	3	3	100.0
20-29	3	5	60.0	5	8	62.5	8	14	57.1
30-39	4	4	100.0	4	7	57.1	8	11	72.7
40-49	1	3	33.3	1	2	50.0	2	5	40.0
50+	3	5	60.0	0	1	0.0	3	6	50.0
Indet.	0	0	na	0	0	na	0	1	0.0
TOTAL	12	18	66.7	10	18	55.6	24	40	60.0

Chi-sq. tests male:female	
Chi-sq.	p
0.468	0.494

Three Age Classes

	male			female			total individuals		
	obs	n	%	obs	n	%	obs	n	%
<30	4	6	66.7	5	8	62.5	11	17	64.7
30-39	4	4	100.0	4	7	57.1	8	11	72.7
40+	4	8	50.0	1	3	33.3	5	11	45.5
Indet.	0	0	na	0	0	na	0	1	0.0
TOTAL	12	18	66.7	10	18	55.6	24	40	60.0

tests for LEH homogneity by age class	
Chi-sq.	p
1.856	0.395
Likelihood ratio	p
1.846	0.397

Table 5.1c. Mean Number of Mandibular Canine LEH[1] Events

Per Individual

	obs	n[2]	mean	t	df	p
Da But Culture	34	21	1.62	**2.518**	43	**0.016**
Metal Period	60	24	2.50			

[1] severe and moderate (grades 1 and 2) LEH only

[2] individuals displaying LEH

Maxillary Canines

Da But Period Sample

Of the 36 individuals with assessable maxillary canines, 69.4% displayed LEH (Table 5.2a). There appears to be a greater prevalence of LEH in females (83.3%) than males (53.3%), but this is not statistically significant. Examining the distribution of LEH by age indicates a slightly lower prevalence in the oldest age class, relative to the two younger classes (Table 5.2a). However, the null hypothesis of LEH prevalence homogeneity by age class is not rejected

Metal Period Sample

Of the 47 individuals with assessable maxillary canines, 53.2% displayed LEH (Table 5.2b). A slightly greater proportion of males have LEH (63.2%) than females (50.0%), although not to a significant degree. There is a pronounced, and significantly ($p<0.01$), uneven distribution of LEH by age class (Table 5.2b). Only a small proportion of the oldest age class (18.2%) exhibits LEH.

Table 5.2a. Da But Period. Maxillary Canine LEH[1] by Age

Five Age Classes

	male			female			total		
	obs	n	%	obs	n	%	obs	n	%
<15	0	0	na	0	0	na	0	1	0.0
15-19	1	1	100.0	1	1	100.0	6	7	85.7
20-29	0	0	na	1	2	50.0	3	4	75.0
30-39	4	7	57.1	3	3	100.0	7	10	70.0
40-49	2	3	66.7	3	4	75.0	6	8	75.0
50+	1	4	25.0	2	2	100.0	3	6	50.0
TOTAL	8	15	53.3	10	12	83.3	25	36	69.4

Chi-sq. tests male:female

Chi-sq.	p
2.700	0.100

Three Age Classes

	male			female			total		
	obs	n	%	obs	n	%	obs	n	%
<30	1	1	100.0	2	3	66.7	9	12	75.0
30-39	4	7	57.1	3	3	100.0	7	10	70.0
40+	3	7	42.9	5	6	83.3	9	14	64.3
TOTAL	8	15	53.3	10	12	83.3	25	36	69.4

tests for LEH homogneity by age class

Chi-sq.	p
0.352	0.839
Likelihood ratio	p
0.353	0.838

Table 5.2b. Metal Period. Maxillary Canine LEH[1] by Age

Five Age Classes

	male			female			total		
	obs	n	%	obs	n	%	obs	n	%
<15	0	0	na	0	0	na	1	2	50.0
15-19	0	0	na	1	1	100.0	2	3	66.7
20-29	5	6	83.3	2	6	33.3	9	16	56.3
30-39	5	5	100.0	6	9	66.7	11	14	78.6
40-49	1	3	33.3	0	1	0.0	1	5	20.0
50+	1	5	20.0	0	1	0.0	1	6	16.7
Indet.	0	0	na	0	0	na	0	1	0.0
TOTAL	12	19	63.2	9	18	50.0	25	47	53.2

Chi-sq. tests male:female

Chi-sq.	p
0.652	0.419

Three Age Classes

	male			female			total		
	obs	n	%	obs	n	%	obs	n	%
<30	5	6	83.3	3	7	42.9	12	21	57.1
30-39	5	5	100.0	6	9	66.7	11	14	78.6
40+	2	8	25.0	0	2	0.0	2	11	18.2
Indet.	0	0	na	0	0	na	0	1	0.0
TOTAL	12	19	63.2	9	18	50.0	25	47	53.2

tests for LEH homogneity by age class

Chi-sq.	p
9.176	**0.010**
Likelihood ratio	p
9.760	**0.008**

Table 5.2c. Mean Number of Maxillary Canine LEH[1] Events Per Individual

	obs	n[2]	mean
Da But Culture	56	25	2.24
Metal Period	45	25	1.80

t	df	p
1.341	48	0.186

[1] severe and moderate (grades 1 and 2) LEH only
[2] individuals displaying LEH

The LEH prevalence by maxillary canines exhibits a pattern which is the reverse of that seen in the mandibular canines. The Da But period sample displays the greater frequency of maxillary canine LEH defects, albeit not significantly so (Chi-sq. 2.248, p=0.134). Further, the Da But period sample also displays a higher mean LEH rate per individual (2.24) than the Metal period sample (1.80), although this is not statistically significant (Table 5.2c).

Maxillary and Mandibular Canines Combined

The unit of analysis in this instance is defined as an individual with an assessable maxillary canine and/or an assessable mandibular canine.

Da But Period Sample

Of the 53 individuals with assessable maxillary and/or mandibular canines, 71.7% displayed LEH (Table 5.3a). A greater proportion of females display LEH (81.3%) than males (63.0%), albeit not significantly so. While the oldest age class has the lowest proportion of individuals with LEH, there is no significant heterogeneity by age class (Table 5.3a).

Metal Period Sample

Of the 55 individuals with assessable maxillary and/or mandibular canines, 67.3% displayed LEH (Table 5.3b). A greater proportion of males (80.0%) had LEH than females (65.2%), albeit not to a statistically significant degree. Again, while there appears to be a lower representation of LEH in the oldest age class, the null hypothesis of age class homogeneity for LEH prevalence is not rejected (Table 5.3b).

The higher prevalence of LEH in the Da But period sample (71.7%) is not significantly greater (Chi-sq. 0.249, p=0.618) than that for the Metal period sample (67.3%). Further, the mean number of LEH events per individual in the Metal period sample (2.86) is not significantly larger than the mean for the Da But period sample (2.37) (Table 5.3c).

To further explore any potential relationship between LEH and age-at-death, the age distribution of individuals with LEH is compared to those without LEH for both samples (Figures 5.1a-5.1c and Figures 5.2a-5.2c). Figure 5.1a displays the age distribution of individuals in the Da But period sample with assessable maxillary or mandibular canines. Examining Figure 5.1b, it is apparent that the age distribution of individuals with LEH closely resembles the age distribution of the entire sample. On the other hand, the age distribution of individuals without LEH is clearly skewed to older age in the Da But period sample (Figure 5.1c).

Figure 5.2a displays the age distribution of individuals in the Metal Period sample with assessable maxillary and mandibular canines. The age distribution of individuals with LEH (Figure 5.2b) closely resembles the age distribution of the whole sample. However, the age distribution of those individuals without LEH is more complex (Figure 5.2c). Over 40% of this sub-sample is aged 20-29 years while just under 40% are aged 40+ years.

Discussion and Conclusions

Ideally the use of all anterior tooth classes is preferable in an examination of the prevalence of LEH. The poor preservation of the incisors in these skeletal series from Vietnam has mitigated against their use in this study. Nonetheless, the long developmental life of the canine (Skinner & Goodman 1992; Malville 1997) and the demonstrated sensitivity of this tooth type to stress (Goodman & Armelagos 1985a) suggest little, if any, information will be lost by restricting the analysis to the canine. However, Condon & Rose's (1992) stricture that the use of a single tooth type limits the study to the range of that particular tooth's sensitivity to stress must be heeded. As previously noted, the window of stress sensitivity provided by the maxillary and mandibular canine lies approximately between 1.5 and 4.5 years of age. However, this window is likely even further attenuated by the observation that most stress events occur in the middle third of all tooth types (Condon & Rose 1992), thus limiting the window to about a year in length (Malville 1997) around the middle of the 1.4 to 4.5 year range.

Although the use of one tooth type, the canine, is appropriate, the use of either the maxillary or mandibular variant alone seems to be contra-indicated. The prevalence of LEH by individual for each sample reversed depending on the type of canine used. This would indicate that neither canine is more sensitive to stress than the other, but rather that there is a range of variability in the sensitivity of the maxillary and mandibular canine. It should be noted also that the apparent differences in inter-sample LEH prevalence were not statistically significant for either canine type. Nonetheless, the Metal period did display a significantly higher mean LEH count per tooth for the mandibular canine at least. Given that this was the only significant inter-sample difference the meaning of this is equivocal. With regard to the other observed differences, albeit not significant, in prevalence between the jaws, this can simply be put down to the vagaries of sampling. By combining the maxillary and mandibular canines most of the inter-sample differences, by overall prevalence, as well as mean number of LEH events per individual, evened out. There were no statistically significant intra-sample sex differences either.

Apart from the observation that Metal period maxillary canine LEH was not distributed evenly by age category to a significant degree, no other marked differences were noted. However, the apparent, albeit non-significant, lower frequency of LEH in the older age class of both samples is intriguing. This is especially so for the Metal sample, although the null hypothesis of homogeneity of LEH prevalence by age class was only rejected for the sample of mandibular canines. A smaller proportion of all aged individuals displaying LEH might suggest increased chances of survival to old age for those who did not suffer the stresses causing LEH. It is unlikely that the lower prevalence of LEH in older individuals is an artefact of LEH events being lost through wear, because extremely worn teeth were not included in the study sample. Lacking data on infant mortality, the so-called paradox of greater levels of EH indicating either increased or decreased population health is not testable here. Notwithstanding this however, when only examining those who survived the LEH sensitive years, it would appear that there is a correlation between survival to older age and absence of LEH in the Metal period sample. It is also worth pointing out that the Metal period sample has a somewhat attenuated age distribution overall (Figure 5.2a). The younger age of the population is thus likely masking any mortality benefits that lower stress levels, as measured by absence of LEH, might have given.

The same pattern, albeit to a more pronounced degree, is apparent in the Da But period sample. There is a clear skewing of the age distribution of the Da But period individuals without LEH to older age (Figure 5.1c).

Table 5.3a. Da But Period. Maxillary and Mandibular Canine LEH[1] by Age

Five Age Classes

	male			female			total		
	obs	n	%	obs	n	%	obs	n	%
<15	0	0	na	0	0	na	0	1	0.0
15-19	1	1	100.0	1	1	100.0	6	7	85.7
20-29	3	4	75.0	3	3	100.0	9	10	90.0
30-39	6	9	66.7	3	3	100.0	9	12	75.0
40-49	2	4	50.0	3	6	50.0	6	11	54.5
50+	5	9	55.6	3	3	100.0	8	12	66.7
TOTAL	17	27	63.0	13	16	81.3	38	53	71.7

Chi-sq. tests male:female

Chi-sq.	p
1.593	0.207

Three Age Classes

	male			female			total		
	obs	n	%	obs	n	%	obs	n	%
<30	4	5	80.0	4	4	100.0	15	18	83.3
30-39	6	9	66.7	3	3	100.0	9	12	75.0
40+	7	13	53.8	6	9	66.7	14	23	60.9
TOTAL	17	27	63.0	13	16	81.3	38	53	71.7

tests for EH homogneity by age class

Chi-sq.	p
2.594	0.273
Likelihood ratio	p
2.648	0.266

Table 5.3b. Metal Period. Maxillary and Mandibular Canine LEH[1] by Age

Five Age Classes

	male			female			total		
	obs	n	%	obs	n	%	obs	n	%
<15	0	0	na	0	0	na	1	2	50.0
15-19	1	1	100.0	1	1	100.0	5	5	100.0
20-29	5	6	83.3	5	9	55.6	12	19	63.2
30-39	5	5	100.0	8	10	80.0	13	15	86.7
40-49	2	3	66.7	1	2	50.0	3	6	50.0
50+	3	5	60.0	0	1	0.0	3	6	50.0
Indet.	0	0	na	0	0	na	0	2	0.0
TOTAL	16	20	80.0	15	23	65.2	37	55	67.3

Chi-sq. tests male:female

Chi-sq.	p
1.162	0.281

Three Age Classes

	male			female			total		
	obs	n	%	obs	n	%	obs	n	%
<30	6	7	85.7	6	10	60.0	18	26	69.2
30-39	5	5	100.0	8	10	80.0	13	15	86.7
40+	5	8	62.5	1	3	33.3	6	12	50.0
Indet.	0	0	na	0	0	na	0	2	0.0
TOTAL	16	20	80.0	15	23	65.2	37	55	67.3

tests for EH homogneity by age class

Chi-sq.	p
4.261	0.119
Likelihood ratio	p
4.408	0.110

Table 5.3c. Mean Number of Maxillary and Mandibular Canine LEH Events Per Individual [1]

	obs	n[2]	mean	t	df	p
Da But Culture	90	38	2.37	1.380	73	0.172
Metal Period	106	37	2.86			

[1] severe and moderate (grades 1 and 2) LEH only
[2] individuals displaying LEH

Moreover, in comparison to the overall Da But period age distribution (Figure 5.1a), the age distribution of individuals with LEH is more normally distributed. These patterns clearly indicate a bias in survival to old age for those without LEH in the Da But period sample.

A number of studies have found a relationship between the presence of LEH and decreased mean age at death (Goodman & Armelagos 1988; Duray 1996; Saunders & Keenleyside 1999). A North Amerindian sample of some 111 individuals from the Dickson Mounds site revealed significant differences in mean age-at-death for two of the three time periods encompassed in the study. The chronologically latest sequence exhibited the largest differences, with the mean age-at-death of individuals without defects being more than seven years older than those with at least one defect and nearly 16 years older than those with two or more defects (Goodman & Armelagos 1988).

Another North American study, examining 143 individuals from the North American Libben site in Ohio, found

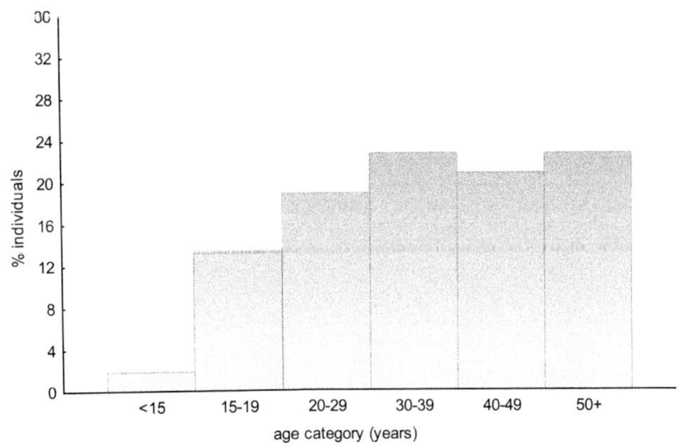

Figure 5.1a. Da But Period: Age Distribution of Individuals with Assessable Canines (n=53)

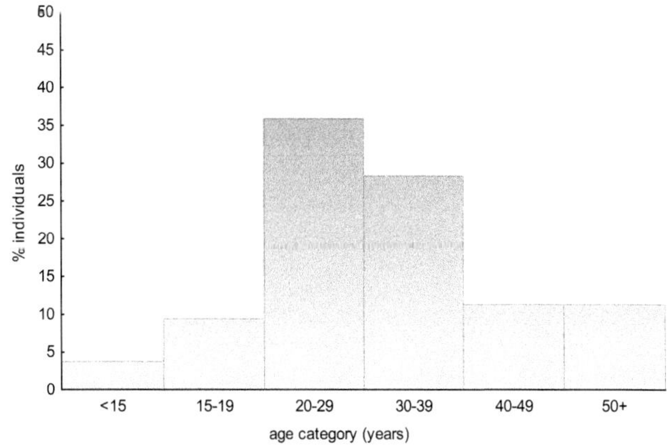

Figure 5.2a. Metal Period: Age Distribution of Individuals with Assessable Canines (n=53)

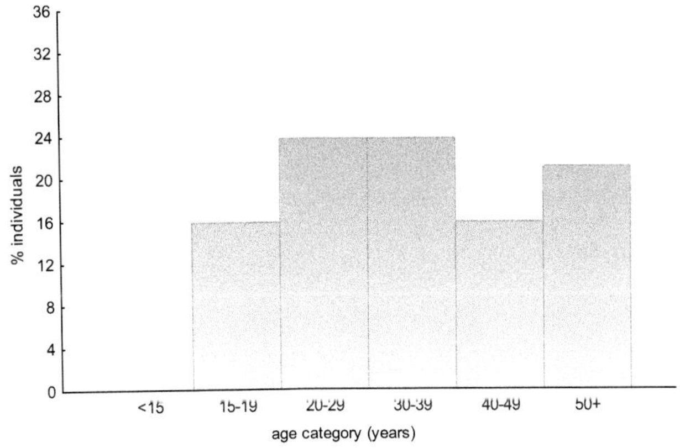

Figure 5.1b. Da But Period Age Distribution of Individuals Displaying Canine Linear Enamel Hypoplasia (n=38)

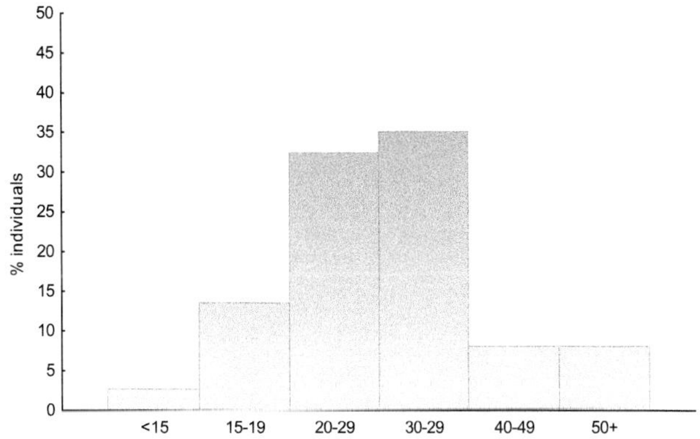

Figure 5.2b. Metal Period: Age Distribution of Individuals Displaying Canine Linear Enamel Hypoplasia (n=37)

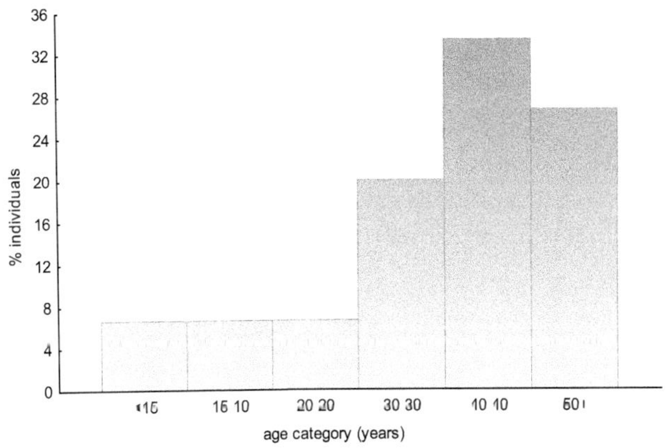

Figure 5.1c. Da But Period: Age Distribution of Individuals Without Canine Linear Enamel Hypoplasia (n=15)

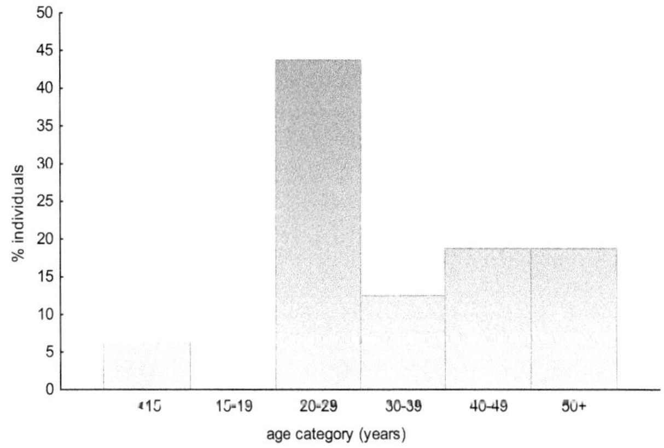

Figure 5.2c. Metal Period: Age Distribution of Individuals Without Canine Linear Enamel Hypoplasia (n=16)

that the mean age-at-death of those without defects was significantly greater, at 5.37 years, than those with some sort of enamel defect (Duray 1996). In a sample of nearly 200 age estimated and age documented individuals from a 19th century Canadian Anglican cemetery, the mean age-at-death was greater for the sub-samples without hypoplastic defects, although the differences were not statistically significant (Saunders & Keenleyside 1999).

Goodman & Armelagos (1988) propose three non-mutually exclusive hypotheses to explain the relationship between increased stress, prevalence of LEH, and decreased life expectancy, or specifically decreased mean age-at-death. The first hypothesis can be termed the 'weaker constitution model' where a likely genetic susceptibility to stress results in childhood stress, the development of LEH, and subsequent adulthood stress and earlier mortality. The second hypothesis can be called the 'wear and tear model' where stress suffered as a child has weakened the individual's ability to deal with stress as an adult thus leading to earlier mortality. The third hypothesis can be termed the 'socially disadvantaged model' where culturally determined factors such as status and differential exposure to nutrition, shelter and so forth will carry on from childhood through adulthood thus leading to earlier mortality. Duray (1996) suggested a fourth hypothesis in which the skeletal assemblage is a sampling of two or more generations over a period of differential resource scarcity and/or presumably varying levels of disease.

Only the latter two models can really be tested in the archaeological record. The model of differential social advantage, or status, can be approached by way of determining status through grave goods for example. The fourth model of inter-generational differences needs fine temporal control of the sample. Neither of these models can be tested using the available Vietnamese data. Moreover, the situation may not be that straightforward even with excellent data quality. Zhou & Corruccini (1998:724) make the important comment that:

> there appears to be a considerable baseline variation both within and between populations in [the] amount of seemingly random hypoplasia formation that may not be confidently attributable to metabolic stress. In other words, the information content of hypoplasias as pathologies is mixed with variable amounts of 'background noise' in given populations. A key would be to determine baseline rates of hypoplasia formation in a given tooth or population, and then to search for evidence of excess rates of hypocalcification in response to specific stressors.

One clear stricture to be drawn is that inter-population comparisons of the prevalence of hypoplasiae are fraught with further complicating unknowns, over and above the usual problems of inter-observer error and differing recording protocols. Nonetheless, it is still worth examining the prevalence of EH displayed in these Vietnamese samples to other samples globally.

The overall prevalence of LEH in the Vietnamese samples is not unusually high relative to some bio-archaeological studies. For instance, very high hypoplasia frequencies have been reported for maxillary canines (98.4%) and mandibular canines (98.2%) in the mid-Holocene Ra's al-Hamra-5 sample from Oman (Mack & Coppa 1992). However, the Vietnamese canine LEH prevalence does appear somewhat elevated with respect to other East and Southeast Asian studies.

East and Southeast Asian analyses of LEH show a great deal of variation in reported EH frequencies. For example, Yamamoto (1992) has examined the prevalence of permanent mandibular canine EH in Japanese from the Jomon period, beginning some 10000 years bp, through to the modern period. The foraging Jomon sample displayed an EH prevalence of 48.1%, the agricultural and foraging Kofun an EH prevalence of 36.4%, the early modern urban Edo an EH prevalence of 65.1%, while the modern sample exhibited an EH prevalence of 39.5%. Yamamoto (1992) cited variations in subsistence, sanitation and population density as the chief aetiological factors involved in the variation in EH prevalence over time.

Tayles (1992) reports 50% of maxillary and 48% of mandibular permanent canines displayed LEH in the 4000 to 3500 years bp coastal site Khok Phanom Di in Thailand. This finding is somewhat intriguing in face of comments by Higham (1989) that the Khok Phanom Di habitat was particularly favourable to its inhabitants. Nelson (1999) reports that 30% of individuals with maxillary canines displayed EH, while 45.9% of individuals with mandibular canines had EH in the 2300 to 1700 years bp site Noen U-Loke in Thailand.

Prevalence of LEH has also been provided by Michele Douglas (pers. comm.) for two sites in Thailand: Non Nok Tha (5500 to 3000 years bp) and Ban Chiang (5600 to 1700 years bp). Frequencies by adult individuals for the maxillary canine, mandibular canine and both canines combined are given. The prevalence of LEH, by individual, is 10.3% for the maxillary canine and 8.0% for the mandibular canine in the Non Nok Tha dental series. For the Ban Chiang series the prevalence of LEH by individual is 15.8% for the maxillary canine and 26.8% for the mandibular canine. Combining maxillary and mandibular canines gives a prevalence of LEH by individual of 11.9% at Non Nok Tha and 29.1% at Ban Chiang.

The chief subsistence economies of the inhabitants of the Thai sites, in particular, seem to display a general pattern with regard to EH prevalence. Khok Phanom Di and Noen U-Loke can both be described as having predominantly agricultural subsistence economies (see Table 3.34, Chapter 3), while the Ban Chiang and Non Nok Tha sites are perhaps best described as mixed (foraging with agricultural and/or horticultural components). The prevalence of EH is somewhat more elevated in the agriculturally oriented sites relative to those with a mixed economy. However, the Japanese Jomon foragers displayed a higher frequency of EH relative to the temporally later agricultural period. Nonetheless, the urban Edo sample displayed the highest level of EH and the individuals comprising this sample would have experienced elevated levels of population density and lowered levels of public sanitation.

Accepting that high levels of canine LEH reflect an agricultural subsistence economy is inconsistent with the archaeological evidence from the Da But period sites at least. However, both populations were likely quite sedentary and factors associated with sedentism, and perhaps population density, may be important in the level of LEH present in these Vietnamese samples. Further, the coastal/estuarine orientation of these communities may have exposed them to high parasitic loads (see Walker 1986).

With respect to a possible positive correlation between sedentism and population density are Webb's (1995) results for prehistoric Australia. Webb (1995) reports permanent maxillary canine LEH frequencies for several regions of the prehistoric Australian continent. LEH frequency was reported to range from a low of 20.7% in the arid, inland Rufus River region (southern Australia) to a high of 45.6% in the fertile southwest corner of the continent (Webb 1995). There was a clear tendency for the frequency of LEH to be low (20.7% to 28.8%) in the arid regions of Australia and higher (37.6% to 45.6%) in the more fertile and more densely populated regions. Webb (1995) favoured differences in weaning customs, timing, between arid and fertile regions to account for the observed differences in LEH frequency. However, it is probably more likely that the clear and marked environmental, climatic and settlement pattern differences, in particular population density, between these regions were more important underlying causes in LEH frequency differences.

Wood *et al.* (1992:356) demonstrate that, by using the prevalence of enamel hypoplasia as a measure of health status, it can be shown that more hypoplasia indicates greater health, or as they state "Better health makes for worse skeletons". The reasoning is that a greater proportion of subadult survivors of stress episodes will exhibit enamel defects as adults due to the differential loss of non-survivors (those that did not have the chance to develop defects) from the sample. In this sense, the possession of an EH defect is a badge of successful passage or a sign of adaptation. This is fine if that is how population health is defined. However, if the health of a sample is defined as the degree of physiological stress it has endured, then high levels of population LEH, or adaptation using Wood *et al.'s* (1992) term, can only be interpreted as indicating poor health.

In this context it is well worth again examining Zhou & Corruccini's (1998) retrospective study of the relationship between EH and famine stress in Chinese.

> [The] death rate rose 135% during the famine years and young children under 6.5 years suffered very high mortality, a significant proportion of highly stressed individuals probably were removed from the possibility of study.
> (Zhou & Corruccini 1998:728)

Notwithstanding this massive infant mortality rate, the survivors of the famine stress, that is those who formed the study cohort for that period, displayed a much higher prevalence of LEH than the pre-famine and post-famine samples. These findings can be interpreted in two ways. The first is that the high frequency of LEH in the famine cohort is a measure of the robusticity of this cohort to stress. Employing Wood *et al.'s* (1992) reasoning suggests that this famine sample is healthy in that it displays a high level of adaptation, as indicated by a high prevalence of LEH, to stress. However, a perhaps more compelling interpretation relates to the observation that this famine cohort comprises the survivors of extreme famine-induced stress. The frequency of LEH in the famine cohort is a direct reflection of the level of stress individuals experienced during the famine years. In other words, the high prevalence of LEH in this cohort indicates the relative poor health of these individuals compared to their less stressed pre and post-famine compatriots.

It would appear from Zhou & Corruccini's (1998) study that high levels of physiological stress will also result in a high prevalence of EH. This appears to be true even in the face of extreme stress, such as famine, and concomitant high levels of mortality. This tends to support the claims of bio-archaeological studies that suggest higher frequencies of population LEH indicate higher levels of population metabolic stress and by inference decreased, rather than increased, levels of general well being or health.

What can be said of the Da But period and Metal period samples is that the overall level of well being of both populations appears to be similar with respect to LEH prevalence at least. How elevated their respective LEH levels are relative to the base-line level for populations from which they are sampled cannot be determined and is ultimately unknowable. Moreover, the specific aetiological factors responsible for these LEH levels must remain unknown. However, it is important to remember that there is some evidence indicating that those individuals without LEH tended to live longer than those individuals with LEH. This would tend to suggest that the causes of LEH had a deleterious affect on the health of these ancient Vietnamese.

Finally, comparing the frequency of LEH in the Da But and Metal period samples to other East and Southeast Asian samples indicates that the Vietnamese prevalence of LEH is somewhat elevated. If the prevalence of EH does correlate with the level of physiological stress a population endured, it would appear that the Vietnamese samples endured somewhat higher levels of stress than did other populations thus far sampled in the tropical Southeast Asian context.

Porotic Hyperostosis

In addition to enamel hypoplasia, porotic hyperostosis is a commonly documented marker of physiological stress in the bio-archaeological record (e.g. Webb 1982; Walker 1986; Palkovich 1987; Stuart-Macadam 1991; Lovell 1997a). The first task of this study is to examine the nature, causes and role of porotic hyperostosis as an indicator of metabolic disturbance. Secondly, the prevalence of porotic hyperostosis, cribra orbitalia in particular, is documented for the Vietnamese skeletal assemblages. The implications of the frequency and distribution of individuals with this condition by sex and age is discussed.

Literature Review

Descriptions of Porotic Hyperostosis

The classic form of porotic hyperostosis, manifesting in the cranial vault, can be seen in Figure 5.3. Ortner & Putschar (1981:258), in describing porotic hyperostosis note:

> The lesions are found on the skull vault, primarily the outer surface of the parietal bones, and on the orbital roof. They consist of porous periosteal bone deposition in these areas of the skull.

Most recent descriptions of these lesions refer to the orbital manifestation as cribra orbitalia, and the vault lesions as porotic hyperostosis, although the latter term is still often used to refer to both conditions. It is also important to note that other definitions of porotic hyperostosis do not necessarily refer specifically to the presence of porous periosteal bone deposition. Manifestations of porotic hyperostosis have been described as ranging from the presence of light scattered fine foramina, to large isolated and confluent foramina forming a trabecular structure, to the most severe form which displays the porous periosteal condition mentioned above (Stuart-Macadam 1985). However, the so-called porous periosteal bone deposition extending past the outer table is perhaps more aptly described as: the result of a continuation of the process of diploetic expansion, subsequent to destruction of the outer table.

A Condition of Childhood

Webb (1982) surveyed the incidence of cribra orbitalia in adults and children from a large number of bio-archaeological studies and noted the universal trend of much higher frequencies in children compared to adults. Webb (1982) suggested that the lower incidence in adults was due to an increased ability to fight under-nutrition, disease and parasitic infection in adults relative to children. However, it had been suggested before this that porotic hyperostosis was a childhood condition facilitated by the thinner and less mineralised bones of children (El-Najjar *et al.* 1975). These writers also suggested that adult lesions represented unhealed lesions that developed in childhood.

Stuart-Macadam (1985) has presented a strong case that porotic hyperostosis is a condition that only develops in childhood and at a period most likely confined to early childhood. Stuart-Macadam (1985) cited physiological evidence that hemopoietic marrow (red cell producing), that fills all marrow cavities in the infant including the orbital and vault spaces, is gradually replaced by non-hemopoietic marrow as the child matures. This change is quite noticeable at about four years of age (Stuart-Macadam 1985). Moreover, the bone of children is more plastic and sensitive to the sorts of pressure changes associated with diploetic expansion of orbital and vault bone. It is erythroid hyperplasia of the hemopoietic marrow that leads to the skeletal changes recognised as porotic hyperostosis. In anaemic adults, physiological responses to the condition are performed without recourse to utilisation of all available marrow space, thus sparing the orbital and vault regions.

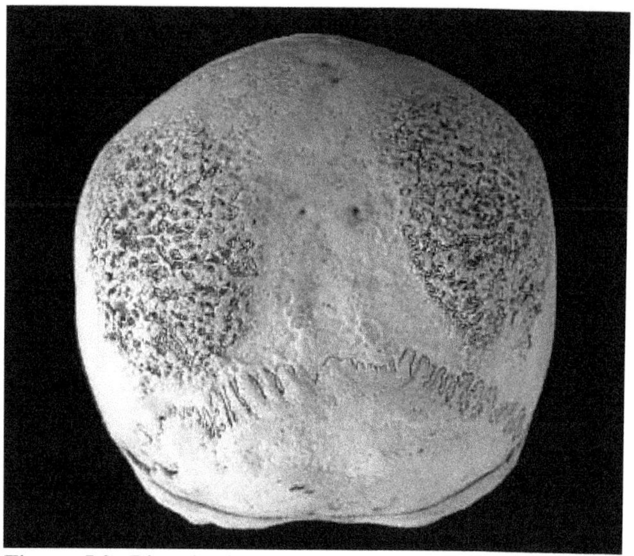

Figure 5.3. Bi-parietal porotic hyperostosis in Nicolas Ferry (Bébé), housed in the Musée de l'Homme (Paris)

Stuart-Macadam (1985) also refers to clinical data that suggests skull responses are limited to childhood. Moreover, Stuart-Macadam (1985) cites a range of anthropological studies demonstrating that porotic hyperostosis was far more common in childhood than adulthood. However, this evidence is not as convincing as the clinical and physiological data, as it is open to other interpretations such as suggested by Webb (1982) above. Nonetheless, there is a reasonable consensus now that porotic hyperostosis is essentially a childhood condition and lesions seen in adults represent the inactive and variously remodelled expression of this. It does not appear possible, at present, to assess the prevalence and severity of anaemia in adults. The situation is analogous to the use of Harris lines, which are formed during subadult growth and development and only exist in varying states of resorption and remodelling in the adult.

Aetiology of Porotic Hyperostosis

Schultz (Schultz & Larsen 1997; Carli-Thiele & Schultz 1997) argues that a number of vitamin deficiency diseases and inflammatory processes can potentially contribute to observed porotic hyperostosis in skeletal samples. However, there has been a general assumption in the bio-archaeological literature that porotic hyperostosis is caused by either acquired or hereditary anaemia (Ortner & Putschar 1981). Stuart-Macadam (1987a, 1987b) specifically tested this view citing clinical and bio-archaeological literature and her own radiographic study in support. While these two studies clearly suggested anaemia as the chief aetiological agent in the development of porotic hyperostosis in general, the specific association with vault lesions still remains unclear. Much of the problem in Stuart-Macadam's (1987a) radiographic study was that very few individuals only displayed vault lesions and so the relationship between these particular lesions and anaemia could not be adequately assessed. Moreover, the link between diploic thickening and anaemia was confounded in one of the skeletal samples (Poundbury) that did not display any significant radiographic differences between individuals with and without lesions. Unfortunately, the contribution of cases to the study by category of porotic severity was not revealed. It would have been useful to examine the relationship between radiographic signs of anaemia in those individuals that only displayed fine and scattered ectocranial porosity of the vault. As it is, Stuart-Macadam (1989) noted that only severe forms of vault lesions were always associated with orbital lesions. Severe in this case was defined as lesions displaying confluent foramina exhibiting a trabecular pattern and/or trabecular out-growths extending past the outer table.

The meaning and significance of ectocranial porosity in the absence of these more severe vault changes is not clear. Mann & Murphy (1990) suggest that porotic hyperostosis can only be scored as present when there is both porosity and thickened bone. They also note that ectocranial porosity, especially concentrating on the parietals, frontal and occipital, is very common in many skulls and on its own cannot be used as a marker of porotic hyperostosis. For the orbits, at least, the presence of varying degrees of porosity is used exclusively to define the presence of porotic hyperostosis (Stuart-Macadam 1985), without direct reference to diploetic thickening, as this is assumed to have led to the condition.

The problem of identifying porotic hyperostosis in skeletal material is all the more acute when vault ectocranial porosity is used in isolation to score the presence of anaemia. A study of an assemblage of 5000 to 4000 years bp Harrapan individuals, from the Indus valley, described a case of anaemia based on the observation of scattered ectocranial porosity on an unidentified cranial fragment (Lovell 1997a). Diploetic thickening could not be confirmed for the fragment. It would seem premature to use vault ectocranial porosity as a criterion for inferring anaemia in skeletal material before more work is done on determining the relationship, if any, between this somewhat ubiquitous vault condition and anaemia.

As suggested, the chief causal agent implicated in porotic hyperostosis is anaemia. Mosely (1974:169) notes that:

> Most of the bone changes that are due to anemia are manifestations of compensatory erythroid hyperplasia of the marrow…In the skull, hyperplasia of the marrow widens the diploic space and thins the outer table. If the erythroid hyperplasia is severe enough, the hyperplastic marrow may perforate the outer table [ectocranial lesions of varying severity] and proliferate under the periosteum [the porous periosteal deposition seen in severe cases], producing a radial spiculation that has come to be known as the 'hair-on-end' pattern.

There does not appear to be any correlation between the severity, or even presence, of skeletal changes and the severity of the underlying anaemia (Moseley 1974; Palkovich 1987). Furthermore, there is a large degree of variability in bone remodelling subsequent to recovery from anaemia (Palkovich 1987).

Ryan (1997:25) notes that iron deficiency anaemia is the chief cause of anaemia and it is the most common nutritional deficiency in the world, with some 50% of children aged below five affected in developing countries. However, Stuart-Macadam (1992:40) has argued that diet only plays a minor role, if any, in the development of iron deficiency anaemia. Two critical pieces of evidence in her argument relate specifically to anthropological studies which display high levels of porotic hyperostosis even in the face of evidence for iron adequate diets and secondly, the nature of the chief source of iron in adult hemopoiesis.

With reference to the first issue, Stuart-Macadam (1992) cites Walker's (1986) study of a marine economy-based California population. While this population potentially had access to a very rich source of dietary iron still, half of the sample of individuals aged 18 years or under displayed cribra orbitalia (Walker 1986). Walker (1986) favoured diarrhoeal disease associated with water pollution problems as the chief aetiological factor in the prevalence of cribra orbitalia, although he also suggested marine-borne parasites played a role. However, it is still very possible for dietary deficiencies to occur in nutritionally rich environments. One potentially very important culturally induced dietary deficiency relates to the employment of food prohibitions or avoidances.

Food taboos, prohibitions and avoidances are common in many societies, present and past (Wilson 1980). Importantly, many of these food prohibitions can occur even in the face of a great variety and abundance of food-stuffs. Furthermore, pregnant and postpartum mothers are commonly the target of these food taboos and avoidances. Because the practice of food prohibitions and avoidances seems to have attracted little attention in discussions of dietary deficiencies in past populations, a number of examples, specifically detailing tropical and Asian instances, are detailed here.

In tropical northern Australia, among the fishing and shellfish collecting Gidjingali of coastal Arnhem Land, it is reported that women are prohibited from eating animal flesh, including most species of shellfish, for several weeks after birth (Meehan 1982:137). Yolngu women, also from Arnhem Land, are likewise affected by a number of food restrictions during pregnancy. Moreover, some dietary restrictions may have to be maintained throughout the growing years of their children (Reid 1983).

Ferro-Luzzi (1980a) reports on a considerable variety of food avoidances observed during pregnancy in Tamilnad, South India. Avoidance was particularly important with respect to fruits and grains, especially papaya and sesame seeds. A wide range of foods are also avoided puerperium and sometimes well beyond (Ferro-Luzzi 1980b). It is also not uncommon for previously non-vegetarian women to become vegetarian for extended periods after birth.

In Malaysia, Wilson (1980) reports severe dietary taboos imposed on mothers during the first 40 days after birth, in an east coast fishing community. Prohibited items include culturally categorised cold foods such as most fruits and vegetables and toxic foods such as fish. Wilson (1980) also notes the variable nutritional status of Malay women in general, given the common dietary practice of inconstant consumption of vegetables and their use of fruit as a snack food only. These puerperium food taboos only serve to further compromise the mother's nutritional status during that vital first month of infant suckling.

Food restrictions or prohibitions are also applied in connection to certain illnesses and can unintentionally exacerbate an existing disease. For example, McKay (1980) writes of a small community in rural Ulu Trengganu, West Malaysia, that practices a range of food prohibitions during illness or as a preventative measure. For instance, xerophthalmia (an eye condition resulting from Vitamin A deficiency that can eventually lead to blindness) is treated by prohibiting citrus fruits, vegetables and papaya. The condition is believed by the community to be due to helminth infestations and so the use of cooling or sour foods is avoided (McKay 1980). Unfortunately, helminth infestations are not affected by the treatment. However the treatment, in this instance food taboos, likely contributes to the high prevalence of xerophthalmia in this community.

Food restrictions, not necessarily taboos, also apply in that children may not have received adequately nutritious diets for no other reason than that many foods were not available in a digestible form, given their age, or were not considered appropriate within the particular culture. In a study of dietary taboos in a Yogyakarta, Java, community, Hull (1986), in addition to noting food taboos for pregnant and postpartum women, also recorded how children's diets were influenced by what foods were deemed appropriate at certain ages. Even by two years of age, and not uncommonly older, children were not receiving an adult diet (Hull 1986:244). Hull (1986) goes on to state that by the first year only 26% were receiving green vegetables, less than 20% were receiving fruits other than bananas, 56% were eating bananas, and only 9% were consuming meat and/or fish. By 18 months only about half the children were consuming green vegetables and fruit while a third were eating meat and/or fish. Even by 2 ½ years some 13% were still not eating green vegetables, 31% were not eating fruit and nearly 20% were not consuming meat and/or fish.

Nutritionally compromised mothers can have a deleterious impact on their suckling children. A direct correlation between anaemic mothers and their children has been demonstrated (see Ryan 1997). A range of food types are targeted by such taboos and over varying lengths of time preceding and following birth. Food restrictions, not necessarily taboos, also apply to children that can considerably delay their access to the full range of foods available to adults. Furthermore, local perceptions of disease and treatment can lead to dietary restrictions that may be not only ineffective but can unwittingly lead to the development of other dietary disorders. The manifestation and importance of these various factors vary in the present and doubtless did so in the past. The potentially immensely important role of such factors is rarely appreciated in bio-archaeological studies.

The second important piece in Stuart-Macadam's (1992) argument, minimising the role of dietary influences in iron-deficiency anaemia, relates to the observation that over 90% of the iron required for red blood cell production comes from the turnover of senescent red blood cells. However, Ryan (1997:29) points out that the picture is very different for infants:

> due to the rapid expansion of the levels of hemoglobin and myoglobin during the first year of life, infants depend on dietary iron to a greater extent than adults, requiring about 30% of iron from their diet.

Low birth weight children require even more iron due to their faster rates of post-natal growth.

As mentioned, there now also appears to be a relationship between iron deficient mothers and increased risk of iron deficiency in infants (Ryan 1997). Previously it was assumed that infants could store sufficient iron despite the anaemic or otherwise status of their mothers. Ryan (1997) suggests much of the evidence for this assumption comes from studies focussing on women who were either not iron deficient or had iron supplements in late pregnancy. Studies that have looked at anaemic mothers have also found their infants to be iron deficient. Further, in a study looking at infants of anaemic mothers up to a year after birth it was found that such children were significantly more likely to become anaemic in their first 12 months. This occurred even after controlling for socioeconomic variables, feeding practices, and morbidity (Colomer *et al.* 1990).

Given the pivotal role of dietary sources of iron in an infant's growth and development, it is also important to realise that only 50% of this dietary iron is available from breast milk. Up until about four to six months infants can draw on their own supplies of stored iron. After six months they require additional sources of dietary iron. This pattern explains why porotic hyperostosis is rarely seen in individuals less than six months old. Anaemia after this time, as assessed through the presence of porotic hyperostosis in skeletal assemblages, cannot be assigned a specific cause. Insufficient levels of dietary iron in foods supplementing breast milk, or the delayed effects of the mother's own iron deficiency, will be indistinguishable. The development of acquired parasitic infections with the introduction of weaning foods will also be subsumed into a complex milieu of aetiological factors. The presence of mild forms of hereditary anaemia, in geographic areas where these are common, will also no doubt help complicate the picture.

Weinberg (1984; 1992; 1999) has been advocating the importance of the iron-withholding hypothesis. Iron-withholding refers to a host's denial of iron to infectious agents and neoplastic cells. Pathogens, such as protozoa and fungi, require the host's iron reserves for their own metabolism and reproduction. Weinberg (1999) has suggested that much evidence points to hosts having developed a complex response, the iron-withholding defence system, to such microbial attacks. Weinberg (Kent *et al.* 1994) suggests that anaemia associated with infection and chronic disease should be seen as an immunological defence against these pathogenic agents rather than a disorder or deleterious response to disease. In the context of this model, Stuart-Macadam (1992:44) states:

> The acceptance of the two concepts, that diet is of little importance in the development of iron deficiency anemia, and that iron deficiency is an adaptive response to stress [high pathogen load], has a profound effect on the interpretation of porotic hyperostosis.

This assertion needs to be treated with caution. In the first instance, it has been pointed out that diet plays a very significant role in maintaining infant iron levels. Further, iron-deficient mothers likely have a significant effect on the iron status of their offspring even up to at least a year after birth. As Stuart-Macadam (1985) has convincingly argued, porotic hyperostosis appears to develop exclusively in young children, not in the adult for whom diet plays a less important role in maintaining adequate iron levels. There are also problems with the iron-withholding hypothesis.

Tests of the iron-withholding hypothesis generally centre on assessing the prevalence of certain infections in samples of populations that have been given iron supplements. If the iron-withholding hypothesis is a realistic model, then infection rates should increase where the iron-withholding system has been compromised. Ryan (1997), in a comprehensive review of the literature investigating this model, concluded that the evidence is ambiguous at best. The best data supporting the hypothesis suggested that iron supplementation leads to an increase in the prevalence of malaria infection in areas where it is present. However, even this study was not conclusive due to the greater difficulty of detecting malarial parasites in iron deficient as relative to iron rich individuals. Ryan (1997) noted that most of these studies were carried out in disadvantaged populations without the possibility of controlling for a range of other factors potentially involved in infection susceptibility. Further, a number of studies support the role of iron supplementation in increasing immunity (Ryan 1997).

It is probably not a particularly useful exercise to attempt to link iron-deficiency anaemia in skeletal populations to any one aetiological agent in a global generalisation of the condition. Because active porotic hyperostosis, the only consistent evidence of iron-deficiency anaemia in skeletal material to date, appears to be restricted to young children, the factors that affect such a segment of the population are the most important to focus on. While dietary iron intake is of minor short-term significance in adults, it plays a significant role in the iron status of young children. The iron status of an infant's mother also appears to play a crucial role in the susceptibility of an infant to developing iron-deficiency anaemia. The influence of parasites likely played a material role in some communities in the past, as they do in the present. Northern Vietnam, for example, is not particularly unique in being host to a range of various protozoa (malaria is an important example), helminths (flukes and worms) and even arthropods (human ticks, fleas and lice for instance). In areas where malaria is endemic, the development of high fixations of individuals heterozygous for various hereditary anaemic conditions no doubt also played an important role. The potential role of the iron-withholding mechanism also needs to be considered although its impact, if any, on the prevalence of iron deficiency anaemia in past populations, let alone living, cannot be gauged at present. In the absence of other evidence pointing to a specific aetiology, such as skeletal changes consistent with a hereditary anaemia, it is perhaps better to treat porotic hyperostosis as a childhood marker for iron-deficiency anaemia of unknown or uncertain cause.

Methods and Materials

Cribra Orbitalia

All cranial material with at least the preservation of the antero-lateral and antero-medial portions of the orbital roof of at least one orbit was assessed. The justification for using the preservation of this portion of the orbit is based on Stuart-Macadam's (1991) finding that these are by far the most commonly affected portions of the orbit. It would seem unlikely that this minimum criterion will impact to any great degree on the prevalence of cribra orbitalia recorded in these samples.

Cribra orbitalia is defined and scored as either open or closed. Open-cribra orbitalia includes perforations of the outer table of the orbital roof. This category does not distinguish between active-cribra orbitalia and remodelled-cribra orbitalia. Figure 5.4 demonstrates an extreme form of open-cribra orbitalia. Figure 5.5 represents the other end of the open-cribra orbitalia spectrum, with its combination of remodelling and slight porosity. Closed-cribra orbitalia

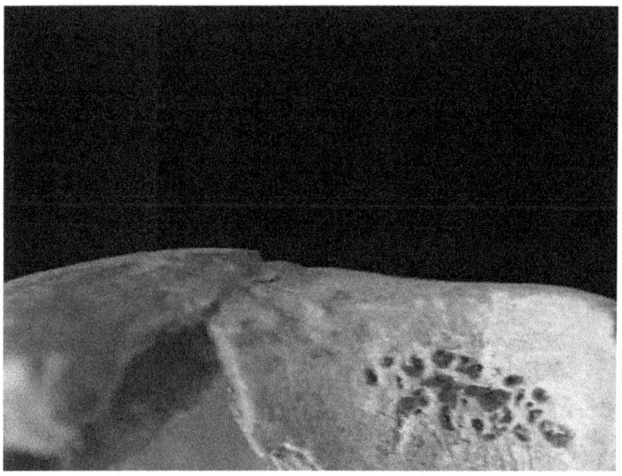

Figure 5.4. 66VQ1M31. Metal Period adult female of indeterminate age. Severe form of open cribra orbitalia (left orbital roof shown)

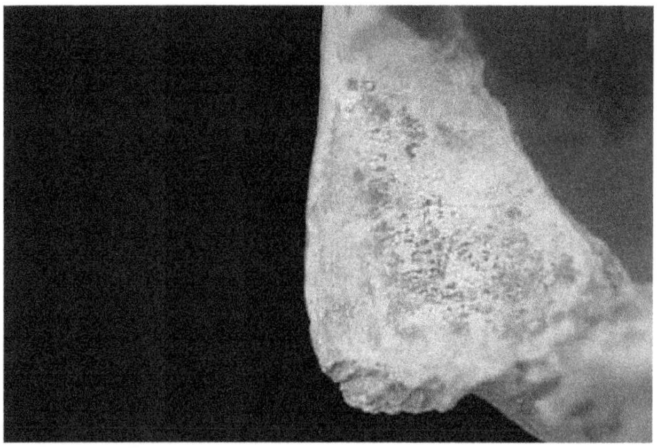

Figure 5.5. 80CCNM5a. Da But Period female, aged 20-29 years. Mild form of open cribra orbitalia with extensively remodelled lesions (right orbital roof shown)

refers to what is believed to be highly remodelled cribra orbitalia, where perforations of the outer table of the orbital roof can no longer be observed. This form is generally easily recognisable by the distinctive worm-like patterning of the outer table of the orbital roof (Figure 5.6). If Stuart-Macadam's (1985) view that cribra orbitalia is a condition that develops exclusively in childhood is correct, then virtually all of the observed cribra orbitalia in this sample was inactive and in various stages of remodelling at death. If such is the case, as appears likely, open and closed-cribra orbitalia here refer to the opposite ends of a remodelling continuum. As such, the categorisation may seem somewhat redundant. However, most researchers only record what is termed here open-cribra orbitalia. If the heavily remodelled expression of this trait is generally ignored, it may explain the large disparity between the prevalence of cribra orbitalia in subadults and adults noted in the previous discussion. It has been pointed out that remodelling can be a slow and drawn out process (Palkovich 1987). The closed-cribra orbitalia category recorded here would seem to represent the terminal stages of remodelling after which the orbital roof will no longer display signs of an event that most probably occurred decades before in some cases.

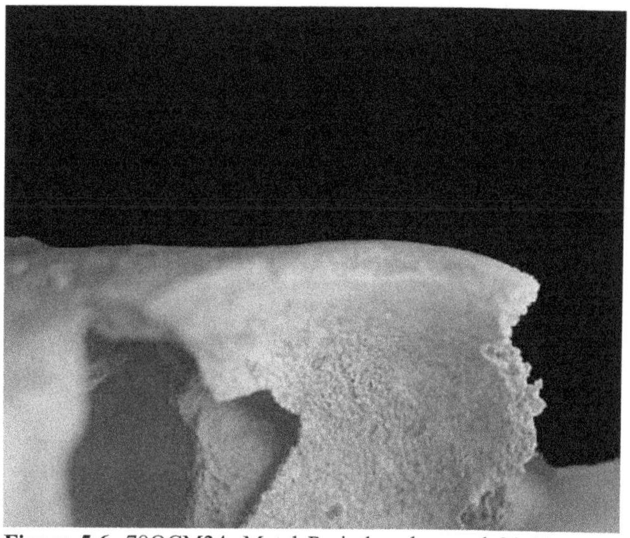

Figure 5.6. 78QCM24. Metal Period male, aged 20-29 years. Extensively remodelled cribra orbitalia displaying distinctive 'worm like' residual scaring (right orbital roof shown)

As was the case for LEH, cribra orbitalia is almost certainly a condition acquired in childhood. Nonetheless, it is necessary to determine if there is a correlation between the age-at-death of adults in these samples displaying evidence of cribra orbitalia acquired in childhood.

Ectocranial Porosity

A preliminary survey of the Vietnamese samples in 1996 revealed that ectocranial porosity was limited to the supero-posterior aspect of the frontal bone, the superior aspect of the parietals, never descending below the temporal line, and the superior aspect of the occipital bone. As such, any individual with cranial remains including these regions was assessed for ectocranial porosity. Ectocranial porosity was defined as the presence of small pits on the outer table of the cranium. Figure 5.7a displays the general distribution and limits of ectocranial porosity, while Figure 5.7b is a close up of ectocranial porosity as defined here.

Results

The distribution of cribra orbitalia was not quantified according to position on the orbital roof. However, most of the lesions, both open and closed, tended to manifest in the antero-lateral and antero-medial portions of the orbital roof. As already suggested, this appears to be the most common site for these lesions (Stuart-Macadam 1991).

Cribra Orbitalia

Da But Period

Examining all manifestations of cribra orbitalia, from large confluent perforations of the outer table to virtually totally remodelled lesions that leave only faint traces in the form of worm track-like scars, 81% of the assessable sample displayed evidence for this condition (Table 5.4a). Only one individual was less than 15 years old, a nine year old child with extensive confluent orbital lesions. Only two other affected individuals were young adults. One,

Table 5.4a. Da But Period. Prevalence of Cribra Orbitalia (Open and Closed)

Five Age Classes

	male			female			total			Chi-sq. tests male:female	
	obs	n	%	obs	n	%	obs	n	%	Chi-sq.	p
<15	0	0	na	0	0	na	1	1	100.0	0.667	0.414
15-19	0	0	na	0	0	na	2	2	100.0		
20-29	2	2	100.0	3	4	75.0	8	9	88.9		
30-39	11	13	84.6	2	4	50.0	13	17	76.5		
40-49	2	3	66.7	6	7	85.7	9	11	81.8		
50+	6	8	75.0	3	4	75.0	9	12	75.0		
Indet.	4	4	100.0	0	0	na	5	6	83.3		
TOTAL	25	30	83.3	14	19	73.7	47	58	81.0		

Three Ages Classes

	male			female			total			tests for cribra homogneity by age class	
	obs	n	%	obs	n	%	obs	n	%	Chi-sq.	p
<30	2	2	100.0	3	4	75.0	11	12	91.7	1.213	0.545
30-39	11	13	84.6	2	4	50.0	13	17	76.5		
40+	8	11	72.7	9	11	81.8	18	23	78.3	Likelihood ratio	p
Indet.	4	4	100.0	0	0	na	5	6	83.3	1.394	0.498
TOTAL	25	30	83.3	14	19	73.7	47	58	81.0		

Table 5.4b. Da But Period. Prevalence of Open Cribra Orbitalia

Three Ages Classes

	male			female			total			tests for cribra homogneity by age class	
	obs	n	%	obs	n	%	obs	n	%	Chi-sq.	p
<30	1	2	50.0	2	4	50.0	7	12	58.3	**7.823**	**0.020**
30-39	3	13	23.1	0	4	0.0	3	17	17.6		
40+	1	11	9.1	2	11	18.2	4	23	17.4	Likelihood ratio	p
Indet.	2	4	50.0	0	0	na	2	6	33.3	**7.181**	**0.028**
TOTAL	7	30	23.3	4	19	21.1	16	58	27.6		

Table 5.4c. Da But Period. Prevalence of Closed Cribra Orbitalia

Three Age Classes

	male			female			total			tests for cribra homogneity by age class	
	obs	n	%	obs	n	%	obs	n	%	Chi-sq.	p
<30	1	2	50.0	1	4	25.0	4	12	33.3	2.658	0.265
30-39	8	13	61.5	2	4	50.0	10	17	58.8		
40+	7	11	63.6	7	11	63.6	14	23	60.9	Likelihood ratio	p
Indet.	2	4	50.0	0	0	na	3	6	50.0	2.679	0.262
TOTAL	18	30	60.0	10	19	52.6	31	58	53.4		

Figure 5.7a. 80CCNM8a. Da But Period male, aged 30-39 years. The usual distribution of EP is confined within the black border

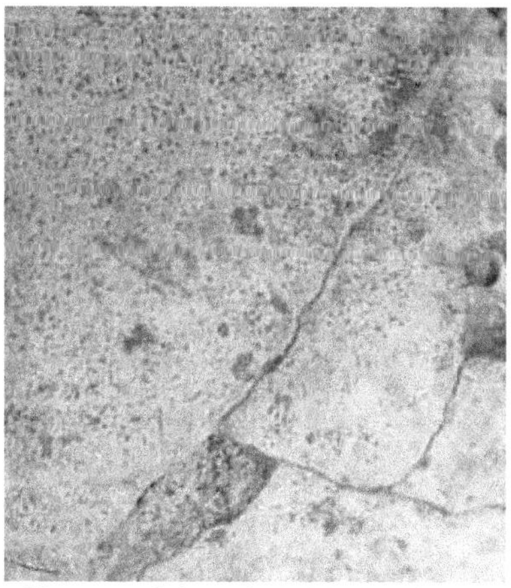

Figure 5.7b. Close up of above individual (left posterior-superior parietal)

Table 5.5a. Metal Period. Prevalence of Cribra Orbitalia (Open and Closed)

Five Age Classes

	male			female			total		
	obs	n	%	obs	n	%	obs	n	%
<15	0	0	na	0	0	na	2	3	66.7
15-19	1	1	100.0	1	2	50.0	5	6	83.3
20-29	6	7	85.7	6	9	66.7	13	17	76.5
30-39	2	5	40.0	7	8	87.5	9	13	69.2
40-49	2	4	50.0	0	2	0.0	2	6	33.3
50+	4	4	100.0	1	1	100.0	5	5	100.0
Indet.	1	1	100.0	3	3	100.0	5	6	83.3
TOTAL	16	22	72.7	18	25	72.0	41	56	73.2

Chi-sq. tests male:female	
Chi-sq.	p
0.000	0.956

Three Age Classes

	male			female			total		
	obs	n	%	obs	n	%	obs	n	%
<30	7	8	87.5	7	11	63.6	20	26	76.9
30-39	2	5	40.0	7	8	87.5	9	13	69.2
40+	6	8	75.0	1	3	33.3	7	11	63.6
Indet.	1	1	100.0	3	3	100.0	5	6	83.3
TOTAL	16	22	72.7	18	25	72.0	41	56	73.2

tests for cribra homogeneity by age class	
Chi-sq.	p
0.744	0.689
Likelihood ratio	p
0.736	0.692

Table 5.5b. Metal Period. Prevalence of Open Cribra Orbitalia

Three Age Classes

	male			female			total		
	obs	n	%	obs	n	%	obs	n	%
<30	4	8	50.0	4	11	36.4	12	26	46.2
30-39	1	5	20.0	0	8	0.0	1	13	7.7
40+	1	8	12.5	0	3	0.0	1	11	9.1
Indet.	0	1	0.0	2	3	66.7	3	6	50.0
TOTAL	6	22	27.3	6	25	24.0	17	56	30.4

tests for cribra homogeneity by age class	
Chi-sq.	p
8.861	**0.012**
Likelihood ratio	p
9.653	**0.008**

Table 5.5c. Metal Period. Prevalence of Closed Cribra Orbitalia

Three Age Classes

	male			female			total		
	obs	n	%	obs	n	%	obs	n	%
<30	3	8	37.5	3	11	27.3	8	26	30.8
30-39	1	5	20.0	7	8	87.5	8	13	61.5
40+	5	8	62.5	1	3	33.3	6	11	54.5
Indet.	1	1	100.0	1	3	33.3	2	6	33.3
TOTAL	10	22	45.5	12	25	48.0	24	56	42.9

tests for cribra homogeneity by age class	
Chi-sq.	p
3.966	0.138
Likelihood ratio	p
4.015	0.134

likely aged 17-19 years, displayed evidence of remodelled cribra orbitalia but without perforations of the orbit while the other, likely aged 13-16 years, displayed remodelling in addition to perforations of the orbital roof.

A slightly greater prevalence of cribra orbitalia is seen in males (83.3%) compared to females (73.7%), but this difference is not statistically significant. The frequency of cribra orbitalia is constant across each of the three age classes. When only those cases of open-cribra orbitalia are examined, 23.3% of the males and 21.1% of females, or a total of 27.6% of the assessable sample are so affected (Table 5.4b). With regard to closed-cribra orbitalia, 30% of the males, 52.6% of the females and 53.4% of the assessable sample are so affected (Table 5.4c). Of the affected sample, 16/47 (34.0%) displayed open-cribra orbitalia.

There is a clear tendency for open-cribra orbitalia to occur in the youngest age class. Moreover, this difference in distribution by age is statistically significant ($p<0.05$). Furthermore, there is also a trend for closed-cribra orbitalia, to be more common in the older age classes. However, the null hypothesis that the distribution is even across all age classes is not rejected.

Metal Period

In the Metal period sample the prevalence of all manifestations of cribra orbitalia is 73.2% (Table 5.5a). Males display a frequency of 72.7% while females are virtually equivalent with 72.0% affected. Included in this sample are nine individuals aged less than 20 years, of which seven display cribra orbitalia. Two of three individuals aged less than 15 years displayed cribra orbitalia. One was an 8-11 year old while the other was

Table 5.6. Prevalence of Anaemia in Modern Northern Vietnam by Age and Geography*

	age (in months)		
	6-11	12-23	24-35
lowland	60.1%	37.7%	7.7%
mountains	65.0%	44.6%	35.1%

Table 5.7. Prevalence of Anaemia in Modern Northern Vietnam by Sex and Locality (children aged 6-35 months)*

	locality			
	lowlands		mountainous	
	urban	rural	urban	rural
male	26.2%	50.0%	43.0%	55.8%
female	24.8%	43.2%	43.7%	55.4%

* Adapted from Nguyen Cong Khanh *et al.* (1995)

aged 4-6 years. Both of these children displayed open lesions, presumably active given their ages. Of the other five individuals aged 15-19 years, four displayed open and one displayed closed-cribra orbitalia.

When only those cases of open-cribra orbitalia are considered, 27.3% of the males and 24.0% of females or a total of 30.4% of the assessable sample was affected (Table 5.5b). With regard to closed-cribra orbitalia, 30% of the males, 52.6% of the females and 53.4% of the assessable sample were affected (Table 5.5c). Of the affected sample, 17/41 (41.5%) displayed open-cribra orbitalia.

As with the Da But period sample, most of the open-cribra orbitalia in the Metal period sample occurs in the youngest age class. This distribution by age is also statistically significant ($p<0.05$). Moreover, closed-cribra orbitalia is more prevalent in the older age classes, although the pattern is not as straightforward as in the Da But period sample. Also, the null hypothesis that the distribution is even across all age classes is not rejected.

Prevalence of Anaemia in Contemporary North Vietnamese Children

A recent study has examined the prevalence and distribution of anaemia in children from rural, urban, mountainous and lowland areas of northern Vietnam (Nguyen Cong Khanh *et al.* 1995). Anaemia was clinically defined as a haemoglobin level less than 11g/dl. Tables 5.6 and 5.7 are adapted from this work (Nguyen Cong Khanh *et al.* 1995). Childhood anaemia in these modern north Vietnamese samples is quite high, especially for infants in the six to 11 month age bracket (Table 5.6). However, the prevalence of anaemia decreases markedly in the second year of life and for lowland children, at least, virtually disappears during the third year of childhood. There are no noticeable differences in the prevalence of anaemia by sex for children aged six months to three years of age (Table 5.7). There does appear to be a greater risk of anaemia in highland communities relative to lowland communities. Moreover, rural populations are more at risk than urban populations. Nguyen Cong Khanh *et al.* (1995) suggested that the very low iron serum levels of anaemic children implicate iron-deficiency anaemia. They also noted a correlation between increased levels of malnutrition and anaemia.

Ectocranial Porosity

Da But Period

As discussed, the aetiological causes of ectocranial porosity (EP) are far from certain. In the Da But period sample it is virtually ubiquitous. Of those individuals with assessable cranial material 67/74 (90.5%) displayed EP. Of those individuals with both assessable orbits and vault material 51/58 (87.9%) displayed EP, while 43/58 (74.1%) displayed both EP and cribra orbitalia.

In individuals with at least a complete calva, the pattern of EP generally followed the distribution seen in Figure 5.7a. EP generally concentrated on the upper frontal near bregma and extended in a strip, usually several centimetres wide, straddling the sagittal suture and often formed two concentrations on the rear of each upper parietal before extending onto the upper occipital. EP was never seen below the temporal line on either the frontal or parietal of any individual.

Metal Period

In the Metal period sample 30/64 (46.9%) assessable crania displayed EP. Of those individuals with both assessable vaults and orbits 30/54 (55.6%) displayed EP, while 22/54 (40.7%) had both EP and cribra orbitalia.

The nature and distribution of EP was quite different for each temporal sample. The more regular and clearly differentiated EP observed in the Da But period cranial material (Figure 5.7a), was extremely rare in the Metal period sample. For the most part, crania from the Metal period sample displayed EP that was scattered and fine. Nonetheless, when present, Metal period EP was limited to that area described for the Da But period sample and displayed in Figure 5.7a. Only one Metal period individual (Figures 5.8a and 5.8b) displayed a very clear instance of localised marked porosity, of the type sometimes described for vault porotic hyperostosis. This individual was of indeterminate sex, aged 15-19 years and also displayed open-cribra orbitalia.

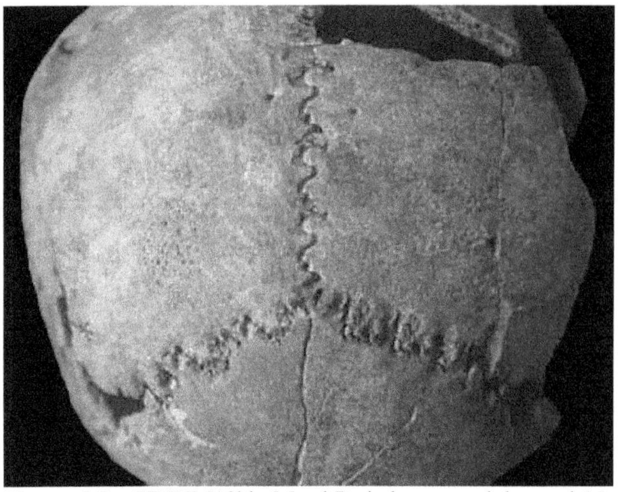

Figure 5.8a. 77NNM18kb. Metal Period young adult, aged 15-19 years. Bi-parietal localised marked porosity

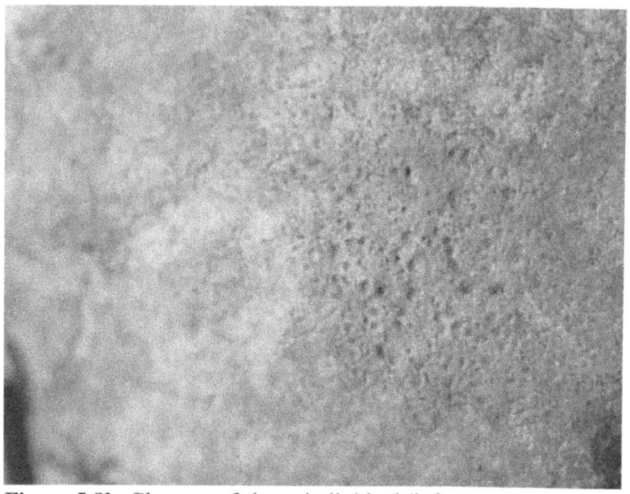

Figure 5.8b. Close up of above individual (left posterior-superior parietal)

Discussion and Conclusions

Understanding the patterning and prevalence of cribra orbitalia is even more difficult than is the case for linear enamel hypoplasia (LEH). The reason is that while LEH leaves a virtually indelible mark on the teeth cribra orbitalia, on the other hand, is prone to remodelling. Ideally, a good sample of subadults is needed for which remodelling would be expected to be much more limited. Such a sample is not available for these time periods in northern Vietnam. Nonetheless, there is an exceptionally high prevalence of cribra orbitalia in both samples. Stuart-Macadam (1991) has suggested that only 50% to 75% of individuals with iron-deficiency anaemia will develop skeletal responses. This would indicate that most individuals in these Vietnamese communities suffered from anaemia as children.

One of the reasons for this high adult prevalence of cribra orbitalia may be the inclusion of individuals displaying extensively remodelled lesions. Widely cited recording protocols, such as Stuart-Macadam's (1985) and Hengen's (1971), require some degree of porosity to be present in the orbital roof before a positive score is recorded. Such definitions would mean that highly remodelled forms, such as the shallow worm-tracks without associated pitting mentioned previously, will not generally be recorded.

Tayles (1992), using Hengen's (1971) recording protocol for cribra orbitalia, reports a prevalence of 54.4% for adults and 76.9% for subadults (all aged less than 12 years) in the Khok Phanom Di sample from Thailand. In the Non Nok Tha assemblage from Thailand, Douglas (1996), using the presence of porosity as a minimum scoring requirement, reports a prevalence of 11.9% for adults and a complete absence of cribra orbitalia in the eight assessable subadults. In the Ban Chiang collection, also from Thailand, Douglas (1996) reports a prevalence of 18.8% for the adults and 40% for the subadults. In comparison, using porosity as a minimum criterion (open-cribra orbitalia), 27.6% of the Da But period sample and 30.4% of the Metal period sample are affected. This places the Vietnamese prevalence of cribra orbitalia somewhere within the centre of the reported range for this condition in pre and early metal period Southeast Asian sites.

The prevalence of cribra orbitalia has been linked to increased mortality in at least one study. In Mittler & Van Gerven's (1994) research on Christian Kulubnarti, North Africa, they report a significant reduction in mean life expectancy for the sample with cribra orbitalia. For the Vietnamese samples under investigation here, however, there does not appear to be any obvious health disadvantage in having cribra orbitalia. When both open and closed-cribra orbitalia are considered, the distribution of affected individuals is even across all age classes for both the Da But period and Metal period samples. The chief differences seen in distribution by age only apply to cribra orbitalia of varying levels of remodelling. More remodelled cribra orbitalia is more common in the older age classes, while less remodelled cribra orbitalia is more prevalent in the younger age classes.

What are the underlying aetiological causes for the high prevalence of cribra orbitalia, all manifestations included, in the Vietnamese populations? As previously discussed, it is very difficult to isolate any one cause or even the varying contributions of multiple, likely synergistic, factors. The only fairly definite common factor in both periods is that these communities were essentially sedentary. This of course does not necessarily imply a significant or even minor agricultural subsistence economy. Nonetheless, the effects of sedentism appear to the one common global factor in the increased prevalence of porotic hyperostosis beginning at the start of the Holocene (Stuart-Macadam 1998). Because these Vietnamese communities were situated in the tropics, with the associated huge potential for parasitic infestation, this factor also needs to be considered. It is also worth noting that coastal and marine based communities can suffer from high parasitic loads (Walker 1986). It is also probable that these populations were affected by hereditary anaemic conditions, such as various forms of thalassemia.

Tayles (1996) has recently demonstrated the likely presence of thalassemia in a number of individuals from Khok Phanom Di in Thailand, thus documenting the extreme antiquity of this genetic condition in Southeast Asia. Tayles (1996) has also noted that 40% of the modern Thai population display forms of the various thalassemia syndromes and haemoglobinopathies. As noted in Chapter 1, reports of the occurrence of malaria in northern Vietnam were documented over 2000 years ago. However, skeletal changes associated with the more severe homozygous forms of hereditary anaemias are so rare that these individuals are unlikely to have been preserved. The situation is further exacerbated by the rarity of subadult material in the Vietnamese skeletal series.

While high pathogen loads and genetic anaemias likely played a role in the high prevalence of cribra orbitalia in Northern Vietnam during the mid-Holocene and early Bronze ages, a causal association cannot be demonstrated. Moreover, the likely important role played by dietary deficiencies in young growing children and the effects of anaemic mothers, also probably experiencing various degrees of parasitic infestation and perhaps genetically derived anaemia, cannot be assessed. However, very high rates of anaemia are reported in the children of living north Vietnamese communities (Nguyen Cong Khanh *et al.* 1995). The high prevalence or cribra orbitalia in Da But period and Metal period northern Vietnam is not remarkable in this context. It is also of interest that Nguyen Cong Khanh *et al.* (1995) report a correlation between increased levels of malnutrition and anaemia for the modern samples.

All the aforementioned factors likely played some role in the anaemia experienced in these samples from Vietnam. The important point to note, however, is that with such a large proportion of people experiencing cribra orbitalia as children, it does not appear to have been a condition associated with any increased level of adult mortality. Nguyen Cong Khanh *et al.'s* (1995) study of anaemia in modern Vietnamese children was not longitudinal. Nonetheless, there is no obvious evidence suggesting that the dramatic decrease in anaemia from the first to second years of life was not indicative of recovery from an event likely originating some time shortly after six months of age. With regard to the ancient Vietnamese samples, even if there was a high mortality rate among subadults with cribra orbitalia, it does not change the observation that most of the adults also had the condition as children. Having cribra orbitalia as a child does not appear to have compromised the health of these individuals as later adults. The implication is that while porotic hyperostosis is a marker of physiological stress in children, it may not have any real dire, and certainly not obvious, implications for the general health of these ancient Vietnamese as adults.

6
CASE STUDIES IN VIETNAMESE PALAEOPATHOLOGY

> First carefully and fully describe what you see including what is occurring and where it appears in the skeleton. An attempt at diagnosis can then be made with the assurance that others will at least have the option of reaching a different diagnostic conclusion on the basis of the descriptive information you have provided should that be appropriate.
> (Ortner 1990: xii)

In a series of case studies this chapter examines pathological conditions identified in the Da But period and Metal period skeletal samples from northern Vietnam. The preceding chapters have dealt with skeletal manifestations of health, or ill-health as it may be, and behaviour from a population oriented perspective. Most of the case studies examined in this chapter represent isolated conditions not amenable to such an analysis. Notwithstanding this, however, the trauma data can be examined from both a case study and population perspective.

The primary aim of the first part of this chapter is to assess the nature, extent and implications of evidence for traumatic lesions in the Vietnamese skeletal assemblages. The chief aim of the second part of this chapter is to document the range of non-traumatic conditions in the Vietnamese samples. This will contribute to the ongoing development of the epidemiological database of disease in tropical and sub-tropical Southeast Asia specifically, and the world generally. Moreover, the non-definitive differential diagnoses of these conditions offer insights into both the range of diseases these early Vietnamese populations may have been exposed to and also the history and development of disease in the region. The implications of the nature and frequency of both traumatic and non-traumatic pathology in ancient northern Vietnam are discussed.

Traumatic Lesions

Healed Fractures

Lovell (1997b:141) defines a fracture as "an incomplete or complete break in the continuity of a bone". There has been a recent trend toward a population approach to fracture analysis (Lopez *et al.* 1995; Berger & Trinkaus 1995; Kilgore *et al.* 1997; Alvrus 1999; Judd & Roberts 1999; Neves *et al.* 1999) with Lovejoy & Heiple's (1981) work being frequently cited as seminal in this regard. However, Lovell (1997b:148) also stresses that proper

> description of an injury is the first step in trauma analysis and is the basis for determining the mechanism, or proximate cause, of the injury. In turn, an understanding of the proximate cause is crucial for the identification of the ultimate cause of trauma, usually behavior.

The identification of pre or peri-mortem fractures is extremely difficult (Lovell 1997b) and only the more likely cases of peri-mortem trauma are examined here. A population approach to fracture analysis in the Vietnamese samples is hampered by both the low overall number of healed fractures and the fragmentary nature of the remains. The latter problem was overcome, to an extent, in the analysis of articular change in Chapter 4. In that chapter one aim was to establish the gross patterning of articular change by joint complex with respect to age and sex. A large number of preserved articular surfaces facilitated that analysis. The analysis of fractures, however, is more dependent on the availability of complete bones. Lovejoy & Heiple (1981) went as far as to exclude all bones from their analysis that were not complete. However, sample frequencies of fractured bone types can be attempted by specifying the denominator as a specific section of bone. Most long bone fractures are reported as being located in the proximal, middle or distal thirds of the bone. These portions can serve as useful denominators when reporting the frequency of fractures per bone type in fragmentary remains.

The purpose here is to document all observed fractures and establish a proximal cause in each instance. Each fractured element was examined both macroscopically and radiologically. The descriptive terminology and definitions expounded by Lovell (1997b) are employed. The frequency of fractures by bone type is then summarised. Finally, the significance of these fractures and fracture patterns, or ultimate causes, is discussed with respect to behaviour, the environment and other reported fracture data in Southeast Asia. A description of the skeletal and dental material available for each individual in the following case studies is presented in Appendix 1.

Healed Fracture Case Studies

Case Study One

80CCNM35a, from Con Co Ngua Pit 2, was male and aged 30-39 years. He was one of only two individuals buried with red ochre in the shape of a gourd. Given the rarity of this practice at Con Co Ngua, this might be interpreted as recognition of some specific status in the community.

The right femur displayed a healed fracture just distal to mid-shaft 163mm superior to the condyles (Figure 6.1). The complete shortened length of the femur, from the tip of the greater trochanter to the end of the condyles, is 379mm. The head was missing post-mortem. The morphology of the fracture suggests a transverse separation of the diaphysis with the proximal fragment being displaced anteriorly and subsequent 0% apposition on healing. There is a 66mm overlap to the proximal and distal ends of the shaft with a slight posterior angulation. There does not appear to be any lateral angulation. Following convention, angulation refers to the position of the distal fragment relative to the midline of the proximal fragment (Lovell 1997b). Furthermore, there does not appear to be any obvious rotation. There is complete union of the both fragments with little remaining callus. The distal end of the proximal fragment is well remodelled and quite smooth, while the proximal end of the distal fragment displays several spicules of bone. The proximal fragment is quite flat antero-posteriorly and displays a reduced *linea aspera*. This is likely due to disruption of muscles and the intermuscular septa in this region.

The preserved distal articular area shows some evidence of articular change. There is a distinct marginal osteophytic ridge to the medial aspect of the medial condyle, the antero-inferior aspect of the intercondylar sulcus and also to the superior aspect of the anterior condyle. Much of the lateral aspect of the lateral condyle has eroded away post-mortem.

Transverse fractures can be produced by either direct or indirect force. Bucholz & Brumback (1991) suggest that the most common proximal cause, or mechanism, of femoral shaft fractures is bending load. A fall from a height is a likely ultimate cause of such a fracture.

In addition to the femoral trauma there are several vertebral irregularities. This individual had the best vertebral preservation in the Da But period sample. There is a likely spondylolysis, separated neural arch, of the fifth lumbar vertebra. The separation is at *pars interarticularis*. However, only the right side can be assessed as the left superior articular facet has been eroded post-mortem. There is no apparent point of breakage inferior to the right superior articular facet. This is possibly a case of congenital separation of the neural arch. Merbs (1996a, 1996b) has argued that a stress fracture is the most common aetiology for spondylolysis. Such a susceptibility to stress fracturing is believed to be related to the mechanics of both habitual bipedalism (Merbs 1996b) and behaviours such as heavy lifting (Merbs 1996a). The most common age of occurrence is early childhood (Merbs 1996b). As such, it is not improbable that the neural arch in this case was separated in youth, with subsequent remodelling over the life of this individual leading to the appearance of a congenital defect.

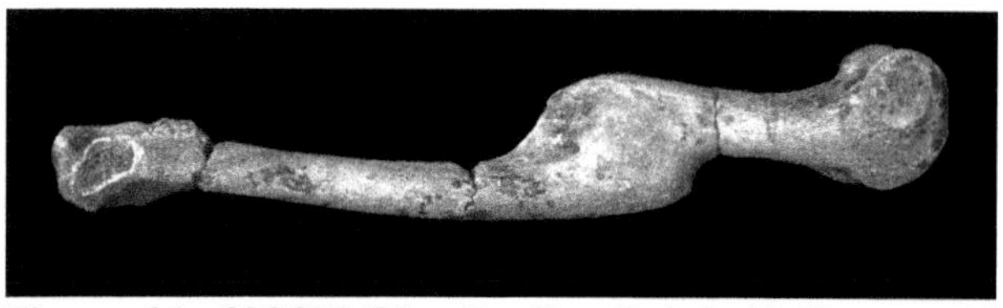

Figure 6.1. 80CCNM35a. Da But Period 40-49 year old male. Healed fracture to middle third of right femur (medial aspect shown)

Centra of two of the preserved lower thoracic vertebrae display enlarged foramina. An extremely enlarged foramen, 9.2mm in diameter, of T9 occurs on the anterior aspect of the centrum. A smooth-floored, shallow channel encircles the anterior aspect of the centrum and the lesion is situated within this. T10 displays a similar channel and several large (~4.0mm diameter) pits along the course of the channel. The foramina are very deep and extend up to 10mm into the centrum (Figure 6.2).

The shallow laterally running channel is likely caused by the anterior external plexus which anastomoses (interconnects) internally, through foramina in the sides of the centrum, and externally with other vertebral veins. The basivertebral veins, which exit through the posterior aspect of the body leaving a distinctive foramen, are known to increase in size with age (Williams & Warwick 1984). It is thus likely that the anterior external plexus does also. All blood vessels, no matter what their size, are susceptible to distension or aneurysm (Aufderheide & Rodrigues-Martin 1998). It is unclear what the normal range of variation is for the size of the vessels comprising the anterior external plexus. As such, it is difficult to determine if this case provides evidence for some underlying condition such as hypertension or anaemia, for example. Apart from closed and highly remodelled cribra orbitalia and localised

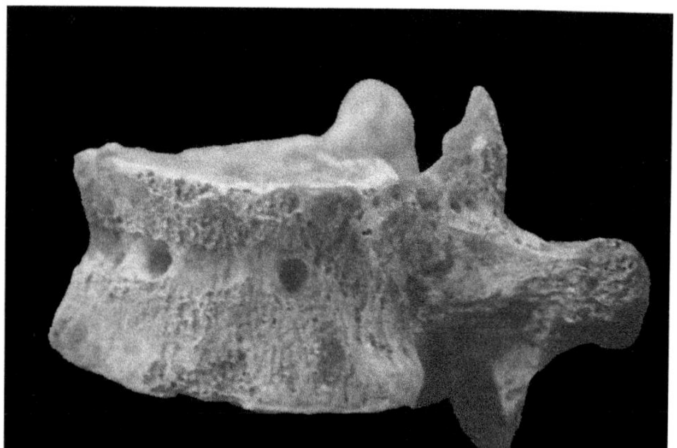

Figure 6.2. 80CCNM35a. Da But Period 40-49 year old male. Latero-anterior aspect of T10

ectocranial porosity of the supero-posterior parietals, no other pathologies were noted.

Case Study Two

80CCNM47a, from Con Co Ngua Pit 1, was a child aged 9 years +/- 24 months, based on dental eruption timing. The right femur displays a healed fracture to approximate mid-shaft, 170mm superior to the distal metaphysis (Figure 6.3). The bone is missing the proximal and distal articular epiphyses, which had not yet fused. This transverse fracture has healed without any obvious rotation. There is a marked 25^0 medial or varus angulation and 20^0 posterior angulation. The shortened length of the diaphysis is 242mm with a distal and proximal shaft overlap of 35mm, as such there is 0% apposition. The proximal fragment has displaced anteriorly. Much of the outer table surrounding the trauma is finely porous, as is most of the antero-medial aspect of the shaft. Porosity is also concentrated on the postero-proximal aspect around the gluteal tuberosity and especially superior to the tuberosity. There is a large, smooth-walled, oval opening some 18.7mm long, supero-inferiorly, situated on the proximal most aspect of distal fragment. This opening forms a communication between the medullary cavity and the exterior of the shaft. The proximal causes of such transverse fractures have been discussed for case one.

Fractures such as presented in this child will eventually remodel with almost no visible trace of the fracture (Steve Baddeley pers. comm.). A radiograph of the fracture showed remodelling of the lateral aspect, that would have eventually led to considerable straightening of the medial angulation. This fracture may have occurred between six months and a year before death (Steve Baddeley pers. comm.). The increased vascularity associated with subadult fractures will lead to increased growth of the bone and can lead to a lengthening of the shaft. This would in part ameliorate the shortening of the shaft due to the overlap of each fragment (Steve Baddeley pers. comm.).

The distal diaphysis displays at least six clear, and approximately equidistant, growth arrest (Harris) lines. All of this individual's permanent incisors had recently erupted and were preserved. Each of the maxillary incisors exhibited two linear hypoplastic bands, while the mandibular incisors were free of any evidence of enamel hypoplasia. The in situ primary dentition, including all mandibular and maxillary molars in addition to the left maxillary deciduous canine, were also free of enamel

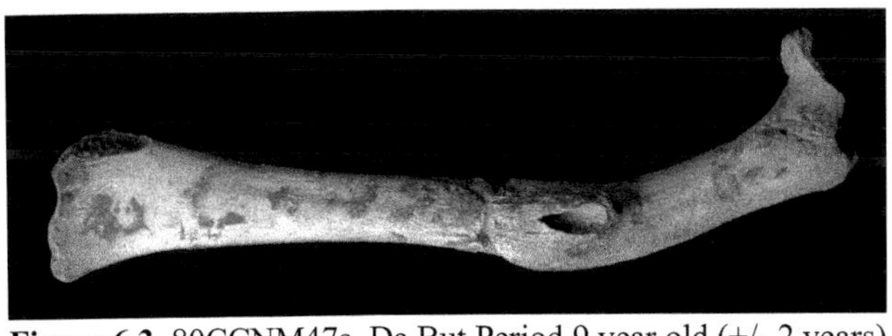

Figure 6.3. 80CCNM47a. Da But Period 9 year old (+/- 2 years) child. Healed fracture to approximate mid-shaft of right femur (posterior aspect shown, proximal end to the right)

defects. The preserved left orbital roof displayed marked open-cribra orbitalia with some evidence of remodelling. It would appear then that this child was experiencing some degree of physiological stress in the years prior to and up until death. The number and spacing of the Harris lines suggests a regular or periodic nature of one of the stressors at least. The porotic hyperostosis, LEH and most of the Harris lines would have developed before the femoral fracture. While the fracture had healed, the child's mobility would have been severely handicapped. The metabolic demands needed to cope with such serious trauma may have critically depleted the already compromised physiological resources of the child, thus contributing to an early death.

The cranium, while quite narrow anteriorly (minimum frontal breadth of 90.1mm, euryon-euryon 129.0mm, glabella-opisthocranion 186.0mm), did not display any obvious additional pathological features. The cranial bones were not abnormally porous and the diploe to table ratio was normal. Further, there was no obvious frontal or parietal bossing. Apart from enamel hypoplasia, the gnathic remains were also completely free of any obvious macroscopic pathology.

Case Study Three

80CCNMX5 is represented by an isolated right femur, of which only the proximal two-thirds was preserved. There is no surviving documentation as to which pit at Con Co Ngua this individual came from. The discriminant function formulae (Chapter 2) classified the bone as female. The only age-at-death information is that she is adult, based on the complete fusion of all proximal epiphyses.

A transverse fracture occurred 122mm distal to the top of the femoral head. There is 0% apposition and the proximal fragment has been displaced anteriorly and overlaps the distal fragment by 44mm (Figure 6.4). The lesser trochanter appears to have stopped the proximal migration of the distal fragment and has subsequently become engulfed within the massively exuberant and complex callus formation. There does not appear to be any obvious rotation of the shaft and only a slight anterior angulation. Both the distal end of the proximal fragment and proximal end of the distal fragment display extensive remodelling and closure of their respective ends. However, a large 'cavern' has formed on the anterior aspect of the femur beneath the overhanging proximal fragment and posteriorly displaced distal fragment (Figure 6.5). This cavern does not communicate with the medullary space of the proximal fragment. The dimensions of the cavern show a maximum internal medial-lateral width of 32.8mm and a maximum internal height of 15.8mm. This cavern does not appear to be the result of infectious activity. Charlotte Roberts (pers. comm.) has suggested that the cavern was likely formed through a soft tissue inclusion during the healing process. An examination of a lateral radiograph of the femur shows clear evidence of periosteal tearing or lifting on the posterior aspect of the distal fragment. This raised periosteal bone can also be seen macroscopically.

As can be seen in Figure 6.4, there is extensive post-mortem erosion of the femoral neck and head. The undamaged portions of the head are free from any articular changes such as osteophytic development, porosity or eburnation. However, it is likely that the extensively eroded portions of the head displayed some degree of articular change in life. The proximal causes of transverse fractures to the femoral diaphysis have been discussed in case one.

Case Study Four

80CCNM54a was a 30-39 year old female from Con Co Ngua Pit 2. There is a well healed steeply oblique (supero-lateral to infero-medial) fracture to the lower end of the distal left humerus. It is difficult to assess the degree of rotation, if any, as only the distal quarter of the bone has been preserved. There appears to be complete or 100% apposition. An examination of a radiograph of the fracture reveals a clear medial angulation of 10^0 degrees. This has resulted in a marked bulge on the antero-lateral aspect of the shaft above the most superior point of the fractured diaphysis (Figure 6.6). This most superior point is just below the post-mortem break, 110mm superior to the distal end of the humerus.

The distal condyles were well preserved and displayed minimal articular change. Mild erosion was evident on the medial most aspect of the anterior portion of the trochlea. Moreover, there was an isolated area of porosity on the lateral most aspect of the posterior face of the trochlea. The proximal half of the left ulna was also preserved but displayed no signs of articular changes.

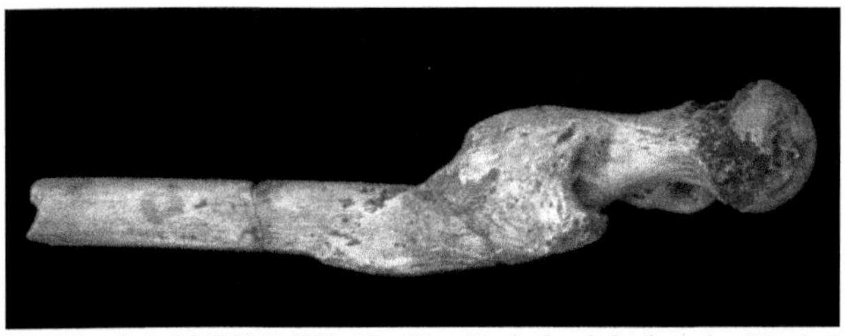

Figure 6.4. 80CCNMX5. Da But Period adult female. Healed fracture to the proximal third of the right femur (medial aspect shown)

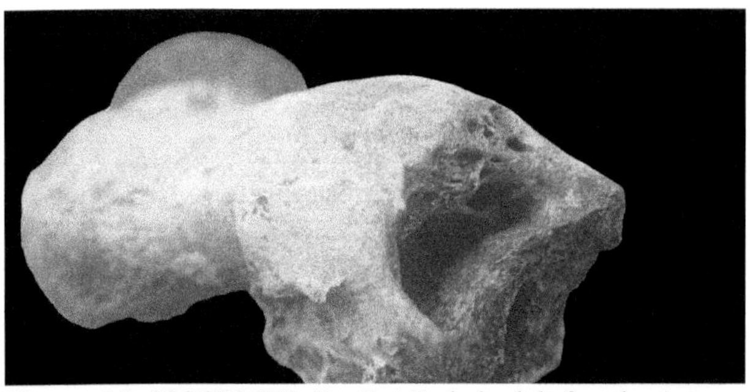

Figure 6.5. 80CCNMX5. Da But Period adult female. Antero-lateral aspect of fracture seen in Figure 6.4. Note the 'cavern' formed as a result of the anterior displacement of the proximal fragment

An oblique fracture will be caused by a force acting at a distance to the site of the fracture. In the humerus, such a fracture is likely produced by a combination of bending (angulation) and torsion (Epps & Grant 1991). An oblique fracture of the humerus can be caused by a fall onto the elbow with the arm abducted (Apley & Solomon 1993).

With respect to other preserved skeletal and dental elements, apart from three distinct linear hypoplastic bands on the preserved left maxillary canine, there were no gnathic pathologies. The left femur displayed a very pronounced *linea aspera* while the *fovea capitis* of each of the preserved proximal femoral heads were very deep (up to 5mm) and wide (up to 16.5mm). The anterior articular face of the left cuboid (for articulation with the fourth and fifth metatarsals) exhibited a slight osteophytic marginal rim. Moreover, the plantar aspect of the distal left first metatarsal displayed clear osteophytic development.

Case Study Five

80CCNM57a was a 20-29 year old male from Con Co Ngua Pit 2. The distal end of the left humeral diaphysis has been fractured in two places. It is unclear if both breaks occurred simultaneously or if some period of time separated them. The maximum shortened length of the humerus is 315mm.

There is a healed steeply oblique (supero-lateral to infero-medial) fracture to the distal diaphysis that encompasses 50.5mm of the length of the distal shaft. There is no obvious rotation but there is a clear lateral displacement of the distal fragment giving an apposition value of approximately 70%. The proximal fragment appears to have slightly overridden, disto-medially, the distal fragment. This has resulted in a slight posterior angulation. A prominent superiorly oriented spur, representing the supero-medial most point of the distal fragment, is apparent at the approximate point where the radial nerve makes its way from the posterior to anterior aspect of the shaft (Figure 6.7).

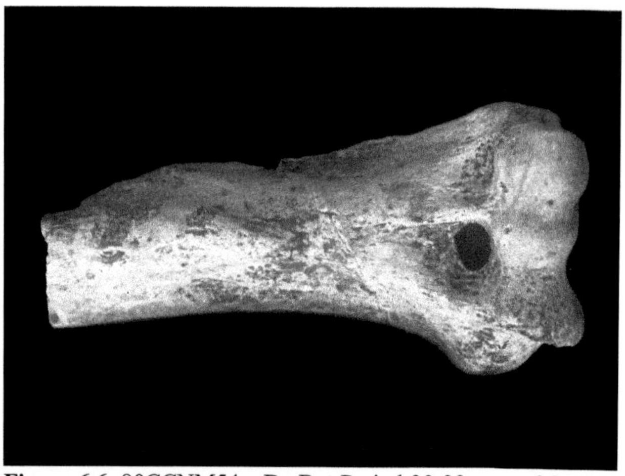

Figure 6.6. 80CCNM54a. Da But Period 30-39 year of female. Well healed steeply oblique fracture to the distal third of the left humerus (anterior aspect shown)

There is a marked medial build up of callus well above the level of the proximal limit of the lower oblique fracture. There does not appear to be any rotation of this fracture either and there is complete apposition. However, there is a very slight anterior angulation. The anterior angulation associated with this more superior fracture and the posterior angulation associated with the more inferior fracture has resulted in a dogs-leg appearance in the distal shaft when examined laterally. The anterior aspect of the more superior fracture displays a series of grooves, running proximo-distally, and pits indicating remodelling around the fracture site. Extensive pitting and long tear-like streaks of bone also appear on the opposite side of the humerus, on the posterior aspect of the more superior fracture.

A large build up of bone is apparent on the lateral epicondyle that extends onto the posterior surface. This build up extends 25mm along the lateral edge running supero-inferiorly. It is 25.5mm wide on the posterior aspect running latero-medially (Figure 6.8). This bony development represents ossification of the origin of the *extensor carpi radialis longus*. It is also possible that part, at least, of the continuation of this ossification onto the posterior aspect of the shaft included the distal most disto-lateral extent of the origin of the medial head of the *triceps*. The medial head of the *triceps* is critically important in all forms of forearm extension. The *extensor carpi radialis longus* is involved with flexion of the fingers and extension of the wrist.

The articular ends of the humerus do not display any obvious pre-mortem articular changes with the exception of a large area of micro-porosity on the dorso-lateral aspect of the proximal head. The preserved articular ends of the left radius (both) and ulna (proximal only) did not display any signs of articular change.

A common neural complication in distal fractures of the humerus is radial nerve palsy or wrist drop (Epps & Grant 1991). Even if the nerve is not lacerated, or even completely severed, during injury it can be contused or worse still become compressed within the fracture callus. The superiorly oriented spur, described in the fractured humerus above, could have served to capture and compress the radial nerve thus leading to neural complications.

The proximal causes of oblique fractures of the humerus have already been discussed with respect to case four. The more superior transverse fracture of this humerus could have been caused by either a direct or indirect force. A tapping fracture, caused by direct force, can result in a transverse fracture. Alternatively, a bending or angulation fracture of the shaft, due to indirect force, can also result in a transverse fracture (Harkess *et al.* 1991). This fracture could have been caused by either a direct blow or a fall. It is unlikely that both fractures occurred at the same time and is probably impossible to determine which fracture occurred first (Steve Baddeley pers. comm.).

With respect to the rest of the skeleton, the mandibular canines displayed multiple LEH bands (four on each canine). No other gnathic pathologies were noted. The orbital roofs were not preserved but extensive and marked orange peel-like ectocranial porosity was seen on frontal and parietal fragments. The preserved T12 and L1 displayed enlarged foramina along the anterior and lateral aspects of each centra, similar but less severe as that noted for 80CCNM35a (case one, Figure 6.2). The only other pathology was a lytic lesion on the left first cuneiform. The lesion presented on the lower centre of the lateral face, that is inferior to the articular facet for the second cuneiform. It is a large, 7.0mm in diameter, circular lesion up to 5.0mm deep. The edges are jagged and the floor is also jagged and irregular. There is no evidence of adjacent porosity.

Case Study Six

95DXM1a is a 30-39 year old female from the Metal period site Dong Xa. This case is unique among the Vietnamese material in being virtually complete. This individual displays a number of pathologies of which only the fracture is dealt with here.

There is a healed fracture to the proximal third of the left radius. Only the proximal two-thirds of the bone is preserved. The fracture angle is steeply oblique (supero-medial to infero-lateral). The apex of the fracture begins

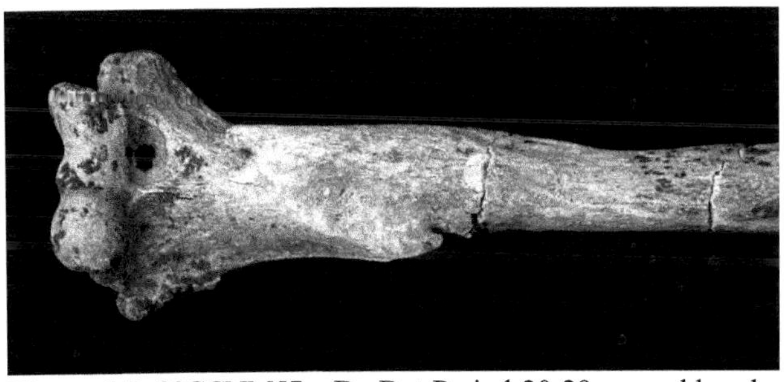

Figure 6.7. 80CCNM57a. Da But Period 20-29 year old male. The distal third of this humerus has been fractured in two places. Slightly superior to the healed steeply oblique fracture is a healed transverse fracture to the humeral shaft (anterior aspect shown)

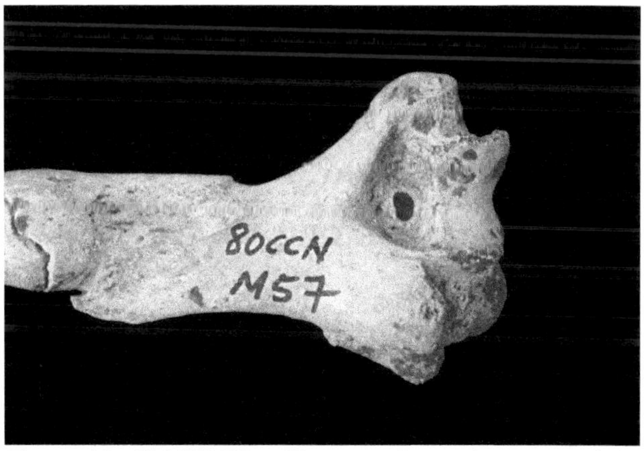

Figure 6.8. 80CCNM57a. Close up of posterior aspect of the left humerus seen in Figure 6.7. Note osseous development of posterior aspect of lateral epicondyle

57.5mm from the top of the proximal head and the entire proximo-distal length of the fracture is 51.1mm. There is considerable rotation with the proximal and distal fragments rotating at 90^0 relative to each other (Figure 6.9). This is caused by supination of the proximal fragment and pronation of the distal fragment when the shaft is fractured distal to the *supinator teres* and proximal to *the pronator teres* (Anderson & Meyer 1984). Because the ends of each fragment have rotated away from each other, there is 0% apposition.

The proximal end of the distal fragment has completely remodelled and closed. There is a 12^0 lateral angulation of the distal fragment relative to the midline of the proximal fragment. A shallow sulcus (20.0 x 5.0mm) runs down the radial shaft on the lateral aspect just below the distal most extent of the fracture. Most of the fracture area is smooth and well healed. There is a slight pointed projection or spur projecting superiorly for 3.0mm on or near the top of the original fracture on the medial face. Also, the posterior face at the fracture site has mild signs of localised porosity and unevenness.

The left ulna displays a clear disto-lateral bowing deformity. Stuart-Macadam *et al.* (1998) have recently reviewed the clinical literature on what are termed acute plastic bowing deformities. They point out that while more common in children, they are seen in adults and the ulna is the most frequently affected bone. Moreover, the condition is more common when associated with a mid-shaft fracture to an accompanying bone.

While the distal articular area of the radius is missing post-mortem, the preserved proximal head displays extensive porosity and erosion particularly to the medial aspect of the head. Moreover, there is extensive porosity and erosion of the biceps insertion including the adjacent area. The proximal head of the left ulna also displays extensive porosity, concentrated on the radial notch. The superior aspect of the notch is very smooth and the usual step and inferior anterior and posterior margins are flush with the adjacent bone. The remaining (non-eroded) face of the radial notch displays eburnation. The distal articular region of the humerus displays slight scattered micro-porosity of the trochlea and capitulum. The same is apparent on the right humerus. Proximally, the left humeral head displays porosity and marginal osteophytic development on the superior aspect of the lesser tubercle (attachment for the *subscapularis*). Both humeri display *pectoralis major* cortical defects.

Anderson & Meyer (1984) suggest that fractures to the proximal two-thirds of the radius alone are not common. This region of the radius is well protected by muscle and seldom is the radius in an anatomical position that would expose it to fracture forces that did not also affect the ulna. It has been suggested that a fall could have resulted in such a fracture (Charlotte Roberts pers. comm.) This fracture would have caused permanent subluxation (dislocation) of the distal radioulnar joint resulting in limited and painful pronation and supination of the forearm (Anderson & Meyer 1984).

Case Study Seven

89MDM2a is a 40-50 year old female from the Metal period site Minh Duc. There is a transverse fracture to the distal third of the ulna with 100% apposition and no apparent rotation, angulation or overlap (Figure 6.10). The apparent normal length of the shaft is 220mm. It is very well healed as was possibly splinted by an unfractured radius (which was not preserved). The medial surface of the callus is irregular while the lateral aspect is very smooth. The distance from the styloid process to the point of fracture is 49.0mm.

The proximal articular area of the ulna displays porosity to the upper lateral aspect of the olecranon. Most of the articular margins of the head display marginal osteophytic development, especially the anterior olecranon and anterior lunar notch. The distal articular end also displays articular changes.

The ventral aspect of distal ulna presents a very irregular development of osteophytic bone. Also, the medial articular extension (for articulation with the ulnar notch of the radius) is irregular and an osteophytic lip has developed on the superior border.

While the left radius was not preserved, the left humerus also displays articular changes. The distal articular region of the humerus displays a marked ridge of osteophytic bone and porosity along the lateral border of the trochlea. This ridge extends onto the posterior aspect as far as the supero-posterior limit of the trochlea. There is also osteophytic lipping of the postero-lateral edge of the trochlea and medial edge of the capitulum. This probably extended to the antero-medial edge of the trochlea, but this region was eroded post-mortem.

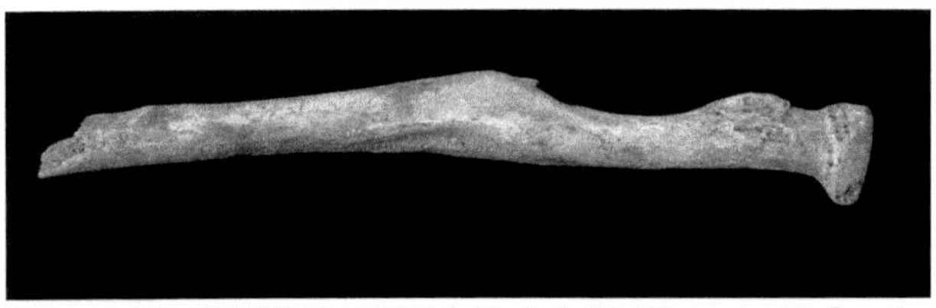

Figure 6.9. 95DXM1a. Metal Period 30-39 year old female. Healed fracture to proximal third of the left radius (anterior aspect shown)

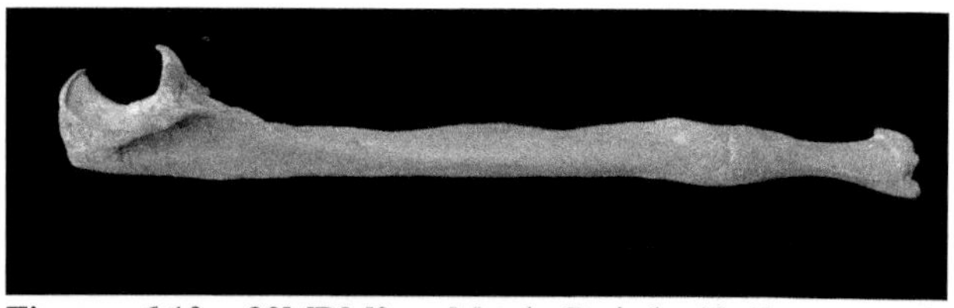

Figure 6.10. 89MDM2a. Metal Period 40-49 year old female. Healed fracture to distal third of the left ulna (medial aspect shown)

The proximal left humeral head also displays signs of articular change. There is extensive porosity to the supero-lateral aspect of the proximal head. Further, the posterior aspect of the greater tubercle appears worn with the appearance of what appears to be a large posteriorly sloping facet, the centre of which displays two porous lesions. This region forms the origin for the *infraspinatus*, one of the rotator cuff complex involved in shoulder stabilisation. The glenoid of the scapula also displayed porosity and marginal osteophytosis.

Anderson & Meyer (1984) point out that a fracture of the distal half of the ulna alone, resulting in minimal displacement, is quite common. This type of fracture is generally caused by some type of direct blow. Pollock *et al.* (1983) note that such fractures can heal minus complications in an average of six to seven weeks without the use of immobilisation.

This individual displayed extensive proliferative joint changes to most of her preserved joints. Of the few surviving vertebrae, T1 displays extensive proliferative and porous lysis of the interarticular facets, while the centra was free from pathology. The surviving T10-12 also displayed interarticular facet osteophytes, with T12 expressing osteophytic development of the ventral edge of the centra as well. While L5 is missing, the sacrum displays extensive proliferative development of the lateral juxta-articular region of the promontory and a combination of proliferative and lytic changes to the superior articular facets. Laterally the sacrum has a strong incurvature, even though female sacra are generally less curved than male sacra. There is also extensive ossification of the *erector spinae (sacrospinalis)* which attaches, in the form of a strong ligament here, along the dorsal median crest of the sacrum (Figure 6.11).

The right os coxa presents a lytic lesion on the supero-lateral aspect of the ischial tuberosity. It is approximately 4.0mm in diameter and 3.0mm deep. There is a distinct localisation of porosity inferior to the lesion but this does not extend superiorly. Internally, the lesion is relatively smooth but with porous walls and floor. This region forms the origin of the *semimembranosus*.

This individual also suffered extensive alveolar destruction of both the mandible and maxilla through the development of numerous alveolar defects of pulpal origin. Figure 6.12 demonstrates the massive anterior maxillary alveolar destruction with over 50% of the original height of the subnasal maxilla lost in parts.

Other Pre-Mortem Trauma Case Studies

Case Study Eight

80CCNMX4 is represented by the proximal half of an isolated right humerus. There is no surviving documentation as to which pit at Con Co Ngua this bone came from. The discriminant function formulae (Chapter 2) classified the bone as male. The only age-at-death information is that it is adult based on the complete fusion of the proximal epiphysis.

The right humerus displays extensive ossification of the *deltoid* insertion. The ossified soft tissue extends the length of the insertion some 17.6mm proximo-distally. The ossification projects outward to a maximum of 8.2mm and is 5.7mm thick at the base or insertion (Figure 6.13).

There is also a clear *pectoralis major* cortical defect extending some 44.5mm along the shaft and expanding to 5.0mm at its widest point. Only the middle section of the groove is porous and this is 18.5mm long. A pronounced ridge also runs parallel to the cortical defect with an associated narrow and shallow sulcus on the medial side of the groove. This ridge is the attachment site for the *latissimus dorsi* muscle. The length of this sulcus is 35.0mm.

This condition of the deltoid can be produced by trauma to the muscle and subsequent ossification of tissues within the resulting haematoma, the deltoid being a common site (Ortner & Putschar 1981). In such instances the condition is referred to as myositis ossificans traumatica. In distinction to this is the hereditary condition myositis ossificans progressiva. This causes extensive ossification of muscle origin and attachment sites (Ortner & Putschar 1981). A much more complete skeleton, than is available for this Vietnamese individual, would be needed to exclude

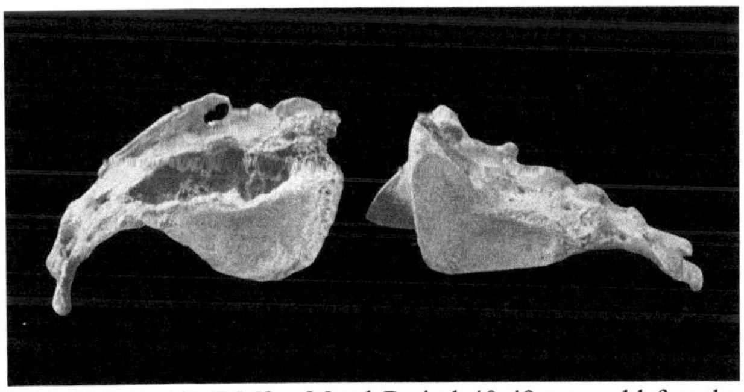

Figure 6.11. 89MDM2a. Metal Period 40-49 year old female. Right lateral aspect of sacrum (on left). Note the marked incurvature and ossification of the *erector spinae.* The comparative sacrum (on the right) is from a 20-29 year old female from the same site

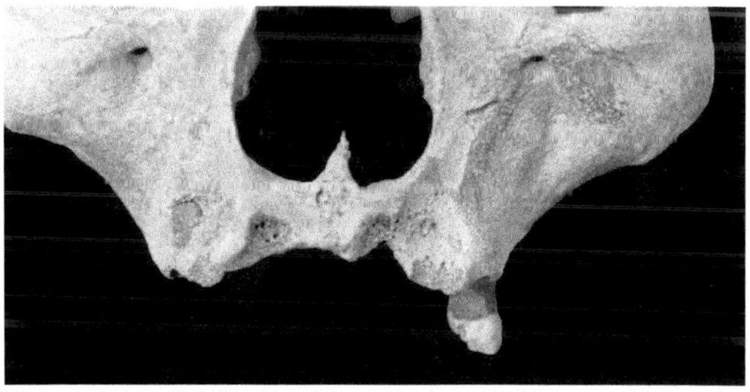

Figure 6.12. 89MDM2a. Metal Period 40-49 year old female. Note extensive maxillary alveolar destruction and numbeous alveolar defects of pulpal origin

this condition.

Case Study Nine
80CCNM1Y was a female aged 40-49 years. It is unknown from which pit at Con Co Ngua she came. The lateral aspect of each mandibular condyle, just inferior to the articular area, displays an oval raised process. The surface of each process is flattened and rough or uneven. In each instance the process displays a definite sharp rim on its anterior aspect. Each process is about 8.5mm high and 6.5mm wide and is raised at least 1.0mm above the surrounding bone (Figure 6.14).

Both the fibrous capsule and lateral temporomandibular ligament attach to this site. This process thus likely represented bilateral ossification of one or both of these ligamentous attachments. It is possible that this resulted from some form of trauma to the mandible, the force of which concentrated on the lateral region of the condyles. In addition, a clear circular section of porosity within a slightly concave depression is apparent on the postero-lateral aspect of each condyle. Such condylar depressions have been described and their possible aetiology reviewed in Chapter 3.

Peri-Mortem Trauma Case Studies

The following cases are only provisionally identified as instances of peri-mortem trauma. Post-mortem/ post depositional factors can often mimic skeletal changes that are otherwise consistent with trauma immediately preceding the death of an individual.

Case Study Ten
The cranial morphology of 66VQ1M31a is consistent with the sex of female. Ecto- and endocranial synostosis of portions of the sagittal and much of the coronal sutures tends to indicate an adult of middle age. She was excavated from the Metal period site Vinh Quang and is represented by a calva only.

The right side of the frontal bone presents a large cut running antero-posteriorly. Anteriorly, the cut begins 31.5mm superior to the middle of the right superior orbital rim. The cut extends posteriorly for a maximum length of 45.0mm (Figure 6.15). The lateral margin of the cut is straight and it extends through the outer table deep into trabecular bone. At its deepest point, at about mid-length, it is 6.2mm deep. On the medial aspect of the cut two fragments of bone have been pried loose or flaked away (Figure 6.16). These likely represent percussion flakes. The larger piece is semi-circular in shape with its anterior margin being 15.0mm posterior to the anterior most extent of the cut, after which it extends 21.5mm posteriorly. The greatest medial-lateral width of this flake is 13.6mm, which is also the maximum width of the entire cut. The second piece, a scalloped lesion, begins at the anterior most extent of the cut and extends 9.0mm posteriorly. The greatest medial-lateral width of this smaller flake is 12.2mm. There is no sign of porosity or remodelling. Moreover, the calva was free of any pathology endocranially.

An abundant array of bronze knives, swords, axes and adzes were available during the Metal period in northern Vietnam. The cut to this woman's skull is consistent with the use of such a sharp implement. Nonetheless, the lack of an extant detailed excavation report means that an excessively enthusiastic excavation technique cannot be ruled out.

Case Study Eleven
80CCNM42a is a 30-39 year old probable female from Con Co Ngua pit 2. The left ilium presents an irregular somewhat circular perforation of the blade. The superior margin of the perforation is 16.5mm from the iliac crest while the posterior margin is 81.5mm posterior to the anterior-superior iliac spine (Figure 6.17). The blade is 7.0mm thick at the site of the perforation. On the lateral aspect of the blade the dimensions of the lesion are 12.5mm supero-inferiorly and 11.2mm antero-posteriorly. On the medial face the perforation is 21.5mm supero-inferiorly and 18.4mm antero-posteriorly at its widest points. The perforation is more irregular on the medial face and there is more damage to the outer table of this face. There are

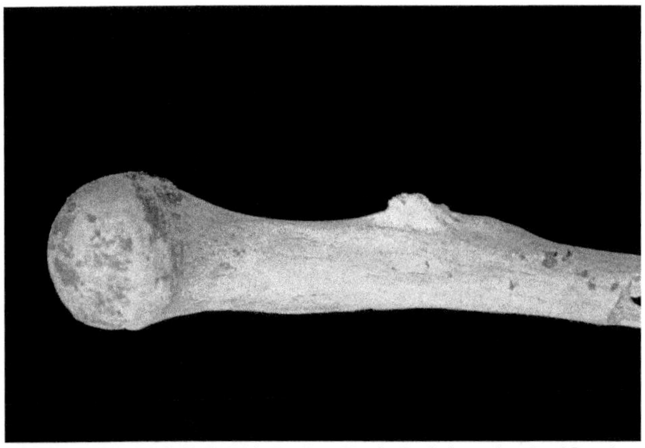

Figure 6.13. 80CCNMX4. Da But Period adult male. Ossification of the right humeral deltoid insertion (medial aspect shown)

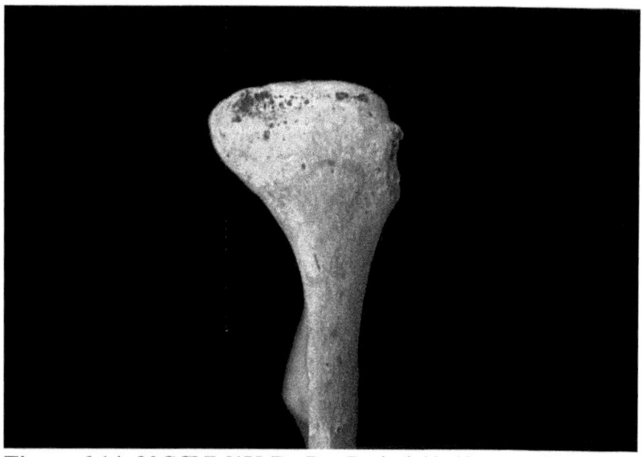

Figure 6.14. 80CCNM1Y. Da But Period 40-49 year old female. Right mandibular condyle (posterior aspect shown). Note development of a raised process on the lateral aspect of the condyle

Table 6.1. Da But Period. Fracture Frequencies by Skeletal Element

	left			right			total		
humerus	n	obs	%	n	obs	%	n	obs	%
proximal third	18	0	0.0	21	0	0.0	39	0	0.0
mid third	36	0	0.0	50	0	0.0	86	0	0.0
distal third	34	2	5.9	35	0	0.0	69	2	2.9
diaphysis[1]	20	2	10.0	25	0	0.0	45	2	4.4
radius	n	obs	%	n	obs	%	n	obs	%
proximal third	32	0	0.0	32	0	0.0	64	0	0.0
mid third	23	0	0.0	27	0	0.0	50	0	0.0
distal third	17	0	0.0	28	0	0.0	45	0	0.0
diaphysis[1]	11	0	0.0	16	0	0.0	27	0	0.0
ulna	n	obs	%	n	obs	%	n	obs	%
proximal third	41	0	0.0	44	0	0.0	85	0	0.0
mid third	20	0	0.0	26	0	0.0	46	0	0.0
distal third	15	0	0.0	13	0	0.0	28	0	0.0
diaphysis[1]	10	0	0.0	11	0	0.0	21	0	0.0
femur	n	obs	%	n	obs	%	n	obs	%
proximal third	33	0	0.0	30	1	3.3	63	1	1.6
mid third	49	0	0.0	41	2	4.9	90	2	2.2
distal third	15	0	0.0	20	0	0.0	35	0	0.0
diaphysis[1]	23	0	0.0	23	3	13.0	46	3	6.5
tibia	n	obs	%	n	obs	%	n	obs	%
proximal third	24	0	0.0	17	0	0.0	41	0	0.0
mid third	29	0	0.0	28	0	0.0	57	0	0.0
distal third	34	0	0.0	26	0	0.0	60	0	0.0
diaphysis[1]	19	0	0.0	10	0	0.0	29	0	0.0

Note: n refers to the number of assessable bone elements
[1]complete diaphysis

also numerous pieces of outer-table fragments wedged into the trabecular bone around the perforation, with some still cemented in place with mud, on the medial face of the blade. This suggests that if this was a low speed projectile (spear?) wound, the projectile entered from the medial face and exited through the buttock.

Apart from articular changes, such as marginal osteophytic activity, no other pathological conditions were present. It is unclear if the iliac perforation is peri or post-mortem. The only likely post-mortem agent is an excavator's tool. This seems unlikely, but not impossible, from conversations I have had with those attending the excavation. Peri-mortem explanations might include some sort of low speed projectile implement such as a spear, or alternatively a fall and subsequent impalement on a sharp projecting object. Kevin Lee (pers. comm.) suggested that

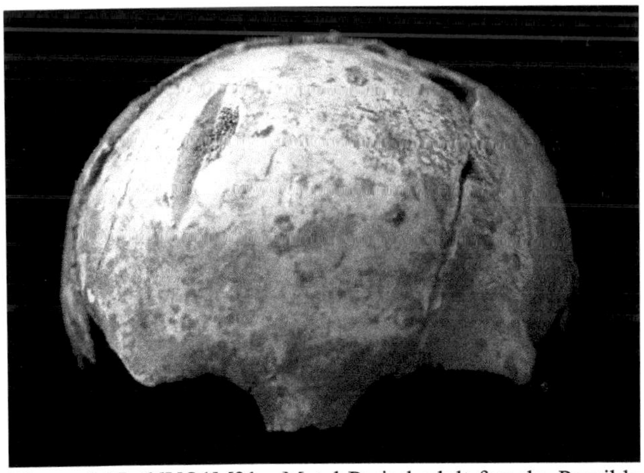

Figure 6.15. 66VQ1M31a. Metal Period adult female. Possible pre-mortem sharp instrument would to right frontal bone

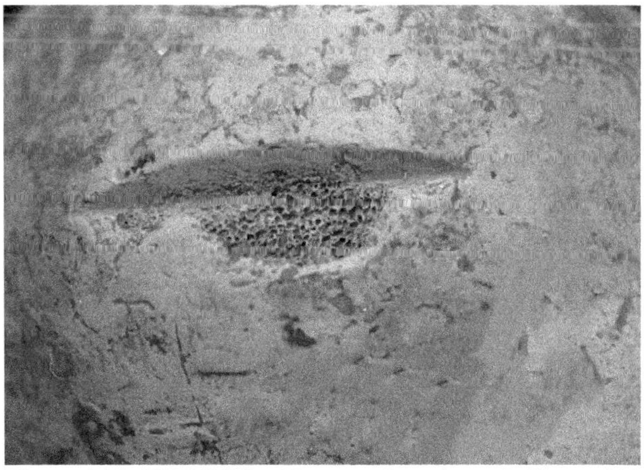

Figure 6.16. 66VQ1M31a: Close up of cut shown in Figure 6.15

Table 6.2. Metal Period. Fracture Frequencies by Skeletal Element

	left			right			total		
humerus	n	obs	%	n	obs	%	n	obs	%
proximal third	13	0	0.0	9	0	0.0	22	0	0.0
mid third	16	0	0.0	13	0	0.0	29	0	0.0
distal third	15	0	0.0	10	0	0.0	25	0	0.0
diaphysis[1]	12	0	0.0	11	0	0.0	23	0	0.0
radius	n	obs	%	n	obs	%	n	obs	%
proximal third	9	1	11.1	13	0	0.0	22	1	4.5
mid third	8	0	0.0	12	0	0.0	20	0	0.0
distal third	6	0	0.0	11	0	0.0	17	0	0.0
diaphysis[1]	5	1	20.0	9	0	0.0	14	1	7.1
ulna	n	obs	%	n	obs	%	n	obs	%
proximal third	15	0	0.0	12	0	0.0	27	0	0.0
mid third	10	0	0.0	10	0	0.0	20	0	0.0
distal third	8	1	12.5	10	0	0.0	18	1	5.6
diaphysis[1]	6	1	16.7	8	0	0.0	14	1	7.1
femur	n	obs	%	n	obs	%	n	obs	%
proximal third	9	0	0.0	10	0	0.0	19	0	0.0
mid third	17	0	0.0	19	0	0.0	36	0	0.0
distal third	10	0	0.0	10	0	0.0	20	0	0.0
diaphysis[1]	11	0	0.0	14	0	0.0	25	0	0.0
tibia	n	obs	%	n	obs	%	n	obs	%
proximal third	9	0	0.0	9	0	0.0	18	0	0.0
mid third	13	0	0.0	11	0	0.0	24	0	0.0
distal third	10	0	0.0	9	0	0.0	19	0	0.0
diaphysis[1]	9	0	0.0	9	0	0.0	18	0	0.0

Note: n refers to the number of assessable bone elements
[1]complete diaphysis

the perforation is not unlike those he has observed, in the capacity of a forensic pathologist, in Aboriginal Australian victims of spearings. This individual was also one of three cases where stag horns were placed next to the corpse (Bui Vinh 1980). The other two individuals interred with stag horns included a 40-49 year old female and a 30-39 year old male.

Case Study Twelve

80CCNM13a is a male aged 30-39 years old from pit 2 of the Con Co Ngua site. The left femur presents a number of tangential slices to the shaft. Anteriorly, there is a slice in the proximal half of the shaft. The slice is 23.0mm long supero-distally and 12.0mm wide. The surface of the of slice presents a number of grooves, separated by 2.0mm intervals, running roughly parallel to axis of shaft. The slice is shallow, only 1.0mm deep, and has a scalloped out appearance. A further slice begins distally to this one and is situated just inferior to the mid-point of the shaft. This slice terminates at a post-mortem break in the shaft.

The anterior aspect of the right femur displays a large groove or channel just superior to the mid-point of the shaft. The channel, 21.0mm wide, extends almost the entire width of the shaft and is 12.0mm wide (supero-inferiorly). At its deepest point it is 5.0mm deep. Examining the femur from its lateral aspect the channel has a very symmetrical 'U' shaped appearance (Figure 6.18). Multiple equidistant groves, 0.5mm apart, running with the channel perpendicular to the axis of the shaft, line its floor.

The posterior aspect of the right femur presents a shallow slice lateral to and at the beginning of *linea aspera.*

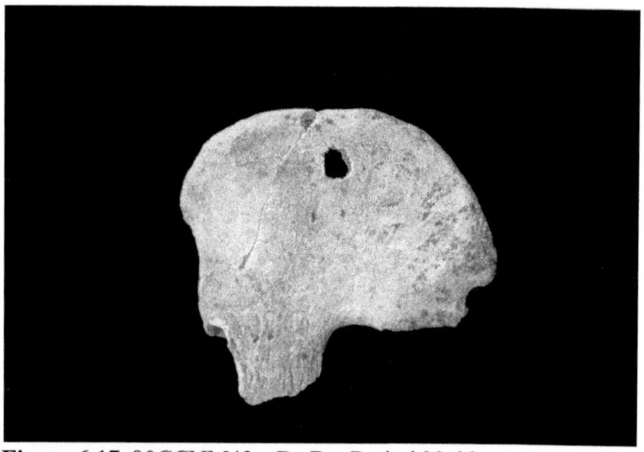

Figure 6.17. 80CCNM42a. Da But Period 30-39 year old female. Possible pre-mortem perforation of the left iliac blade (lateral aspect shown)

Table 6.3. Major Long Bone Fracture Frequencies in East Asia

	Ban Chiang[1]			Non Nok Tha[1]			Ban Na Di[2,4]			Khok Phanom Di[3]			Nong Nor[4]			Ban Lum Khao[4]			Yoshigo[5]		
	n	obs	%	n	obs	%	n	obs	%	n	obs	%	n	obs	%	n	obs	%	n	obs	%
complete humerus							16	0	**0.0**	104	0	**0.0**	n/a	0	**0.0**	n/a	0	**0.0**	83	0	**0.0**
humerus distal third	65	1	**1.5**	96	1	**1.0**															
humerus diaphysis	72	1	**1.4**																		
complete radius							18	0	**0.0**	102	0	**0.0**	13	0	**0.0**	48	3	**6.3**	105	2	**1.9**
radius diaphysis	90	2	**2.2**																		
radius proximal third	70	1	**1.4**																		
radius distal third	54	1	**1.9**	105	1	**1.0**															
complete ulna							10	0	**0.0**	74	0	**0.0**	9	2	**22.2**	40	3	**7.5**	91	0	**0.0**
complete femur							13	1	**7.7**	85	0	**0.0**	17	0	**0.0**	37	0	**0.0**	55	0	**0.0**
femur mid third	82	1	**1.2**																		
complete tibia							17	0	**0.0**	90	0	**0.0**	n/a	0	**0.0**	n/a	0	**0.0**	73	0	**0.0**
tibia mid third				125	2	**1.6**															
complete fibula							4	0	**0.0**							29	1	**3.4**	45	0	**0.0**
fibula mid third				85	1	**1.2**															

Note: n refers to the total number of assessable elements (left and right)
[1] Douglas (1996)
[2] Wiriyaromp (1984)
[3] Tayles (1992)
[4] Domett (1999)
[5] Nakai *et al.* (1999)

The slice is wide lateral to *linea aspera* and narrows as it travels distally and medially toward the *linea aspera*. Another deep cut starts just above the mid-point of the shaft directly through *linea aspera*. This slice terminates at a post-mortem break. A further slice occurs distal to the shaft's mid-point. Its surface presents many equidistant grooves and the slice is deeper proximally.

The only other skeletal element to present any sort of possible pathology was the right pubis. There is a possible pre-mortem lytic lesion on the supero-lateral aspect of the right pubis. The lesion is roughly circular, 8.4mm in diameter, and extends very deep into the trabecular bone. It has an irregular uneven edge and is considerably larger internally. There is erosion of the outer table superior to the lesion but no sign of any porosity or remodelling. The lesion was dirt filled when originally observed. This lesion would have formed in the region of the origin of the *adductor longus*. This muscle in involved in flexion and extension of the knee as well as medial rotation of the thigh (Williams & Warwick 1984).

The aetiology of the femoral slices and the gouge or channel is unclear. Animal gnawing is a possibility, although an examination of the literature failed to reveal any examples of animal gnawing that would produce the deep, symmetrical channel on the anterior aspect of the right femur. The slices are reminiscent of the sorts of tangential slices a sharp bladed implement, such as a sword, might produce from glancing blows to the femur. Such slices, cranial at least, have been recorded in European medieval battle victims and have been attributed to such glancing blows from a sword-like or axe-like weapon (Manchester 1983). However, again it is hard to conceive how the deep channel may have been formed by such actions. These femoral modifications only occur in this individual and only on these two bones. If animal gnawing were the aetiological agent it might have been expected that other bones and individuals would have demonstrated similar modifications. If a weapon or implement of some nature was involved, it could not have been metal as this is a pre-metal period site.

The Prevalence of Trauma

Da But Period and Metal Period Samples

Healed fractures for Vietnam are summarised by frequency in Table 6.1 and Table 6.2. The denominator in each case refers to a specific portion of the bone. Because all of the Vietnamese fractures occurred on the shaft, the total number of preserved diaphyses is also given for each bone. Despite reasonable sample sizes for each of the three sections of the ulna, radius and tibia in the Da But period sample (Table 6.1), no fractures were recorded. The frequency of fractures to the humerus is 2.9%, using the number of humeri with an assessable distal third as the denominator. Of the three fractures to the femur, two occurred in adults while one of the mid-shaft fractures occurred in a nine year old child. Using complete diaphyses as the denominator gives a fracture frequency of 3/46 (6.5%) for the femur. The adult femoral fracture frequency, again using the diaphysis as the denominator, is 2/45 (4.4%).

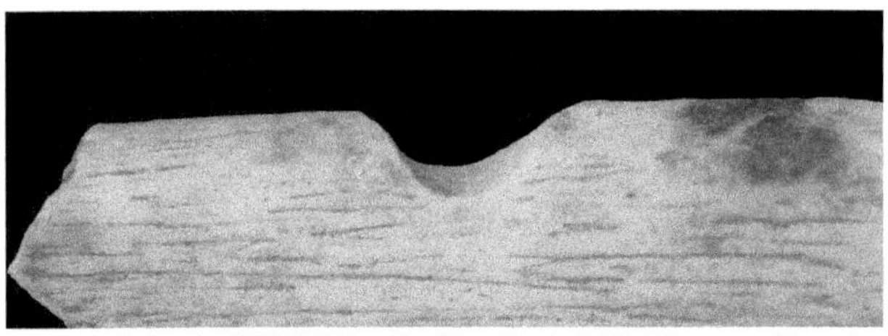

Figure 6.18. 80CCNM13a. Da But Period 30-39 year old male. Possible pre-mortem cut to the anterior aspect of the right femoral shaft, just proximal to the mid-point of the diaphysis (medial aspect shown)

The denominators are rather small for the Metal period sample (Table 6.2). It is very difficult to determine how representative the fracture frequency is. Only two fractures for the entire sample were recorded and both in the forearms. It is perhaps also relevant that these fractures belonged to individuals with good levels of preservation.

Other East Asian Sites

Table 6.3 summarises major long bone fracture rates, the number of healed fractures per total number of assessable elements, for six skeletal series from Thailand and one series from a Japanese Late-Jomon (3400 to 2400 years bp) period site. The denominator for Ban Na Di and Khok Phanom Di is based on the number of complete long bones used in stature estimation for each of these studies. The Ban Na Di skeletal assemblage included 32 individuals aged over 15 years in a variable state of preservation (Wiriyaromp 1984). The preservation of the Khok Phanom Di has been described as excellent (Tayles 1992). Neither Ban Na Di nor Khok Phanom Di displayed any fractures to major long bones. Wiriyaromp (1984) does, however, note one healed first metatarsal fracture. Tayles (1992) reports healed fractures of two clavicles, two metacarpals and a metatarsal.

Two healed radial fractures (Table 6.3) are reported for the Jomon period site (Nakai *et al.* 1999). However, this site included numerous healed fractures of smaller bones: 2/65 clavicles (3.1%); 6/511 metacarpals (1.2%); 1/798 hand phalanges (0.1%); 2/124 tali (1.6%); 1/82 calcani (1.2%); 2/606 metatarsals (0.3%); and 6/468 foot phalanges (1.3%) (Nakai *et al.* 1999).

Douglas (1996) recorded five healed fractures of major long bones for Ban Chiang (Table 6.3) including one femoral fracture, one humeral shaft fracture and an avulsion fracture of a humeral medial epicondyle. In addition there were healed fractures to 9/558 (1.6%) ribs; 1/39 (2.6%) hamates; 2/296 metacarpals (0.7%); 1/237 metatarsals (0.4%); 3/55 cervical vertebrae (5.5%); 1/7 hyoids (14.3%); and 2/49 frontal bones (4.1%).

For Non Nok Tha, Douglas (1996) recorded five major long bone healed fractures (Table 6.3). These included a supracondylar fracture of a humerus, two midshaft tibial fractures, one of which included the fibula and a fracture of the distal radius. In addition there were healed fractures to 3/455 ribs (0.4%); 1/34 hamates (2.9%); 1/75 metacarpals (1.3%); 2/69 cervical vertebrae (2.9%); 2/80 parietals (2.5%); and 4/90 clavicles (4.4%).

Domett (1999) notes fractures to two ulnae (22.2%), one clavicle (5.6%), a foot phalanx and the body of a mandible for Nong Nor, Thailand. For Ban Lum Khao, Thailand, the majority of fractures affected the forearms with three radii (6.3%) and three ulnar (7.5%) fractures (Domett 1999). Other Ban Lum Khao fractures included a clavicle (2.6%), a fibula (3.4%) and a possible broken nose (Domett 1999). For Ban Na Di a fractured third metatarsal and a single femoral (7.7%) fracture are reported (Wiriyaromp 1984, Domett 1999).

Discussion and Conclusions

While the proximal causes of the Vietnamese fractures have been discussed, within each case study, the ultimate cause of each fracture is difficult to approach. The patterning and frequency of fractures in Vietnam is not particularly informative for the Metal period in particular. Until further data are available, it is tentatively suggested that Metal period behaviour was not associated with a great deal of risk from fractures. The forearm fractures are entirely consistent with misadventure. The potential sharp implement wound to the frontal bone of case ten represents the only apparent example of interpersonal violence in the Metal period. The lack of further such examples of interpersonal violence might be interpreted as suggesting that this was not a particularly common behaviour during this period in northern Vietnam. This is however, not consistent with historical and material culture evidence for this period. The skeletal assemblage representing the Metal period has been drawn from a number of village sites. The nature of warfare in the Metal period may have been such that villages were not foci of conflict.

The frequency and patterning of trauma in the Da Dut period sample is quite different. While this sample is quite fragmentary, there are numerous well preserved sections of most major long bones, with the exception of the fibula. Moreover, the bones displaying fractures are among the most robust bones in the body. It is worth pointing out that as serious as these fractures would have been, they could all have healed without any sort of intervention, such as splinting (Steve Baddeley pers. comm.). However, this does not mean that some sort of treatment was not attempted.

The types of fractures sustained by these individuals from Con Co Ngua could all have occurred as a result of some form of fall. The isolated case of myositis ossificans traumatica (case eight) and the mandibular condyle pathologies (case nine) also represent instances of significant trauma. Including the potential case of peri-mortem trauma to the hip (case eleven) indicates that 8/94 (8.5%) of individuals from Con Co Ngua suffered some form of major trauma.

It is difficult to determine what type of behaviour would have predisposed this population to such levels of risk. As highlighted in Chapter one, the region was home to numerous large animals, some of which may have been hunted by this community. The elevation of the landscape was quite varied within the immediate environs of Con Co Ngua. This would have provided the opportunity for misadventure at least. What does seem clear is that the risk of major long bone trauma, femoral and humeral fractures, was greater at Con Co Ngua than in any other skeletal assemblage thus far examined in Southeast and East Asia. Isolating specific behaviours predisposing this particular population to traumatic injury can only be highly speculative.

Non-Traumatic Lesions

The identification of specific diseases in archaeological specimens is not straightforward or even possible in many instances. Miller *et al.* (1996:224) suggest that impediments to diagnosis in archaeological material include: "the paucity of well-documented, clinically diagnosed skeletal samples, and the difficulty in finding skeletal abnormalities or patterns of abnormalities that are unique to individual disease entities". For the most part, bioarchaeologists seek comparative and diagnostic data in the clinical literature. However, disease pathology as assessed and described in living individuals is not necessarily compatible with what is observed in a defleshed skeleton (Ortner 1991). Miller *et al.* (1996:225) state that:

> Accurate palaeopathological diagnosis depends on identification of characteristic bony changes in an expected distribution ('constellation'). This requirement is not met by early cases and only by a minority of fully developed cases. Modern material teaches that the 'classic' morphologic expression of a given disease process represents, in fact, the narrow tails of the normal distribution.

Moreover, Tayles (1996:22) has noted that:

> A difficulty with using the evidence described and illustrated in clinical reports as a basis for making a diagnosis from dry bones is that clinical reports tend to describe the more extreme cases, where the evidence is radiographically visible. As a consequence of this, many textbooks describe these extreme skeletal reactions as if they should be expected in any individual suffering from a disease.

Neither disease pathogens nor their hosts are adaptively or evolutionarily stagnant. Ortner (1991) points out that there is a tendency for infectious diseases to increasingly manifest as chronic conditions over the length of their evolutionary existence. The implication of this is that changes in the frequency of specific diseases over time may be due more to alterations in their virulence or expression rather than the proportion of the population affected. Nonetheless, it seems reasonable to suggest that the identification of disease using higher order classifications (eg., metabolic, circulatory, neoplastic etc.) may be accurately achieved (Miller *et al.* 1996). Moreover, the increase in publications seeking to specifically differentiate between otherwise grossly similar (as expressed skeletally) conditions greatly assists in the more accurate diagnosis of disease in dry bones. At the very least, such works help limit differential diagnoses to manageable levels.

Because the pathologies present in the Vietnamese material have not been systematically described and recorded to date, this forms the chief aim of this section. Pathological elements, and in certain cases all of the skeletal material representing an individual, were examined both macroscopically and radiologically. This is however, not only an exercise in describing these conditions and attempting to delimit a range of potential or likely causes. It is also important to address the implications of the nature and extent of disease in these samples. What are the implications for the environment-host-culture dynamic with respect to the observed types and frequency of skeletally sensitive disease in the Vietnamese samples? Has the nature and frequency of disease, defined as skeletally sensitive conditions, changed over time in Vietnam? Can such a question be practically addressed? Such issues are explored in the discussion section following the descriptions and differential diagnoses of the case studies.

Non-Traumatic Lesion Case Studies

A description of the skeletal and dental material available for each individual in the following case studies is presented in Appendix 1.

Case Study Thirteen

78QCM3a, a male aged 30-39 years, from the Metal period site Quy Chu presents articular lesions. The condition affects both knee joints but the right femur and left tibia are the only bones representing the knees to have survived. The inferior aspect of the right femoral distal condyle manifests a circular lesion cut in and around the articular surface. This lesion is doughnut-like in shape with a groove around the outer edge and a ballooning in the centre, with a pit in the very centre. The overall diameter of the lesion is 16.5 mm (Figure 6.19).

The distal right femur displays moderate development of an osteophytic rim on the inner, medial and lateral aspects of the posterior condyles as well as to the anterior aspect of the condyles. Moreover, a slight surface build up of bone on the central-anterior aspect of the dorsal face of the distal condyles is also evident. The femoral head also displays a very large and deep *fovea capitis*.

The left tibia presents a roughly circular lesion on the central-posterior aspect of the articular area of the proximal lateral condyle (Figure 6.20). The lesion displays a slightly porous perimeter and slopes down into a central pit. The lesion is somewhat similar in appearance to the right femoral lesion, but this one on the tibia does not balloon out in a doughnut-like fashion. In fact, it more resembles the negative impression you would expect from the femoral lesion pushing into the proximal tibia. The overall diameter of this lesion is 9.7mm. The tibia also exhibits marked enthesophytic development of the soleal (popliteal).

No other pathologies were observed for this individual. However, the orbits displayed extensively remodelled open-cribra orbitalia and the only preserved canine, left maxillary, displayed two enamel hypoplastic bands. There were also numerous pacchionian pits endocranially. This individual had a very pronounced nuchal crest, large mastoids and pronounced canine juga.

Differential Diagnosis

An extensive search failed to locate any specific conditions completely compatible with the lesions described for this case. The following conditions are tentatively forwarded as possibilities only.

Osteochondritis Dissecans

Ortner & Putschar (1981:242) describe osteochondritis dissecans as "a type of aseptic necrosis characterized by the formation of a small, in profile triangular, sequestrum consisting of articular cartilage on one surface and necrotic subchondral cancellous bone". The necrotic bone can eventually become detached to form a loose osteocartilaginous body within the joint. The case under consideration here may be representative of the early stages of this condition. Ortner & Putschar (1981) note that the condition displays a major predilection for the knee, although generally manifesting on the lateral aspect of the medial condyle. Further, they suggest that this is a disease of adolescents and young adults.

Synovial Haemangioma

These are described as benign vascular tumours of the joint by Aufderheide & Rodrígues-Martín (1998). The most common site is the knee. These authors also note that intralesional calcification is common. However, synovial haemangioma manifests unilaterally and is monoarticular, thus tending to exclude it from further consideration. Further, bony changes to the joint are not discussed for this particular condition.

Chondrocalcinosis

This is a pseudogout, also termed calcium phosphate

deposition disease. The condition is characterised by the "deposition of dihydrate calcium pyrophosphate crystals in the articular tissues with calcification within the cartilage" (Aufderheide & Rodrígues-Martín 1998:114). Bony change associated with the condition is not discussed.

Articular Chondromas
This condition is also referred to as dysplasia epiphysealis hemimelica as well as Trevor's disease (Resnick & Niwayama 1981). These same authors describe the condition as a form of cartilage proliferation that can involve several epiphyses but on one side of the body. It is said to be commonly observed on the talus, distal femoral and distal tibial epiphyses (Resnick & Niwayama 1981). The bony changes associated with the condition are not discussed. That this condition appears to only manifest unilaterally would tend to exclude it from further consideration.

Case Study Fourteen
80CCNM74a was a male aged 50+ years old from Con Co Ngua pit 2. The right humerus presents a large spherical lesion on the posterior aspect of the lateral epicondyle. The lesion has completely resorbed the posterior face of the condyle and extends to, and encroaches on, the postero-lateral border of the capitulum. The lesion would also have encroached on and probably obliterated the origin of the *anconeus*, a muscle that assists the *triceps* in extension of the elbow (Williams & Warwick 1984). The interior walls are smooth with some porosity. They are discoloured black, as if painted with tar. There is no sign of vascularity surrounding the lesion (Figure 6.21). The lesion is 16.8mm in diameter and is 8.0mm at its deepest, which is the centre of the lesion. The regular shape and steeply sloping smooth interior walls of the lesion are what would be expected were a hot marble pressed into soft clay. The preserved fragment of the proximal humeral head does not display anything unusual. A radiographic examination of all the skeletal material attributed to this individual failed to reveal any further lesions.

In addition to the lesion there is evidence of a ridge of osteophytic bone along the lateral border of the trochlea and some minor evidence of porosity on the antero-medial aspect of the trochlea. The right proximal radius was too eroded post-mortem to assess articular change. However, the distal radial articular area, medial aspect for articulation with the lunate, displays a raised and very porous plateau of osteophytic bone. There is eroded lipping of the posterior border of the distal articular facet and extensive lipping on the anterior border as well. The styloid process has been destroyed post-mortem. No carpal bones were preserved.

Only the proximal three-quarters of the right ulna is preserved. The proximal head is very rugged in appearance with a marked development of the supinator and interosseous crests. There is also a marked spur on the olecranon representing ossification of the *triceps brachii* insertion. The trochlea notch is well developed and there is extensive marginal lipping to the superior and medial edges of the olecranon (semi lunar notch), all around the radial notch and anterior and medial aspects of the coronoid process. The highest point on the floor of the coronoid process also exhibits localised porosity.

The left arm also displayed considerable evidence for articular change. While the proximal humerus exhibited porosity only, the distal articular end displayed marginal osteophytes on the superior border of the anterior aspects of the trochlea and capitulum, extending onto the medio-posterior and latero-posterior edges of the trochlea and capitulum respectively. An osteophytic ridge runs medio-laterally along the ventral aspect of the capitulum. There is also osteophytic development on the anterior aspect of the lateral edge of the trochlea. An extensive build up of exuberant bone has developed within the olecranon fossa and is concentrated on the medial-inferior portion, but fills most of the floor of the fossa. The articular regions of the left radius were not preserved. The left ulna was missing the distal articular region. Proximally, the left ulna displayed osteophytic development to a much lesser degree than the right. There was slight intermittent marginal lipping to the radial notch and antero-superior section of the olecranon. The few preserved phalanges displayed marked medial

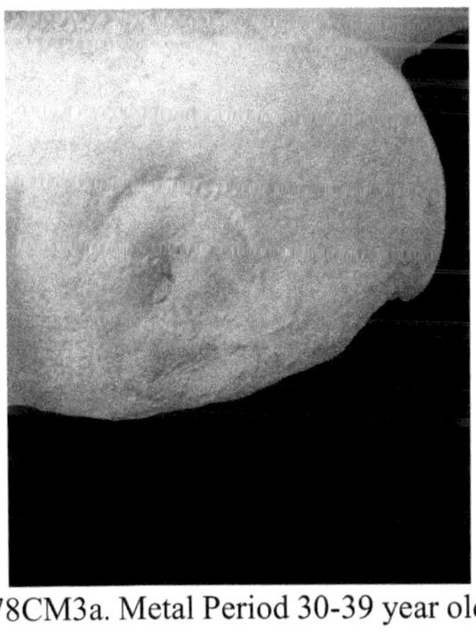

Figure 6.19. 78CM3a. Metal Period 30-39 year old male. Lesion presenting on the inferior aspect of the distal lateral condyle of the right femur (inferior aspect shown)

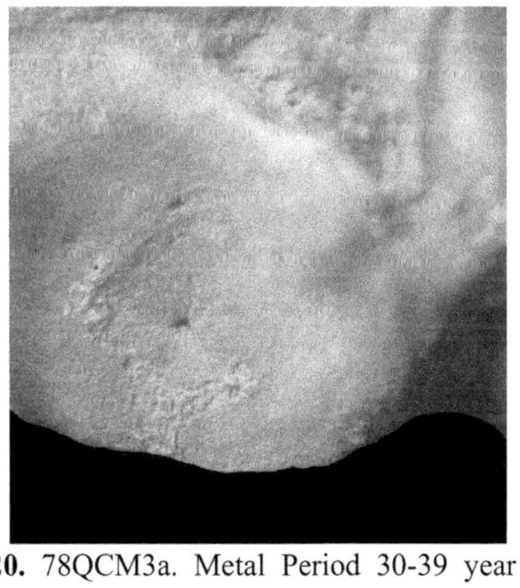

Figure 6.20. 78QCM3a. Metal Period 30-39 year old male. Lesion presenting on the proximal lateral condyle of the left tibia (superior aspect shown)

and lateral ridging to the palmar aspect of the diaphyses.

Other preserved skeletal elements displayed articular changes also. The left clavicle has a well developed rhomboid fossa, while the right clavicle presents a distinct conoid joint process. Both glenoids of the scapulae displayed marginal osteophytic development. The distal condyles (proximal heads not preserved) of the right femur displayed marginal lipping while the left did not. The left distal femur did, however, present a lesion, 5.0mm in diameter, inferior to the lateral epicondyle. There are numerous large pits anterior and superior to the lesion. This possibly represents trauma or stress to the collateral ligament.

While the articular portions of both tibiae were preserved, the proximal articular areas were too eroded post-mortem to assess for articular changes. The distal ends were free of obvious changes however. Both patellae were available with the posterior aspect of the left displaying localised porosity to the supero-lateral aspect of the medial facet. There was also porosity and osteophytic development in the mid to inferior central section of the medial facet. The limited number of vertebrae were too eroded post-mortem to assess. Of the ribs, four left and one right articular section were preserved. Two of these ribs displayed considerable lipping and porosity of the facets.

In summary, this individual displayed extensive marked osteophytic and porous changes to most preserved articular regions. These changes are consistent with the advanced age of the individual. However, the right arm was affected by these changes to a greater degree than the left.

Differential Diagnosis

Chondroblastoma

Aufderheide & Rodrígues-Martín (1998) point out that 98% of epiphyseal lesions are due to osteomyelitis, chondroblastoma or giant cell tumour. The lesion under consideration here bears no morphological resemblance to bony changes consistent with osteomyelitis and this condition can be excluded from further discussion.

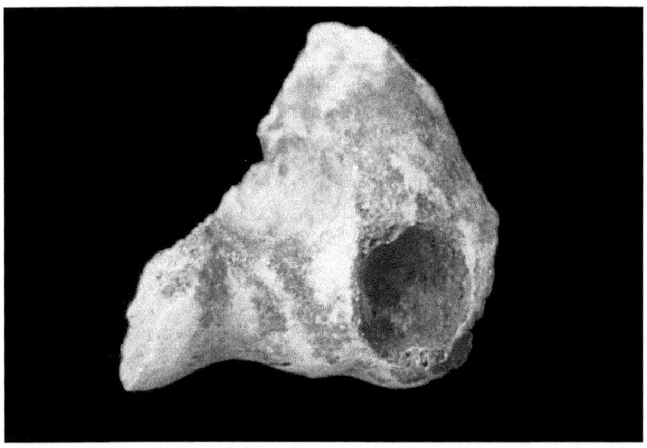

Figure 6.21. 80CCNM74a. Da But Period 50+ year old male. Large spherical lesion presenting on the posterior aspect of the lateral epicondyle. Right humerus (posterior aspect shown)

Chondroblastoma is a rare benign tumour accounting for only 1% of all bone tumours (Aufderheide & Rodrígues-Martín 1998; Mazabraud 1998). These tumours only form in epiphyseal and apophyseal sites and some three-fifths of documented modern cases appear in males (Mazabraud 1998). Mazabraud (1998) suggests that chondroblastomas are more frequently found in younger individuals with some 70% manifesting in the second decade. Resnick & Niwayama (1981) cite the greatest incidence of this tumour occurring during the third and fourth decades. The most common sites include the humeral and femoral heads (Resnick & Niwayama 1981; Mazabraud 1998). It is also common in the greater trochanter, cranium, ribs, iliac bones, os calais (Mazabraud 1998) and proximal tibia (Resnick & Niwayama 1981). However, Aufderheide & Rodrígues-Martín (1998) suggest that this tumour is extremely rare in the skull. They also note that other rare sites include the hands, feet, flat bones in general and spine. None of the aforementioned sources make explicit reference to the frequency of this tumour in the distal humerus.

The tumour can range in size from 1 to 5cm, has lytic foci, may or may not destroy the outer cortex with expansion and often displays surrounding periostitis (Aufderheide & Rodrígues-Martín 1998). Rothschild & Martin (1993) also note that a periosteal reaction is usually present even if very subtle. However, Resnick & Niwayama (1981) suggest that periostitis of surrounding bone occurs in only about a third of cases. They also note that morphologically, the lesion is well defined and either spheroid or oval. Furthermore, calcification within the lesion can occur about half the time.

The size, epiphyseal location, well-defined spherical morphology, lack of periosteal reaction and apparent expansion and subsequent destruction of the outer cortex with respect to the lesion under consideration is consistent with a chondroblastoma. The age-at-death of this individual has been estimated at 50+ years and while this tumour is most common from the 20s through 40s, it can occur at any age (Mazabraud 1998). As noted above, this is a benign tumour, probably asymptomatic during the life of the individual, but these tumours have the potential to lead to considerable joint complications (Mazabraud 1998).

Pigmented Villonodular Synovitis

Pigmented Villonodular Synovitis produce "monoarticular large, well-demarcated periarticular erosions with sclerotic margins" (Rosthchild & Martin 1993:183). The condition is related to giant cell tumour (Rosthchild & Martin 1993), other names include giant cell tumour of the tendon sheath and benign synovioma (Aufderheide & Rodrígues-Martín 1998). While the most common site is the knee, followed by the hip, ankle and foot, other sites include the wrist, shoulder, TMJ and elbow (Rosthchild & Martin 1993). This benign neoplasm has its highest incidence between the ages of 20 and 40 years (Aufderheide & Rodrígues-Martín 1998). When discussing a differential diagnosis in considering chondroblastoma, Mazabraud (1998) suggests that based on radiological evidence alone villanodular synovitis is a possibility. Moreover, Ortner & Putschar (1981:435) state that a "solitary lesion of this type could, in the dry bone, be indistinguishable from an epiphysial tumour (chondroblastoma or enchondroma)".

Juxta-Articular Cyst
This is not a tumour, but rather a cyst of unclear aetiology. Mazabraud (1998) indicates that the aetiology could include necrobiosis or a cyst related to vascular disturbance in a weight bearing area. Another possibility includes increased synovial fluid pressure. Mazabraud (1998) points out that for the most part the aetiology of this cyst is idiopathic. It is generally asymptomatic and is thus only noted by chance in living populations. The frequency of the condition is therefore unknown. The highest incidence of the cyst, or its detection at least, occurs between 25-55 years (Mazabraud 1998). While the cyst can arise anywhere it is most commonly found in the wrist followed by the ankle, elbow, proximal femur, acetabulum and proximal tibia (Mazabraud 1998). Radiographic imaging reveals a clear round or oval morphology localised in the epiphysis close to the articular cartilage (Mazabraud 1998).

Giant Cell Tumour
Giant cell tumours are more common than chondroblastomas, in that they are responsible for between 5-7% of all bone tumours and 20% of benign tumours (Mazabraud 1998). These tumours are also more common in later life with the highest incidence in the third and fourth decades (Mazabraud 1998). Approximately half the documented cases manifest in the knee, followed by the lower radius, sacrum and then lower ulna and fibula (Mazabraud 1998). While about three-quarters manifest near the ends of long bones they are confined to metaphyseal or metaphyseo-epiphyseal sites (Mazabraud 1998). Steinbock (1976), on the other hand, states that giant cell tumours, or osteoclastomas, nearly always have an epiphyseal location but can occasionally spread to include the metaphysis. It is possible that Steinbock (1976) has conflated the two separate conditions of giant cell tumor and chondroblastoma. Another name for chondroblastoma is giant cell chondromatous tumour of Codman (Aufderheide & Rodrígues-Martín 1998), which may explain the confusion. If it is accepted that giant cell tumours are only metaphyseal or metaphyseo-epiphyseal in location, then the strictly epiphyseal lesion under consideration here is unlikely to represent a giant cell tumour.

Other Possibilities
Steinbock (1976) suggests that radiologically, a fibrosarcoma of the epiphyseal region can resemble a giant cell tumour. The age of occurrence is also similar for giant cell tumours. However, Steinbok (1976) also notes that most examples display a periosteal reaction, something that is absent in the lesion under consideration. Mazabraud (1998) suggests in his differential diagnosis for chondroblastoma, that sarcomas such as chondrosarcoma of chondroblastic type with epiphyseal localization and epiphyseal osteosarcoma need to be excluded. However, this can only be done histologically. Moreover, the bony changes left by such sarcomas are not detailed.

Case Study Fifteen
70DCM1a was a child aged 5 years +/- 16 months, from the Metal period site Duong Co. This individual presents a lytic lesion on the rear, left parietal at the level of lambda. It is 43.0mm from lambda to the centre of the lesion and 20.0mm from the centre of the lesion straight down to the left lambdoid suture. The lesion is slightly oval in shape, 7.4mm antero-posteriorly and 5.7mm medio-laterally. It has a very regular perimeter but its edges are irregular or bumpy on close inspection (x10 magnification). The lesion penetrates the inner table in the most postero-inferior aspect of the floor of the lesion. The inner table perforation is 2.3mm in diameter and is not surrounded in porosity or any obvious discolouration of the bone. However, it is possible that this perforation of the inner table is a post-mortem artefact. The lesion penetrates the outer table and diploe with an average depth of 3.0mm. The floor of the lesion appears slightly remodelled or sclerotic. A pale but distinct ring of discoloured bone encircles the lesion, which is wider inferiorly than superiorly. The pale ring of bone is dotted with numerous tiny pits indicating some degree of vascularity in life (Figure 6.22).

With the exception of open-cribra orbitalia, no other pathologies were present. None of the preserved primary dentition (only the deciduous maxillary molars were preserved) displayed enamel hypoplasia. Radiologically, the cranium appeared to be free of any other pathological conditions.

Differential Diagnosis
Without accompanying postcranial material it will be perhaps impossible to isolate the cause of the lesion in this case. Nonetheless, given the young age of this individual several possibilities can be mooted.

Langerhan's Cell Histiocytosis
Langerhan's cell histiocytosis comes under the classificatory umbrella of reticuloendothelial disorders (Ortner & Putschar 1981). This condition, formerly referred to as histiocytosis X, is sub-classified with respect to three forms: localised (eosinophilic granuloma of bone); disseminated acute or subacute (Letterer-Siwe's disease); and disseminated chronic (multifocal eosinophilic granuloma or Hand-Schuller-Christian disease) (Mazabraud 1998). The localised form mainly affects older children and young adults while the disseminated acute form affects children under two years of age (Mazabraud 1998). The skeletal lesions associated with the disseminated chronic form, occurring mainly in later childhood and adolescence, can also be associated with diabetes insipidus and exophthalmos (eyeball protrusion) (Mazabraud 1998).

Ortner & Putschar (1981) point out that while the bony lesions in each form are similar, they can be differentiated with respect to distribution. The skull is the most commonly affected region with the localised form, eosinophilic granuloma of bone, commonly presenting as an isolated round or oval defect with a bevelled edge (Ortner & Putschar 1981). The other two forms manifest cranially with multiple lytic foci. Ortner & Putschar (1981) go on to say that periosteal reactive bone is not usually seen, even in cases where both tables have been destroyed, although a periosteal reaction can occur (Rothschild & Martin 1993). Further, this form is described as self-limiting and benign (Ortner & Putschar 1981), with the lesions able to regress and eventually disappear (Mazabraud 1998). The lesion in this case was not bevelled and there was clear evidence of increased vascularity or periosteal reaction surrounding the lesion. It is not clear if bevelled edges are a crucial diagnostic feature of this condition. Moreover, the morphology of a healing or regressing lesion is not clearly described either.

Epidermoid Cyst

The following discussion is based on Mazabraud (1998). The epithelial tissue itself can be introduced by way of trauma and as such these cysts are commonly found in bones close to the skin and favour the cranium, jaws and phalanges. Radiologically, they are described as forming a clear cavity and can be surrounded by a sclerotic layer. Cranially, they can present as well circumscribed areas of osteolysis that can destroy both tables. These cysts can be found at any age and can even develop from an inclusion in the first stages of foetal development. Such cysts can disappear suddenly and continuous growth is said to be rare.

Tuberculosis

It is necessary to include tuberculosis as a possible source of this lesion as when it does manifest in the skull it is generally in young children rather than adults (Ortner & Putschar 1981). In a survey of the cranial distribution of tuberculous lesions, Strauss (1933 cited in Ortner & Putschar 1981) noted that the vast majority occurred in either the frontal or parietal. Moreover, most of the victims were less than 10 years old. However, Ortner & Putschar (1981) point out that such lesions in children are often multiple. Moreover, the lesion in this case is only superficially similar to Ortner & Putschar's (1981) description of a tuberculous cranial lesion. These authors note that both the inner and outer tables are perforated and the inner table generally displays greater destruction than the outer table. This is not the case for this child and is unclear if the small perforation of the inner table is a post depositional artefact or a true communication between the endo and ectocranium.

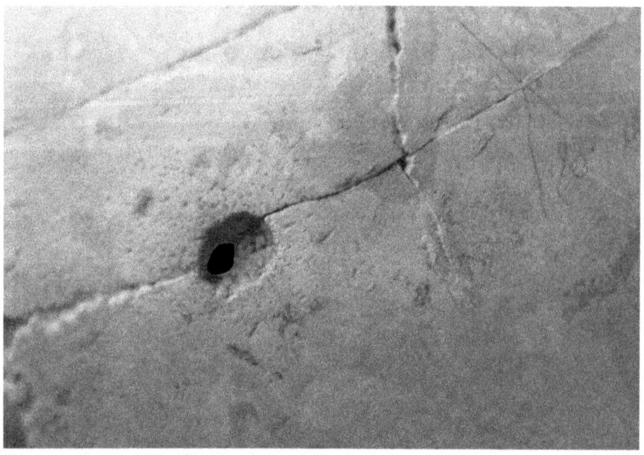

Figure 6.22. 70DCM1a. Metal Period 5 year (+/- 16 months) old child. Lesions presenting on posterior aspect of left parietal bone (lambdoid suture to bottom left and dorsal aspect of cranium to top of photo)

Metastatic Neuroblastoma

In their discussion of tuberculous cranial lesions, Ortner & Putschar (1981) suggest that with children a metastatic neuroblastoma must also be considered with respect to cranial lesions. This is described as "a malignant tumor of the adrenal or of the sympathetic chain in infants and children, usually under 3 years of age" (Ortner and Putschar 1981:395). As with tuberculosis a more certain diagnosis would require the post cranial skeleton which should exhibit extensive periostitis of the long bones. Nonetheless, Ortner & Putschar (1981) suggest that osteoblastic reactions are more common, something which is not a feature of this case.

Other Possibilities

Mycotic or fungal infections can produce lytic lesions in bone, although it would appear almost impossible to differentiate the various forms based on bony morphology alone (Ortner & Putschar 1981). There are three possibilities described by Ortner & Putschar (1981), among several that can affect bone, with global distributions that may be relevant here. Histoplasmosis, while being most common in North America is distributed world-wide. Bony lesions are not common but the African form (histoplasmosis duboishii) can cause multiple round lesions in the skeleton including the cranium. Sporotrichosis is described by Ortner & Putschar (1981) as principally a skin or soft tissue infection, the organism entering the body through a cut or abrasion. The organism mainly affects children and young adults. Bony lesions manifest in only about 10% of cases, can affect most of the skeleton but favour the cranium and tibia. However, Ortner & Putschar (1981) note that the lesion is generally periosteal rather than lytic. Cryptococcosis can also lead to well circumscribed lytic lesions in most bones including the cranium (Ortner & Putschar 1981). Additionally, two other fungal infections may need to be considered that are known to both occur in Southeast Asia. Mycetoma, both bacterial and fungal vectors, can spread by way of lymphatic or hematogenous pathways to eventually cause extensive and destructive osteomyelitis (Chandler & Watts 1987). Penicilliosis marneffei, believed to be endemic to Southeast and Far East Asia, is also known to produce osteolytic lesions (Chandler & Watts 1987). Fungal infections are discussed in more detail for case study twenty.

Case Study Sixteen

76NNM2c is a probable male aged 40-49 years from the Metal period site Nui Nap. The prefix 78 on the scale in the following photographs is an error and should read 76. There is a large lytic lesion just superior to the external auditory meatus on the left temporal bone (Figure 6.23). The structures of the external auditory meatus are nearly completely lost post-mortem, all that remains is the superior border of the meatus including most of the TMJ fossa. The lesion is positioned just superior, and slightly posterior, to the external auditory meatus. The antero-inferior border of the lesion encroaches on the external superior border of the meatus. In fact, it is situated directly over the usual position of the suprameatal triangle The lesion is almost perfectly circular, 9.0mm in diameter and intrudes approximately 9.0mm at its deepest point. The floor of the lesion is composed of trabecular bone and the sides of the lesion are also irregular or rough.

Scattered porosity surrounds the lesion and there is a localised area of slightly raised, very porous, bone on the posterior aspect of the lesion. This likely represents mild and localised osteitis or periostitis. The superior edge of the lesion is quite abrupt in that it does not gently slope into the cavity. The inferior border is slightly damaged, probably post-mortem.

The lesion possibly communicated with the mastoid antrum, but is difficult to be certain due to post-mortem breakage and damage to the petrous process. The petrous portion was preserved, albeit damaged post-mortem, and was missing the roof of the carotid canal along most of its length. It is difficult to determine whether this is post or pre-mortem damage.

In addition to this lesion, both orbits display extensive remodelled cribra orbitalia. Endocranially, there is extensive pacchionian, or arachnoid, pitting running from mid-frontal to vertex all along sagittal plane. Very deep meningeal grooves run parallel to the coronal suture with the right one, 15.0mm in diameter, ending in a large pacchionian lacuna near bregma. Numerous pits are present within the left meningeal groove and tributaries.

Dentally, this individual presented a severely worn dentition, with eight teeth displaying attrition induced pulp chamber exposure. Due to this level of wear there were numerous alveolar defects of pulpal origin

Differential Diagnosis

It is difficult to determine if this lesion is strictly otological in origin. Because there is no obvious communication between the lesion and the internal architecture of the ear, there are a number of possibilities that need to be discussed in a differential diagnosis.

Mastoiditis

This condition is discussed in more detail in the following two cases, which display destruction of the mastoid process. As noted, there is probably a lack of communication between this lesion and the inner ear. This would suggest the origin of the lesion is not to be found in the eustachian tube. A superficially similar lesion is illustrated by McKenzie & Brothwell (1967), in a case reported on by Elliot Smith and Wood Jones in 1910. As with the Vietnamese case, the lesion was located on the suprameteal triangle. However, in that instance the lesion communicated with the antrum, thus suggesting the likelihood of an ear infection. It also needs to be noted that mastoiditis is more a condition of young children (Kemink *et al.* 1993). The lesion under consideration here has clear signs of recent vascular activity, suggesting it is not simply an unremodelled lesion of childhood.

Madura Skull

Yagi *et al.* (1998:68) indicate that mycetoma "should be considered in the differential diagnosis of mastoid bone pathology in the tropics". They report on a case study, a 22 year old Sudanese female, with mycetoma of the mastoid. The condition resulted in bony destruction of the mastoid. Yagi *et al.* (1998) suggest that the fungus was probably introduced into the ear by way of an ear-cleaning implement. Typically, the pathogen is introduced via traumatic inoculation.

Chondroblastoma

This condition is discussed in detail for case study fourteen. It is not uncommon in the cranio-facial region and the temporal bone is a very common site when only considering the skull (Mazabraud 1998). While the morphology of this lesion is described for the long bones (Mazabraud 1998), it is unclear if there are any significant differences in a cranial manifestation.

Aneurysmal Cyst

This condition is rarer than giant cell tumour but more common than chondroblastoma. While it is most commonly found in long bone metaphyses, other sites include the vertebrae (posterior arch more so than the body), ribs, pelvis and skull, especially the temporal bone and mandible (Mazabraud 1998). The dry bone morphology of such a cranial focus of this cyst is unclear from the literature.

Figure 6.23. 76NNM2c. Metal Period 40-49 year old probable male. Lytic lesion, with associated slight periostitis, situated over the suprameatal triangle of the left temporal bone (anterior of cranium to left of photo)

Neoplasm
Friedmann (1974) points out that neoplasms of the ear are not uncommon. Further, benign cysts frequently form between the angle of the pinna and mastoid region (Friedmann 1974). Such lesions can readily become infected. Another possibility is a spread of a tumour or abscess from the pinna to the adjacent temporal region. Wright (1997) suggests that such can occur in the case of basal cell carcinoma, a non-metastasising neoplasm, originally of the pinna. A true metastasis of a carcinoma is also a possibility.

Langerhan's Cell Histiocytosis
Langerhan's cell histiocytosis, also discussed previously, needs some consideration. Solitary eosinophil granulomas are often localised in the temporal bone (Friedmann 1974) and the temporal bone is commonly the first affected area in Langerhan's cell histiocytosis (Brookes & Booth 1997). In their own study of 22 cases of ear or temporal bone histiocytosis X, McCaffrey & McDonald (1979) noted that while the majority of cases occurred in children under 10 years of age, there was, however, a later small peak in occurrence between 30-39 years of age.

Other Possibilities
Epidermoid cyst, discussed previously, needs to be included in a differential diagnosis.

Case Study Seventeen
77NNM10ka was a male, aged 20-29 years, from the Metal period site Nui Nap. This case presents extensive destruction of the mastoid process of the left temporal bone. Much of the outer table of the mastoid is completely eroded leaving visible the network of small honey comb-like sinuses. There are several more isolated holes in the superior region of the process just inferior to the most posterior extension of the zygomatic crest, the supramastoid crest. There are at least three confluent lesions in this region ranging in size from 4.5mm to 5.8mm in diameter (Figure 6.24). Each has regular smooth sloping edges. There is also a lesion at the tip of the process with a similar smooth regular edge. Adjacent to this lesion there is another tiny lesion, also with smooth edges. There is no sign of pitting or vascularity associated with any of the lesions.

The latero-posterior aspect of the condyle of the mandible has been eroded. There is also the remnant of a circular lesion on the dorso-lateral edge of the condyle. The lesion is 2.2mm in diameter with smooth edges and walls. This may be related to the extensive lytic destruction of the left mastoid.

Both orbits displayed clear evidence of highly remodelled cribra orbitalia. Moreover, orange peel-like ectocranial porosity affected the rear frontal, and parietals above the temporal lines, while much of the occipital was missing so could not be assessed. Endocranially, there were pacchionian pits on the left parietal adjacent to the sagittal suture. In fact, there is a large, 40.0mm diameter, depression just distal to bregma and bisected by the sagittal suture. This is probably a large, shallow pacchionian lacuna. This lacuna contains several clusters of smaller pacchionian pits. Such extensive granulation activity is unusual in such a young individual (Mann & Murphy 1990).

This individual was free from alveolar defects of pulpal origin and carious lesions, but the maxillary canines and incisors displayed enamel hypoplasia. The anterior teeth exhibited a deep red betel nut stain.

Differential Diagnosis
Wells (1967) illustrated a case of pseudopathology that was originally assessed as mastoiditis. The outer table of the mastoid region is rather thin and easily eroded post-mortem thus revealing the numerous air cells. The lesions in the Vietnamese case appear actual rather than artefactual. Extensive bony destruction caused by mastoiditis will no doubt sufficiently weaken the structure of the mastoid to induce rapid post-mortem destruction. The combined processes of pre and post-mortem destruction may be difficult to separate in dry specimens. If it is accepted that the destruction of this mastoid is in part due to pre-mortem pathology, the most likely candidate is mastoiditis.

Middle Ear Infection
Otitis media is an extremely common condition affecting nearly all children at some stage (Morris 1998). Such middle ear inflammatory conditions are categorised with respect to the underlying disease process as well as location in the living individual (Kemink *et al.* 1993). They can range from an "incidental finding of fluid in the middle ear space to chronic bacterial infection of the mastoid and middle ear cavities" (Morris 1998). It has long been observed that there is a close association between the high frequency of respiratory disease and high prevalence of middle ear disease in many Australian Aboriginal Communities (Morris 1998). In reviewing the literature on the aetiology of middle ear infection, Morris (1998) notes that there is also an apparent but somewhat equivocal link between the frequency of middle ear disease and socioeconomic conditions. Further, some evidence, albeit inconclusive, points to a genetic or racial link to middle ear disease prevalence.

Studies of mastoid infection in skeletal remains have focussed on population level radiographic examination of mastoid air cell morphology (Gregg *et al.* 1965; Titche *et al.* 1981). The mastoid's air cells are susceptible to pathological alteration during development, in childhood, as a result of ear infection. This is not the case with adults however.

Other Possibilities
There are a number of other possibilities that need to be considered. An epidermoid cholesteatoma can lead to extensive bony destruction and can occur at any age (Friedmann 1974). Sarcoidosis of the ear can lead to the

development of single or multiple lesions and xanthoma of the ear can imitate the effects of chronic mastoiditis (Friedmann 1974). However, it is not clear how the effects of these conditions will present in dry bone specimens. Finally, it has been suggested that mycetoma must be considered in a differential diagnosis of mastoid pathology in tropical regions (Yagi *et al.* 1998).

Case Study Eighteen

78NNM7ka was a 30-39 year old female from the Metal period site Nui Nap. Three small irregularly shaped lesions, approximately 2-3.0mm in diameter, are present on the mastoid process of the right temporal bone. Two are situated on the superior aspect of the mastoid, while the other is on the mid-anterior aspect (Figure 6.25). Each lesion has a smooth margin that slopes into the interior of the cavity. They were all filled with dirt prior to examination. Each lesion displays at least one tiny foramina-like perforation of its floor leading into the interior of the mastoid. The internal walls of each lesion are steep and smooth. There is also an area of erosion of the outer table at the tip of the mastoid process exposing a honey-comb of bone, the mastoid sinuses. There is a large cavity on the mid-posterior border of the mastoid that has sharp edges and appears to represent clear post-mortem damage. There is a further small pit, with smooth irregular edges, just posterior to the mastoid and positioned equidistant between the height of the two superior and one inferior lesions on the mastoid.

Endocranially, the petrous portion is completely missing post-mortem. There is a cavity with smooth irregular edges posterior to the base of the petrous process. The dimensions of this cavity are 12.0mm by 6.5mm at the widest point. The inferior aspect presents a combination of smooth walls and exposed trabecular bone with no sign of pitting. It is unclear if this is post-mortem damage or related to the mastoid lesions.

With the exception of incisor and canine enamel hypoplasia no other pathological conditions were detected. The anterior teeth are darkly stained by way of betel nut.

Differential Diagnosis

See previous discussion of case study eighteen.

Case Study Nineteen

89MDM7a was a 20-29 year old female from the Metal period site Minh Duc. There is a circular, albeit slightly jagged, lytic lesion situated just anterior to the coronal suture and just lateral to bregma, high on the right side of the frontal bone (Figure 6.26). The floor and sides are irregular or rough. The lesion appears to be punched-out but does not penetrate through to the inner table. The lesion itself is 7.5mm in diameter and 3.0mm deep. A ring of concentrated pitting completely surrounds the edges of the lesion and averages 2.5mm in width. There is another band of slightly darker, smoother and shinier bone also surrounding the lesion, which ranges from a minimum width of 5.5mm to a maximum width of 12.0mm. The only difference between this outer-ring and the rest of the bone of the frontal is that this bone appears 'wet' or 'lacquered' in contrast to the matt appearance of the adjacent bone and rest of the skull for that matter. This outer ring is nearly completely confined to the frontal bone, but there is some slight overlap onto the left parietal just lateral to bregma. There is no sign of remodelling and the lesion was likely active at the time of death.

There is also a small, 4.2mm in diameter, button osteoma on the right frontal tuber. The individual had extensive open-cribra orbitalia and associated remodelling in both orbits. Two isolated areas of mild ectocranial porosity are present on the rear of both parietals, either side of the sagittal suture, that do not extend onto the occipital.

The right humerus presents a mild case of periostitis. It manifests as an elongated island of raised bone some 16.4mm long, 5.2mm wide and 1-2.0mm high, on the postero-lateral-superior aspect of the deltoid tuberosity. There is extensive porosity surrounding this piece of bark-like bone with the surface of the new bone itself being quite dense. Alternatively, this may simply represent mild ossification of this portion of the *deltoid* muscle insertion. This would be consistent with other signs of mild

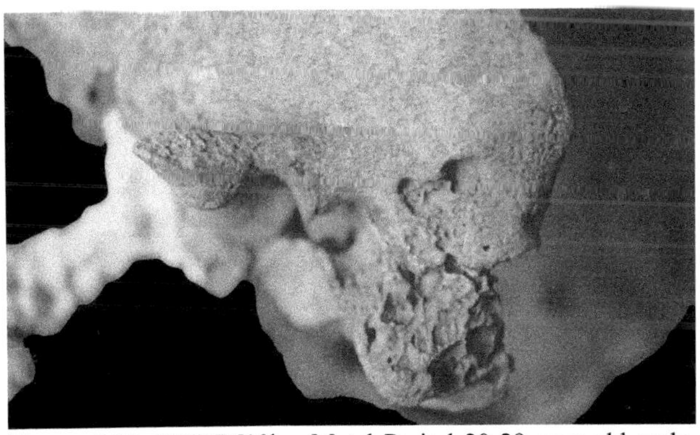

Figure 6.24. 77NNM10ka. Metal Period 20-29 year old male. Note the well circumscribed lesions to the superior portion of the left mastoid and massive destruction of the lower aspect of the mastoid (anterior of cranium to the left of photo)

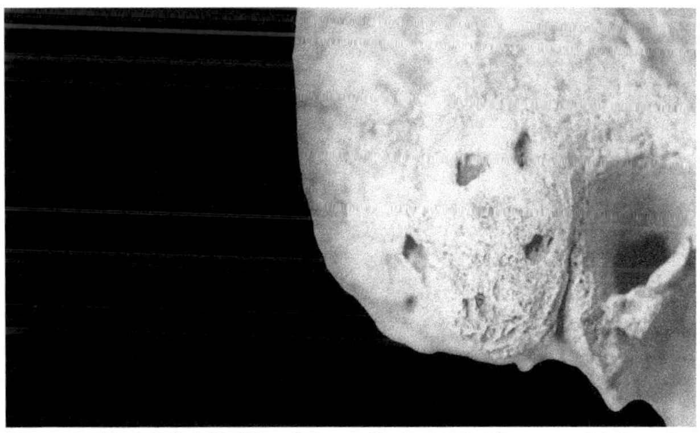

Figure 6.25. 78NNM7ka. Metal Period 30-39 year old female. Note well circumscribed lesions exposing sinuses of the right mastoid

osteophytic development of the elbow and knee joints.

She did not display any oral pathology, nor evidence for enamel hypoplasia. However, this is the only Metal period individual to display orange-red staining to all exposed surfaces of all teeth. This may be evidence of habitual betel nut chewing, in distinction to the other individuals where stains appear only on the labial aspect of anterior teeth. She also exhibits an unusual lingual maxillary central incisor wear pattern. The aetiology of such lingual task wear facets is discussed in Chapter 3.

Differential Diagnosis

As with Case fifteen, this individual presents a single lytic cranial lesion. Very little post-cranial material has survived making the isolation of a particular or even narrow range of causes difficult.

Langerhan's Cell Histiocytosis

If this case does represent this particular condition, it is most likely the localised form termed an eosinophilic granuloma. The age of this individual is consistent with the age group most susceptible to this condition, that is both adolescents and young adults. However, similar to case study fifteen, the lesion here is not bevelled and displays clear signs of active vascularity prior to death. Again, it is not clear that these features are sufficient to exclude this condition from further consideration.

Metastatic Tumour

Primary malignant carcinomas are known to metastasise to the cranium, among other sites. Ortner & Putschar (1981) point out that such metastases are by definition malignant. In differentiating between malignant and benign lesions Rothschild & Martin (1993:167) suggest that "bone destruction with well-defined margins, easily separated from surrounding normal bone, is generally a sign of a relatively benign process, independent of the smoothness or shagginess of the bone margins". As such, the lesion under consideration here would be regarded as benign thus excluding it from further consideration as a metastatic lesion. The significance of the co-occurrence of a button osteoma in this individual is difficult to gauge. The aetiology of such bone masses is varied and includes mechanical, vascular, inflammatory, neoplastic (Capasso 1997) and probably even genetic (Ruggieri *et al.* 1998) factors. The presence of this osteoma cannot be used to infer a neoplastic status for the lytic lesion.

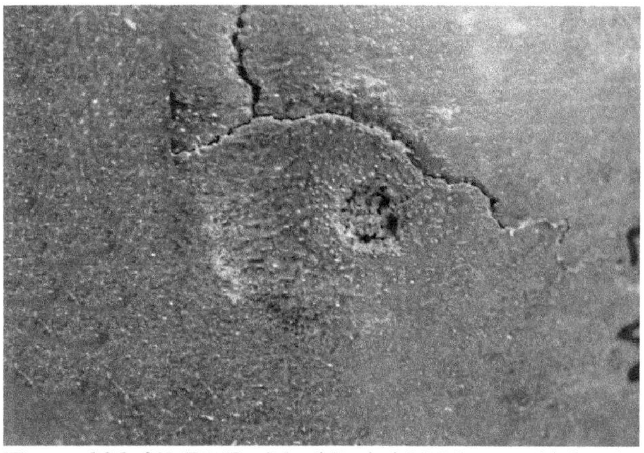

Figure 6.26. 89MDM7a. Metal Period 20-29 year old female. Lytic lesion penetrating the outer table of the frontal bone, just forward and to the left of bregma (anterior of cranium to bottom of photo)

Other Possibilities

Epidermoid cysts have been discussed already. The dry bone morphology of such cysts and the observation that they can occur at any age means that such a condition must be considered here. As noted previously, chondroblastomas are known to form in cranial bones although their dry bone morphology has not been described for such a location. A mycotic aetiology also needs to be considered. Fungal possibilities discussed previously, include sporotrichosis, histoplasmosis, cryptococcosis, mycetoma and penicilliosis. The fungal disease penicilliosis marneffei is dicussed in greater detail with respect to case study twenty. It is a condition endemic to Southeast Asia and has been observed to cause osteolytic lesions in the cranial vault (Yayanetra *et al.* 1984).

Case Study Twenty

95DXM1a, a 30-39 year old female, has already been examined with respect to a healed radial fracture (see case study six). The vertebral column, including the sacrum, was completely preserved. The right inferior interarticular facet of T3 has been completely destroyed. There is extensive porosity surrounding the lesion, or former site of the articular facet. Porosity of the ventral centrum and major destruction of the anterior centrum, which may be post-mortem, is also evident. The left transverse process is also missing but it is unclear if this is pre-mortem.

The corresponding right superior facet of T4 is also virtually completely eroded away with only porous, slightly remodelled trabecular bone remaining. There is extensive erosion of the dorsal and dorso-anterior centrum and the tips of all processes. There is pre-mortem erosion of the left superior interarticular facet, which corresponds to slight erosion of the T3 left inferior facet.

The eighth thoracic vertebra displays erosion of the dorsal, left lateral and much of the ventral aspect of the centrum. It is unclear if this is post-mortem or related to the clear pre-mortem lesions on T9 and T10. Given that there is pre-mortem destruction of the left lateral aspect of the centrum, it is likely that this is related to the cause of the T9 and T10 lesions.

A large scooped out lesion presents in the left lateral aspect of the T9 centrum. The lesion is approximately 12.0mm in diameter with a maximum depth of 10.0mm (Figure 6.27). It is circular and internally spherical with the walls and floor being exposed porous trabecular bone. There is also considerable erosion of the anterior and antero-ventral aspects of the centrum.

The left latero-posterior aspect of the T10 centrum presents a large scalloped lesion (Figure 6.27). The lesion extends as far posteriorly as the base of the left lateral transverse process. It is 16.7mm long, 12.2mm high (deep) and is circular except for the posterior extension. The lesion is roughly spherical internally, exposing porous trabecular bone. There is also erosion of the upper anterior aspect of the centrum and an oval hollow in the antero-dorsal right lateral portion of the centrum.

The fifth lumbar vertebra presents a case of spondylolysis, bilaterally separated neural arch. The separation occurred at the *pars interarticularis*. The L5 is preserved as a centrum, with both lateral transverse processes and superior articular facets, and separated left and right sides of the neural arch. Bony spurs project inferiorly where the neural arch would normally have joined inferiorly. There is no indication of clear breakage and remodelling. However, the area where the arch would normally join is smooth, but irregular with a moderate degree of porosity. The dorsal centrum of L5 displays a slight shallow depression, which is probably a Schmorl's node.

While most of the left lateral aspect of the sacrum is missing post-mortem, there is a clear neural arch defect affecting the lower three sacral vertebrae. It is likely that there was a complete neural arch defect involving all sacral vertebrae. With the exception of the aforementioned vertebrae, no other evidence of pathology or articular change, such as osteophytosis, was observed in any other vertebral element.

The left os coxa presents a shallow lesion on the medial face of the superior ischial ramus (Figure 6.28). It is oval in shape and measures 19.0mm supero-inferiorly and 11.5mm antero-posteriorly. For the most part it has smooth sloping sides, although there is a sharp border superiorly due to pressure and fracturing of this region. The floor of the lesion is stained a shiny dark brown compared to the adjacent matt yellowish bone. There are two distinct shallow pits within the crater of the lesion. There is no evidence of porosity or remodelling and it seems purely lytic in nature.

There is also a localised and discoloured circular area, 14.5mm in diameter, in the superior region of the iliac fossa. There are three small foramina within the discoloured area, the limits of which are sharply defined. Nonetheless, the surface of the bone appears the same as adjacent bone in texture. The discolouration is the same as that found in the floor of the aforementioned ischial lesion. The *iliacus* covers the area where this lesion is located. Further, it might also be relevant that the left iliac fossa cradles the terminal extension of the descending colon.

A radiographic examination of all skeletal material attributed to this individual failed to detect any further pathologies. With the exception of articular changes, noted in relation to the fractured radius, the only other changes seen were generalised micro-porosity of most joint surfaces. This individual was also free from dental caries although incisor, canine and premolar enamel hypoplasia was present. Extensively remodelled closed-cribra orbitalia was observed in the left orbit but was absent in the right. Finally, the medio-inferior aspect of the left clavicle presented a deep, up to 4.0mm, crater-like fossa at the attachment site of the costoclavicular ligament. Mann & Murphy (1990) suggest this is a common finding. In this instance in manifested unilaterally.

Differential Diagnosis

The differential diagnosis in this case is somewhat complicated due to the number and range of pathologies. At least two likely quite separate conditions are apparent in this case and as such the congenital or developmental pathologies of the bifid sacrum and L5 spondylolysis are discussed separately. The remaining lesions are treated as related manifestations of a more systemic condition. This assumption may not be warranted, but at least has the advantage of limiting the differential diagnosis to a manageable level.

Tuberculosis

Ortner & Putschar (1981:145) point out that vertebral "tuberculosis is, in practically all the clinical and autopsy series, the most common and most characteristic lesion". They go on to note that the vast majority of cases involve at least two, rarely more, adjacent vertebral bodies. Isolated involvement of portions of vertebrae other than the centra are said to be rare. Ortner & Putschar (1981) also suggest that development of paravertebral abscesses is a common complicating factor in vertebral tuberculosis. While the lesions described in this instance, in the two adjacent vertebrae T9 and T10, do not display the flaring bony extensions described by Ortner & Putschar, the lesions are grossly similar to those illustrated in their Figure 201 (Ortner & Putschar 1981:147). Such abscesses extend inferiorly and can initiate further sites of infection. The well-circumscribed region of dark-stained cortex in the iliac fossa of the left ilium may be indicative of the

Figure 6.27. 95DXM1a. Metal Period 30-39 year old female. Note destruction of the cortex of T8 (left of photo) and large lesions to the lateral aspects of T9 and T10 (left lateral view of the lower thoracic vertebrae shown)

presence in life of such an (psoas?) abscess. Moreover, the ischial lesion of the same innominate is consistent with the further downward extension of such an abscess, or alternatively an additional secondary site of infection. Moreover, the morphology of the ischial lesion is consistent with pressure from a soft tissue abscess rather than a bone specific site of infection.

The isolated destruction of the right hand side interarticular facets between T3 and T4 is, however, not consistent with the general pattern of vertebral tuberculosis. Ortner & Putschar (1981:145) state that isolated "tuberculous foci in posterior elements of vertebrae are extremely rare". Nevertheless, in a later paper Ortner (1991) provides an example of posterior element involvement in a case of juvenile skeletal tuberculosis.

Vertebral Osteomyelitis

Sapico & Montgomerie (1990) suggest that pyogenic vertebral osteomyelitis and vertebral tuberculosis are difficult to distinguish. It is unclear from the literature if the skeletal lesions of these conditions can be differentiated. Ortner & Putschar (1981) suggest that vertebral osteomyelitis only accounts for 2% of all cases of osteomyelitis. Further, in the majority of cases they note that only one vertebra is affected. Further, is should be noted that the spinous process and neural arch are often affected, which is in contrasts to the pattern seen in tuberculosis (Ortner & Putschar 1981).

Actinomycosis

Another bacterial condition that needs to be considered is actinomycosis. Crank *et al.* (1982) suggest that actinomycosis needs to be considered in a differential diagnosis when considering tuberculous, blastomycotic and coccidioidal infections. Unlike tuberculosis, actinomycosis can affect any part of a vertebra, although the centra is the preferred site. Crank *et al.* (1982:167) note however, "[w] hen the thoracic spine is involved the posterior ends of the corresponding ribs also are likely to be affected". There was no rib involvement in 95DXM1a however. Further, Crank *et al.* (1982:167) describe actinomycotic vertebral lesions as presenting "an irregular, saw-tooth appearance".

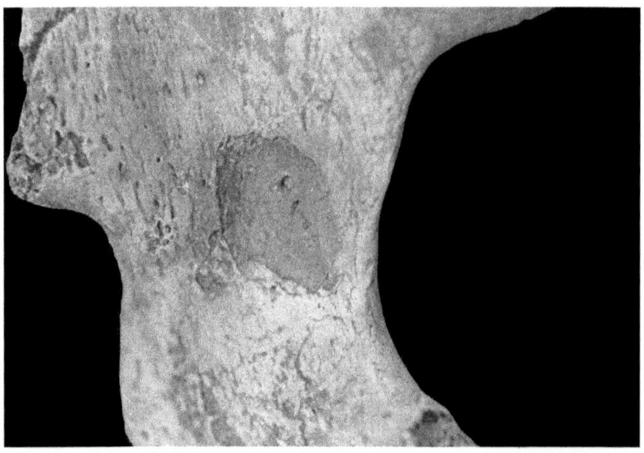

Figure 6.28. 95DXM1a. Well circumscribed, smooth floored lesion on the medial aspect of the left, upper ischial ramus

Moreover, reactive new bone is usually present. Neither of these conditions were a feature of this Vietnamese case.

Mycotic Infection

Clinically, the fungal infections blastomycosis (Frean *et al.* 1993; Guler *et al.* 1995) and coccidioidomycosis (Wesselius *et al.* 1977; McGahan *et al.* 1979) have been reported to imitate the symptoms, including skeletal, of tuberculosis. A recent paper by Hershkovitz *et al.* (1998) provides a set of criteria, principally based on lesion morphology, to differentiate among blastomycosis, tuberculosis and metastatic cancer. While the descriptions of the fungal lesions are based on only one case, it is nonetheless an important first step in differentiating between these three conditions in dry bone.

Neither blastomycosis nor coccidioidomycosis is endemic to Southeast Asia. Nonetheless, the skeletal lesions described for case study twenty are not inconsistent with descriptions of the morphology and distribution of lesions seen in cases of blastomycosis and coccidioidomycosis. As so few fungal diseases have been described with respect to their specific bony manifestations, it is worthwhile examining the skeletal presentation of these two fungal conditions in more detail as a clue to the nature of the condition affecting case twenty.

The lesions described for both T9 and T10 can be categorised as having a space-occupying appearance. This is a characteristic of both metastatic carcinoma and blastomycosis, but not tuberculosis. However, Hershkovitz *et al.* (1998:53) note that in the case of metastatic cancer the space-occupying appearance is due to "the formation of an expansile bone displacing mass" with displacing used "as a term to denote bone resorption and reformation at the outer edge of the tumor mass". This does not describe the appearance of the thoracic lesions in this instance. With respect to tuberculosis, Hershkovitz *et al.* (1998) describe the lesions as excavations that not uncommonly coalesce. Blastomycotic lesions, on the other hand, are generally isolated. They also point out that blastomycosis does not spare the posterior vertebral elements, as is the case in tuberculosis. However, a recent review of eight cases of vertebral blastomycosis (Saccente *et al.* 1998) reported that infection can spread though the disk spaces to include adjacent vertebrae. Moreover, the manifestation of contiguous vertebral abscesses is a distinctive feature of the disease. Paravertebral or psoas abscesses are not infrequently observed. Saccente *et al.* 1998 also point out that infection can affect non-adjacent vertebrae by tracking through the anterior longitudinal ligament.

The morphology of the ischial lesion is very different to those of the thoracic vertebrae. However, Hershkovitz *et al.* (1998) described a blastomycotic lesion of a humerus as smooth and elliptical. The morphology of this humeral lesion described by Hershkovitz *et al.* (1998) is at least consistent with that of the innominate lesion in the Vietnamese case. Furthermore, the hypothesised iliac fossa lesion and this ischial lesion were posited to be consistent with a psoas abscess when discussing tuberculosis. The observation that paravertebral and psoas abscesses are also a feature of vertebral blastomycosis (Saccente *et al.* 1998) is then also relevant here.

Coccidioidomycosis is a fungal condition that affects the skeleton in some 20% of disseminated cases (Dalinka & Greendyke 1971). Osteolytic lesions are known to affect most bones with the axial skeleton being affected in 59% of cases in a recent study (Zeppa *et al.* 1996). The most common type of lesion is described as having a punched-out, or well demarcated, appearance (McGahan *et al.* 1979; Zeppa *et al.* 1996) and is more frequently seen in long bones and flat bones. It is not uncommon to observe sclerotic margins in these lesions (Zeepa *et al.* 1996). Zeepa *et al.* (1996) also describe the presence of more permeative lesions that not uncommonly affect the vertebral bodies. Osteolytic lesions of the vertebrae can affect the body, pedicles and transverse processes (Dalinka & Greendyke 1971). Furthermore, contiguous involvement of the ribs has also been observed to occur (Dalinka & Greendyke 1971).

The discussion of case study fifteen suggested six possible fungal infections that potentially involve the bony tissues. Of these, histoplasmosis can likely be ruled out as extensive bony involvement is only seen in the form specific to Africa. Sporotrichosis can also likely be eliminated, as vertebral involvement has not been reported (Hershkovitz *et al.* 1998). Crytococcosis has a high propensity to affect both the vertebrae and innominate. This condition cannot be ruled out at this stage. The specific appearance of cryptococcosis in dry bones is unknown. A further fungal condition to consider is penicilliosis marneffei.

The fungal agent responsible for penicilliois marneffei, *Penicillium maneffei*, is a dimorphic fungus endemic to Southeast Asia and southern China (Louthrenoo *et al.* 1994). Deng & Connor (1985) report on eight cases from Guang Xi province, southwestern China, discovered between 1964 and 1973. The age of affected individuals ranged between 11 and 40 years. The majority of infected cases were farmers. The course of the illness ranged between 2 months and 3 years. All cases were described as "disseminated, progressive, and fatal" (Deng & Connor 1985:323). Deng & Connor (1985) point out that two species of bamboo rat living in the region could be potential reservoirs of the pathogen. They note that rats are regularly trapped and eaten by local inhabitants. There are also Tang dynasty (AD 618-906) references to the trapping of rats by the indigenes of the region (Schafer 1967:53). Given the extensive liver and intestinal involvement in their five autopsied cases, they suggest ingestion of the pathogen by way of eating infected rats is a likely cause of human infection. However, in a later paper Deng *et al.* (1986) suggests that it is unlikely that the bamboo rat is instrumental in human infection by the fungus. Deng *et al.* (1986) surmise that an as yet unknown environmental factor is responsible for both rat and human infection.

While Deng & Connor (1985) note bone marrow involvement in all their autopsied cases, no mention is specially made of osteolytic lesions. However, in a study of five cases reported on from Thailand two individuals displayed osteolytic lesions (Yayanetra *et al.* 1984). In one case, a male aged 35 years, osteolytic lesions to the proximal femur, greater trochanter, clavicle, scapula, sternum, ribs and big toe were described. In the other case, female aged 50 years, an ostelytic lesion to the right frontal bone was described. More recently Louthrenoo *et al.* (1994) have examined eight individuals from Thailand manifesting penicilliosis marneffei. They describe the osteolytic response thus:

> The osteomyelitis manifested as multiple osteolytic lesions with minimal sclerotic margin involving flat bones, long bones and small bones of the hands and feet. There was no periosteal reaction. One patient (case 2) who had cervical spine involvement had an osteolytic lesion at the posterior portion of the axis.
> (Louthrenoo *et al.* 1994:1148)

Louthreno *et al.* (1994) go on to suggest that the distribution of osteolytic lesions seen in these instances of penicilliosis marneffei have been reported for other mycotic infections, as such they are not unique to this particular condition.

Metastatic Carcinoma

This condition has been considered in the preceding discussion of fungal disease. The lesions in this instance do not appear to be consistent with a diagnosis of metastatic carcinoma.

Other Possibilities

Perhaps the only other condition that could produce lesions even superficially similar to those seen in T9 and T10 is an aortic aneurysm. Kelley (1979) suggests that a differential diagnosis of vertebral aneurysmal erosions should include tuberculosis and metastatic carcinoma. While the location, left lateral, of the T9 and T10 lesions is consistent with an abdominal aneurysm, the overall morphology is not. Such aneurysms leave smooth depressions in the wall of the vertebrae (Kelley 1979). In this Vietnamese case the lesions have more of a space-occupying appearance suggestive of a granulamatous mass, rather than an expanding artery. Even if it were accepted that an aneurysm caused the lower thoracic lesions, the remaining lesions still need to explained.

Congenital/Developmental Features

It is not entirely improbable that the bifid sacrum, spondylolysis of L5 and the pedestaled anterior and medial articular facets of both calcanei form a related group of congenital defects. However, Charles Merbs (pers. comm.) has suggested that in this Vietnamese case the L5 spondylolysis is more likely due to trauma than congenital separation of the neural arch. Merbs suggests that although both conditions, bifid sacrum and L5 spondylolysis, have different aetiologies their occasional co-occurrence can be readily explained. Merbs argues that Spina bifida of the sacrum weakens the arch of L5 to the extent that it eventually causes spondylolysis, or fracturing due to repeated microtrauma.

Case Study Twenty-One

80CCNM67a was a young female aged between 15-25 years from Con Co Ngua pit 2. 80CCNM67b was a 7-8 lunar months old foetus found buried with the aforementioned female. The female did not display any obvious skeletal pathology. However, fragments of the parietal bones displayed ectocranial porosity and the only preserved canine, right maxillary, presented three bands of enamel hypoplasia.

Very few subadult remains were recovered from Con Co Ngua. Their fragility and low mineral content predispose them to rapid breakdown in the ground. 80CCNM67b is the only example of an individual aged less than five years old to be recovered from the site. It is not unlikely that these foetal remains survived internment because they were protected for some time within the womb of the mother at burial. It is not improbable that 80CCNM67a was the mother and that she died in the third trimester of her pregnancy. The pelvic region was too poorly preserved to assess for potential obstetrical problems.

Discussion and Conclusions

Earlier in this chapter it was found that the Da But period sample clearly displayed a higher frequency of major traumatic lesions relative to the temporally later Metal period sample. Moreover, the level of serious trauma in the Da But period appears to be greater than for any comparable archaeological assemblage in Southeast Asia. In this section, examining non-traumatic lesions, only one of the nine case studies reviewed derived from the Da But period sample. Moreover, the differential diagnosis for this single Da But period individual appears to favour some form of tumour. Of the remaining eight case studies, one (case thirteen) displays a joint pathology of unclear aetiology, while the remaining seven cases include an infectious disease in their respective differential diagnoses. However, two of these cases may very well have suffered from a reticuloendothelial disorder (cases fifteen and nineteen). Overall, 7/96 (7.3%) of the individuals in the Metal period sample displayed evidence of non-traumatic lesions potentially of bacterial or fungal origin. This is likely an underestimate, as the denominator includes individuals represented by very little skeletal material. The Da But period sample, even with its greater level of skeletal preservation, does not provide a single case of possible infectious disease. What these findings mean with reference to the relative health and well being of the populations, each of these samples represents, is unclear.

It would be simplistic, and potentially quite inaccurate, to state that the Metal period sample suffered poorer health based on the demonstrated greater prevalence of non-traumatic pathology. Ortner (1991) points out that it is more common for skeletal responses to infectious disease to manifest later in the disease process.

> This means that the individual with the disease must survive the initial, often acute, phase of infection before the skeletal manifestations occur. This implies a good immune response and a relatively healthy individual, in contrast with a person having the same disease who dies before the bone tissue is affected by the disease process.
> (Ortner 1991:10)

Nonetheless he also notes that "such an immune response is not as effective as one that successfully rids the body of the infectious organism during the early stages of the disease" (Ortner 1991:10).

This point has further implications when considering some diseases that generally only take hold in individuals who are either debilitated or have a depressed immune response. For example, the fungal infection penicilliosis marneffei is such a disease that also includes an osteolytic response. A high incidence of such lesions in an archaeological sample (assuming the condition could be accurately diagnosed) would indicate a high level of debilitation or lowered immune response for the population as a whole. In this instance, to assume a good immune response from the hosts would be an inaccurate interpretation of the data.

The immediate implication here is that there is no simple relationship between the sample frequency of lesions per se and the level of population health. This is true regardless of the demographic profile of the sample. The underlying condition or disease causing the lesions must be identified before the overall population health implications can be assessed. Of course, isolating a specific disease is not straightforward and may never be a practicable aim in many instances. This inability to isolate specific aetiological agents responsible for the lesions in the Metal period sample makes it extremely difficult to infer the relative health of the sample based on the frequency of osteolytic lesions alone. At least four of the case studies included fungal diseases in their differential diagnoses. As such, this may be more indicative of debilitation and/or decreased immunocompetence in the Metal period, than evidence of a healthy immune response to a chronic infection.

It also needs to be considered that if the Metal period evidence is pointing to higher rates of infectious disease, it may simply be due to changes in the human/pathogen vector dynamic. For example, there is equivocal evidence that the Vietnamese bamboo rat plays an important role in the epidemiology of penicilliosis marneffei. Higher population density and/or altered living conditions in the Metal period may have led to an increased human exposure to such disease carrying vectors. A decline in sanitation may also be implicated in the potential incidence of ear infection in the Metal period as well. Even if case study twenty turns out to be tuberculosis, rather than a fungal condition, it would still be consistent with the hypothesis of increased population density and decreased levels of sanitation. An increased prevalence of tuberculosis is closely tied with both increased levels of crowding and poor levels of nutrition. The earliest historical references to tuberculosis in China place it at 4700 years bp, although clear descriptions of symptoms do not appear until 2400 years bp (Johnston 1993). However, one confounding factor, not mentioned in such historical research, is that fungal infections, blastomycosis for example, have been known to imitate the symptoms of tuberculosis (Guler *et al.* 1995; Frean *et al.* 1993). As such, it is possible that fungal conditions, rather than tuberculosis, are being described in early Chinese historical accounts.

Numerous studies have presented evidence supporting a positive correlation between increases in the frequency of infectious disease and changes in settlement patterns and/or subsistence economy. This correlation has been reported for Eastern Europe (Angel 1984); South Asia (Kennedy 1984); Japan (Suzuki 1991); Australia (Webb 1995); North America (Cassidy 1984; Cook 1984; Goodman *et al.* 1984; Larsen 1984; Perzigian *et al.* 1984; Rose *et al.* 1984; Eisenberg 1991; Milner 1991); Central America (Norr 1984); and South America (Ubelaker 1984; 1994). Interestingly, a review of research carried out in Western Europe failed to find an unequivocal increase in infectious disease from the Mesolithic to Neolithic periods (Meiklejohn *et al.* 1984; Meiklejohn & Zvelebil 1991). However, part of the explanation for this may lie in the observation that trends were already evident in Western European Mesolithic samples toward increased population density (Meiklejohn & Zvelebil 1991). A study of pre-agricultural and agricultural samples from Southwest Asia found a decrease in observed infectious disease frequency over time (Rathbun 1984). However, it was proposed that the adoption of agriculture likely saw an increase in acute disease and as such is not represented in the skeletal samples analysed (Rathbun 1984).

A number of researchers have highlighted the link between the changing settlement patterns, adoption of sendentism and increased population density, rather than the adoption or intensification of agriculture per se, and the increase in infectious disease prevalence (Allison 1984; Cook 1984; Armelagos 1990; Larsen 1984, 1995). It has also been demonstrated that contact between hitherto separate cultures can have dire health consequences (Larsen & Milner 1994). Increases in the frequency of chronic infectious disease as observed skeletally, over and above historically documented marked increases in acute infectious disease, have been reported in post-contact North America (Stodder 1994) and South America (Ubelaker 1994). Moreover, clear evidence for a decline in post-contact Mayan health, but without evidence for chronic infectious disease, has been interpreted as being due to the weeding-out effects of high mortality from acute infection (Cohen *et al.* 1994). Furthermore, Suzuki (1991) has suggested that the first appearance and subsequent high frequency of chronic non-specific infection and specific infectious disease, such as tuberculosis, in post-Jomon Japan can be attributed to more than subsistence changes and increasing population density alone. Perhaps more importantly, this increase in chronic and specific infectious disease can be understood in the context of major human migrations into Japan from the Korean peninsula during the third to seventh centuries AD (Suzuki 1991). This is potentially a significant factor with respect to the Vietnamese situation during the Metal period. As detailed in Chapter one, northern Vietnam was invaded and conquered by the Han Chinese in 186 BC. Although, systematic colonising efforts did not begin until the first century AD. Nonetheless, this socio-culturally tumultuous period of time encompasses the temporal range of the Metal period sample analysed in this study.

In summary it appears that the Metal sample, in stark contrast to the Da But period sample, displays the skeletal manifestations of what is potentially infectious disease. The possible implications of this are:

(1) Increased contact with bacterial and/or fungal vectors in the Metal period. This could have been facilitated by changes to the environment-host-pathogen dynamic. Two examples include increasing population density; land clearing/cultivation. Alternatively, or in addition, contact and colonising efforts by the Han Chinese may have contributed to the introduction and/or spread of infectious disease.
(2) Higher levels of debilitation and/or decreased levels of immunocompetence in the Metal period.
(3) The evolution of pathogens over time, leading to more chronic rather than acute human responses to the same pathogens.

A combination of these factors could also have occurred. The first two factors are both related to life style changes potentially associated with changes in subsistence economy and/or settlement patterns. In the case of mycotic infections, the fungal vector of which cannot be transmitted between humans, reduced immunocompetence, in association with a change in the environment-host-pathogen dynamic, is a potential scenario. The last factor, pathogen evolution, would indicate a long regional history for the pathogens involved. However, without being able to isolate the lesions to specific pathogens it will be extremely difficult to isolate the specific and likely interrelated reasons for the higher prevalence of non-traumatic lesions in the Metal period, relative to the temporally previous Da But period.

7
CONCLUSIONS

Research into the health and skeletally sensitive behaviours of mid-Holocene and Bronze Age inhabitants of the northern portion of what is now the Socialist Republic of Vietnam has not been undertaken previous to this study. The primary aim of this research was to both describe and interpret bioarchaeological aspects of health and behaviour using curated skeletal and dental material from two temporally separate periods.

Northern Vietnam was, and still is, an extremely rich and complex region topographically, climatically, florally and faunally. Human occupation extends at least as far back as 30,000 years ago and perhaps even much further. Communities have very likely been operating within marine and riverine contexts for a considerable period of time prior to the development of the first archaeologically visible coastal communities, which include the Da But period, in the mid-Holocene. This coastal and marine orientation continued and flourished, ultimately giving birth to the socio-political sophistication of the Metal period. The rich social structure and incredible material cultural wealth of this time was most likely also contingent upon a riverine and maritime economic focus. It is within these environmental and socio-cultural contexts that interpretations of the skeletal material have been set, and an understanding of human biological responses to the Southeast Asian sub-tropical environment-culture dynamic been facilitated. The principal findings of this study are now brought together and examined with respect to the various markers of health and behaviour, residential mobility and subsistence.

Health Conclusions and Implications

Oral Health

Metal period oral health is poorer relative to the Da But period. The Metal period sample had greater levels of caries and pulp exposure. Metal period individuals also suffered from more occlusal problems as indicated by the frequency of hypodontia, malaligned and impacted teeth. Metal period individuals suffered from more occlusal problems despite their jaw robusticity being greater, and tooth size smaller, than in the Da But period. While the Da But period displayed higher levels of ante-mortem tooth loss and interproximal grooves, these traits are more related to cultural practices. For instance, the patterning of much of the evidence for ante-mortem tooth loss, mainly affecting incisors, indicates the probability that deliberate tooth ablation occurred in the Da But period. Few observations of interproximal grooves were directly associated with other pathologies and likely represented the habit of picking the teeth.

Behavioural differences between each sample are also implicated in the poorer oral health of the Metal period sample with respect to pulp exposure. A great deal of evidence points to the use of the anterior dentition in industrial activities during the Metal period. For example, in the Metal period pulp exposure is concentrated anteriorly, alveolar defects are not limited to the posterior teeth as in the Da But period, task wear facets are virtually exclusive to the Metal period and the anterior to posterior tooth wear gradient is much steeper.

Another clear behavioural difference between each period involves the use of betel nut, for the first time, in the Metal period. The health benefits, or otherwise, regarding the use of this plant are equivocal. It seems likely that most individuals displaying stained teeth had their anterior dentition deliberately dyed using betel nut compounds. It is unclear, however, if betel nut was habitually chewed for its narcotic and medicinal attributes.

There are limited, and no statistically significant, oral health differences between the sexes in the Da But period. Those differences that do occur relate to a greater frequency of caries, tooth malalignment and interproximal grooves in females. Males displayed a greater frequency of pulp exposure. More female tooth malalignment is not unexpected given smaller female jaw size. The tendency for more females to be affected by caries is potentially related to dietary differences between the sexes and/or different eating habits, such as the spacing between meals. For instance, frequent snacking can increase the risk of tooth decay.

Intra-sample differences are greater in the Metal period, and several significantly so, by the tooth count reporting method. Females exhibit greater amounts of ante-mortem tooth loss, caries and impacted teeth. Males have more pulp exposure, chipped teeth, rotated teeth and task wear facets. The possible reasons for an increased frequency in female caries have been mentioned with respect to the Da But period. The female contribution to Metal period occlusal problems is disproportionately large. Those traits affecting relatively more males are for the most part related to using the teeth industrially, and a division of labour is thus indicated.

Articular Health

The Da But period sample has a greater frequency of articular change than the Metal period sample. Other differences in the articular health of each sample concern the distribution of osteoarthritis (OA) by sex and joint. An important aetiological factor to consider with regard to elevated levels of Da But period OA is the older age structure of this sample. This study has found a clear positive correlation between increasing risk of OA and increasing age-at-death. Other factors, than age alone, have contributed to the frequency of articular change in each sample. Males displayed a higher frequency of articular change in the Da But period, while females evidenced more change in the Metal period. This is despite the observation that the Da But sample had a greater proportion of older females, while the Metal period had a greater proportion of older males. It is likely that a synergistic host of intrinsic and extrinsic factors, perhaps including behaviour, underlie these differences.

As mentioned above, this study also revealed differences in the patterning of articular change between each sample. The spine displayed the greatest level of change in the Da But sample, but shared this position with the elbow in the Metal period. In the Da But sample the elbow was the second most affected joint. Aside from age, which is positively correlated with an increased frequency of elbow OA in these samples, the effects of a predisposition of the elbow to bone forming, osteophytosis, processes must be considered. Other potential aetiological factors include higher levels of elbow injury, which can contribute to secondary OA, and also perhaps behaviour. However, there is little evidence to support the wear-and-tear hypothesis being significant in the aetiology of OA. Elbow OA is consistently reported to be high in many bioarchaeological studies. This is further evidence in support of a biological, rather than behavioural, aetiology for increased levels of OA in this complex joint. It is more difficult to sustain a global behavioural model to explain these results. However, an evolutionary model based on less than optimal elbow design needs to be tested.

A higher frequency of knee and hip articular change was found in the Metal period sample. The knee was the second, after the spine and elbow, and the hip the fourth most affected joint in the Metal period. The same aetiological factors relevant to the elbow should also be considered here. Even in the Da But sample, over a fifth of assessable individuals displayed knee changes. In general, the prevalence of knee OA in archaeological assemblages tends to be high. As with the elbow, this would implicate a biologically oriented aetiology, rather than a behavioural one. Less than a tenth of the Da But sample suffered from hip changes while over a quarter of the Metal period sample did. Both ancestral and behavioural factors have been suggested in explanations of the aetiology of modern hip disease. The evidence regarding the genetic homogeneity of the Da But and Metal period assemblages is equivocal. Nonetheless, it is intriguing that differences in hip change between each sample may be indicative of ancestral and/or behavioural differences.

The chief importance of this study of articular change is that it has revealed patterns of OA generally consistent with that seen globally, both in the past and in recent times. The level and patterning of articular change, between each sample and compared to other studies world-wide, can be explained with respect to the influences of age, sex, ancestry and perhaps behaviour. The implications of the frequency and patterning of articular change for the overall health of these samples is equivocal. However, given the extremely small level of serious articular change, joint surface eburnation, it would seem that the articular health of these samples was relatively good.

Physiological Stress in Childhood

Two indicators of non-specific stress suffered during childhood were examined in this study, linear enamel hypoplasia (LEH) and porotic hyperostosis. The implications for the health of these individuals as adults, rather than children, was investigated.

The frequency of canine LEH was similar for both samples, and somewhat high in comparison to other Southeast Asian assemblages. It is also evident that there is a bias toward survival to older age for those individuals without evidence of canine LEH in both dental series. These high levels of LEH indicate, at face value, elevated levels of physiological stress in the children of these past populations. The frequency of LEH in a sample does not indicate the proximate causes of the condition. A synergistic interplay between poor nutrition and disease plays a major aetiological role in the manifestation of this trait. Regardless of the specific mechanics of the relationship between childhood stress and later adult health, it is clear that those children suffering stress to such a degree as to manifest enamel defects had an attenuated lifespan relative to those that did not develop LEH.

The prevalence of porotic hyperostosis, whether manifesting as ectocranial porosity or cribra orbitalia, was also relatively high for both samples. The skeletal scars of iron deficiency anaemia in childhood appear so common as to suggest that most of these adults suffered this condition as children. While the frequency of cribra orbitalia was evenly distributed by age, more remodelled forms of cribra orbitalia were seen in older individuals while less remodelled forms were predominantly found in younger people. This is consistent with the view that cribra orbitalia is a condition manifesting in childhood, the skeletal scars of which differentially remodel as the individual ages.

The aetiological pathways leading to anaemia are numerous and include poor nutrition and disease. In sub-tropical regions the potential aetiological role of parasitism is enormous. It is also believed that elevated levels of porotic hyperostosis can be partly attributed to high parasitic loads in coastal/marine oriented communities. The haemoglobinopathy thalassemia, and by inference malaria, is also known to have been present in Southeast Asia over 4000 years ago. A range of variables including genetic blood conditions, malaria and other parasites in addition to the effects of population density, sanitation and diet, including food taboos and restrictions, potentially played a role in the prevalence of this condition in both samples. Nonetheless, despite the high prevalence of childhood anaemia, a situation also seen in modern Vietnam, and in contrast to the LEH results, there is no obvious impact on the health of adults experiencing anaemia as children.

Trauma

Over 7% of the Da But period sample displayed evidence of major trauma compared to 3% in the Metal period. Putting this in its regional context, of nine Southeast Asian skeletal series reported on to date, only four humeral fractures have been identified, with two of these occurring in the Da But assemblage. Also, three of the five femoral fractures reported on in the region derive from the Da But series. In terms of overall health, the Da But period experienced a lesser level of well being, with respect to traumatic injury, than the Metal period and arguably the Southeast Asian region as a whole. Life appears to have been associated with a greater potential for misadventure and dangerous injury in this temporally earlier period. Nonetheless, those individuals who did suffer from serious fractures appear to have survived for some period after a traumatic event that likely received minimal, if any, treatment. This would indicate a robust level of general health at least.

Non-Traumatic Lesions

Seven case studies were examined that potentially implicate infectious disease in the Metal period. No evidence was found for the presence of infectious disease in the Da But period sample. Differential diagnoses of the Metal period case studies included fungal diseases in several instances. Not only does this provide an insight into the palaeoepidemiology of mycotic disease in Southeast Asia, but it also suggests decreased levels of immunocompetence in the Metal period sample. Fungal diseases do not generally affect otherwise healthy individuals. If this proposition is accepted, the implication is for generally poorer health in the Metal period than in the Da But period. Skeletal lesions can, in some instances, be interpreted in terms of host resistance to disease, and by implication good health. In the case of fungal diseases, which tend to affect those with decreased resources for resistance, it cannot be argued that more lesions equate with better health. At the least, the Metal period sample was subject to infectious diseases of a nature that also affected skeletal tissues. There is a complete absence for such conditions in the temporally earlier Da But period.

Stature

Da But period height is greater in terms of both mean and maximum range than Metal period height both for combined sex samples and separately by sex. If the Da But and Metal period samples are for the most part genetically homogenous, the shorter attained stature in the Metal period supports the general findings of poorer Metal period health. In particular, reduced Metal period stature is consistent with the first evidence for chronic infectious disease occurring in the Metal period.

Behavioural Conclusions and Implications

Residential Mobility

There is human biological evidence for a low degree of residential mobility in both the Da But and Metal period samples. The prevalence of canine LEH was elevated for each sample relative to other Southeast Asian assemblages examined. While there is a general global trend for the prevalence of LEH to increase with moves from foraging to agriculture and agricultural intensification, it would appear that other aetiological factors are more important in Vietnam. One such factor is the potential link between increased prevalence of LEH and sedentism and associated variables such as population density and sanitation. Northern Vietnam is not alone in Southeast Asia in this respect either. Two Southeast Asian sites, Khok Phanom Di and Noen U-Loke, that displayed both low levels of residential mobility and agricultural subsistence bases, also exhibited rather elevated levels of canine LEH.

The prevalence of porotic hyperostosis is also high in each sample. Globally, increases in the prevalence of porotic hyperostosis appear to be associated with the adoption of sedentism. Proximal causes of porotic hyperostosis have been mentioned above. Particularly important, with respect to the relationship between sedentism and porotic hyperostosis, are the effects of increased population density and decreased levels of sanitation.

As noted, infectious disease makes its first appearance in northern Vietnam during the Metal period. Numerous studies make reference to the association between increasing levels of infectious disease and either change in subsistence orientation and/or change toward a more sedentary residential pattern. The biological evidence for low levels of residential mobility in the Metal period, particularly markers of physiological stress and infectious disease, are consistent with the picture provided by archaeological and historical data.

The lack of evidence for infectious disease in the Da But period could be interpreted in terms of relatively greater levels of residential mobility compared with the Metal period. But the evidence from childhood physiological stress does not support this. The Da But sample displays much more evidence for healed trauma than the Metal period. In the case of healed femoral fractures, the affected individuals would have had to have been immobilised for several months at least and suffered permanent disability for the remainder of their lives. A certain degree of residential sedentism is thus implied. The extent of the Da But cemetery site Con Co Ngua and the relatively large number of inhumations also indicate some level of sedentary behaviour. Moreover, the evidence for pig and buffalo domestication in Da But contexts also suggests that levels of residential mobility were not high.

Subsistence

Descriptions of oral health are useful in investigations of subsistence orientations. The Da But period sample displayed an extremely low prevalence of caries and ante-mortem tooth loss. Ante-mortem tooth loss was also attrition, rather than caries, induced in the main. The frequency of other oral pathologies such as alveolar defects of pulpal origin, tooth malalignment and tooth impactions were also low. On the other hand, a high proportion of individuals displayed calculus deposits. The remarkably good oral health of this sample is consistent with a foraging subsistence base. A relatively high protein intake in the diet, a characteristic of pre-agricultural communities, is consistent with both the low frequency of caries and the high frequency of calculus. While diets high in protein or carbohydrates cannot be differentiated by calculus levels alone, diets high in protein will protect the teeth from acidogenic bacteria by increasing oral pH levels. Further protection from caries would be expected from the fluoride effects of a marine-oriented diet. The archaeological evidence points to an estuarine/marine subsistence orientation for the Da But period. The oral health of the sample is certainly consistent with these data.

The Da But period sample was also free from evidence of any lesions that could be interpreted as indicating infectious disease. There is mounting bioarchaeological evidence that the adoption of agriculture and/or agricultural intensification is often associated with an increase in infectious disease. This further supports the view maintained here that the chief subsistence focus of the Da But period was foraging.

The Metal period sample also displayed a low frequency of caries. In a regional context, the Metal period prevalence of caries was equal third lowest out of twelve Southeast Asian dental assemblages studied to date. Caries played a minimal role in the aetiology of pulp chamber exposure in this sample and an extremely low proportion of carious lesions affected the dentin. It was argued in this study that regardless of total carious lesion frequency, it would be expected that a greater contribution to the total lesion count would come from dentin lesions if grains were a significant component of the diet. The implication is that grain products did not form a significant part of the diet in the Metal period sample.

There is an increase in Metal period mandibular robusticity relative to the Da But period. Part of the reason for this increase may be due to increased use of the dentition in industrial activities, mentioned earlier. The marked robusticity of Metal period jaws could also be an indication that softer foods, such as cooked grains, did not form a significant part of their diet.

If the initial appearance of infectious disease during the Metal period is connected with the adoption of agriculture or intensification in agricultural production, it supports the view that Han political control of the region during this time led to the use of northern Vietnam as a rice bowl. The initiation and/or intensification of activities associated with rice agriculture, such as land clearing, earth movement, ploughing, planting and so forth, would create optimal conditions for an increased risk from infectious agents. Fungal disease has been implicated in a number of Metal period instances of infectious disease. The spores and secondary vectors of mycotic conditions would be particularly problematic to these early sub-tropical agriculturalists. If the local inhabitants did engage in agriculture and grow crops for Han export, rather than local consumption, it explains both the high level of infectious disease and relatively good oral health.

Another factor, other than adoption and intensification of agriculture or factors associated with sedentism, could have contributed to the observation of infectious disease in the Metal period. The evidence is also consistent with information on massive Han migration into the region at this time. New people, as well as new pathogens, may have entered the region at the same time. Increased migration also likely led to increased population density, which is associated with increased risk of the spread of disease.

The biological evidence suggests that the subsistence orientation in the Metal period was primarily foraging, with a small contribution from grains being possible. This view is consistent with the archaeological and historical data that indicate an extremely important riverine/maritime economic, including subsistence, focus. There is no reason to believe that the formation of complex semi-autonomous states, such as the Metal period in northern Vietnam represents, could not have developed in the context of aquatic and marine based subsistence economies. Indeed, such a development would appear to be represented by the Vietnamese Metal period.

EPILOGUE

As noted in the Preface, it has been almost 20 years since the data for this monograph was collected, with supporting literature extending up until the year 2000. In the intervening 15 to 20 years my views have changed quite considerably with respect to some of the findings reported here, while in other areas I still hold to my original views. Being closely involved in both promoting and synthesising bioarchaeological research in the Asia-Pacific region over the last decade or so has also altered my opinions and approaches somewhat (e.g. Oxenham and Tayles 2006, Pechenkina and Oxenham 2013, Oxenham and Buckley in press-c). One of the most important changes in perspective, with respect to this monograph at least, is associated with the nature of the Da But period, as well as dates for the emergence of the Neolithic and Bronze periods in Southeast Asia.

The Da But period is now known to be part of a geographically extensive pre-Neolithic, albeit pottery using, hunter-gatherer culture that was spread throughout northern Vietnam and southern China from at least c. 7000 BP (see Oxenham and Buckley in press-b). The Neolithic itself, referring to "communities that grew domesticated crops and raised domesticated animals" (Oxenham and Buckley in press-a), emerges quite rapidly in Southeast Asia after c. 4300 BP, giving rise to the Bronze Age c. 3000 BP, followed by the Iron Age at c. 2400 BP (see Higham *et al.* 2011). While no Neolithic cemetery sites were known for northern Vietnam in the late 1990s, bioarchaeologically the early Neolithic is now well represented by Man Bac (Oxenham *et al.* 2008, Oxenham *et al.* 2009, Oxenham *et al.* 2011). Indeed, the timing, nature and processes associated with the emergence of the Neolithic in Southern Vietnam have also recently been brought to light (e.g. Bellwood *et al.* 2011, Oxenham *et al.* 2015). Within the context of this new dating framework the very early dates reported for Ban Chiang and Non Nok Tha (see page 53 of the monograph) are now defunct.

We currently know considerably more regarding the genetic, or ancestral, background of the Da But and Metal period samples than was the case 20 years ago. On page 21 of the monograph I suggested the data were equivocal with regard to this issue. However, extensive subsequent research indicates that the Da But sample is essentially directly descended from the original colonists of the region, some 50,000 years ago, and show a clear Australo-Melanesian cranio-facial morphology in distinction to Metal Age populations which have a clear East Asian cranio-facial morphology (e.g. Matsumura and Oxenham 2014). The population history of the region has been recently summarised in Oxenham and Buckley (in press-b). The quite distinct genetic backgrounds of the Da But and Metal period samples does not necessarily affect any of the health and behavioural findings in the monograph, however, differences in stature (see page 49) are now believed to be more a function of the genetic backgrounds of these samples and not related to a decline in stature over time as a response to changes in diet and/ or disease loads.

My views on oral health have also developed over the intervening years (e.g. Oxenham *et al.* 2002b, Oxenham 2006, Oxenham *et al.* 2006). Moreover, a somewhat significant change in opinion relates to very recent interpretations of oral health in Southeast Asia. In the monograph I essentially subscribe to a dietary (or subsistence) model of oral health outcomes. While the dietary model still holds prominence in the mainstream literature today, work by one of my PhD students (Anna Willis) has convinced me that a fertility model provides much more promise in explaining major trends in oral health from the pre-Neolithic, Neolithic and Bronze/Iron Ages in Southeast Asia, and indeed, with respect to much of the rest of the world (see Willis and Oxenham 2013). Notwithstanding, this is not to say that I have totally abandoned the view that diet plays a role in oral health, I simply now believe that diet is simply one of a range of contributing factors, with the fertility model accounting for the lion's share of oral health variability in this regard.

A number of minor points and issues will be dealt with in the remainder of the epilogue. In discussing the ostensible correlation between low caries rates and high occlusal wear, I referred to Costa's (1980) findings regarding high caries rates and high wear among the Ipiutak of Alaska (see pages 53-4). Subsequently, I have argued (Oxenham and Matsumura 2008) that the Ipiutak series was somewhat unique in that the community targeted terrestrial game, as opposed to the marine resources hunted and collected by other populations in the region that, incidentally, also displayed lower caries rates. We also proposed that the Ipiutak may have been utilising the carbohydrate rich (and caries inducing) moss filled stomach contents of the caribou they preferentially hunted. In the context of the same discussion of influences on caries rates, the use of betel nut in Metal period Vietnam was explored. A detailed study of the Metal period teeth was subsequently undertaken where it was demonstrated that the tooth staining was deliberate and likely carried out for aesthetic purposes, thus somewhat reducing any caries prophylactic properties the betel nut itself may have had (Oxenham *et al.* 2002a).

It was suggested in the monograph that "[a]rchaeological evidence indicates a marine dietary component in both periods, although the relative contribution of marine resources to the diet is unknown" (page 93). This assertion needs to be tempered with subsequent dietary stable isotopic work which suggests that at least one of the Metal period sites, Nui Nap, may have had a significant rice contribution to their diet, and not such an important marine input (Bower *et al.* 2006, Willis and Oxenham in press). Further, the putative important marine focus of the Da But period population is not borne out by the same isotopic study, albeit using a very small sample (Bower *et al.* 2006). A recent new excavation of the Da But Con Co Ngua site I undertook in 2013 indicates that terrestrial fauna, especially water buffalo, had a significant part to play in the Da But diet (Rebecca Jones pers. comm.).

In the context of the chapter looking at osteoarthritis in ancient Vietnam, condylar depressions are reviewed (e.g. pages 118 and 156). These are variably expressed depressions on the posterior aspect of the mandibular condyle. A follow up study of these depressions concluded they were, indeed, not associated with osteoarthritis, but would appear to be a type of cortical defect (see Oxenham and Whitworth 2006).

Regarding the aetiology of cribra orbitalia, the literature has been somewhat active over recent years. Walker *et al.* (2009) have suggested that iron deficiency anaemia should not be considered as a potential candidate for the condition, a position which has been shown to be based on a misunderstanding of the actual haematopoietic response to a range of anaemic conditions (Oxenham and Cavill 2010). It is clear that much more work needs to be done with respect to understanding the aetiology of cribra orbitalia, and porotic hyperostosis, but iron deficiency anaemia is still very much a prime candidate in any differential diagnosis associated with these skeletal lesions.

When looking at Vietnamese examples of palaeopathology, case 12 (see page 204) prompted a discussion of a series of somewhat unusual gouges and slices to the long bones of one individual. While an obvious reason for these modifications was unclear at the time, we now know that porcupines were present at the Da But period site Con Co Ngua and may have been responsible for some post-mortem damage to the human remains. Moreover, a recent post-excavation season looking at the 160+ burials excavated in the 2013 season indicates widespread and systematic post-mortem mutilation of most adults as part of an elaborate funerary treatment. In terms of the demography of the Da But period, it was noted that very few subadults were available for assessment. The 2013 excavation of Con Co Ngua uncovered a significant proportion of subadults, including a number of perinates. While the examination of these new remains is in its very early stages, minimally it is now clear that previous excavation and curation techniques were responsible for the dearth of Da But period subadults available to be examined in this monograph.

Finally, when I undertook this study of pre-Neolithic Con Co Ngua from the Da But period and the collective assemblage of Metal period sites, a Neolithic series from the region was unavailable, and indeed, had not yet been found. This gap was filled through a series of excavations between 1999 and 2007 (Oxenham *et al.* 2011), at the site of Man Bac, located on the southern edges of the Red River delta. Man Bac has been pivotal in our attempts to bridge the large temporal and socio-cultural gap between the pre-Neolithic and early Metal period of northern Vietnam. Notwithstanding, the chief findings of the "*Bioarchaeology of Ancient Northern Vietnam*" has weathered the test of time, new excavations and new insights quite well.

BIBLIOGRAPHY

Ascadi, G.Y. and J. Nemeskeri 1970. *History of Human Life Span and Mortality*. Budapest: Akademiai Kiado.

Akazawa, T. 1986a. Hunter-gatherer adaptations and the transition to food production in Japan. In M. Zvelebil (ed) *Hunters in Transition: Mesolithic Societies of Temperate Eurasia and their Transition to Farming*. Cambridge: Cambridge University Press, pp:151-165.

Akazawa, T. 1986b. Regional variation in procurement systems of Jomon hunter-gatherers. In T. Akazawa and C.M. Aikens (eds) *Prehistoric Hunter-Gatherers in Japan*. Tokyo: University of Tokyo Press, pp:73-89.

Akpata, E.S. and E.O. Akinrimisi 1977. Antibacterial activity of extracts from some African chewing sticks. *Oral Surgery, Oral Medicine and Oral Pathology* 44 (5):717-722.

Alexander, C.J. 1989. Relationship between the utilisation profile of individual joints and their susceptibility to primary osteoarthritis. *Skeletal Radiology* 18:199-205.

Allison, M.J. 1984. Paleopathology in Peruvian and Chilean populations. In M.N. Cohen & G.J. Armelagos (eds) *Paleopathology at the Origins of Agriculture*. Orlando: Academic Press, pp:515-529.

Altman, R., E. Asch, D. Bloch, D. Borenstein, K. Brandt, W. Christy, T.D. Cooke, R. Greenwald, M. Hochberg, D. Howell, D. Kaplan, W. Koopman, S. Longley, III, H. Mankin, D.J. McShane, T. Medsger, Jr., R. Meenan, W. Mikkelsen, R. Moskowitz, W. Murphy, B. Rothschild, M. Segal, L. Sokoloff and F. Wolfe 1986. Development of criteria for the classification and reporting of osteoarthritis. Classification of osteoarthritis of the knee. *Arthritis and Rheumatism* 29 (8):1039-1049.

Alvrus, A. 1999. Fracture patterns among the Nubians of Semna South, Sudanese Nubia. *International Journal of Osteoarchaeology* 9:417-429.

Anderson, D.L., G.W. Thompson and F. Popovich 1976. Age of attainment of mineralization stages of the permanent dentition. *Journal of Forensic Sciences* 21:191-200.

Anderson, L.D. and F.N. Meyer 1984. Fractures of the shafts of the radius and ulna. In C.A. Rockwood, Jnr., D.P. Green and R.W. Bucholz (eds) *Fractures in Adults*. Philadelphia: J.B. Lippincott, pp:679-737.

Angel, J.L. 1984. Health as a crucial factor in the changes from hunting to developed farming in the Eastern Mediterranean. In M.N. Cohen and G.J. Armelagos (eds) *Paleopathology at the Origins of Agriculture*. Orlando: Academic Press, pp:51-73.

Angel, J.L., J.M. Suchey, M.Y. Iscan and M.R. Zimmerman. 1986. Age at death estimated from the skeleton and viscera. In M.R. Zimmerman and J.L. Angel (eds) *Dating and Age Determination of Biological Materials*. London: Croom Helm, pp:179-220.

Anon 1994. Position of the American Dietetic Association: the impact of fluoride on dental health. *Journal of the American Dietetic Association* 12:1428-1431.

Apley, A.G. and L. Solomon 1993. *Apley's System of Orthopaedics and Fractures*. Oxford: Butterworth-Heinemann.

Armelagos, G.J. 1990. Health and disease in prehistoric populations in transition. In A.C. Swedlund and G.J. Armelagos (eds) *Disease in Populations in Transition: Anthropological and Epidemiological Perspectives*. New York: Bergin and Garvey, pp:127-144.

Aufderheide, A.C. and C. Rodriguez-Martin 1998. *The Cambridge Encyclopedia of Human Paleopathology*. New York: Cambridge University Press.

Aykroyd, R.G., D. Lucy, A.M. Pollard and T. Solheim. 1997. Technical note: regression analysis in adult age estimation. *American Journal of Physical Anthropology* 104:259-265.

Baetsen, S., P. Bitter and T.J.D. Bruint Jes 1997. Hip and knee osteoarthritis in an eighteenth century urban population. *International Journal of Osteoarchaeology* 7:628-630.

Banerjee, K.K. and B.B. Agarwal. 1998. Estimation of age from epiphyseal union at the wrist and ankle joints in the capital city of India. *Forensic Science International* 98 (1-2):31-39.

Barrett, J.H. 1936. The union of certain epiphyses in a mixed female population in Rangoon. *Journal of Anatomy* 70:432-434.

Bass, W.M. 1987. *Human Osteology: A Laboratory and Field Manual*. Columbia, Missouri: Missouri Archaeological Society.

Beckett, S. and N.C. Lovell 1994. Dental disease evidence for agricultural intensification in the Nubian C-Group. *International Journal of Osteoarchaeology* 4:223-240.

Bedford, M.E., K.F. Russell, C.O. Lovejoy, R.S. Meindl, S.W. Simpson and P.L. Stuart-Macadam. 1993. Test of the multifactorial aging method using skeletons with known ages-at-death from the

Grant collection. *American Journal of Physical Anthropology* 91:287-297.

Bellwood, P., M. F. Oxenham, C. H. Bui, K. D. Nguyen, A. Willis, C. Sarjeant, P. Piper, H. Matsumura, K. Tanaka, N. Beavan-Athfield, T. Higham, Q. M. Nguyen, N. K. Dang, K. T. K. Nguyen, T. H. Vo, N. B. Văn, T. K. Q. Trần, P. T. Nguyen, F. Campos, Y. Sato, L. C. Nguyen and N. Amano 2011. An Son and the Neolithic of Southern Vietnam. *Asian Perspectives* 50: pp 144-174.

Benfer, R. and D.S. Edwards. 1991. The principal axis method for measuring rate and amount of dental attrition: estimating juvenile or adult tooth wear from unaged adult teeth. In M.A. kelley and C.S. Larsen (eds) *Advances in Dental Anthropology*. New York: Wiley-Liss, pp:325-340.

Berger, T.D. and E. Trinkaus 1995. Patterns of trauma among the Neandertals. *Journal of Archaeological Science* 22:841-852.

Bermudez de Castro, J.M. 1989. Third molar agenesis in human prehistoric populations of the Canary Islands. *American Journal of Physical Anthropology* 79:207-215.

Bermudez de Castro, J.M., J.L. Arsuaga and P.J. Perez 1997. Interproximal grooving in the Atapuerca-SH hominid dentitions. *American Journal of Physical Anthropology* 102:369-376.

Black, T.K.III. 1978. A new method for assessing the sex of fragmentary skeletal remains: femoral shaft circumference. *American Journal of Physical Anthropology* 48:227-232.

Blakey, M.L. and G.J. Armelagos. 1985. Deciduous enamel defects in prehistoric Americans from Dickson Mounds: prenatal and postnatal stress. *American Journal of Physical Anthropology* 66:371-380.

Blakey, M.L., T.E. Leslie and J.P. Reidy 1994. Frequency and chronological distribution of dental enamel hypoplasia. *American Journal of Physical Anthropology* 95:371-383.

Blondiaux, J., A. Cotton, C. Fontaine, C. Hanni, A. Bera and R.M. Flipo 1997. Two Roman and Medieval cases of symmetrical erosive polyarthropathy from Normandy: anatomico-pathological and radiological evidence for rheumatoid arthritis. *International Journal of Osteoarchaeology* 7:451-466.

Bombin, M. 1990. Transverse enamel hypoplasia on teeth of South African Plio-Pleistocene hominids. *Naturwissenschaften* 77:128-129.

Boucher, B.J., S.W. Ewen and J.M. Stowers 1994. Betel nut (*Areca catechu*) consumption and the induction of glucose intolerance in adult CD1 mice and their F1 and F2 offspring. *Diabetologia* 37 (1):49-55.

Bower, N. W., Y. Yasutomo, M. F. Oxenham, L. C. Nguyen and K. T. Nguyen 2006. Preliminary reconstruction of diet at a Neolithic site in Vietnam using Stable Isotope and BA/SR analyses. *Bulletin of the Indo-Pacific Prehistory Association* 26: pp 79-85.

Brace, C.L., K.R. Rosenberg & K.D. Hunt. 1987. Gradual change in human tooth size in the late Pleistocene and post-Pleistocene. *Evolution* 41 (4):705-720.

Brace, C.L., S.L. Smith and K.D. Hunt 1991. What big teeth you had Grandma! Human tooth size, past and present. In M.A. Kelley and C.S. Larsen (eds) *Advances in Dental Anthropology*. New York: Wiley-Liss, pp:33-57.

Brickley, M. and T. Waldron 1998. Relationship between bone density and osteoarthritis in a skeletal population from London. *Bone* 22 (3):279-283.

Bridges, P.S. 1991. Degenerative joint disease in hunter-gatherers and agriculturalists from the southeastern United States. *American Journal of Physical Anthropology* 85:379-391.

Bridges, P.S. 1992. Prehistoric arthritis in the Americas. *Annual Review of Anthropology* 21:67-91.

Brookes, G.B. and J.B. Booth 1997. Diseases of the temporal bone. In A.G. Kerr (ed.) *Scott-Brown's Otolaryngology*. Oxford: Butterworth-Heinemann, pp: 3/15/1-3/15-52.

Brothwell, D.R. 1981. *Digging up Bones: The Excavation, Treatment and Study of Human Skeletal Remains*. Ithaca: Cornell University Press.

Brown, P. 1987. Pleistocene homogeneity and Holocene size reduction: the Australian human skeletal evidence. *Archaeology in Oceania* 22:41-67.

Brown, T. and S. Molnar 1990. Interproximal grooving and task activity in Australia. *American Journal of Physical Anthropology* 81:545-553.

Bucholz, R.W. and R.J. Brumback 1991. Fractures of the shaft of the femur. In C.A. Rockwood, Jnr., D.P. Green and R.W. Bucholz (eds) *Fractures in Adults*. Philadelphia: J.B. Lippincott, pp:1653-1723.

Buckwalter, J.A. and N.E. Lane 1997. Athletics and Osteoarthritis. *The American Journal of Sports Medicine* 25 (6):873-881.

Bui Van Liem and Pham Lun Quan 1990. Excavation at Minh Duc, the boat coffin site (Ha Tay province). *New Archaeological Discoveries in 1990*:84-91.

Bui Van Liem and Pham Quoc Quan 1991. Boat coffin burial ground at Minh Duc. *Khao Co Hoc* 1991 (2):56-63.

Bui Vinh 1980. Bao Cao Khai Quat, Di Tich Van Hoa Da But, Dia Diem Khao Co Hoc: Con Co Ngua (Thanh-Hoa). [Cultural Vestiges at Da But Archaeological Site at Con Co Ngua (Thanh Hoa)]. Unpublished Excavation Report, Institute of Archaeology, Hanoi, Vietnam.

Bui Vinh 1991. The Da But Culture in the Stone Age of Vietnam. *Bulletin of the Indo-Pacific Prehistory Association* 10:127-131.

Bui Vinh 1994. On the post-Hoabinhian period in the Stone Age of Vietnam. *Vietnam Social Sciences* 5 (43):35-39.

Bui Vinh 1996. Da But Culture in the post-Hoabinhian in Vietnam. *Vietnamese Studies* 120 (2):23-29.

Buikstra, J.E. and J.H. Mielke. 1985. Demography, diet, and health. In R.I. Gilbert, Jr. and J.H. Mielke (eds) *The Analysis of Prehistoric Diets*. Orlando: Academic Press, pp:359-422.

Buikstra, J.E. and D.H. Ubelaker 1994. *Standards for Data Collection from Human Skeletal Remains*. Arkansas Archaeological Survey Research Series No. 44.

Burr, D.B., R.B. Martin, M.B. Schaffler, R.D. Jurmain, E.J. Harner and E.L. Radin 1983. Osteoarthrosis: sex-specific relationship to osteoporosis. *American Journal of Physical Anthropology* 61:299-303.

Burr, D.B. and M.S. Schaffler 1997. The involvement of subchondral mineralized tissues in osteoarthrosis: quantitative microscopic evidence. *Microscopy Research and Technique* 37:343-357.

Burris, B.G., D.D.S. Harris and E.F. Harris. 1998. Identification of race and sex from palate dimensions. *Journal of Forensic Sciences* 43 (5):959-963.

Calcagno, J.M. and K.R. Gibson 1991. Selective compromise: evolutionary trends and mechanisms in hominid tooth size. In M.A. Kelley and C.S. Larsen (eds) *Advances in Dental Anthropology*. New York: Wiley-Liss, pp:59-76.

Canniff, J.P. and W. Harvey 1981. The aetiology of oral submucous fibrosis: the stimulation of collagen synthesis by extracts of areca nut. *International Journal of Oral Surgery* 10 (Suppl. 1):163-167.

Capasso, L. 1997. Osteoma: palaeopathology and phylogeny. *International Journal of Osteoarchaeology* 7:615-620.

Carli-Thiele, P. and M. Schultz 1997. Microscopic differential diagnosis of so called cribra orbitalia - a contribution to the etiology of orbital porotic hyperostosis. *American Journal of Physical Anthropology* (abstract) Supp. 24: 88.

Cassidy, C.M. 1984. Skeletal evidence for prehistoric subsistence adaptation in the central Ohio river valley. In M.N. Cohen and G.J. Armelagos (eds) *Paleopathology at the Origins of Agriculture*. Orlando: Academic Press, pp:307-345.

Cawson, R.A. 1991. *Essentials of Dental Surgery and Pathology*. Edinburgh: Churchill Livingston.

Chamberlain, A.T. and B.A. Wood 1985. A reappraisal of variation in hominid mandibular corpus dimensions. *American Journal of Physical Anthropology* 66:399-405.

Chandler, F.W. and J.C. Watts 1987. *Pathologic Diagnosis of Fungal Infections*. Chicago: ASCP Press.

Chang, M.C., M.Y. Kuo, L.J. Hahn, C.C. Hsieh, S.K. Lin and J.H. Jeng 1998. Areca nut extract inhibits the growth, attachment and matrix protein synthesis of cultured human gingival fibroblasts. *Journal of Periodontology* 69 (10):1092-1097.

Chapman, K., Z. Mustafa, C. Irven, A.J. Carr, K. Clipsham, A. Smith, J. Chitnavis, J.S. Sinsheimer, V.A. Bloomfield, M. McCartney, O. Cox, L.R. Cardon, B. Sykes and J. Loughlin 1999. Osteoarthritis-susceptibility locus on chromosome 11q, detected by linkage. *American Journal of Human Genetics* 65 (1):167-174.

Chapman, N.E.M. 1997. Evidence for Spanish influence on activity induced musculoskeletal stress markers at Pecos Pueblo. *International Journal of Osteoarchaeology* 7:497-506.

Chung, W.C. and B.C. Ko 1976. Treatment of Taenia saginata infection with mixture of areca nuts and pumpkin seeds. *Chung Hua Min Kuo Wei Sheng Wu Huseh Tsa Chih* 9 (1-2):31-5.

Churchill, S.E. and A.G. Morris 1998. Muscle marking morphology and labour intensity in prehistoric Khoisan Foragers. *International Journal of Osteoarchaeology* 8:390-411.

Clarke, N.G. 1990. Periodontal defects of pulpal origin: evidence in early Man. *American Journal of Physical Anthropology* 82:371-376.

Clarke, N.G. and R.S. Hirsch 1991a. Physiological, pulpal, and periodontal factors influencing alveolar bone. In M.A. Kelley and C.S. Larsen (eds) *Advances in Dental Anthropology*. New York: Wiley-Liss, pp:241-266.

Clarke, N.G. and R.S. Hirsch 1991b. Tooth dislocation: the relationship with tooth wear and dental abscesses. *American Journal of Physical Anthropology* 85:293-298.

Clarke, N.G., S.E. Carey, W. Srikandi, R.S. Hirsch and P.I. Leppard 1986. Periodontal disease in ancient populations. *American Journal of Physical Anthropology* 71:173-183.

Cockburn, A., H. Duncan and J.M. Riddle 1979. Arthritis, ancient and modern: guidelines for field workers. *Henry Ford Hospital Medical Journal* 27 (1):74-79.

Cohen, M.N. K. O'Connor, M. Danforth, K. Jacobi and C. Armstrong 1994. Health and death at Tipu. In C.S. Larsen and G.R. Milner (eds) *In the Wake of Contact: Biological Responses to Conquest*. New York: Wiley-Liss, pp:121-133.

Colomer, J., C. Colomer, D. Gutierrez, A. Jubert, A. Nolasco, J. Donat, R. Fernandez-Delgado, F. Donat and C. Alvarez-Dardet 1990. Anaemia during pregnancy as a risk factor for infant iron deficiency: report from the Valencia Infant Anaemia Cohort (VIAC) study. *Paediatric and Perinatal Epidemiology* 4 (2):196 204.

Comuzzie, A.G. and D.G. Steele 1989. Enlarged occlusal surface on first molars due to severe attrition and hypercementosis: examples from prehistoric coastal populations. *American Journal of Physical Anthropology* 78:9-15.

Condon, K. and J.C. Rose 1992. Intertooth and intratooth variability in the occurrence of developmental enamel defects. In A.H. Goodman and L.L. Capasso (eds) *Recent Contributions to the Study of Enamel Developmental Defects. Journal of Paleopathology, Monographic Publications* 2:61-77.

Cook, D.C. 1984. Subsistence and health in the lower Illinios valley: osteological evidence. In M.N. Cohen & G.J. Armelagos (eds) *Paleopathology at the Origins of Agriculture*. Orlando: Academic Press, pp:235-269.

Cook, D.C. and J.E. Buikstra 1979. Health and differential survival in prehistoric populations: prenatal dental defects. *American Journal of Physical Anthropology* 51:649-664.

Corruccini, R.S. 1991. Anthropological aspects of orofacial and occlusal variations and anomalies. In M.A. Kelley and C.S. Larsen (eds) *Advances in Dental Anthropology*. New York: Wiley-Liss, pp:295-323.

Costa, R.L. 1980. Incidence of caries and abscesses in archaeological Eskimo skeletal samples from Point Hope and Kodiak, Alaska. *American Journal of Physical Anthropology* 52:501-514.

Costa, R.L. 1986. Determination of age at death: dentition analysis. In M.R. Zimmerman and A.J. Lawrence (eds) *Dating and Age Determination of Biological Materials*. London: Dover, N.H:Croom Helm, pp:249-269.

Crank, R.N., M. Sundaram and J.B. Shields 1982. Case Report 197. *Skeletal Radiology* 8:164-167.

Creamer, P. and M.C. Hochberg 1997. *Osteoarthritis*. The Lancet 350:503-510.

Cucina, A. and M.Y. Iscan 1997. Assessment of enamel hypoplasia in a high status burial site. *American Journal of Human Biology* 9:213-222.

Cushnaghan, J. and P. Dieppe 1991. Study of 500 patients with limb joint osteoarthritis. I. Analysis by age, sex, and distribution of symptomatic joint sites. *Annals of the Rheumatic Diseases* 50:8-13.

Cutress, T.W. and G.W. Suckling 1982. The assessment of non-carious defects of enamel. *International Dental Journal* 32 (2):117-122.

Dalinka, M.K. and W.H. Greendyke 1971. The spinal manifestations of coccidioidomycosis. *Journal of the Canadian Association of Radiologists* 22:93-99.

Dang Phuong Kiet and Bui Vu Huy 1995. Breast-feeding practice of mothers in Central High Plateaus and Hau Giang. *Vietnamese Studies* 4 (118):66-76.

Dave, B.J., A.H. Trivedi and S.G. Adhvaryu 1992. In vitro genotoxic effects of areca nut extract and recoline. *Journal of Cancer Research and Clinical Oncology* 118 (4):283-288.

De Crespigny, R. 1990. *Generals of the South: the Foundation and Early History of the Three Kingdoms State of Wu*. Canberra: Australian National University, Faculty of Asian Studies, New Series No. 16.

De Crespigny, R. 1996. South China in the Han period. In F.D. Bulbeck and N. Barnard (eds) *Ancient Chinese and Southeast Asian Bronze Age Cultures: The Proceedings of a Conference Held at the Edith and Joy London Foundation Property, Kiola, NSW, 8-12 February 1988*. Taipei: SMC Publishing, pp:759-768.

De Crespigny, R. n.d. [Unpublished Translation of Book 86, Hou Han Shu (Chronicles of the Later Han)].

De Miranda, C.M., C.W. van Wyk, P. van der Bijl and N.J. Basson 1996. The effect of areca nut on salivary and selected oral microorganisms. *International Dental Journal* 46 (4):350-356.

Deacon, A., K. Bennell, Z.S. Kiss, K. Crossley and P. Brukner 1997. Osteoarthritis of the knee in retired, elite Australian Rules footballers. *The Medical Journal of Australia* 166 (4):187-190.

Deng, Z. and D.H. Connor 1985. Disseminated penicilliosis cased by *Penicillium marneffei*: report of eight cases and differentiation of the causative organism from *Histoplasma capsulatum*. *American Journal of Clinical Pathology* 84:323-327.

Deng, Z., M. Yun and L. Ajello 1986. Human penicilliosis marneffei and its relation to the bamboo rat (*Rhizomys pruinosus*). *Journal of Medical and Veterinary Mycology* 24:383-389.

Dequeker, J., L. Mokassa, J. Aerssens and S. Boonen 1997. Bone density and local growth factors in generalized osteoarthritis. *Microscopy Research and Techniques* 37:358-371.

Di Bennardo, R. and J.V. Taylor. 1979. Sex assessment of the femur: a test of a new method. *American Journal of Physical Anthropology* 50:63-638.

Di Bennardo, R. and J.V. Taylor.1983. Multiple discriminant function analysis of sex and race in the post cranial skeleton. *American Journal of Physical Anthropology* 61:305-314.

Di Vella, G., C.P. Campobasso, M. Dragone and F. Introna Jr. 1994. Skeletal sex determination by scapular measurements. *Bollettino Societa Italiana Biologia Sperimentale* 70 (12):299-305.

Dias, G. and N. Tayles 1997. 'Abscess cavity'-a misnomer. *International Journal of Osteoarchaeology* 7:548-554.

Dieppe, P. 1990. Osteoarthritis: a review. *Journal of the Royal College of Physicians of London* 24 (4):262-267.

Dieppe, P. and J.M. Rogers 1993. Skeletal pathology of rheumatic disorders. In D.M. McCarty and W.J. Koopman (eds) *Arthritis and Allied Conditions*. Philadelphia: Lea and Febiger, pp:9-16.

Dittrick, J. and J. M. Suchey. 1986. Sex determination of prehistoric Central Californian skeletal remains using discriminant analysis of the femur and humerus. *American Journal of Physical Anthropology* 70:3-9.

Doherty, M. and B. Preston 1989. Primary osteoarthritis of the elbow. *Annals of the Rheumatic Diseases* 48 (9):743-747.

Doherty, M., I. Watt and P. Dieppe 1983. Influence of primary generalised osteoarthritis on development of secondary osteoarthritis. *The Lancet* 2:8-11.

Domett, K.M. 1999. Health in Late Prehistoric Thailand. Unpublished PhD Thesis, University of Otago, New Zealand.

Donnelly, S.M., S.M. Hens, N.L. Rogers and K.L. Schneider 1998. Technical note: a blind test of mandibular ramus flexure as a morphologic indicator of sexual dimorphism in the human skeleton. *American Journal of Physical Anthropology* 107:363-366.

Douglas, M. 1996. Paleopathology in Human Skeletal Remains from the Pre-Metal, Bronze and Iron Ages, Northeastern Thailand. Unpublished PhD Thesis, University of Hawaii.

Duray, S.M. 1996. Dental indicators of stress and reduced age at death in prehistoric native Americans. *American Journal of Physical Anthropology* 99:275-286.

Eisenberg, L. 1991. Interpreting measures of community health during the Late Prehistoric period in Middle Tennessee: a biocultural approach. In H. Bush and M. Zvelebil (eds) *Health in Past Societies: Biocultural Interpretations of Human Skeletal Remains in Archaeological Contexts*. BAR International series 567, pp:115-127.

El-Najjar, M.Y., B.Lozoff and D.J.Ryan 1975. The paleoepidemiology of porotic hyperostosis in the American Southwest: radiological and ecological considerations. *American Journal of Roentgenology, Radium Therapy and Nuclear Medicine.* 125 (4):918-924.

Elvin Lewis, M. 1983. The antibiotic and healing potential of plants used for teeth cleaning. In L. Romanucci-Ross, D.E. Moerman and L.R. Tancredi (eds) *The Anthropology of Medicine: From Culture to Method.* Massachusetts: Bergin and Garvey, pp:201-220.

Epps, C.H.Jr. and R.E. Grant 1991. Fractures of the shaft of the humerus. In C.A. Rockwood, Jnr., D.P. Green and R.W. Bucholz (eds) *Fractures in Adults*. Philadelphia: J.B.Lippincott, pp:843-869.

Fairgrieve, S.I. And T.S. Oost. 1995. On a test of the multifactorial aging method by Bedford et al. (1993). *American Journal of Physical Anthropology* 97:83-87.

Fang, C., S. Liu, Z. Sheng, Z. Li and F. Tang 1997. Study of the cytotoxic and DNA damaging effects in oral mucosal fibroblasts by areca nut extract. *Hunan I Ko Hsueh Pao* 22 (2):105-108.

Federation Dentaire International 1982. An epidemiological index of developmental defects of dental enamel (DDE Index). *International Dental Journal* 32 (2).159-167.

Feldesman, M.R. and R.L. Fountain 1996. "Race" specificity and the femur/stature ratio. *American Journal of Physical Anthropology* 100:207-224.

Feldesman, M.R., J.G. Kleckner and J.K. Lundy 1990. Femur/stature ratio estimates of stature in mid- and late-Pleistocene fossil hominids. *American Journal of Physical Anthropology* 83:359-372.

Felson, D.T., N.N. Couropmitree, C.E. Chaison, M.T. Hannan, Y. Zhang, T.E. McAlindon, M. LaValley, D. Levy and R.H. Myers 1998. Evidence for a Mendelian gene in a segregation analysis of generalized radiographic osteoarthritis: the Framingham Study. *Arthritis and Rheumatism* 41 (6):1064-1071.

Fergusson, C.M. 1987. The aetiology of osteoarthritis. *Postgraduate Medical Journal* 63:439-445.

Ferro-Luzzi, G.E. 1980a. Food avoidances of pregnant women in Tamilnad. In J.R.K. Robson (ed.) *Food, Ecology and Culture: Readings in the Anthropology of Dietary Practices*. New York: Gordon and Breach Science Publishers, pp:101-108.

Ferro-Luzzi, G.E. 1980b. Food avoidances during the puerperium and lactation in Tamilnad. In J.R.K. Robson (ed.) *Food, Ecology and Culture: Readings in the Anthropology of Dietary Practices.* New York: Gordon and Breach Science Publishers, pp:109-117.

Flynn, M. 1977. Black teeth: a primitive method of caries prevention in Southeast Asia. *Journal of the American Dental Association* 95:96-97.

Frayer, D.W. 1991. On the etiology of interproximal grooves. *American Journal of Physical Anthropology* 85:299-304.

Frayer, D.W. and M.D.Russell 1987. Artificial grooves on the Krapina Neanderthal teeth. *American Journal of Physical Anthropology* 74:393-405.

Frean, J., L. Blumberg and M. Woolf 1993. Disseminated blastomycosis masquerading as tuberculosis. *Journal of Infection* 26 (2):203-206.

Friedmann, I. 1974. *Pathology of the Ear.* Oxford: Blakewell Scientific Publications.

Funayama, M., Y. Aoki, T. Kudo and K. Sagisaka. 1986. Sex determination of the human skull based upon line drawing from roentgen cephalograms. *Tohoku Journal of Experimental Medicine* 149 (4):407-416.

Galera, V., D.H. Ubelaker and L.C. Hayek. 1998. Comparison of macroscopic cranial methods of age estimation applied to skeletons from the Terry collection. *Journal of Forensic Sciences* 43 (5):933-939.

Gelber, A.C., M.C. Hochberg, L.A. Mead, N.Y. Wang, F.M. Wigley and M.J. Klag 1999. Body mass index in young men and the risk of subsequent knee and hip osteoarthritis. *The American Journal of Medicine* 107:542 548.

Geyh, M.A., H.R. Kudrass and H. Streif 1979. Sea level changes during the late Pleistocene and Holocene in the straits of Malacca. *Nature* 278: 441-443.

Gilbert, B.M. 1973. Misapplication to females of the standard for aging the male os pubis. *American Journal of Physical Anthropology* 38:39-40.

Gilbert, B.M. and T.W. McKern. 1973. A method for aging the female Os pubis. *American Journal of Physical Anthropology* 38:31-38.

Giles, E. 1964. Sex determination by discriminant function analysis of the mandible. *American Journal of Physical Anthropology* 22:129-135.

Giles, E. 1970. Discriminant function sexing of the human skeleton. In T.D. Stewart (ed.) *Personal Identifi-*

cation in Mass Disasters. Washington: National Museum of Natural History, Smithsonian Institution, pp: 99-109.

Giles, E. and O. Elliot 1963. Sex determination by discriminant function analysis of crania. *American Journal of Physical Anthropology* 21:53-68.

Gomez-Pellico, L. and F.J. Fernandez-Camacho. 1992. Biometry of the anterior border of the human hip bone: normal values and their use in sex determination. *Journal of Anatomy* 181 (Pt 3):417-422.

Goodman, A.H and G.J. Armelagos 1985a. Factors affecting he distribution of enamel hypoplasias within the human permanent dentition. *American Journal of Physical Anthropology* 68:479-493.

Goodman, A.H. and G.J. Armelagos 1985b. The chronological distribution of enamel hypoplasia in human permanent incisor and canine teeth. *Archives of Oral Biology* 30 (6):503-507.

Goodman, A.H. and G.J. Armelagos 1988. Childhood stress and decreased longevity in a prehistoric population. *American Anthropologist* 90:936-944.

Goodman, A.H. and J.C. Rose 1991. Dental enamel hypoplasias as indicators of nutritional status. In M.A. Kelley and C.S. Larsen (eds) Advances in Dental Anthropology. Wiley-Liss: New York, pp:279-293.

Goodman, A.H. and L.L. Capasso (eds) 1992. *Recent Contributions to the Study of Enamel Developmental Defects*. Journal of Paleopathology, Monographic Publications 2.

Goodman, A.H., G.J. Armelagos and J.C. Rose 1980. Enamel hypoplasias as indicators of stress in three prehistoric populations from Illinios. *Human Biology* 52 (3):515-28.

Goodman, A.H., J. Lallo, G.J. Armelagos & J.C. Rose 1984. Health changes at Dickson Mounds, Illinios (AD 950-1300). In M.N. Cohen & G.J. Armelagos (eds) Paleopathology at the Origins of Agriculture. Orlando: Academic Press, pp:271-305.

Goodman, A.H. L.H. Allen, G.P. Hernandez, A. Amador, L.V. Arriola, A. Chavez and G.H. Pelto 1987. Prevalence and age at development of enamel hypoplasias in Mexican children. *American Journal of Physical Anthropology* 72:7-19.

Goodman, A.H., D.L. Martin, C.P. Klein, M.S. Peele, N.A. Cruse, L.R. McEwen, A. Saeed and B.M. RobinsonIn 1992. Cluster bands, Wilson bands and pit patches: histological and enamel surface indicators of stress in the Black Mesa Anasazi population. In A.H. Goodman and L.L. Capasso (eds) *Recent Contributions to the Study of Enamel Developmental Defects*. Journal of Paleopathology, Monographic Publications 2:115-127.

Gorman, C. 1971. The Hoabinhian and after: subsistence patterns in Southeast Asia during the late Pleistocene and early Recent periods. *World Archaeology* 2 (3):300-320.

Green, S.B., N.J. Salkind and T. Akey 1997. *Using SPSS for Windows: Analysing and Understanding Data.* New Jersey: Prentice Hall.

Gregg, J.B., J.P. Steel and S. Clifford 1965. Ear disease in skulls from the Scully burial site. *Plains Anthropologist* 10 (30):233-239.

Guler, N. A. Palanduz, U. Ones, A. Ozturk, A. Somer, N. Salman and I. Yalcin 1995. Progressive vertebral blastomycosis mimicking tuberculosis. *Pediatric Infectious Disease Journal* 14 (9):816-818.

Ha Van Phung 1984. Nap Mount (Thanh Hoa Province): new perceptions. *Khao Co Hoc* 1984 (4):27-40.

Ha Van Phung and Phan Ninh Huyen 1978. Excavations at Nap Mount site (Than Hoa). *New Archaeological Discoveries in 1978*.

Ha Van Tan 1985. Prehistoric pottery in Viet Nam and its relationships with Southeast Asia. *Asian Perspectives* 26 (1):135-146.

Ha Van Tan. 1997. The Hoabinhian and before. *Bulletin of the Indo-Pacific Prehistory Association* 16 (3):35-41.

Hacking, P., T. Allen and J. Rogers 1994. Rheumatoid arthritis in a Medieval skeleton. *International Journal of Osteoarchaeology* 4:251-255.

Hager, L.D. 1996. Sex differences in the sciatic notch of great apes and modern humans. *American Journal of Physical Anthropology* 99:287-300.

Haqq, C.M. and P.K. Donahoe. 1998. Regulation of sexual dimorphism in mammals. *Physiological Reviews* 78 (1):1-33.

Hardie, J. and K. Ahmed. 1995. The Miswak as an aid in oral hygiene. *Journal of the Philippine Dental Association* 47 (1):33-38.

Harkess, J.W., W.C. Ramsey and J.W. Harkess 1991. Principles of fractures and dislocations. In C.A. Rockwood, Jnr., D.P. Green and R.W. Bucholz (eds) *Fractures in Adults*. Philadelphia: J.B.Lippincott, pp:1-180.

Harris, E.F. and T.A. Rathbun 1991. Ethnic differences in the apportionment of tooth sizes. In M.A. Kelley and C.S. Larsen (eds) *Advances in Dental Anthropology*. Wiley-Liss: New York, pp:121-142.

Hartnady, P. and J.C. Rose 1991. Abnormal tooth-loss patterns among Archaic-period inhabitants of the lower Pecos region, Texas. In M.A. Kelley and C.S. Larsen (eds) *Advances in Dental Anthropology.* New York: Wiley-Liss, pp:267-278.

Hattab, F.N. 1997. Meswak: the natural toothbrush. *Journal of Clinical Dentistry* 8 (5):125-129.

Haun, S.J. 2000. Brief communication: a study of the predictive accuracy of mandibular ramus flexure as a single morphologic indicator of sex in an archaeological sample. *American Journal of Physical Anthropology* 111:429-432.

Hawkey, D.E. 1998. Disability, compassion and the skeletal record: using musculoskeletal stress markers (MSM) to construct an osteobiography from early New Mexico. *International Journal of Osteoarchaeology* 8:326-340.

Hayden, B., M. Eldridge, A. Eldridge and A. Cannon 1985. Complex hunter-gatherers in interior British Columbia. In T.D. Douglas and J.A. Brown (eds) Prehistoric Hunter-Gatherers: The Emergence of Cultural Complexity. Orlando: Academic Press, pp:181-199.

Hengen, O.P. 1971. Cribra orbitalia: pathogenesis and probable etiology. *Homo* 22: 57-75.

Hershkovitz, I., B. Latimer, O. Dutour, L.M. Jellema, S. Wish-Baratz, C. Rothschild and B.M. Rothschild. 1997. Why do we fail in aging the skull from the sagittal suture? *American Journal of Physical Anthropology* 103:393-399.

Hershkovitz, I. B.M. Rothschild, O. Dutour and C. Greenwald 1998. Clues to recognition of fungal origin of lytic skeletal lesions. *American Journal of Physical Anthropology* 106:47-60.

Hickey, G.C. 1982. *Sons of the Mountains: Ethnohistory of the Vietnamese Central Highlands to 1954.* New Haven: Yale University Press.

Higham, C.F.W. 1989. *The Archaeology of Mainland Southeast Asia.* Cambridge: Cambridge University Press.

Higham, C.F.W. 1996. *The Bronze Age of Southeast Asia.* Cambridge: Cambridge University Press.

Higham, C. F. W., T. Higham, R. Ciarla, K. Douka, A. Kijngam and F. Rispoli 2011. The Origins of the Bronze Age of Southeast Asia. *Journal of World Prehistory* 24(4): pp 227-274.

Higham, C.F.W. and R. Thosarat 1998. Noen U-Loke and the Implications for the Origins of Early States. Paper presented at the 16th Congress of the Indo-Pacific Prehistory Association, Melaka, Malaysia 1st to 7th July 1998.

Hildebolt, C.F. and S. Molnar 1991. Measurement and description of periodontal disease in anthropological studies. In M.A. Kelley and C.S. Larsen (eds) *Advances in Dental Anthropology.* Wiley-Liss: New York, pp:225-240.

Hildebolt, C.F., S. Molnar, M. Elvin-Lewis and J.K. McKee 1988. The effect of geochemical factors on the prevalences of dental diseases for prehistoric inhabitants of the state of Missouri. *American Journal of Physical Anthropology* 75:1-14.

Hildebolt, C.F., M. Elvin-Lewis, S. Molnar, J.K. McKee, M.D. Perkins and K.L. Young 1989. Caries prevalences among geochemical regions of Missouri. *American Journal of Physical Anthropology* 78:79-92.

Hill, C.A. 2000. Technical note: evaluating mandibular ramus flexure as a morphological indicator of sex. *American Journal of Physical Anthropology* 111:573-577.

Hillson, S. 1996. *Dental Anthropology.* New York: Cambridge University Press.

Hillson, S. and S. Bond 1997. Relationship of enamel hypoplasia to the pattern of tooth crown growth: a discussion. *American Journal of Physical Anthropology* 104:89-103.

Hinton, R.J. 1981. Form and patterning of anterior tooth wear among aboriginal human groups. *American Journal of Physical Anthropology* 54:555-564.

Hinton, R.J. and D.S. Carlson 1979. Temporal changes in human temporomandibular joint size and shape. *American Journal of Physical Anthropology* 50:325-331.

Hirsch, R. 1996. Association of hand and knee osteoarthritis: evidence for a polyarticular disease subset. *The Journal of the American Medical Association* 55:25-29.

Hoaglund, F.T., C.S. Oishi and G.G. Gialamas 1995. Extreme variations in racial rates of total hip arthroplasty for primary coxarthrosis: a population-based study in San Francisco. *The Rheumatic Diseases* 54:107-110.

Hoaglund, F.T., A.C.M.C.Yau and W.L.Wong 1973. Osteoarthritis of the hip and other joints in Southern Chinese in Hong Kong. *The Journal of Bone and Joint Surgery* 55A (3):545-557.

Hocart, C.H. and B. Fankhauser 1996. Betel nut residues in archaeological samples of human teeth from the Mariana Islands. *Experientia* 52: 281-285.

Hochberg, M.C., N.E. Lane, A.R. Pressman, H.K. Genant, J.C. Scott and M.C. Nevitt 1995. The association of radiographic changes and osteoarthritis of the hand and hip in elderly women. *Journal of Rheumatology* 22 (12):2291-2294.

Hochberg, M.C., M. Lethbridge-Cejku, W.W. Scott Jnr., C.C. Plato and J.D. Tobin 1993. Obesity and osteoarthritis of the hands in women. *Osteoarthritis Cartilage* 1 (2):129-135.

Holland, T.D. 1991. Sex assessment using the proximal tibia. *American Journal of Physical Anthropology* 85:221-227.

Holland, T.D. 1992. Estimation of adult stature from fragmentary tibias. *Journal of Forensic Sciences* 37 (5):1223-1229.

Holland, T.D. 1995. Brief communication: estimation of adult stature from the calcaneus and talus. *American Journal of Physical Anthropology* 96:315-320.

Holman, D.J. and K.A. Bennett. 1991. Determination of sex from arm bone measurements. *American Journal of Physical Anthropology* 84:421-426.

Holmgren, J. 1980. *Chinese Colonisation of Northern Vietnam: Administrative Geography and Political Development in the Tongking Delta, First to Sixth Centuries A.D.* Canberra: Australian National University.

Hough, A.J. Jnr. 1993. Pathology of osteoarthritis. In D.J. McCarty and W.J. Koopman (eds) *Arthritis and Allied Conditions.* Philadelphia: Lea and Febiger, pp:1699-1721.

Hoyme, L.E. 1957. The earliest use of indices for sexing pelves. *American Journal of Physical Anthropology* 15:537-546.

Huch, K., K.E. Kuettner and P. Dieppe 1997. Osteoarthritis in ankle and knee joints. *Seminars in Arthritis and Rheumatism* 26 (4):667-674.

Hull, V.J. 1986. Dietary taboos in Java: myths, mysteries and methodology. In L. Manderson (ed.) *Shared Wealth and Symbol: Food, Culture and Society in Oceania and Southeast Asia.* Cambridge: Cambridge University Press, pp:237-258.

Hutchinson, D.L. and C.S. Larsen 1997. Stressed to the max? Physiological perturbation in the Krapina Neandertals. *Current Anthropology* 38:904-914.

Hutterer, K.L. 1983. The natural and cultural history of Southeast Asian agriculture: ecological and evolutionary considerations. *Anthropos* 78:169-212.

Hutton, C.W. 1987. Generalised osteoarthritis: an evolutionary problem? *The Lancet* 1 (8548):1463-1465.

Ikehara-Quebral, R. and M.T. Douglas 1997. Cultural alteration of human teeth in the Mariana Islands. *American Journal of Physical Anthropology* 104:381-391.

Indrayana, N.S., J. Glinka and S. Mieke. 1998. Mandibular ramus flexure in an Indonesian population. *American Journal of Physical Anthropology* 105:89-90.

Ingold, T. 1980. *Hunters, Pastoralists and Ranchers: Reindeer Economies and their Transformations.* Cambridge: Cambridge University Press.

Inokuchi, J. H. Okabe, T. Yamauchi, A. Nagamatsu, G. Nonaka and I. Nishioka 1986. Antihypertensive substance in seeds of *Areca catechu L. Life Sciences* 38 (15):1375-1382.

Inoue, M., T. Inoue, Y. Fushimi and K. Okada. 1992. Sex determination by discriminant function analysis of lateral cranial form. *Forensic Science International* 57:109-117.

Inoue, K., S. Hukuda, M. Nakai, K. Katayama and J. Huang 1999. Erosive peripheral polyarthritis in ancient Japanese skeletons: a possible case of rheumatoid arthritis. *International Journal of Osteoarchaeology* 9:1-7.

Institute of Archaeology HS 55 n.d. Report of Thieu Duong Site Excavation (1960-1961). Unpublished Report, Institute of Archaeology, Hanoi, Vietnam.

Institute of Archaeology HS 63 n.d. Report of Duong Co Site Excavation. Unpublished Report, Institute of Archaeology, Hanoi, Vietnam.

Institute of Archaeology HS 85 n.d. Report of Vinh Quang Site Excavation. Unpublished Report, Institute of Archaeology, Hanoi, Vietnam.

Institute of Archaeology HS 224 n.d. Report of Minh Duc Excavation 1989 (Ha Tay Province). Unpublished Report, Institute of Archaeology, Hanoi, Vietnam.

Institute of Archaeology HS 251 n.d. Report of Nap Mount Site Excavation (Thanh Hoa 1977). Unpublished Report, Institute of Archaeology, Hanoi, Vietnam.

Introna, F.Jr., G. Di Vella, C.P. Campobasso and M. Dragone 1997. Sex determination by discriminant analysis of calcanei measurements. *Journal of Forensic Sciences* 42 (4):725-728.

Iscan, M.Y. and P. Miller-Shaivitz. 1984a. Discriminant function sexing of the tibia. *Journal of Forensic Sciences* 29 (4):1087-1093.

Iscan, M.Y. and P. Miller-Shaivitz. 1984b. Determination of sex from the tibia. *American Journal of Physical Anthropology* 64:53-57.

Iscan, M.Y. and D.H. Shi. 1995. Sexual dimorphism in the Chinese femur. *Forensic Science International* 74 (1-2):79-87.

Iscan, M.Y., M. Yoshino and S. Kato. 1994. Sex determination from the tibia: standards for contemporary Japan. *Journal of Forensic Sciences* 39 (3):785-792.

Iscan, M.Y., S.R. Loth, C.A. King, D.H. Shi and M. Yoshino. 1998. Sexual dimorphism in the humerus: a comparative analysis of Chinese, Japanese and Thais. *Forensic Science International* 8 (1-2):17-29.

Iwamoto,M., K. Uchino, T. Toukairin, K. Kawaguchi, T. Tatebayashi, H. Ogawara and Y. Tonosaki 1991. The growth inhibition of *Streptococcus mutans* by 5'-nucleotidase inhibitors from *Areca catechu L. Chemical and Pharmaceutical Bulletin* 39 (5):1323-1324.

Jantz, R.L., D.R. Hunt and L. Meadows 1995. The measure and mismeasure of the tibia: implications for stature estimation. *Journal of Forensic Sciences* 40 (5):758-761.

Jeng, J.H., W.H. Lan, L.J. Hahn, C.C. Hsieh and M.Y. Kuo 1996. Inhibition of the migration, attachment, spreading growth and collagen synthesis of human gingival fibroblasts by arecoline, a major areca alkaloid, in vitro. *Journal of Oral Pathology and Medicine* 25 (7):371-375.

Jeng, J.H., L.H. Hahn, B.R. Lin, C.C. Hsieh, C.P. Chan and M.C. Chang 1999. Effects of areca nut, inflorescence piper beetle extracts and arecoline on cytotoxicity, total and unscheduled DNA synthesis in cultured gingival keratinocytes. *Journal of Oral Pathology and Medicine* 28 (2):64-71.

Ji, L., K. Terazawa, T. Tsukamoto and K. Haga. 1994. Estimation of age from epiphyseal union degrees of the sternal end of the clavicle. *Hokkaido Igaku Zasshi* 69 (1):104-111.

Johnston, W.D. 1993. Tuberculosis. In K.F. Kiple (ed.) *The Cambridge World History of Human Disease.* Cambridge: Cambridge University Press, pp:1059-1068.

Jones, A. and M. Doherty 1995. ABC of rheumatology: osteoarthritis. *British Medical Journal* 310:457-460.

Jonsson. H. and S.T. Valtysdottir 1995. Hypermobility features in patients with hand osteoarthritis. *Osteoarthritis Cartilage* 3 (1):1-5.

Judd, M.A. and C.A. Roberts 1999. Fracture trauma in a Medieval British farming village. *American Journal of Physical Anthropology* 109:229-243.

Jurmain, R.D. 1980. The pattern of involvement of appendicular degenerative joint disease. *American*

Journal of Physical Anthropology 53:143-150.

Jurmain, R. 1990. Paleoepidemiology of a central Californian prehistoric population from CA-Ala-39: dental disease. *American Journal of Physical Anthropology* 81:333-342.

Jurmain, R.D. 1999. *Stories from the Skeleton: Behavioral Reconstruction in Human Osteology*. Australia: Gordon and Breach Publishers.

Jurmain, R.D. and L. Kilgore 1995. Skeletal evidence of osteoarthritis: a palaeopathological perspective. *Annals of the Rheumatic Diseases* 54:443-450.

Katz, D. and J.M. Suchey. 1989. Race differences in pubic symphyseal ageing patterns in the male. *American Journal of Physical Anthropology* 80:167-72.

Katzenberg, M.A., D.A. Herring and S.H. Saunders 1996. Weaning and infant mortality: evaluating the skeletal evidence. *Yearbook of Physical Anthropology* 39:177-199.

Katzenstein, P.L. R.W. Moskowitz, E.J. Weaver, C.J. Malemud, M.N. Pathria, S.A. Jimenez and D.J. Prockop 1990. Genetic linkage of a polymorphism in the type II procollagen gene (COL2A1) to primary osteoarthritis associated with mild chondrodysplasia. *The New England Journal of Medicine* 322 (8):526-531.

Kelley, M.A. 1979. Skeletal changes produced by aortic aneurysms. *American Journal of Physical Anthropology* 51:35.38.

Kelley, M.A. and C.S. Larsen (eds) 1991. *Advances in Dental Anthropology*. New York: Wiley-Liss.

Kelley, M.A., D.R. Levesque and E. Weidl 1991. Contrasting patterns of dental disease in five early northern Chilean groups. In M.A. Kelley and C.S. Larsen (eds) *Advances in Dental Anthropology*. Wiley-Liss: New York, pp:203-213.

Kemink, J.L., J.K. Niparko and S.A. Telian 1993. Mastoiditis. In K.F. Kiple (ed.) *The Cambridge World History of Human Disease*. Cambridge: Cambridge University Press, pp:865-871.

Kennedy, K.A.R. 1984. Growth, nutrition and pathology in changing paleodemographic settings in South Asia. In M.N. Cohen and G.J. Armelagos (eds) *Paleopathology at the Origins of Agriculture*. Orlando: Academic Press, pp:169-192.

Kennedy, K.A.R. 1989. Skeletal markers of occupational stress. In M.Y. Iscan and K.A.R. Kennedy (eds) *Reconstruction of Life from the Skeleton*. New York: Wiley-Liss, pp:129-160.

Kent, S., E.D. Weinberg and P. Stuart-Macadam 1994. The etiology of anaemia of chronic disease and infection. *Journal of Clinical Epidemiology and Infection* 47 (1):23-33.

Kerley, E.R. 1970. Estimation of skeletal age: after about age 30. In T.D. Stewart (ed.) *Personal Identification in Mass Disasters*. National Museum of Natural History, Smithsonian Institution, Washington, pp: 57-70.

Kerr, L.D. 1998. Arthritis of the forefoot. *Clinical Orthopaedics and Related Research* 349:20-27.

Kieser, J.A., J. Moggi-Cecchi and H.T. Groeneveld. 1992. Sex allocation of skeletal material by analysis of the proximal tibia. *Forensic Science International* 56 (1):29-36.

Kilgore, L., R. Jurmain and D. van Gerven 1997. Palaeoepidemiological patterns of trauma in a Medieval Nubian skeletal population. *International Journal of Osteoarchaeology* 7:103-114.

King, C.A. 1997. Osteometric assessment of 20th century skeletons from Thailand and Hong Kong (China). Unpublished MA Thesis, Florida Atlantic Unversity.

King, C.A., M.Y. Iscan and S.R. Loth. 1998. Metric and comparative analysis of sexual dimorphism in the Thai femur. *Journal of Forensic Sciences* 43 (5):954-958.

Kleinbaum, D.G. and L.L. Kupper 1978. *Applied Regression Analysis and Other Multivariable Methods*. Massachusetts: Duxbury Press.

Klepinger, L.L., D. Katz, M.S. Micozzi and L. Caroll. 1992. Evaluation of cast methods for estimating age from the os pubis. *Journal of Forensic Sciences* 37 (3):763-770.

Knüsel, C.J., S. Goggel and D. Lucy 1997. Comparative degenerative joint disease of the vertebral column in the Medieval monastic cemetery of the Gilbertine Priory of St. Andrew, Fishergate, York, England. *American Journal of Physical Anthropology* 103:481-495.

Konradsen, L., E.M. Hansen and L. Sndergaard 1990. Long distance running and osteoarthrosis. *American Journal of Sports Medicine* 18 (4):379-381.

Koski, K. 1996. Mandibular ramus flexure-indicator of sexual dimorphism? *American Journal of Physical Anthropology* 101:545-546.

Krasse, B. 1985. The cariogenic potential of foods-a critical review of current methods. *International Dental Journal* 35:36-42.

Kreitner, K.F., F.J. Schweden, T. Riepert, B. Nafe and M. Thelen. 1998. Bone age determination based on the study of the medial extremity of the clavicle. *European Radiology* 8 (7):1116-1122.

Kreitner, K.F., F. Schweden, H.H. Schild, T. Riepert and B. Nafe. 1997. Computerized tomography of the epiphyseal union of the medial clavicle: an auxiliary method of age determination during adolescence and the 3rd decade of life? *Rofo Fortschritte Aud Dem Gebiete Der Rontgenstrahlen Und Der Neuen Bildgebenden Verfahren* 166 (6):481-486.

Krogman, W.M. 1962. *The Human Skeleton in Forensic Medicine*. Springfield, Illinois: C. C. Thomas.

Krogman, W. and M.Y. Iscan 1986. *The Human Skeleton in Forensic Medicine*. Springfield, Illinois: C.C.Thomas.

Kujala, U.M., J. Kapiro and S. Sarna 1994. Osteoarthritis of weight bearing joints of lower limbs in former elite male athletes. *British Medical Journal* 308:231-235.

Lampl, M. and F.E. Johnston. 1996. Problems in the aging of skeletal juveniles: perspectives from maturation assessments of living children. *American Journal of Physical Anthropology* 101:345-355.

Lane, N.E., H.H. Jones, W.H. Marshall Jnr., P.D. Wood and J.F. Fries 1986. Long-distance running, bone density and osteoarthritis. *Journal of the American Medical Association* 255 (9):1147-1151.

Lane, N.E., J.W. Oehlert, D.A. Bloch and J.F. Fries 1998. The relationship of running to osteoarthritis of the knee and hip and bone mineral density of the lumbar spine: a nine year longitudinal study. *Journal of Rheumatology* 25 (2):334-341.

Larsen, C.S. 1983. Behavioural implications of temporal change in cariogenesis. *Journal of Archaeological Science* 10:1-8.

Larsen, C.S. 1984. Health and disease in prehistoric Georgia: the transition to agriculture. In M.N. Cohen & G.J. Armelagos (eds) *Paleopathology at the Origins of Agriculture*. Orlando: Academic Press, pp:367-392.

Larsen, C.S. 1985. Dental modifications and tool use in the Western Great Basin. *American Journal of Physical Anthropology* 67:393-402.

Larsen, C.S. 1995. Biological changes in human populations with agriculture. *Annual Review Anthropology* 24:185-213.

Larsen, C.S. 1997. *Bioarchaeology: Interpreting Behavior from the Human Skeleton*. Cambridge: Cambridge University Press.

Larsen, C.S. and G.R.Milner 1994. Bioanthropological perspectives on postcontact transitions. In C.S. Larsen and G.R. Milner (eds) *In the Wake of Contact: Biological Responses to Conquest*. New York: Wiley-Liss, pp:1-8.

Larsen, C.S., R. Shavit and M.C. Griffin 1991. Dental caries evidence for dietary change: an archaeological context. In M.A. Kelley and C.S. Larsen (eds) *Advances in Dental Anthropology*. Wiley-Liss: New York, pp:179-202.

Lourandos, H. 1997. Continent of Hunter-Gatherers: New Perspectives in Australian Prehistory. Cambridge: Cambridge University Press.

Lawrence, J.S. and M. Sebo 1980. The geography of osteoarthrosis. In G. Nuki (ed.) *The Aetiopathogenesis of Osteoarthrosis*. Kent: Pitman Medical Publishing Co Ltd, pp:155-183.

Leverett, D.H. 1982. Fluorides and the changing prevalence of dental caries. *Science* 217:26-30.

Li, X.C., P. van der Bijl and C.D. Wu 1998. Binaphthalenone glycosides from African chewing sticks, Diospyros lycioides. *Journal of Natural Products* 61 (6):817-820.

Lieverse, A.R. 1999. Diet and the aetiology of dental calculus. *International Journal of Osteoarchaeology* 9:219-232.

Lim, K.K., J. Rogers, L. Shepstone and P.A. Dieppe 1995. The evolutionary origins of osteoarthritis: a comparative skeletal study of hand disease in 2 primates. *Journal of Rheumatology* 22 (11):2132-2134.

Lim, K.K.T., M.J. Kessler, K.P.H. Pritzker, J.E. Turnquist and P.A. Dieppe 1996. Osteoarthritis of the hand in nonhuman primates: a clinical, radiographic, and skeletal survey of Cayo Santiago rhesus macaques. *Journal of Medical Primatology* 25:301-308.

Linossier, A., M. Gajardo and J. Olavarria 1996. Paleomicrobiological study in dental calculus: *Streptococcus mutans*. *Scanning Microscopy* 19 (4):1005-1013.

Littleton, J. and B. Frohlich 1993. Fish-eaters and farmers: dental pathology in the Arabian Gulf. *American Journal of Physical Anthropology* 92:427-447.

Liu, T.Y., C.L. Chen and C.W. Chi 1996. Oxidative damage to DNA induced by areca nut extract. *Mutation Research* 367 (1):25-31.

Liu, W. 1989. Sex determination of Chinese femur by discriminant function. *Journal of Forensic Sciences* 34 (5):1222-1227.

Lopez, M.C.B., S.A.J. Brobeil and J.A.O. Vallet 1995. Traumatisms in Bronze Age Settlements in the Iberian peninsula: Agar culture. In Paleopathology Association. European Members Meeting (eds) *Proceedings of the IXth European Meeting of the Paleopathology Association: (Barcelona, 1st - 4th Setember, 1992)*. Barcelona : Museu d'Arqueologia de Catalunya, pp:65-72.

Loth, S.R. and M.Henneberg. 1996. Mandibular ramus flexure: a new morphological indicator of sexual dimorphism in the human skeleton. *American Journal of Physical Anthropology* 99:473-485.

Loth, S.R. and M. Henneberg 1998. Mandibular ramus flexure is a good indicator of sexual dimorphism. *American Journal of Physical Anthropology* 105:91-92.

Louthrenoo, W., K. Thamprasert and T. Sirisanthana 1994. Osteoarticular penicilliosis marneffei. A report of eight cases and review of the literature. *British Journal of Rheumatology* 33:1145-1150.

Lovejoy, C.O. 1985. Dental wear in the Libben population: its functional pattern and role in the determination of adult skeletal age at death. *American Journal of Physical Anthropology* 68:47-56.

Lovejoy, C.O. & K.G. Heiple 1981. The analysis of fractures in skeletal populations with an example from the Libben site, Ottowa county, Ohio. *American Journal of Physical Anthropology* 55:529-541.

Lovejoy, C.O., R.S. Meindl, R.P. Mensforth and T.J. Barton. 1985a. Multifactorial determination of skeletal age at death: a method and blind tests of its accuracy. *American Journal of Physical Anthropology* 68:1-14.

Lovejoy, C.O., R.S. Meindl, T.R. Pryzbeck and R.P. Mensforth. 1985b. Chronological metamorphosis of the auricular surface of the ilium: a new method for the determination of adult skeletal age at death. *American Journal of Physical Anthropology* 68:15-28.

Lovell, N.C. 1989. Test of Phenice's technique for determining sex from the os pubis. *American Journal of Physical Anthropology* 79:117-120.

Lovell, N.C. 1997a. Anaemia in the ancient Indus valley. *International Journal of Osteoarchaeology* 7:115-123.

Lovell, N.C. 1997b. Trauma analysis in paleopathology. *American Journal of Physical Anthropology* 40:139-170.

Lovell, N.C. and A.A. Dublenko 1999. Further aspects of fur trade life depicted in the skeleton. *International Journal of Osteoarchaeology* 9:248-256.

Lu, C.T., S.J. Lan, C.C. Hsieh, M.J. Yang, Y.C. Ko, C.C.T sai and Y.Y. Yen 1993. Prevalence and characteristics of areca nut chewers among junior high school students in Changhua county, Taiwan. *Community Dentistry and Oral Epidemiology* 21 (6):370-373.

Lu, G.D. and J.Needham 1993. Diseases of antiquity in China. In K.F. Kiple (ed.) *The Cambridge World History of Human Disease*. Cambridge: Cambridge University Press, pp:345-354.

Luder, H.U. 1997. Frequency and distribution of articular tissue features in adult human mandibular condyles: a semiquantitative light microscope study. *The Anatomical Record* 248:18-28.

Luder, H.U. 1998. Age changes in the articular tissue of human mandibular condyles from adolescence to old age: a semiquantitative light microscope study. *The Anatomical Record* 251:439-447.

Lukacs, J.R. 1989. Dental paleopathology: methods for reconstructing dietary patterns. In M.Y. Iscan and K.A.R. Kennedy (eds) *Reconstruction of Life from the Skeleton*. New York: Wiley-Liss, pp:261-286.

Lukacs, J.R. 1992. Dental paleopathology and agricultural intensification in South Asia: new evidence from Bronze Age Harappa. *American Journal of Physical Anthropology* 87:133-150.

Lukacs, J. R. 1995. The 'caries correction factor': a new method of calibrating dental caries rates to compensate for antemortem loss of teeth. *International Journal of Osteoarchaeology* 5:151-156.

Lukacs, J.R. & J.N. Pal 1993. Mesolithic Subsistence in north India: inferences from dental attributes. *Current Anthropology* 34 (5):745-765.

Lukacs, J.R. and R.F. Pastor 1988. Activity-induced patterns of dental abrasion in prehistoric Pakistan: evidence from Mehrgarh and Harappa. *American Journal of Physical Anthropology* 76:377-398.

Lukacs, J.R., D.H. Retief and J.F. Jarrige 1985. Dental disease in prehistoric Baluchistan. *National Geographic Research* 1 (2):184-197.

Luo, Y.C. 1995. Sex determination from the pubis by discriminant function analysis. *Forensic Science International* 74 (1-2):89-98.

Maat, G.J.R., R.W. Mastwijk and E.A. van der Velde 1997. On the reliability of non-metrical morphological sex determination of the skull compared with that of the pelvis in the Low Countries. *International Journal of Osteoarchaeology* 7:575-580.

Mack, M.E. and A. Coppa 1992. Frequency and chronological distribution of enamel hypoplasias from the Ra's al-Hamra-5 (RH5) skeletal collection (Oman). In A.H. Goodman and L.L. Capasso (eds) *Recent Contributions to the Study of Enamel Developmental Defects*. Journal of Paleopathology, Monographic Publications 2:131-141.

McCaffrey, T.V. and T.J. McDonald 1979. Histiocytosis of the ear and temporal bone: review of 22 cases. *The Laryngoscope* 89:1735-1742.

McGahan, J.P., D.S. Graves and P.E.S. Palmer 1979. Coccidioidal spondylitis. *Radiology* 136: 5-9.

McKay, D.A. 1980. Food, illness and folk medicine: insights from Ulu Trengganu, West Malaysia. In J.R.K. Robson (ed.) *Food, Ecology and Culture: Readings in the Anthropology of Dietary Practices*. New York: Gordon and Breach Science Publishers, pp:61-66.

McKenzie, W. and D.R. Brothwell 1967. Disease in the ear region. In D.R. Brothwell and A.T. Sandison (eds) *Diseases in Antiquity: A Survey of the Diseases, Injuries, and Surgery of Early Populations*. Springfield, Illinois: C. C. Thomas, pp:464-473.

McKern, T.W. and T.D. Stewart 1957. Skeletal age changes in young American males. U.S. Army Quartermaster Research and development Command, Technical Report EP-45.

MacLaughlin, S.M. and M.F. Bruce. 1985. A simple univariate technique for determining sex from fragmentary femora: its application to a Scottish short cist population. *American Journal of Physical Anthropology* 67:413-417.

MacLaughlin, S.M. & M.F. Bruce. 1986. The sciatic notch/acetabular index as a discriminator of sex in European skeletal remains. *Journal of Forensic Sciences* 31 (4):1380-1390.

MacLaughlin, S.M. and M.F. Bruce. 1990. The accuracy of sex identification in European skeletal remains using the Phenice characters. *Journal of Forensic Sciences* 35 (6):1384-1392.

Maetzel, A., M. Makela, G. Hawker and C. Bombardier 1997. Osteoarthritis of the hip and knee and occupational exposure: a systematic overview of the evidence. *Journal of Rheumatology* 24:1599-1607.

Maher, R., A.J. Lee, K.A. Warnakulasuriya, J.A. Lewis and N.W. Johnson 1994. Role of areca nut in the causation of oral submucous fibrosis: a case-control study in Pakistan. *Journal of Oral Pathology and Medicine* 23 (2):65-69.

Mair, F., J. Rogers, K. Lim, and P. Dieppe 1998 (abstract). Patterns of joint involvement in osteoarthritis: support for the evolutionary hypothesis. *British Journal of Rheumatology* 37 (1): 155.

Malville, N.J. 1997. Enamel hypoplasia in ancestral Puebloan populations from Southwestern Colorado:1. Permanent dentition. *American Journal of Physi-*

cal Anthropology 102:351-367.
Manchester, K. 1983. *The Archaeology of Disease*. Bradford: The University of Bradford.
Mankin, J. and K.D. Brandt 1997. Pathogenesis of osteoarthritis. In W.N. Kelley, E.D. Harris, S. Ruddy and C.B. Sledge (eds) *Textbook of Rheumatology*. Philadelphia: W.B. Saunders Co, pp:1369-1382.
Manly, B.F.J. 1994. *Multivariate Statistical Methods: A Primer*. London: Chapman and Hall.
Mann, R.W. and S.P. Murphy 1990. *Regional Atlas of Bone Disease: A Guide to Pathological and Normal Variation in the Human Skeleton*. Springfield: C. C. Thomas.
Marti, B., M. Knobloch, A. Tschopp, A. Jucker and H. Howald 1989. Is excessive running predictive of degenerative disease? Controlled study of former elite athletes. *British Medical Journal* 299:91-93.
Masset, C. 1982. *Estimation de l'ager au Deces par les Sutures Craniennes*. These de Sciences Naturelles, Multigraphiee, Universitie Paris VII.
Masset, C. 1989. Age estimation on the basis of cranial sutures. In M.Y. Iscan (ed.) *Age Markers in the Human Skeleton*. Springfield, Illinois: C.C.Thomas, pp:71-103.
Matsui, H., M. Shimizu and H. Tsuji 1997. Cartilage and subchondral bone interaction in osteoarthrosis of human knee joint: a histological and histomorphometric study. *Microscopy Research and Technique* 37:333-342.
Matsumura, H., L.C. Nguyen, K.T. Nguyen and T. Anezaki (2001). Dental morphology of the Early Hoabinian, the Neolithic Da But and the metal Age Dong Son culture people in Vietnam. *Zeitschrift fur Morphologie und Anthropologie* 83(1):59-73.
Matsumura, H. and M. F. Oxenham 2014. Demographic transitions and migration in prehistoric East/ Southeast Asiathrough the lens of nonmetric dental traits. *American Journal of Physical Anthropology* 155: pp 45-65.
Mazabraud, A. 1998. *Pathology of Bone Tumours: Personal Experience*. Berlin: Springer-Verlag.
Meehan, B. 1982. *Shellbed to Shell Midden*. Canberra: Australian Institute of Aboriginal Studies.
Meiklejohn, C. and M. Zvelebil 1991. Health status of European populations at the agricultural transition and implications for the adoption of farming. In H. Bush and M. Zvelebil (eds) *Health in Past Societies: Biocultural Interpretations of Human Skeletal Remains in Archaeological Contexts*. BAR International series 567, pp:129-145.
Meiklejohn, C. C. Schentag, A. Venema and P. Key 1984. In M.N. Cohen & G.J. Armelagos (eds) *Paleopathology at the Origins of Agriculture*. Orlando: Academic Press, pp:75-100.
Meindl, R.S. and C.O. Lovejoy. 1985. Ectocranial suture closure: a revised method for the determination of skeletal age at death based on the lateral-anterior sutures. *American Journal of Physical Anthropology* 68:57-66.
Meindl, R.S., C.O. Lovejoy and R.P. Mensforth. 1983. Skeletal age at death: accuracy of determination and implications for human demography. *Human Biology* 55 (1):73-87.
Meindl, R.S., K.F. Russell and C.O. Lovejoy. 1990. Reliability of age at death in the Hamann-Todd collection: validity of subselection procedures used in blind tests of the summary age technique. *American Journal of Physical Anthropology* 83:39-357.
Meindl, R.S., C.O. Lovejoy, R.P. Mensforth and L.D. Carlos. 1985a. Accuracy and direction of error in sexing of the skeleton: implications for paleodemography. *American Journal of Physical Anthropology* 68:79-85.
Meindl, R.S., C.O. Lovejoy, R.P. Mensforth and R.A. Walker. 1985b. A revised method of age determination using the Os pubis, with a review and tests of accuracy of other current methods of pubic symphyseal aging. *American Journal of Physical Anthropology* 68:29-45.
Meindl, R.S., M.E. Bedford, K.F. Russell, C.O. Lovejoy and S.W. Simpson. 1995. Testing the test of the multifactorial aging method: a reply to Fairgrieve and Oost. *American Journal of Physical Anthropology* 97:85-87.
Mensforth, R.P. and C.O. Lovejoy. 1985. Anatomical, physiological, and epidemiological correlates of the aging process: a confirmation of multifactorial age determination in the Libben skeletal population. *American Journal of Physical Anthropology* 68:87-106.
Merbs, C.F. 1983. *Patterns of Activity-Induced Pathology in a Canadian Inuit Population*. Archaeological Survey of Canada, Paper No. 119. Ottowa: National Museums of Canada.
Merbs, C.F. 1996a. Spondylolysis of the sacrum in Alaskan and Canadian Inuit skeletons. *American Journal of Physical Anthropology* 101:357-367.
Merbs, C.F. 1996b. Spondylolysis and spondylolisthesis: a cost of being an erect biped or a clever adaptation? *Yearbook of Physical Anthropology* 39:201-228.
Merida-Velasco, J.R., J.F. Rodriguez-Vazquez, J.A. Merida-Velasco, I. Sanchez-Montesinos, J. Espin-Ferra and J. Jimenez-Collado 1999. Development of the human temporomandibular joint. *The Anatomical Record* 255:20-33.
Meulenbelt, I., C. Bijkerk, H.S. Miedema, F.C. Breedveld, A. Hofman, H.A. Valkenburg, H.A. Pols, P.E. Slagboom and C.M. van Duijin 1998. A genetic association study of the IGF-1 gene and radiological osteoarthritis in a population-based cohort study (the Rotterdam Study). *Annals of the Rheumatic Diseases* 57 (6):371-374.
Miles, A.E.W. 1958. The assessment of age from the dentition. *Proceedings of the Royal Society of Medicine* 51:1057.
Miles, A.E.W. 1963a. The dentition in the assessment of individual age in skeletal material. In D.R. Broth-

well (ed.) *Dental Anthropology*. London: Pergamon Press, pp:191-209.

Miles, A.E.W. 1963b. Dentition in the assessment of age. *Journal Dental Research* 42:255.

Miles, A.E.W. 1999. A five-grade categorization of age-related change in the acromio-clavicular remains of early 19th century Londoners of known sex and age. *International Journal of Osteoarchaeology* 9:83-101.

Miller, E., B.D. Ragsdale and D.J. Ortner 1996. Accuracy in dry bone diagnosis: a comment on palaeopathological methods. *International Journal of Osteoarchaeology* 6:221-229.

Milner, G.R. 1991. Health and cultural change in the Late Prehistoric American bottom, Illinois. In M.L. Powell, P.S. Bridges and A.M.W. Mires (eds) *What Mean These Bones? Studies in Southeastern Bioarchaeology.* Tuscaloosa: University of Alabama Press, pp:52-68.

Milner, G.R. and C.S.Larsen 1991. Teeth as artifacts of human behavior: intentional mutilation and accidental modification. In M.A. Kelley and C.S. Larsen (eds) *Advances in Dental Anthropology*. Wiley-Liss: New York, pp:357-378.

Mittler, D.M. and D.P. van Gerven 1994. Developmental, diachronic, and demographic analysis of cribra orbitalia in the Medieval Christian populations of Kulubnarti. *American Journal of Physical Anthropology* 93:287-297.

Mittler, D.M., D.P. van Gerven, S.G. Sheridan and R. Beck 1992. The epidemiology of enamel hypoplasia, cribra orbitalia, and subadult mortality in an ancient Nubian population. In A.H. Goodman and L.L. Capasso (eds) *Recent Contributions to the Study of Enamel Developmental Defects*. Journal of Paleopathology, Monographic Publications 2:143-150.

Moller, I.J., J.J. Pindborg and I. Effendi. 1977. The relation between betel chewing and dental caries. *Scandanavian Journal of Dental Research* 85 (1):64-70.

Molleson, T. 1989. Seed preparation in the Mesolithic: the osteological evidence. *Antiquity* 63:356-362.

Molleson, T. 1994. The eloquent bones of Abu Hureyra. *Scientific American*, August 1994:59-65.

Molnar, S. 1972. Tooth wear and culture: a survey of tooth functions among some prehistoric populations. *Current Anthropology* 13 (5):511-526.

Molnar, S. and I. Molnar 1985a. Observations of dental diseases among prehistoric populations of Hungary. *American Journal of Physical Anthropology* 67:51-63.

Molnar, S. and I. Molnar 1985b. The incidence of enamel hypoplasia among the Krapina Neandertals. *American Anthropologist* 87:536-549.

Molnar, S., J.K. McKee and I. Molnar 1983. Measurements of tooth wear among Australian Aborigines: 1. Serial loss of the enamel crown. *American Journal of Physical Anthropology* 61:51-65.

Moore, W.F. 1993. Dental caries in Britain from Roman times to the nineteenth century. In C.G. Geissler and D.F. Oddy (eds) *Food, Diet and Economic Change Past and Present*. Leicester: Leicester University Press, pp:50-61.

Moorrees, C.F.A., E.A. Fanning and E.E. Hunt. 1963. Age variation of formation stages for ten permanent teeth. *Journal of Dental Research* 42 (suppl):255-263.

Morris, P.S. 1998. Improving the Medical Management of Otitis Media and other Chronic Bacterial Respiratory Diseases in Rural and Remote Australian Aboriginal Children: A Systematic Approach. Unpublished PhD Thesis, University of Sydney, Australia.

Moseley, J.E. 1974. Skeletal changes in the anemias. *Seminars in Roentgenology* 9 (3):169-184.

Moskowitz, R.W. 1993. Clinical and laboratory findings in osteoarthritis. In D.M. McCarty and W.J. Koopman (eds) *Arthritis and Allied Conditions*. Philadelphia: Lea and Febiger, pp:1735-1760.

Murray, K.A. and S.A. Murray 1989. Computer software for hypoplasia analysis. *American Journal of Physical Anthropology* 78:277-278.

Murray, K.A. and T. Murray. 1991. A test of the auricular surface aging technique. *Journal of Forensic Sciences* 36 (4):1162-1169.

Murti, P.R., P.C. Gupta, R.B. Bhonsle, D.K. Daftary, F.S. Mehta and J.J. Pindborg 1990. Effect on the incidence of oral submucous fibrosis of intervention in the areca nut chewing habit. *Journal of Oral Pathology and Medicine* 19 (2):99-100.

Murti, P.R., R.B. Bhonsle, P.C. Guptam D.K. Daftary, J.J. Pindborg and F.S. Mehta 1995. Etiology of oral submucous fibrosis with special reference to the role of areca nut chewing. *Journal of Oral Pathology and Medicine* 24 (4):145-152.

Nagar, Y., C. Taitz and R. Reich 1999. What can we make of these fragments? Excavation at 'Mamilla' Cave, Byzantine period, Jerusalem. *International Journal of Osteoarchaeology* 9:29-38.

Nair, U.J., M. Friesen, I. Richard, R. MacLennan, S. Thomas and H. Bartsch 1990. Effect of lime composition on the formation of reactive oxygen species from areca nut extract in vitro. *Carcinogenesis* 11 (12):2145-2148.

Nakai, M., K. Inoue and S. Hukuda 1999. Healed bone fractures in a Jomon skeletal population from the Yoshigo shell mound, Aichi Prefecture, Japan. *International Journal of Osteoarchaeology* 9:77-82.

Nawrocki, S.P. 1995. Taphonomic processes in historic cemeteries. In A.L. Grauer (ed.) *Bodies of Evidence: Reconstructing History Through Skeletal Analysis*. New York: Wiley-Liss, pp:49-66.

Nayeem, F. 1992. The mandible-an analysis. *Acta Anatomica* 145 (2):132-137.

Neiberger, E.J. 1990. Enamel hypoplasia: poor indicators of dietary stress. *American Journal of Physical*

Anthropology 82:231-233.
Nelsen, K.M. 1999. The Dental Health of the People from Noen U-Loke. A Prehistoric Iron Age Site in Northeast Thailand. Unpublished M.A. Thesis, University of Otago, New Zealand.
Neves, W.A., A.M. Barros and M.A. Costa 1999. Incidence and distribution of postcranial fractures in the prehistoric population of San Pedro de Atacama, Northern Chile. *American Journal of Physical Anthropology* 109:253-258.
Newton, P.M., V.C. Mow, T.R. Gardner, J.A. Buckwalter and J.P. Albright 1997. The effect of lifelong exercise on canine articular cartilage. *The American Journal of Sports Medicine* 25 (3):282-287.
Nguyen Cong Khanh, Y. Hofvander, Duong Ba Truc, Dao Ngoc Dien and Hoang Van Son 1995. Child anaemia in some areas of North Vietnam. *Vietnamese Studies* 118 (4):19-28.
Nguyen Dong Chi 1965. About bronze kinds of artefacts in Dong Son sites and Thieu Duong site. *History Review* 61:35-41.
Nguyen Khac Su 1996. The Sonvian in the prehistory of Vietnam and Southeast Asia. *Vietnamese Studies* 120 (2):7-22.
Nguyen Kach Su 1997. In the framework of the maritime culture of Vietnam: the prehistoric maritime culture of the northeast. *Vietnamese Studies* 123 (1):87-116.
Nguyen Khac Su and Nguyen Lan Cuong (eds) 1997. *Dan So Va Dan Cu Thoi Tien Su Viet Nam*. Hanoi: Institute of Archaeology.
Nguyen Kim Thuy. 1990. Ancient Human Skeletons at Con Co Ngua. *Khao Co Hoc* (1990) 3:37-48.
Nguyen Kim Thuy 1998. Human remains from Con Con Ngua: a Da But culture site. Paper presented at the 16th Congress of the Indo-Pacific Prehistory Association, Melaka, Malaysia 1-7 July 1998.
Nguyen Lan Cuong 1985. Two precious ancient crania discovered in the west of Thanh Hoa Province. *Vietnam Social Sciences* 2:125-129.
Nguyen Lan Cuong 1992. A reconsideration of the chronology of hominid fossils in Vietnam. In T. Akazawa, K. Aoki and T. Kimura (eds) *The Evolution and Dispersal of Modern Humans in Asia*. Tokyo: Hokusen-Sha, pp:321-335.
Nguyen Lan Cuong 1996. *Anthropological Characteristics of Dong Son Population in Vietnam*. Hanoi: Social Sciences Publishing House.
Nguyen Trong Dieu 1995. *Geography of Vietnam: Natural, Human, Economic*. Hanoi: The Gioi Publishers.
Nguyen Van Buu and Diep Dinh Hoa 1987. Bronze artefacts in Thieu Duong site, Than Hoa Province. *New Archaeological Discoveries in 1987*.
Nigam, P and A.B. Srivastava. 1990. Betel chewing and dental decay. *Federation of Operative Dentistry* 1 (1):36-38.
Noor, A.M. 1987. Quantitative analysis of areca catechu (betel) nut flavanols (tannins) in relation to oral submucous fibrosis. *Dental Journal of Malaysia* 9 (2):29-32.
Norr, L. 1984. Prehistoric subsistence and health status of coastal peoples from the Panamanian Isthmus of Lower Central America. In M.N. Cohen & G.J. Armelagos (eds) *Paleopathology at the Origins of Agriculture*. Orlando: Academic Press, pp:463-490.
Oishi, C.S., F.T. Hoaglund, L. Gordon and P.D. Ross 1998. Total hip replacement rates are higher among Caucasians than Asians in Hawaii. *Clinical Orthopaedics and Related Research* 353:166-174.
Oliveria, S.A., D.T. Felson, P.A. Cirillo, J.I. Reed and A.M. Walker 1999. Body weight, body mass index and incident symptomatic osteoarthritis of the hand hip and knee. *Epidemiology* 10 (2):161-166.
Olivier, G. 1966. Craniometrie des Indochinois. *Bulletins et Memoires de la Societe d'Anthropologie de Paris* 9 (11):67-90.
Olsen, J.W. and R.L.Ciochon 1990. A review of the evidence for postulated Middle Pleistocene occupations in Viet Nam. *Journal of Human Evolution* 19:761-788.
Olsson, B. 1978. Efficiency of traditional chewing sticks in oral hygiene programs among Ethiopian children. *Community Dentistry and Oral Epidemiology* 6 (3):05-109.
Ortner, D.J. 1990. Introduction. In R.W. Mann and S.P. Murphy, *Regional Atlas of Bone Disease: A Guide to Pathological and Normal Variation in the Human Skeleton*. Springfield: C. C. Thomas, pp:ix-xii.
Ortner, D.J. 1991. Theoretical and methodological issues in paleopathology. In D.J. Ortner and A.C. Aufderheide (eds) *Human Paleopathology: Current Synthesis and Future Options*. Washington: Smithsonian Institution Press, pp:5-11.
Ortner, D.J. and W.G.J.Putschar 1981. *Identification of Pathological Conditions in Human Skeletal Remains*. Washington: Smithonian Institution Press.
Oxenham, M. F. 2006. Biological responses to change in prehistoric Viet Nam. *Asian Perspectives* 45(2): pp 212-239.
Oxenham, M. F. and H. R. Buckley in press-a. Meta-themes in the bioarchaeology of the Asia pacific region. *The Routledge Handbook of Bioarchaeology in Southeast Asia and the Pacific Islands*. M. F. Oxenham and H. Buckley. London, Routledge.
Oxenham, M. F. and H. R. Buckley in press-b. The population history of Mainland and Island Southeast Asia. *The Routledge Handbook of Bioarchaeology in Southeast Asia and the Pacific Islands*. M. F. Oxenham and H. Buckley. London, Routledge.
Oxenham, M. F. and H. R. Buckley, Eds. in press-c. *The Routledge Handbook of Bioarchaeology in Southeast Asia and the Pacific Islands*. London, Routledge.
Oxenham, M. F. and I. Cavill 2010. Porotic hyperostosis and cribra orbitalia: the erythropoietic response

to iron-deficiency anaemia. *Anthropological Science* 118(3): pp 199-200.

Oxenham, M. F., C. Locher, L. C. Nguyen and K. T. Nguyen 2002a. Identification of Areca catechu (Betel Nut) Residues on the Dentitions of Bronze Age Inhabitants of Nui Nap, Northern Vietnam. *Journal of Archaeological Science* 29(9): pp 909-915.

Oxenham, M. F. and H. Matsumura 2008. Oral and physiological paleohealth in cold adapted peoples: Northeast Asia, Hokkaido. *American Journal of Physical Anthropology* 135(1): pp 64-74.

Oxenham, M. F., H. Matsumura, K. Domett, K. T. Nguyen, K. D. Nguyen, L. C. Nguyen, D. Huffer and S. Muller 2008. Health and the experience of childhood in late Neolithic Viet Nam. *Asian Perspectives* 47(2): pp 190-209.

Oxenham, M. F., H. Matsumura and K. D. Nguyen 2011. *Man Bac: The excavation of a Neolithic site in Northern Vietnam. The biology*. Canberra, ANU Epress.

Oxenham, M. F., L. C. Nguyen and K. D. Nguyen 2006. The oral health consequences of the adoption and intensification of agriculture in Southeast Asia. *Bioarchaeology of Southeast Asia*. M. Oxenham and N. Tayles. Cambridge, Cambridge University Press:pp 263-289.

Oxenham, M. F., L. C. Nguyen and K. T. Nguyen 2002b. Oral health in Northern Vietnam: Neolithic through metal periods. *Bulletin of the Indo-Pacific Prehistory Association* 22(6): pp 121-134.

Oxenham, M. F., P. J. Piper, P. Bellwood, C. H. Bùi, K. T. K. Nguyen, Q. M. Nguyen, F. Campos, C. Castillo, R. Wood, C. Sarjeant, N. Amano, A. Willis and J. Ceron 2015. Emergence and Diversification of the Neolithic in Southern Vietnam: Insights From Coastal Rach Nui. *The Journal of Island and Coastal Archaeology* 10.1080/15564894.2014.980473: pp 1-30.

Oxenham, M. F. and N. Tayles, Eds. 2006. *Bioarchaeology of Southeast Asia*. Cambridge, Cambridge University Press.

Oxenham, M. F., L. Tilley, H. Matsumura, L. C. Nguyen, K. T. Nguyen, K. D. Nguyen, K. Domett and D. Huffer 2009. Paralysis and severe disability requiring intensive care in Neolithic Asia *Anthropological Science*(2): pp 107-112.

Oxenham, M. F. and J. Whitworth 2006. Frequency, location, morphology and aetiology of osseous mandibular condylar concavities. *International Journal of Osteoarchaeology* 16(6): pp 517-527.

Palkovich, A.M. 1987. Endemic disease patterns in paleopathology: porotic hyperostosis. *American Journal of Physical Anthropology* 74:527-537.

Parker, K. 1998. The Early Neolithic Cemetery at Con Co Ngua, Northern Vietnam: Nonmetric Cranial Variation and Relationships. Unpublished M.A. Thesis, The Northern Territory University, Australia.

Perry, M. 1980. *Medicinal Plants of East and Southeast Asia: Attributed Properties and Uses*. Cambridge: The MIT Press.

Pechenkina, K. and M. F. Oxenham 2013. *Bioarchaeology of East Asia: Movement, Contact, Health*. Gainesville, University Press of Florida.

Perzigian, A.L., P.A. Tench and D.J. Braun 1984. Prehistoric health in the Ohio river valley. In M.N. Cohen & G.J. Armelagos (eds) *Paleopathology at the Origins of Agriculture*. Orlando: Academic Press, pp:347-366.

Peterson, J. 1998. The Natufian hunting conundrum: spears, atlatls, or bows? Musculoskeletal and armature evidence. *International Journal of Osteoarchaeology* 8:378-389.

Pham Minh Huyen and Diep Dinh Hoa 1981. The position of Thieu Duong site (Thanh Hoa) in the Dong Son culture. *Khao Co Hoc* 1981 (2):19-34.

Phan Hai Linh 1998. Blackened teeth: a comparison of Vietnamese and Japanese customs. *Vietnam Cultural Window* 6 & 7:38-41.

Phenice, T.W. 1969. A newly developed visual method of sexing the Os pubis. *American Journal of Physical Anthropology* 30:297-302.

Pietrusewsky, M. 1987. Multivariate comparisons of recently excavated Neolithic human crania from Thanh Hoa Province, Socialist Republic of Vietnam. Paper presented at the American Association of Physical Anthropologists Meetings in New York, April 14 1987.

Pietrusewsky, M., Y.Y. Li, X.Q. Shao and Q.Q. Nguyen 1990. Modern and near modern populations of Asia and the Pacific: a multivariate craniometric interpretation. Paper presented at the University of Tokyo Symposium: the Evolution and Dispersal of Modern Humans in Asia, November 14-17, 1990.

Pollock, F.H., A.M. Pankovich, J.J. Prieto and M. Lorenz 1983. The isolated fracture of the ulnar shaft. Treatment without immobilizaion. *Journal of Bone and Joint Surgery* (American Volume) 65 (3):339-342.

Pons, J. 1955. The sexual diagnosis of isolated bones of the skeleton. *Human Biology* 27:12-21.

Potsch-Schneider, L., R. Endris and H. Schmidt. 1985. Discriminant analysis of the mandible for sex determination. *Zeitshrift fur Rechtsmedizin* 94 (1):21-30.

Powell, M.L. 1985. The analysis of dental wear and caries for dietary reconstruction. In R.I. Gilbert and J.H. Mielke (eds) *The Analysis of Prehistoric Diets*. Orlando: Academic Press, pp:301-338.

Powell, M.L. 1988. *Status and health in Prehistory: A Case Study of the Moundville Chiefdom*. Washington: Smithsonian Institution Press.

Pun, Y.L., R.W. Moskowitz, S. Lie, W.R. Sundstrom, S.R. Block, C. McEwen, H.J. Williams, J.F. Bleasel, D. Holderbaum and T.M. Haggi 1994. Clinical correlations of osteoarthritis associated with a single-base mutation (arginine 519 to cysteine) in type II procollagen gene: a newly defined pathogenesis. *Arthritis and Rheumatism* 37 (2):264-269.

Radin, E.L. and I.L. Paul 1970. Does cartilage compliance reduce skeletal impact loads: the relative force-attenuating properties of articular cartilage, synovial fluid, periarticular soft tissues and bone. *Arthritis and Rheumatism* 13 (2):139-144.

Radin, E.L. and R.M. Rose 1986. Role of subchondral bone in the initiation and progression of cartilage damage. *Clinical Orthopaedics and Related Research* 213:34-40.

Radin, E.L., I.L. Paul and M.J. Tolkoff 1970. Subchondral bone changes in patients with early degenerative joint disease. *Arthritis and Rheumatism* 13 (4):400-405.

Radin, E.L., H.G. Parker, J.W. Pugh, R.S. Steinberg, I.L. Paul and R.M. Rose 1973. Response of joints to impact loading, III: relationship between trabecular microfractures and cartilage degeneration. *Journal of Biomechanics* 6:51-57.

Radin, E.L., M.G. Ehrlich, R. Chernack, P. Abernethy, I.L. Paul and R.M. Rose 1978. Effect of repetitive impulsive loading on the knee joints of rabbits. *Clinical Orthopaedics and Related Research* 131:288-293.

Rathbun, T.A. 1984. Skeletal pathology from the Paleolithic through the Metal Ages in Iran and Iraq. In M.N. Cohen & G.J. Armelagos (eds) *Paleopathology at the Origins of Agriculture*. Orlando: Academic Press, pp:137-167.

Reale, B., D. Marchi and S.M.B. Tarli 1999. A case of diffuse idiopathic skeletal hyperostosis (DISH) from a Medieval necropolis in southern Italy. *International Journal of Osteoarchaeology* 9:369-373.

Reichart, P and F. Gehring 1984. Streptococcus mutans and caries prevalence in Lisu and Karen of northern Thailand. *Journal of Dental Research* 63 (1):56-58.

Reid, J. 1983. *Sorcerers and Healing Spirits*. Canberra: Australian National University Press.

Resnick, D. and G. Niwayama 1981. *Diagnosis of Bone and Joint Disorders: With Emphasis on Articular Abnormalities*. Philadelphia: Saunders.

Richards, L.C. 1984. Tooth wear and temporomandibular joint change in Australian Aboriginal populations. *American Journal of Physical Anthropology* 82:377-384.

Richards, L.C. and S.L. Miller 1991. Relationships between age and dental attrition in Australian aboriginals. *American Journal of Physical Anthropology* 84:159-164.

Rogers, J. and P. Dieppe 1993. Ridges and grooves on the bony surfaces of osteoarthritic joints. *Osteoarthritis and Cartilage* 1:167-170.

Rogers, J. and P. Dieppe 1994. Is tibiofemoral osteoarthritis in the knee joint a new disease? *Annals of the Rheumatic Diseases* 53:612-613.

Rogers, T. and S. Saunders. 1994. Accuracy of sex determination using morphological traits of the human pelvis. *Journal of Forensic Sciences* 39 (4):1047-1056.

Rogers, J. and T. Waldron 1995. *A Field Guide to Joint Disease in Archaeology*. New York: John Wiley.

Rogers, J., K. Lim and L. Shepstone 1995. The evolutionary origins of osteoarthritis of the hand. A comparative skeletal study of two primates (abstract). *American Journal of Physical Anthropology*, Supp. 20:182-183.

Rogers, J., L. Shepstone and P. Dieppe 1997. Bone formers: osteophyte and enthesophyte formation are positively associated. *Annals of the Rheumatic Diseases* 56:85-90.

Rose, J.C., B.A. Burnett, M. Nassaney & M.W. Blaeuer 1984. Paleopathology and the origins of maize agriculture in the lower Mississippi Valley and Caddoan culture areas. In M.N. Cohen & G.J. Armelagos (eds) *Paleopathology at the Origins of Agriculture*. Orlando: Academic Press, pp: 393-424.

Rothschild, B.M. 1993. Skeletal paleopathology of rheumatic diseases: the sub-homo connection. In D.J. McCarty and W.J. Koopman (eds) *Arthritis and Allied Conditions*. Philadelphia: Lea and Febiger, pp.:3-7.

Rothschild, B.M. 1997. Porosity: a curiosity without diagnostic significance. *American Journal of Physical Anthropology* 104:529-533.

Rothschild, B.M. and L.D. Martin 1993. *Paleopathology: Disease in the Fossil Record*. Boca Raton: CRC Press.

Rothschild, B.M., K.A. Turner and M.A. DeLuca 1988. Symmetrical erosive peripheral polyarthritis in the Late Archaic period of Alabama. *Science* 241:1498-1501.

Rothschild, B.M., B. Arriaza, R.J. Woods and O. Dutour 1999. Spondyloarthropathy identified as the etiology of Nubian erosive arthritis. *American Journal of Physical Anthropology* 109:259-267.

Rotimi, V.O. and H.A. Mosadomi 1987. The effect of crude extracts of nine African chewing sticks on oral anaerobes. *Journal of Medical Microbiology* 23 (1):55-60.

Ruggieri, M. V. Pavone, A. Polizzi, P. Smilari, G. Magro, M. Merino and P.H. Duray 1998. Familial osteoma of the cranial vault. *British Journal of Radiology* 71 (842):225-228.

Ryan, A.S. 1997. Iron-deficiency anemia in infant development: implications fro growth, cognitive development, resistance to infection, and iron supplementation. *Yearbook of Physical Anthropology* 40:25-62.

Saccente, M. R.S. Abernathy, P.G. Pappas, H.R. Shah and R.W. Bradsher 1998. Vertebral blastomycosis with paravertebral abscess: report of eight cases and review of the literature. *Clinical Infectious Diseases* 26 (2):413-418.

Sahni, D., I. Jit, Neelam, S. Suri. 1998. Time of fusion of

the basisphenoid with the basilar part of the occipital bone in Northwest Indian subjects. *Forensic Science International* 98 (1-2):41-5.

Santini, A., M. Land and G.M. Raab 1990. The accuracy of simple ordinal scoring of tooth attrition in age assessment. *Forensic Science International* 48:175-184.

Sapico, F.L. and J.Z. Montgomerie 1990. Vertebral Osteomyelitis. *Infectious Disease Clinics of North America* 4 (3):539-550.

Sapp, J.P., L.R. Eversole and G.P. Wysocki 1997. *Contemporary Oral and Maxillofacial Pathology*. St. Louis: Mosby.

Saunders, S., C. DeVito, A. Herring, R. Southern and R. Hoppa. 1993. Accuracy tests of tooth formation age estimations for human skeletal remains. *American Journal of Physical Anthropology* 92:173-188.

Saunders, S.R. and A. Keenleyside 1999. Enamel hypoplasia in a Canadian historic sample. *American Journal of Human Biology* 11:513-524.

Schafer, E.H. 1967. *The Vermilion Bird: T'ang Images of the South*. Berkeley: University of California Press.

Schamschula, R.G., B.L. Adkins, D.E. Barmes and G. Charlton 1977. Betal chewing and caries experience in New Guinea. *Community Dentistry and Oral Epidemiology* 5 (6):284-286.

Schamschula, R.G., B.L. Adkins, D.E. Barmes, G. Charlton and B.G. Davey 1978. *WHO Study of Dental Caries Etiology in Papua New Guinea*. Geneva: World Health Organization.

Schneider, K.N. 1986. Dental caries, enamel composition and subsistence among prehistoric Amerindians of Ohio. *American Journal of Physical Anthropology* 71:95-102.

Schulter-Ellis, F.P. and L.C. Hayek. 1988. Sexing North American Eskimo and Indian innominate bones with the acetabulum/pubic index. *Journal of Forensic Sciences* 33 (3):697-708.

Schulter-Ellis, F.P., D.J. Schmidt, L.A. Hayek and J. Craig. 1983. Determination of sex with a discriminant analysis of new pelvic bone measurements: Part 1. *Journal of Forensic Sciences* 28 (1):169-180.

Schultz, M. and C.S.Larsen 1997. Porotic hyperostosis in Spanish Florida: nature and etiology of a frequently observed phenomenon. *American Journal of Physical Anthropology* (abstract) Supp. 24:206.

Schwartz, J.H. 1995. *Skeleton Keys: An Introduction to Human Skeletal Morphology, Development, and Analysis*. New York: Oxford University Press.

Scott, E.C. 1979a. Principal axis analysis of dental attrition data. *American Journal of Physical Anthropology* 51:203-212.

Scott, E.C. 1979b. Dental wear scoring technique. *American Journal of Physical Anthropology* 51:213-218.

Sealy, J.C., M.K. Patrick, A.G. Morris and D. Alder 1992. Diet and dental caries among later Stone Age inhabitants of the Cape Province, South Africa. *American Journal of Physical Anthropology* 88:123-134.

Seidemann, R.M., C.M. Stojanowski and G.H. Doran. 1998. The use of the supero-inferior femoral neck diameter as a sex assessor. *American Journal of Physical Anthropology* 107 (3):305-313.

Shepstone, L., J. Rogers, J. Kirwan and B. Silverman 1999. The shape of the distal femur: a palaeopathological comparison of eburnated and non-eburnated femora. *Annals of the Rheumatic Diseases* 58 (2):72-78.

Siebert, G. and K. Trautner 1985. Fluoride content of selected human food, pet food and related materials. *Zeitschrift fur Ernahrungswissenschaft* 24:54-66.

Sinor, P.N., P.C. Gupta, P.R. Murti, R.B. Bhonsle, D.K. Daftary, F.S. Mehta and J.J. Pindborg 1990. A case-control study of oral submucous fibrosis with special reference to the etiologic role of areca nut. *Journal of Oral Pathology and Medicine* 19 (2):94-98.

Sjovold, T. 1990. Estimation of stature from long bones utilizing the line of organic correlation. *Human Evolution* 5:431-447.

Skinner, M. 1996. Developmental stress in immature hominines from late Pleistocene Eurasia: evidence from enamel hypoplasia. *Journal of Archaeological Science* 23:833-852.

Skinner, M. and A.H. Goodman 1992. Anthropological uses of developmental defects of enamel. In S.R. Saunders and M.A. Katzenberg (eds) *Skeletal Biology of Past Peoples: Research Methods*. New York: Wiley-Liss, pp:153-174.

Slemenda, C., K.D. Brandt, D.K. Heilman, S. Mazzuca, E.M. Braunstein, B.P. Katz and F.D. Wolinsky 1997. Quadriceps weakness and osteoarthritis of the knee. *Annals of Internal Medicine* 127:97-104.

Smith, B.H. 1984. Patterns of molar wear in hunter-gatherers and agriculturalists. *American Journal of Physical Anthropology* 63:39-56.

Smith, B.H. 1991. Standards of human tooth formation and dental age assessment. In M.A. Kelley and C.S. Larsen (eds) *Advances in Dental Anthropology*. New York: Wiley-Liss, pp:143-168.

Snaith, M.L. 1996. On the one hand and on the other. *Lancet* 348:907-908.

Solomon, L. 1997. Clinical features of osteoarthritis. In W.N. Kelley, F.D. Harris, S. Ruddy and C.B. Sledge (eds) *Textbook of Rheumatology*, Philadelphia: W.B.Saunders Co., pp:1383-1393.

Song, H.W., Z.Q. Lin and J.T. Jia. 1992. Sex diagnosis of Chinese skulls using multiple stepwise discriminant function analysis. *Forensic Science International* 54 (2):135-140.

Sowers, M., L. Lachance, D. Jamadar, M.C. Hochberg, B. Hollis, M. Crutchfield and M.L. Jannausch 1999. The associations of bone mineral density and bone turnover markers with osteoarthritis of the hand in pre-and perimenopausal women. *Arthritis and Rheumatism* 42 (3):483-489.

Spector, T.D., F. Cicuttini, J. Baker, J. Loughlin, D. Hart and J. Hopper 1996. Genetic influences on osteoarthritis in women: a twin study. *British Medical Journal* 312:940-948.

Sreebny, L.M. 1983. Cereal availability and dental caries. *Community Dentistry and Oral Epidemiology* 11(3):148-155.

St. Hoyme, L.E. and M.Y. Iscan. 1989. Determination of sex and race: accuracy and assumptions. In M.Y. Iscan and K.A.R. Kennedy (eds) *Reconstruction of Life from the Skeleton.* New York: Wiley-Liss, pp:53-93.

Steele, D.G. 1976. The estimation of sex on the basis of the talus and calcaneus. *American Journal of Physical Anthropology* 45:581-588.

Steinbock, R.T. 1976. *Paleopathological Diagnosis and Interpretation.* Springfield, Illinois: C. C. Thomas.

Stewart, T.D. 1958. The rate of development of vertebral osteoarthritis in American whites and its significance in skeletal age identification. *The Leech* 28:144-151.

Stewart, T.D. 1979. *Essentials of Forensic Anthropology.* Springfield, Illinois: C.C.Thomas.

Steyn, M and M.Y. Iscan 1997. Sex determination from the femur and tibia in South African whites. *Forensic Science International* 90:111-119.

Stirland, A.J. 1991. Diagnosis of occupationally related paleopathology: can it be done? In D.J. Ortner and A.C. Aufderheide (eds) *Human Paleopathology: Current Synthesis and Future Options.* Smithsonian Institution Press: Washington, pp:40-47.

Stirland, A.J. 1998. Musculoskeletal evidence for activity: problems for evaluation. *International Journal of Osteoarchaeology* 8:354-362.

Stirland, A.J. and T. Waldron 1997. Evidence for activity related markers in the vertebrae of the crew of the Mary Rose. *Journal of Archaeological Science* 24:329-335.

Stodder, A.L.W. 1994. Bioarchaeological investigations of protohistoric Pueblo health and demography. In C.S. Larsen and G.R. Milner (eds) *In the Wake of Contact: Biological Responses to Conquest.* New York: Wiley-Liss, pp:97-107.

Stojanowski, C.M. 1999. Sexing potential of fragmentary and pathological metacarpals. *American Journal of Physical Anthropology* 109:245-252.

Stuart-Macadam, P.L. 1985. Porotic hyperostosis: representative of a childhood condition. *American Journal of Physical Anthropology* 6:391-398.

Stuart-Macadam, P.L. 1987a. A radiographic study of porotic hyperostosis. *American Journal of Physical Anthropology* 74:511-520.

Stuart-Macadam, P.L. 1987b. Porotic hyperostosis: new evidence to support the anemia theory. *American Journal of Physical Anthropology* 74:521-526.

Stuart-Macadam, P.L. 1989. Porotic hyperostosis: relationship between orbital and vault lesions. *American Journal of Physical Anthropology* 80:187-193.

Stuart-Macadam, P.L. 1991. Anaemia in Roman Britain: Poundbury Camp. In H. Bush and M. Zvelebil (eds) *Health in Past Societies: Biocultural Interpretations of Human Skeletal Remains in Archaeological Contexts.* BAR International Series 567:101-113.

Sturt-Macadam, P.L. 1992. Porotic hyperostosis: a new perspective. *American Journal of Physical Anthropology* 87:39-47.

Stuart-Macadam, P.L. 1998. Iron deficiency anemia: exploring the difference. In A.L. Grauer and P. Stuart-Macadam (eds) *Sex and Gender in Paleopathological Perspective.* Cambridge: Cambridge University Press, pp:45-63.

Stuart-Macadam, P.L., B. Glencross and M. Kricun 1998. Traumatic bowing deformities in tubular bones. *International Journal of Osteoarchaeology* 8:252-262.

Suchey, J.M. 1979. Problems in the aging of females using the Os pubis. *American Journal of Physical Anthropology* 51:467-70.

Suchey, J.M., D.V. Wiseley & D. Katz. 1986. Evaluation of the Todd and McKern-Stewart methods for aging the male os pubis. In K.J. Reichs (ed.) *Forensic Osteology: Advances in the Identification of Human Remains.* Springfield: C.C.Thomas.

Suga, S. 1992. Hypoplasia and hypomineralization of tooth enamel. In A.H. Goodman and L.L. Capasso (eds) *Recent Contributions to the Study of Enamel Developmental Defects.* Journal of Paleopathology, Monographic Publications 2:269-292.

Sundqvist, K. and R.C. Grafstrom 1992. Effects of areca nut on growth, differentiation and formation of DNA damage in cultured human buccal epithelial cells. *International Journal of Cancer* 52 (2):305-310.

Sundqvist, K., Y. Liu, J. Nair, H. Bartsch, K. Arvidson and R.C. Grafstrom 1989. Cytotoxic and genotoxic effects of areca-related compounds in cultured human buccal epithelial cells. *Cancer Research* 49(19):5294-5298.

Sutherland, L.D. and J.M. Suchey. 1991. Use of the ventral arc in pubic sex determination. *Journal of Forensic Sciences* 36 (2):501-511.

Suzuki, T. 1991. Paleopathological study on infectious disease in Japan. In D.J. Ortner and A.C. Aufderheide (eds) *Human Paleopathology: Current Synthesis and Future Options.* Smithsonian Institution Press: Washington, pp:128-139.

Taitz, C. 1999. Osteophytosis of the cervical spine in South African Blacks and Whites. *Clinical Anatomy* 12:103-109.

Tayles, N. 1992. The people of Khok Phanom Di: Health as Evidence of Adaptation in a Prehistoric Southeast Asian Population. Unpublished PhD Thesis, University of Otago, New Zealand.

Tayles, N. 1996. Anemia, genetic diseases, and malaria in prehistoric mainland Southeast Asia. *American Journal of Physical Anthropology* 101:11-27.

Tayles, N. 1997. Tooth ablation in prehistoric Southeast Asia. *International Journal of Osteoarchaeology* 6:333-345.

Taylor, K.W. 1983. *The Birth of Vietnam*. Berkeley: University of California Press.

Titche, L.L., S.W. Coulthard, R.D. Wachter, A.C. Thies and L.L. Harries 1981. Prevalence of mastoid infection in prehistoric Arizona Indians. *American Journal of Physical Anthropology* 56:269-273.

Tjia, H.D. 1996. Sea-level changes in the tectonically stable Malay-Thai Peninsula. *Quarternary International* 31 (1):95-101.

Todd, T.W. 1920. Age changes in the pubic bone: I. The white male pubis. *American Journal of Physical Anthropology* 3:467-470.

Todd, T.W. 1921. Age Changes in the pubic bone: II. Pubis of male Negro-White hybrid; III. The pubis of the white female; IV. The pubis of the female Negro-White hybrid. *American Journal of Physical Anthropology* 4:1-70.

Tomenchuck, T. and J.T. Mayhall. 1979. A correlation of tooth wear and age among modern Igloolik Eskimos. *American Journal of Physical Anthropology* 51:67-78.

Turner, C.G. II. 1979. Dental anthropological implications of agriculture among the Jomon people of central Japan. *American Journal of Physical Anthropology* 51:619-636.

Turner, C.G. II. and I.M.C. Machado 1983. A new dental wear pattern and evidence for high carbohydrate consumption in a Brazilian archaic skeletal population. *American Journal of Physical Anthropology* 61:125-130.

Turner, C.G. II, C.R. Nichol and G.R. Scott 1991. Scoring procedures for key morphological traits of the permanent dentition: the Arizona State University dental anthropology system. In M.A. Kelley and C.S. Larsen (eds) *Advances in Dental Anthropology*. New York: Wiley-Liss, pp:13-31.

Ubelaker, D.H. 1984. Prehistoric human biology of Ecuador: possible temporal trends and cultural correlations. In M.N. Cohen & G.J. Armelagos (eds) *Paleopathology at the Origins of Agriculture*. Orlando: Academic Press, pp:491-513.

Ubelaker, D.H. 1987. Estimating the age at death from immature human skeletons: an overview. *Journal of Forensic Sciences* 32 (5):1254-1263.

Ubelaker, D.H. 1989. *Human Skeletal remains: Excavation, Analysis, Interpretation*. Washington: Taraxacum.

Ubelaker, D.H. 1994. The biological impact of European contact in Ecuador. In C.S. Larsen and G.R. Milner (eds) *In the Wake of Contact: Biological Responses to Conquest*. New York: Wiley-Liss, pp:147-160.

van Dijk, C.N., L.S.L. Lim, A. Poortman, E.H. Strubbe and R.K. Marti 1995. Degenerative joint disease in female Ballet dancers. *The American Journal of Sports Medicine* 23 (3):295-300.

van Houte, J. 1994. Role of micro-organisms in caries etiology. *Journal of Dental Research* 73 (3):672-681.

van Saase, J.L.C.M., J.P. Vandenbroucke, L.K.J. van Romunde and H.A. Valkenburg 1988. Osteoarthritis and obesity in the general population: a relationship calling for an explanation. *The Journal of Rheumatology* 15:1152-1158.

van Saase, J.L.C.M., L.K.J. van Romunde, A. Cats, J.P. Vandenbroucke and H.A. Valkenburg 1989. Epidemiology of osteoarthritis: Zoetermeer survey. Comparison of radiological osteoarthritis in a Dutch population with that in 10 other populations. *Annals of the Rheumatic Diseases* 48:271-280.

van Wyk, C.W., I. Stander, A. Padayachee and A.F. Grobler-Rabie 1993. The areca nut chewing habit and oral squamous cell carcinoma in South African Indians. A retrospective study. *South African Medical Journal* 83 (6):425-429.

van Wyk, C.W., A. Olivier, C.M. de Miranda, P. van der Bijl and A.F. Grobler-Rabie 1994. Observations of the effect of areca nut extracts in oral fibroblast proliferation. *Journal of Oral Pathology and Medicine* 23 (4):145-148.

Vilain, E. and E.R. McCabe 1998. Mammalian sex determination: from gonads to brain. *Molecular Genetics and Metabolism* 65 (2):74-84.

Vingard, E., L. Alfredsson and H. Malchau 1998. Osteoarthritis of the hip in women and its relationship to physical load from sports activities. *The American Journal of Sports Medicine* 26 (1):78-82.

Vingard,E., L. Alfredsson, I. Goldie and C. Hogstedt 1993. Sports and osteoarthritis of the hip: an epidemiologic study. *The American Journal of Sports Medicine* 21 (2):195-201.

Vo Hung and Nguyen Lan Cuong 1979. Ancient people at Nap Mount. *New Archaeological Discoveries in 1979*: 86-91.

Waldron, H.A. 1997. Association between osteoarthritis of the hand and knee in a population of skeletons from London. *Annals of the Rheumatic Diseases* 56:116-118.

Waldron, H.A. and M. Cox 1989. Occupational arthropathy: evidence from the past. *British Journal of Industrial Medicine* 46:420-42.

Waldron, T. 1985. DISH at Merton Priory: evidence for a new occupational disease? *British Medical Journal* 291:1762-1763.

Waldron, T. and J. Rogers 1991. Inter-observer variation in coding osteoarthritis in human skeletal remains. *International Journal of Osteoarchaeology* 1:49-56.

Walker, P.L. 1986. Porotic hyperostosis in a marine-dependent California Indian population. *American Journal of Physical Anthropology* 69:35-354.

Walker, P. L., R. R. Bathurst, R. Richman, T. Gjerdrum and V. A. Andrushko. 2009. The causes of porotic hyperostosis and cribra orbitalia: A reappraisal of the iron-deficiency-anemia hypothesis. *American Journal of Physical Anthropology* 139(2): pp 109-125.

Walker, P.L. and J.M. Erlandson 1986. Dental evidence for prehistoric dietary change on the northern Channel Islands, California. *American Antiquity* 51 (2):375-383.

Walker, P.L. and B.S. Hewlett. 1990. Dental health, diet and social status among central African foragers and farmers. *American Anthropologist* 92 (2):383-398.

Walker, P.L., G. Dean and P. Shapiro. 1991. Estimating age from tooth wear in archaeological populations. In M.A. Kelley and C.S. Larsen (eds) *Advances in Dental Anthropology*. Wiley-Liss: New York, pp:169-178.

Walters, I., M. Oxenham, Nguyen Lan Cuong and Nguyen Kim Thuy (in prep). Holocene dates for Hoabinhian human remains from Thanh Hoa Province, Vietnam.

Washburn, S.L. 1948. Sex differences in the pubic bone. *American Journal of Physical Anthropology* 6:199-208.

Webb, S. 1982. Cribra orbitalia: a possible sign of anaemia in pre- and post-contact crania from Australia and Papua New Guinea. *Archaeology in Oceania* 17 (3):148-156.

Webb, S. 1995. *Palaeopathology of Aboriginal Australians: Health and Disease Across a Hunter-Gatherer Continent*. Cambridge: Cambridge University Press.

Weinberg, E.D. 1984. Iron withholding: a defence against infection and neoplasia. *Physiological Reviews* 64 (1):65-102.

Weinberg, E.D. 1992. Iron depletion: a defence against intracellular infection and neoplasia. *Life Sciences* 50 (18):1289-1297.

Weinberg, E.D. 1999. The role of iron in protozoan and fungal infectious diseases. *Journal of Eukaryotic Microbiology* 46 (3):231-238.

Wells, C. 1967. Pseudopathology. In D.R. Brothwell and A.T. Sandison (eds) *Diseases in Antiquity: A Survey of the Diseases, Injuries, and Surgery of Early Populations*. Springfield, Illinois: C. C. Thomas, pp:5-19.

Wesselius, L.J., R.J. Brooks and E.P. Gall 1977. Vertebral coccidioidomycosis presenting as Pott's disease. *Journal of the American Medical Association* 238 (13):1397-1398.

White, T.D. and P.A. Folkens. 1991. *Human Osteology*. San Diego: Academic Press.

Wienker, C.W. 1984. Sex determination from human skeletal remains: a case of mistaken assumption. In T.A. Rathbun and J.E. Buikstra (eds) *Human Identification : Case Studies in Forensic Anthropology*. Springfield, Illinois: C.C.Thomas, pp:229-243.

Williams, S.A., R.M. Summers, I.A. Ahmed and M.J. Prendergast 1996. Caries experience, tooth loss and oral health-related behaviours among Bangladeshi women resident in West Yorkshire. *Community Dental Health* 13 (3):150-156.

Willis, A. and M. F. Oxenham 2013. The neolithic demographic transition and oral health: The Southeast Asian experience. *American Journal of Physical Anthropology* 152(2): pp 197-208.

Willis, A. and M. F. Oxenham in press. Reconstructing Diet at An Son and Hoa Diem: Implications for Understanding Southeast Asian Subsistence Patterns. *The Routledge Handbook of Bioarchaeology in Southeast Asia and the Pacific Islands*. M. F. Oxenham and H. Buckley. London, Routledge.

Wilson, C.S. 1980. Food taboos of childbirth: the Malay example. In J.R.K. Robson (ed.) *Food, Ecology and Culture: Readings in the Anthropology of Dietary Practices*. New York: Gordon and Breach Science Publishers, pp:67-74.

Winter, G.B. 1990. Epidemiology of dental caries. *Archives of Oral Biology* 35 (Suppl).:1S-7S.

Wiriyaromp, W. 1984. The Human Skeletal Remains from Ban Na Di: Patterns of Birth, Health and Death in Prehistoric North East Thailand. Unpublished Master of Arts Thesis, University of Otago, New Zealand.

Wood, B. 1991. *Koobi Fora Research Project, Volume 4, Hominid Cranial Remains*. Oxford: Clarendon Press.

Wood, J.W., G.R. Milner, H.C. Harpending and K.M. Weiss 1992. The osteological paradox. *Current Anthropology* 33 (4):343-370.

Wood, L. 1996. Frequency and chronological distribution of linear enamel hypoplasia in a North American colonial skeletal sample. *American Journal of Physical Anthropology* 100:247-259.

Wright, D. 1997. Diseases of the external ear. In A.G. Kerr (ed.) *Scott-Brown's Otolaryngology*. Oxford: Butterworth-Heinemann, pp:3/6/1-3/6/20.

Xu, X.H., H.P. Philipsen, N.G. Jablonski, K.P. Pang and J.Z. Zhu. 1992. Age estimation from the structure of adult human teeth: review of the literature. *Forensic Science International* 54:23-28.

Yagi, H.I., A.H. Fahal, A.F. Gadir, A.M. el Hassan 1998. Mycetoma of the mastoid bone. *Transactions of the Royal Society of Tropical Medicine and Hygiene* 92 (1):68.

Yamamoto, M. 1992. Secular trends of enamel hypoplasia in Japanese from the prehistoric to modern period. In A.H. Goodman and L.L. Capasso (eds) *Recent Contributions to the Study of Enamel Developmental Defects*. Journal of Paleopathology, Monographic Publications 2:231-238.

Yayanetra, P., P. Nitiyanant, L. Ajello, A.A. Padhye, S. Lolekha, V. Atichartakarn, P. Vathesatogit, B. Sathaphatayavongs and R. Prajaktam 1984. Penicilliosis marneffei in Thailand: report of five human cases. *American Journal of Tropical Medicine and Hygiene* 33 (4):637-644.

Yesner, D.R. 1980. Maritime hunter-gatherers: ecology and prehistory. *Current Anthropology* 21 (6):727-750.

Zeppa, M.A., A. Laorr, A. Greenspan, J.P. McGahan and L.S. Steinbach 1996. Skeletal coccidioidomycosis: imaging findings in 19 patients. *Skeletal Radiology* 25:337-343.

Zhou, L.M. and R.S. Corruccini 1998. Enamel hypoplasias related to famine stress in living Chinese. *American Journal of Human Biology* 10:723-733.

Zivanovic, S. 1983. A note on the effect of asymmetry in suture closure in mature human skulls. *American Journal of Physical Anthropology* 60:431-435

APPENDIX

Skeletal and dental material were entered into a database developed by the author using Microsoft Access 97. The original Vietnamese individual skeleton identification codes were retained, although alphabetical suffixes were added to each individual in order to track isolated unidentified skeletal elements. Individuals are identified by a unique combination of site code, identification number and identification letter.

Skeletal Preservation of Chapter 6 Case Studies

Case 1 80CCNM35a

Skull: calva, mandible but missing left condyle.
Clavicle: left lateral half, right medial half.
Scapula: acromion, coracoid, glenoid complex.
Rib: 4 left and 7 right articular processes.
Humerus: left proximal head and distal metaphysis; right complete.
Radius: left proximal head and distal quarter; right complete.
Ulna: right complete except missing small section from proximal third of diaphysis.
Hand: right M1; left M1 distal half, M3 proximal three quarters; I second row phalanx.
Vertebra: axis, T5-T10, L1-L5; first and second sacral vertebra plus superior articular facets.
Femur: left proximal three quarters; right distal three quarters.
Tibia: left complete; right proximal two thirds.
Fibula: right distal articular portion and two small diaphyseal fragments.
Talus: left complete.
Calcaneus: left and right complete.
Other foot: left cuboid; right first cuneiform, M1, M3 proximal third.
Innominate: left and right fragments of acetabula, ilia and ischia.

Case Two 80CCNM47a

Skull: calva; complete maxilla; mandible extends from the left condyle to the right permanent M1.
Femur: complete right diaphysis (non-fusion of all epiphyses).

Case Three 80CCNMX5

Femur: proximal two-thirds of right.

Case Four 80CCNM54a

Skull: 3 small fragments of cranium including the left TMJ; left maxilla from I1-M2; left mandibular body C-M3 alveoli.
Rib: 3 small non-articular fragments.
Humerus: left distal quarter.
Radius: left proximal third of diaphysis.
Ulna: left and right proximal halves.
Hand: left M1-M5; right M1 proximal half; 3 proximal row phalanges.
Femur: left proximal two-thirds; right proximal head.
Tibia: left distal articular fragment.
Fibula: left proximal and distal articular areas and some fragments of diaphysis.
Talus: complete left and right.
Calcaneus: complete left.
Other foot: left cuboid, navicular, M1, M3, M5.

Case Five 80CCNM57a

Skull: fragmented cranial remains representing less than 50% of the cranium, including frontal, right parietal and left temporal; mandibular body from left M3 to right M2 alveolus.
Rib: 5 small non-articular fragments.
Humerus: left complete.
Radius: left complete except missing small section of mid diaphysis.
Ulna: left proximal quarter.
Vertebra: fragments of axis (dens and superior facets), C3 centrum, T12 near complete, L1 centrum
Femur: left middle two-thirds.
Tibia: distal fragment of left (including medial and portion of anterior articular region), small diaphyseal fragments (siding uncertain).
Fibula: medial section of left distal articular process, diaphyseal fragments.
Talus: left complete; right missing postero-medial section.
Calcaneus: left postero-plantar fragment.
Other foot:
Innominate: section of left ilium including anterior and inferior superior iliac spines and anterior section of acetabulum; section of right ischium and posterior acetabulum.

Case Six 95DXM1a

Skull: complete cranium, maxilla and mandible.

Clavicle: left diaphysis and acromial end; right complete.
Sternum: manubrium and sternal body.
Scapula: left and right coracoid./acromion/glenoid fossa complexes.
Rib: 23 articular process preserved with most including medial thirds of shaft and 13 fragments including central and distal region of the rib.
Humerus: complete left and right.
Radius: left proximal two-thirds; right complete.
Ulna: left proximal three-quarters; right complete except missing styloid process.
Hand: left navicular, lunate, capitate, hamate, M1-M5; right lesser and greater multangulars, lunate, hamate, M1-M5; 8 proximal row phalanges, 1 mid row phalange, 3 distal row phalanges.
Vertebra: complete C1-L5; sacrum, but much of the left lateral portion eroded away post mortem (coccyx not preserved).
Femur: left and right complete.
Tibia: left and right complete.
Fibula: left and right complete.
Talus: left and right complete.
Calcaneus: left and right complete.
Other foot: left first, second and third cuneiforms, navicular, M1 (distal two-thirds), first proximal and distal phalanges; right first, second and third cuneiforms, cuboid, navicular, M1 (proximal quarter, M2-M3, first distal phalanx.
Innominate: left and right complete except that both missing pubic symphyses and superior pubic rami.

Case Seven 89MDM2a

Skull: virtually complete cranium; complete maxilla; complete mandible.
Sternum: sternal body.
Clavicle: complete left.
Scapula: left and right complete.
Rib: 8 left and 8 right, complete except for some erosion of sternal ends.
Humerus: complete left.
Radius: complete right.
Ulna: complete left.
Vertebra: complete atlas, T1, T2, T11, T12, L4; complete sacrum (missing coccyx).
Femur: complete right (proximal head very eroded).
Tibia: left distal four-fifths.
Fibula: left diaphysis.
Calcaneus: complete left.
Innominate: right and left complete (but right pubic symphysis missing).

Case Eight 80CCNMX4

Humerus: right proximal half.

Case Nine 80CCNM1Y

Skull: calva; left maxillary arcade C-M2, right arcade I1 alveolus to P4; left mandibular condyle and all of the right side from C posteriorly.
Humerus: fragment of distal metaphysis (siding uncertain).
Femur: isolated and very eroded proximal head (siding uncertain).
Foot: left navicular.

Case Ten 66VQ1M31a

Skull: calva.

Case Eleven 80CCNM42a

Skull: calva; right maxilla from P3 to M3, left maxilla from M2-M3; right mandible complete from right I1 alveolus.
Clavicle: right sternal end; fragment of left acromial end.
Scapula: left band right glenoid fossae.
Rib: 6 left and I right articular processes (and 11 diaphyseal fragments).
Humerus: near complete right (missing proximal metaphysis); complete left.
Radius: right distal quarter.
Ulna: left proximal quarter; right proximal two-thirds.
Hand: left capitate, M1, first proximal phalanx; right M1; 3 proximal row phalanges; 2 mid row phalanges.
Vertebra: complete axis, T1-T11, L1-L5.
Femur: left proximal three-quarters; right complete except missing small section of distal diaphysis.
Tibia: left complete except malleolus missing; right proximal two-thirds.
Fibula: left and right distal articular fragments.
Talus: complete right.
Other Foot: left first cuneiform, cuboid, M1-M4, M5 (proximal half), first proximal row phalanx; right first cuneiform, navicular, M1, M4, M2-M3 and M5 (proximal articular ends).
Innominate: complete left and right ilia.

Case Twelve 80CCNM13a

Clavicle: near complete right (erosion of lateral extremity).
Vertebra: complete axis and C6.
Femur: near complete left (missing small section of distal diaphysis); complete right.
Patella: left and right complete.
Tibia: near complete left (malleolus eroded away); complete right.
Fibula: left and right distal three-quarters.
Talus: left and right complete.
Calcaneus: left and right complete.
Other foot: left second cuneiform, navicular, cuboid; right first, second and third cuneiforms, navicular, cuboid, M1, M3, M5, first proximal row phalanx; 2 proximal row phalanges, 1 distal row phalanx.
Innominate: left ischium and acetabular fragments,; right pubis.

Case Thirteen 78QCM3a

Skull: calva; most of the maxilla including complete

alveolar process; complete body and left ascending process but both condyles missing.
Clavicle: left lateral third.
Vertebra: axis, 3 cervical centra.
Femur: right complete.
Tibia: left proximal quarter and distal half.

Case Fourteen 80CCNM74a

Skull: calva; maxillary arcade from left M1-I1, right arcade from C-M1; left mandibular body from P3-M3 and right body from P3-M1.
Clavicle: left complete; right diaphysis.
Scapula: fragment of superior glenoid and coracoid; complete right glenoid and acromion.
Rib: 4 left and 1 right articular fragments and 18 other small fragments.
Humerus: left complete; right distal condyles and fragment of proximal head.
Radius: proximal third left diaphysis; right nearly complete but missing distal metaphysis.
Ulna: left proximal three-quarters; right proximal head.
Hand: left M3 proximal two-thirds; proximal two-thirds of right M1 and M3; 4 proximal row phalanges.
Vertebra: dens and left facets of atlas, 2 thoracic centra, L5 centrum, sacral promontory.
Femur: left distal three-quarters, right distal articular condyles and proximal third of diaphysis.
Patella: left and right complete.
Tibia: left proximal head and distal third; right almost complete but missing small section of distal metaphysis.
Fibula: distal articular section of right and section of mid diaphysis.
Innominate: fragment of posterior right ilium including portion of acetabulum and complete auricular area.

Case Fifteen 70DCM1a

Skull: left side of calva but includes right orbit and large section of right occipital; much of maxilla from permanent M1-M1; fragment of left mandibular body including unerupted M1 and dm2 alveolus, posterior left fragment including condyle.

Case Sixteen 76NNM2c

Skull: calva including face; most of the maxilla preserved although much damage to alveolar process; complete mandible.
Clavicle: left medial half; right medial three-quarters.
Humerus: left distal three-quarters; right distal three-quarters.
Radius: left distal three-quarters; right proximal quarter and mid diaphysis third.
Ulna: left proximal and distal thirds of diaphysis; right complete except missing styloid process.
Hand: right navicular and capitate, M2 proximal half, M3 distal third; 3 proximal row phalanges, 2 mid row phalanges.

Vertebra: left articular facets of atlas, left half of axis, C5-6 centra.
Femur: left proximal two-thirds diaphysis; right proximal two-thirds diaphysis.
Patella: left complete.
Innominate: left auricular area including greater sciatic notch, fragment of superior acetabulum; right superior acetabular fragment.

Case Seventeen 77NNM10ka

Skull: much of cranium but missing base; most of maxilla but right arcade missing posterior to M1; complete mandible.

Case Eighteen 78NNM7ka

Skull: much of cranium but missing base; complete alveolar process of maxilla; most of mandible but missing left ascending ramus and condyle.

Case Nineteen 89MDM7a

Skull: complete cranium; complete maxilla; complete mandible.
Humerus: near complete right but major post mortem damage to lateral head and both tubercles.
Radius: complete right.
Vertebra: complete sacrum.
Femur: left proximal three quarters; complete except for major poste mortem damage to medial and lateral portions of distal condyles.
Tibia: left diaphysis; proximal three-quarters.
Fibula: left distal three-quarters; right complete.
Innominate: left complete; right complete except missing pubic symphysis.

Case Twenty 95DXM1a

See Case Six

Case Twenty One 80CCNM65a & 65b

80CCNM65a
Skull: numerous (>60) small cranial fragments including portions of frontal, parietals and temporals; fragment of left maxillary alveolar process I1-C alveoli, fragment of right from C-M2.
Clavicle: left complete; right sternal half.
Scapula: left glenoid/coracoid complex.
Rib: >60 fragments including 6 left and 7 right articular sections.
Humerus: left distal third; right distal two-thirds.
Radius: right proximal two-thirds.
Ulna: right proximal three-quarters.
Hand: left navicular and first proximal phalanx; right navicular, hamate, M1 and first proximal phalanx; 3 proximal row phalanges, 1 mid row phalanx.

Vertebra: cervical centrum, 4 upper and 1 lower thoracic centra, 4 lumbar centra; fragment of 13 superior and inferior thoracic and lumbar interarticular facets, fragment of sacral promontory.
Calcaneus: left complete.
Other foot: left first cuneiform, M1 and M5; right navicular; 1 proximal row phalanx.
Innominate: left auricular area, superior ischium and infero-posterior acetabular fragment; right auricular area, ischium and fragment of mid acetabulum.

80CCNM67b
Humerus: complete right diaphysis.
Femur: proximal and distal thirds of diaphysis.
Tibia: complete left diaphysis.